Analysing Design Thinking: Studies of Cross-Cultural Co-Creation

Analysing Design Thinking: Studies of Cross-Cultural Co-Creation

Bo T. Christensen

Department of Marketing, Copenhagen Business School, Copenhagen, Denmark

Linden J. Ball

School of Psychology, University of Central Lancashire, Preston, UK

Kim Halskov

Department of Digital Design and Information Studies, Aarhus University, Aarhus, Denmark

CRC Press
Taylor & Francis Group
Boca Raton London New York Leiden

CRC Press is an imprint of the
Taylor & Francis Group, an **informa** business

A BALKEMA BOOK

CRC Press/Balkema is an imprint of the Taylor & Francis Group, an informa business

© 2017 Taylor & Francis Group, London, UK

Typeset by MPS Limited, Chennai, India
Printed and bound in Great Britain by CPI Group (UK) Ltd, Croydon, CR0 4YY

Library of Congress Cataloging-in-Publication Data

Names: Christensen, Bo T., editor. | Ball, Linden J., 1963– editor. | Halskov, Kim, editor.
Title: Analysing design thinking : studies of cross-cultural co-creation / Bo T. Christensen, Department of Marketing, Copenhagen Business School, Copenhagen, Denmark, Linden J. Ball, School of Psychology, University of Central Lancashire, Preston, UK, Kim Halskov, Department of Digital Design and Information Studies, Aarhus University, Aarhus, Denmark.
Description: Leiden, The Netherlands : CRC Press/Balkema, [2017] | Includes bibliographical references and index.
Identifiers: LCCN 2017018992 (print) | LCCN 2017020464 (ebook) | ISBN 9781315208169 (ebook) | ISBN 9781138632578 (hardcover : alk.paper)
Subjects: LCSH: Industrial design—Social aspects. | Intercultural communication. | Teams in the workplace.
Classification: LCC TS171.4 (ebook) | LCC TS171.4 .A524 2017 (print) | DDC 745.2—dc23
LC record available at https://lccn.loc.gov/2017018992

Published by: CRC Press/Balkema
                Schipholweg 107C, 2316 XC Leiden, The Netherlands
                e-mail: Pub.NL@taylorandfrancis.com
                www.crcpress.com – www.taylorandfrancis.com

ISBN: 978-1-138-63257-8 (Hbk)
ISBN: 978-1-315-20816-9 (eBook)
ISBN: 978-1-138-74844-6 (Pbk)

Table of Contents

Acknowledgments

This book is based on video recordings of real-life design processes, so as is customary and appropriate with the credit ordering in any film production, we start with the lead actors: The editors would like to extend their sincere gratitude to the design team, Abby, Ewan, and Kenny, who allowed us to closely observe and trace their design processes across months, work settings and continents.

DTRS11 was designed on the foundations of previous DTRS meetings – especially those organized by Janet McDonnell and Peter Lloyd (DTRS7), and Robin Adams (DTRS10). They helped pave the road for the DTRS data-sharing format, which we built upon and tried to extend here and there. A huge thank you is due to the 28 research teams who spent an enormous amount of time analysing the dataset, producing the findings that make up the following chapters, and contributing to the DTRS11 symposium with inspiring talks and discussions.

The editors would like to thank Innovation Fund Denmark for supporting this research in a grant to the research project 'Creativity in Blended Interaction Spaces' (Grant reference: CIBIS 1311-00001B). Innovation Fund Denmark invests in cultivating and translating ideas, knowledge and technology for the benefit of Danish Society. DTRS11 has also received generous funding from the Carlsberg Foundation, The Danish Council for Independent Research, and Otto Mønsteds Fond. In addition, Designmuseum Danmark supported DTRS11 by providing The Banquet Hall for the symposium anniversary dinner.

DTRS11 and this book would not have been possible without the enthusiastic and hardworking team of student assistants and research assistants who participated in the extensive data collection. We would like to thank our student assistant, David Simon Lindø Sørensen, for his diligent, professional approach when spending months with the design team mounting cameras on a day-to-day basis and collecting data of high-quality. We are also very grateful to our other student assistants, Emil Risum Brøgger, Lærke Cecilie Anbert and Simon Carøe Aarestrup, for their extensive effort in transcribing the many hours of video data and for making transcriptions and videos presentable and shareable. A special thanks to Isak Nord Mirdal for designing the book cover.

Finally, following the customary logic of film credits, we finish with the most important person behind the scenes: Sille Julie J. Abildgaard has held the whole production together with her dedication, hard work and organizing skills. We've received multiple comments on orchestrating a smooth-running data-sharing process, and we need to say that is all thanks to Sille, and her work on everything from data collection, designing e-infrastructure, drafting of the technical report, data sharing, peer-review, right down to event planning at the symposium. Thank you, Sille!

Bo T. Christensen, Linden J. Ball & Kim Halskov

Preface: The History of the Design Thinking Research Symposium

Nigel Cross

The symposia series was initiated by Nigel Cross with Norbert Roozenburg and Kees Dorst at Delft University of Technology, The Netherlands, in 1991, with what was initially expected to be a one-off international meeting on 'Research in Design Thinking'. A proposal for another meeting was developed by Kees Dorst and Henri Christiaans and this was also held in Delft, in 1994, focused on the use of protocol analysis as a research tool for analysing design activity. This became known as the 'Delft Protocols Workshop'. For the first time in design research, a common data set (videotapes of both individual and team design activity) was provided to researchers around the world, for their own analyses, presented at the workshop. The content and format of that meeting were felt by the participants to be so good as to warrant more of the same.

A third meeting was held at the Istanbul Technical University, Turkey, in 1996, on the topic of descriptive models of design, and the fourth meeting was held at the Massachusetts Institute of Technology, Boston, USA, in 1999, on the topic of design representation. It was there that the organisers introduced the term 'Design Thinking Research Symposium' as the generic title for the series. The fifth meeting was again in Delft, in 2001, on the topic of design in context, and developing an interdisciplinary approach to studying design in a broader social context.

The sixth symposium, at the University of Technology, Sydney, Australia, 2003, returned to somewhere near the focus of the original meeting in Delft in 1991, on the nature and the nurture of expert performance in design. This workshop meeting again brought together a relatively small, international group of active researchers. Throughout this series of symposia, this workshop format has been found to be a successful way of synthesising the contributions of an international community, of reporting current work, and of identifying and promoting necessary further research.

A seventh meeting DTRS7 on analysing design meetings was held at Central St. Martin's College, University of the Arts, London in 2007. This was again a small, focused workshop meeting, and again providing researchers worldwide with a common data set for analysis – this time with video recordings of meetings within architectural and engineering product design teams.

Although not a DTRS event, a related meeting adopting the same principle of analysing a common data set was held as a National Science Foundation Workshop at the University of California, Irvine, USA, in February 2010, on 'Studying Professional Software Design'. The data provided were video recordings of pairs of software designers tackling the same design task.

The eighth DTRS meeting, 'Interpreting Design Thinking' was again held in Sydney, Australia, in October 2010, and invited contributions linking design to other disciplines. This meeting acknowledged the growing role of design thinking in business, industry, social services and elsewhere.

DTRS9 'Articulating Design Thinking' was held at the University of Northumbria, Newcastle-upon-Tyne, England, in April 2012. Contributors to the meeting analysed different responses to a given design task related to inclusive design.

The tenth meeting, held at Purdue University, Indiana, USA, in October 2014, focused on analysis of design review meetings.

DTRS topics and previous publications:

DTRS1	**Research in Design Thinking** *Delft University of Technology, The Netherlands in 1991* Cross, N., Dorst, K. & Roozenburg, N. (eds.) (1992). *Research in Design Thinking*. Delft, The Netherlands: Delft University Press.
DTRS2	**Delft Protocols Workshop** *Delft University of Technology, The Netherlands in 1994* Dorst, K. (ed.) (1995). Analysing Design Activity. Special issue of *Design Studies*. Vol. 16, no. 2. Cross, N., Christiaans, H., & Dorst, K. (eds.) (1996). *Analysing Design Activity*. Chichester, UK: John Wiley & Sons
DTRS3	**Descriptive Models of Design** *Istanbul Technical University, Turkey in 1996* Akin, Ö. (ed.) (1997) Descriptive Models of Design. Special issue of *Design Studies*. Vol. 18, no. 4. Akin, Ö. (ed.) (1998) Models of Design, special issue of *Automation in Construction*. Vol. 7, no. 2/3.
DTRS4	**Design Representation** *Massachusetts Institute of Technology, Boston, USA in 1999* Goldschmidt, G. & Porter, W.L. (eds.) (2000). Visual Design Representation. Special issue of *Design Studies*. Vol. 21, no. 5. Goldschmidt, G. & Porter, W.L. (eds.) (2001). Design Representation. Special issue of *Automation in Construction*. Vol. 20, issue 6. Goldschmidt, G. & Porter, W.L. (eds.) (2004). *Design representation*. London: Springer Verlag.
DTRS5	**Design in Context** *Delft University of Technology, The Netherlands in 2001* Lloyd, P. & Christiaans, H. (eds.) (2001). *Designing in Context*. Delft, The Netherlands: Delft University Press. Lloyd, P. (2003). Designing in Context. Special issue of *Design Studies*. Vol. 24, no. 3.

DTRS6	The Nature and the Nurture of Expert Performance in Design
	University of Technology, Sydney, Australia in 2003
	Cross, N. & Edmonds, E. (2003). *Expertise in Design, Creativity and Cognition Press*. Sydney, Australia: University of Technology.
	Cross, N. (ed.) (2004). Expertise in Design. Special Issue of *Design Studies*. Vol. 25, no. 5.

DTRS7	Analyzing Design Meetings
	Central St. Martin's College, University of the Arts, London, England in 2007
	McDonnell, J. & Lloyd, P. (eds.) (2009). *About: Designing – Analysing Design Meetings*. London, UK: Taylor & Francis.
	McDonnell, J. & Lloyd, P. (eds.) (2009). Analysing Design Conversations. Special issue of *CoDesign*. Vol. 5, no. 1.
	Lloyd, P. and McDonnell, J. (eds.) (2009). Values in the Design Process. Special issue of *Design Studies*. Vol. 29, no. 2.

SPSD	Studying Professional Software Design
	University of California, Irvine, USA in 2010
	Petre, M., van der Hoek, A. & Baker, A. (eds.) (2010). Studying Professional Software Design. Special issue of *Design Studies*. Vol. 31, no. 6.

DTRS8	Interpreting Design Thinking
	University of Technology, Sydney, Australia in October 2010
	Stewart, S. (ed.) (2011). Interpreting Design Thinking. Special issue of *Design Studies*. Vol. 32, no. 6.

DTRS9	Articulating Design Thinking
	University of Northumbria, Newcastle-upon-Tyne, England in April 2012
	Rodgers, P. (ed.) (2012). *Articulating Design Thinking*. Faringdon, UK: Libri Publishing.
	Rodgers, P. (ed.) (2013). *Articulating Design Thinking*. Special issue of *Design Studies*. Vol. 34, no. 4.

DTRS10	Design Review Conversations
	Purdue University, Indiana, USA in October 2014
	Adams, R. & Siddiqui (2016). *Analyzing Design Review Conversations*. Purdue University Press.
	Adams, R., McMullen, S., & Fosmire, M. (eds.) (2016). Co-Designing Review Conversations. Special issue of *CoDesign*. Vol. 12, no. 1-2.
	Adams, R., Cardella, M., & Purzer, S. (eds.) (2016). Design Review Conversations. Special issue of *Design Studies*, vol. 45, Part A.

DTRS11   Cross-Cultural Co-Creation
Copenhagen Business School, Frederiksberg, Denmark in
November 2016
Christensen, B. T., Ball, L. J. & Halskov, K. (2017). *Analysing*
Design Thinking: Studies of Cross-Cultural Co-Creation. Leiden:
CRC Press/Taylor & Francis.
Ball, L. J. & Christensen, B. T. (in preparation). Designing in the Wild.
Special issue of *Design Studies*
Halskov, K. & Christensen, B. T. (in preparation). Designing across
Cultures. Special issue of *Co-Design.*

Chapter 1

Introduction: Shared Data in Design Research

Bo T. Christensen, Linden J. Ball & Kim Halskov

ABSTRACT

The Design Thinking Research Symposium (DTRS) series, of which this book is part, is an interdisciplinary symposium series linking international academics with a shared interest in design thinking and design studies coming from a diversity of disciplines, including psychology, anthropology, linguistics, philosophy, architecture, and design studies. The series provides an international forum for pioneering and state-of-the-art research on design thinking that is focused on the study of design practice from various perspectives. The 25 year history of the DTRS series is also a story of almost 25 years of shared datasets in design thinking research. This data-sharing approach was initiated in the seminal 'Delft Protocol Workshop' (now also labelled DTRS2), which was organized by Kees Dorst, Nigel Cross and Henri Christiaans at Delft University of Technology in 1994 (Cross, Christiaans, & Dorst, 1996; Dorst, 1995) and was based around verbal protocol data collected from professional designers in a controlled context. Subsequently, two more DTRS events have involved shared data. DTRS7, organized by Janet McDonnell and Peter Lloyd, involved professional designers (architects and engineers) working in their natural habitats (Lloyd & McDonnell, 2009; McDonnell & Lloyd, 2009a, 2009b), and DTRS10, organized by Robin Adams, involved design review conversations in a design education setting (Adams & Siddiqui, 2016; Adams, Cardella, & Purzer, 2016; Adams, McMullen, & Fosmire, 2016).

At the DTRS11 25th anniversary dinner, Kees Dorst remarked in his celebratory comments that sharing design data is, first and foremost, about academic generosity. For past DTRS organizers, the labour involved in collecting and distributing the data has certainly been substantial. However, the beneficiaries of that labour are not solely restricted to the design research teams involved in the shared data analyses, but extend to the wider academic community as the receivers of the resulting publications. The three previous shared datasets have had a huge impact on the design research literature, with the resulting 3 book publications (not counting all the ensuing journal publications) attracting many hundreds of citations. But outside of design research it remains the case that shared datasets with video data are extremely rare in the humanities, social sciences and technical sciences. Nonetheless, global trends towards so-called 'Open Science' clearly indicate that the sharing of video data holds substantial research potential (Adams, Radcliffe, & Fosmire, 2016).

I   OPEN SCIENCE

Partly spurred on by what has been dubbed the 'reproducibility crisis' in science (Baker, 2016), scholars across many disciplines have recently been pushing towards more research openness, leading to an Open Science 'manifesto' (Munafó *et al.*, 2017). Efforts to increase scientific openness aim centrally at improving the *transparency* of all aspects of the research process because this is viewed as being crucial to making science more reproducible whilst also bringing benefits to research efficiency. To increase the reproducibility of research results, the Open Science agenda sets great store on the value of research teams sharing data, methods and materials so as to facilitate data re-analysis and follow-on replication studies. It seems evident that past DTRS efforts involving shared datasets have already contributed as ground-breaking 'first-mover' cases to this Open Science agenda.

The reproducibility of research results is, however, not the only significant benefit to arise from the sharing of data such as the DTRS datasets. This is because the nature of these datasets also allows for a multitude of different research methods to be applied in their analysis. Furthermore, the fact that the data are 'open-ended' – in the sense that their collection is not restricted to addressing a single, specific research question – allows for both inductively-oriented researchers to explore possible new theoretical angles, while simultaneously allowing for deductively-oriented researchers to test at least some theoretical design models against real-life design cases.

In this latter respect, the shared DTRS video data possess qualities resembling those emphasized as being central to design objects. Sketches, prototypes or similar design objects in-the-making are uncertain, ambiguous, re-frameable, contextually shiftable, generally open to exploration and interpretation, and basically embody qualities that provide creative potential, as captured by dominant theories of design and creativity (e.g., Dorst & Cross, 2001; Finke, Ward & Smith, 1995; Schön & Wiggins, 1992). For the individual designer such qualities allow for continual re-interpretation and object back-talk; in a collaborative setting these qualities ensure a multitude of potentially distinct perspectives being taken on the same shared object of study.

Such qualities of design objects are well-known to designers and design researchers, and shared video data of design processes can be utilized with many of the same types of benefits for the sharing parties. The shared video data thus constitute a common focal-unit of attention for all involved in the symposium, whilst also allowing for individual perspective-taking in terms of particular researchers or research groups diving into theoretically or empirically derived points of interest. In these ways the shared data facilitate discussions among the participating design researchers across the application of a variety of research methods and theoretical units of analysis. In DTRS11 moreover, such dialogue also extended to the practicing design team itself, which served not only as the object of study but also as a partner engaging in discussion and debate at the symposium.

The Open Science agenda seems to pursue mainly singular truths, by focusing narrowly on making sure that data are replicable and reproducible, which itself involves disentangling false negative and false positive results from true effects. But the variant of Open Science sought in the DTRS datasets has helped illustrate the multitude of different (typically complementary and non-competing) conclusions that may be drawn even when starting with the same data at the outset and then subjecting these data to distinct methods in order to answer a range of different research questions. In this respect, Open Science in the present volume values *open-endedness* in the types of research questions the data may be subjected to, in addition to pursuing the goal of allowing for reproducibility of the attained results when addressing a specific research question with either the same or different methods.

2 WHAT TO CAPTURE IN SHARED DESIGN DATASETS?

In acknowledging that shared design data serve the purpose of allowing for a number of methodological approaches in the study of a range of research questions, careful consideration of what to capture seems warranted. The dataset should be rich enough to allow for individual sampling thereof for specific research purposes, and flexible enough to allow for a multitude of methodological approaches to be applied. But on the other hand, the dataset should remain manageable in order to avoid drowning in data complexities leading to *analysis-paralysis*, where the research teams end up spending too much time trying to gain an overview and general understanding of the content at the expense of being able to commence theoretically meaningful analyses. The dataset should also be manageable enough for the research teams to be able to finalize their research projects in time for the symposium, which is, of course, vital in order for participating researchers to utilize the collaborative potential in the shared data serving as a common ground for understanding other research perspectives. All DTRS events involving shared data have approached the question of what data to capture in a similar manner, that is, by honing in on design encounters involving *verbalization* in order to be able to study design activity and cognition. But the purpose of the data collection has differed greatly across symposia, as has methodological considerations on the balance between rich versus manageable datasets.

The DTRS2 Delft Protocol Workshop focused on one particular research method, verbal protocol analysis, aiming to uncover the mysterious cognitive processes involved in expert designing (Cross, Dorst, & Christiaans, 1996). The organizers collected data from five experiments involving professional designers at XeroxPARC who were working for two hours either individually or in teams of three on designing a fastening device that would allow a given backpack to be fastened onto a mountain bike. Two of these experiments were eventually shared in the dataset, involving one individual and one team-based protocol, being selected under consensual considerations of the inherent interestingness of the process and the recording quality. The lone designer worked under think-aloud instructions, where he concurrently verbalized what was going through his short-term memory while designing, whereas the design team was not given such instructions. One important observation stemming from DTRS2 concerns the disadvantages of asking for think-aloud verbalizations, notably how such instructions to enforce concurrent verbalizations may change behavior and cognitive performance (Lloyd, Lawson, & Scott, 1995; see also Davies, 1995; Schooler, Ohlsson, & Brooks, 1993, and for a contrasting position see Ericsson & Simon, 1993). Perhaps as a consequence of these concerns the shared datasets in later DTRS events have all utilized non-forced, naturally-occurring design dialogue in team settings as a window into designing (see also Christensen & Ball, 2014; Christensen & Schunn, 2007).

Ten years on from DTRS2 and the Delft Protocol Workshop, DTRS7 sought to share data collected in naturalistic settings, by recording a total of approximately

7 hours of video from an engineering design team working on designing a new toy using a thermal print-head, and from architectural meetings concerning the design of a crematorium. Unlike DTRS2, the purpose was not narrowly set upon protocol analysis, although the videos were shared along with meeting transcripts. Rather, a multitude of research methods were used to analyse the datasets, including a host of qualitative approaches that utilized the 'single-case' nature of the videos, allowing for inductive research approaches.

The subsequent dataset collected for DTRS10 involved for the first time a central theoretical frame guiding the data collection in that a multitude of 'design review conversations' in educational settings were recorded across six different design-related disciplines (i.e., industrial design at the bachelor and graduate level, entrepreneurship, mechanical design, choreography, and service learning design). The dataset offered comparisons between design domains, while allowing for generalizability and reproducibility by having an extensive number of observations. As a consequence, the dataset was very large, including approximately 25 hours of transcribed video. Although the unit of analysis covering all videos (i.e., design review conversations) made the dataset highly suitable for quantitative approaches, the dataset was also successful in allowing for qualitative approaches, whereby researchers dived into the analysis of smaller parts of the dataset. The extensive data collection, however, also had the side-effect that many research teams ended up focusing on only a small subset of the data in order to avoid drowning in data and to ensure that submission deadlines were met.

Outside of the context of the DTRS series there have been two more data-sharing projects involving design. The 2010 workshop 'Studying Professional Software Design', organized by André van der Hoek, Marian Petre and Alex Baker, involved three pairs of very experienced professional software designers being given a two-hour artificial software design task to solve in a controlled setting (Baker, van der Hoek, Ossher, & Petre, 2011; Petre & van der Hoek, 2013; Petre, van der Hoek, & Baker, 2010). A 2012 workshop entitled 'Investigating Design Thinking of a Complex Multidisciplinary Design Team in a New Media Context', organized by Newton D'souza (D'souza, 2016), involved a comprehensive 20 hour dataset collected over a five day period that focused on a multidisciplinary student design team tasked with creating innovative new concepts for a greeting card company.

3   THE PRESENT DATASET

The decision as to what to study focally for the present dataset was based on careful consideration of what we thought were learning points from the previous shared data events summarized above. Again, naturally-occurring design dialogue – as opposed to enforced verbalization – was viewed as being essential to constitute the backbone of the dataset. It was also deemed to be important for the dataset to allow for the application of a multitude of analysis methods involving differing theoretical lenses so as to lend itself to both deductive and inductive research approaches. In terms of the size of the dataset, we opted for 10–15 hours of video in order to ensure that the data would be manageable in the given timeframe, and could serve as an empirical grounding that would ensure cross-fertilization and shared understanding at the symposium event. As in DTRS7, we wanted to explore professional designers working in their normal context, using their customary tools and interacting with collaborators and co-creators. We wanted to continue the shift away from simple and artificial experimental settings depriving designers of their standard tools, collaborators, co-creators and environments, and instead allow them to work as they normally would. The guiding principle in designing the data collection was *to extend beyond the timeframes and boundaries that had been previously studied, by focusing on a design team traced over time and context, in all of its complexities in the wild.*

Previous shared datasets have sought to capture a limited timeframe of design activity, typically stretching over a few hours per team. Our ambition was to go beyond this restriction by tracing designing longitudinally on a day-to-day basis over the course of several months. The rationale behind this was twofold. First, it would allow for a different set of possible research questions to be established concerning design development. Second, by extending the contextual boundaries, we sought to bring into focus the multitude of important actors in design, from stakeholders to users, all of whom play a potentially important role in co-creating an effective design solution. Already in DTRS7, a stakeholder focus had been possible in the architectural dataset, but for DTRS11 we wanted to capture organizational designing, bringing to the fore organizational stakeholders – who are simultaneously colleagues and clients – as well as the end-users. The present dataset follows a core design team as they cross a number of boundaries in their everyday designing: boundaries between departments within their own organization; boundaries with the outside marketplace in interaction with lead-users; and boundaries between countries in travelling to another continent to conduct cross-cultural user studies.

The final, shared dataset consisted of more than 15 hours of video and audio recordings, including transcriptions. In addition, we distributed a technical report (see Christensen & Abildgaard, 2017; Chapter 2 of this volume) covering details of the project, organisation, the design team's background and summaries of video content.

4 DEBATING SHARED DATA

The shared data aided communication across academic divides, methodological approaches, and researcher/practitioner gaps. The focal data helped facilitate comprehension and discussion of results that were outside the normal theoretical lens of the individual researchers and helped maintain empirical grounding in situations when the theoretical abstractions could have been otherwise incomprehensible. That said, the shared data do not in themselves compensate fully for a lack of theoretical background knowledge when discussing across academic divides, and discussions at DTRS11 appeared most fruitful when similar theoretical constructs were tackled in relation to the same data but using distinct methods.

During the data-sharing process, research teams varied in whether they approached the data mainly inductively or deductively. Early on a number of more deductively oriented researchers expressed to the organizers that they experienced difficulties in matching their theoretical interests to the nature of the data, which meant that they struggled to get to an analysis framework based on the shared data. While a shared dataset does allow for a multitude of perspectives being applied, any single case study does not afford an opportunity to examine every possible aspect of design processes, and the present dataset, despite offering new, extended angles to analyse, also had limitations. A single dataset is *not* infinitely flexible in terms of the analytical approaches that it permits and will, by extension, limit the range of analyses in addition to limiting the generalizability of the conclusions that can be drawn. The question was even raised by a couple of research teams as to whether the data could be characterized as reflecting *design* at all. This raises the issue of where to orient (and where not to orient) the camera when claiming to capture design. Further, questions were raised about whether sharing datasets was 'enough' to warrant symposium coherence, or whether additionally a shared theoretical purpose (as in DTRS10) or a shared methodological purpose (as in DTRS2) was additionally needed.

One underlying tension in these debates seems to be whether the shared dataset is seen as offering mainly new, theoretical angles to explore inductively or whether the shared dataset is mainly seen as one way to test deductively or reproduce findings from existing design models. These important questions were seen as being crucial ones to explore further when looking forward to the future of shared data in design research. Therefore, a panel debate was established during the DTRS11 event at Copenhagen Business School, in order to both look backwards and forwards at the usage of shared data in design research. The participants in the panel were three co-organizers, one from each of the past DTRS shared research events: Kees Dorst, Peter Lloyd and Robin Adams. In addition, Gabriella Goldschmidt was asked to join due to having voiced concerns about the usage of shared data in design research, in order to balance the panel. And finally, 'Ewan' – the design team leader from the design team captured in the DTRS11 dataset – also took part.

The debate was lively and extremely informative in capturing opposing views and key points for reflection in relation to future shared data events. We believe the lively debate is best communicated by maintaining the verbal format, and hence we have transcribed it and present below a central exchange with limited editing:

Gabriella Goldschmidt: "What has come to be called DTRS2, which Kees described to us, came to being, the purpose was [...] to look at the methodology of protocol analysis and see if we could come to some conclusions as to how it could best be used, what its limitations were, etc., etc. [...] Protocol analysis at that time, most of you are not old enough to remember, I mean we're talking about 1994, right? Some of you were just barely born at this point. Protocol analysis emerged at that time as a very powerful and suitable tool for the analysis of design activities, but there were obvious limitations, as there are to any methodology. So, this was a very big opportunity to look at things in great detail and therefore be able to craft some conclusion. I attended that workshop and I thought it was wonderful. And the shared data were wonderful. However, there's a "but". I think that if the only purpose of a shared dataset is to be able to look at something in great detail in a manner of ways that other people could understand because they've been through the same dataset, it's just not enough, because any dataset that you're going to choose is going to leave some people out, because either they're not related to it in terms of their core disciplines or they're not sufficiently interested, etc., etc. [...] Shared datasets are not the only and exclusive way, to research and to analyze design activities or design sessions or design thinking. And here I must share with you something that in my mind is relevant to what we talked about. I take credit for the term "DTRS". The first three workshops that are now called DTRS1-2-3 had different titles at the time. And when it was time to do the fourth, which I was responsible for, I said, "Let's put an umbrella term over this", and we called it "Design Thinking Research Symposium" – but this was before the term "design thinking" became the buzz word that it is today, or the buzz phrase that it is today. It was not meant to be a methodology, it was certainly not of interest to the business and management community at that time. It was just a way of talking about how designers in different disciplines think. Nowadays things are very different, and design thinking has been appropriated by other disciplines, and includes many other things in addition to a design problem. And no offence meant, I think that the dataset that we see here is not design. It's pre-design. You know, you can respond to me and tell me what you think about that. It's pre-design. To me as a layperson it's closer to market research than to anything else. And I would like to make a distinction between what designers do, very often in teams, and what is the design – let's call it "design coalitions". Many people other than designers are involved in projects, which are basically design projects. Same as if you go to design schools nowadays and you look at who teaches in them. There are people who are not designers who teach in design schools. They're very relevant to design, but they're not designers. They may be psychologists, they may be sociologists, etc., etc. The original intent of this series was to look at what designers do. What we've heard here is really interesting, but I don't think it is exclusive to design. It could have been the planning of a, I don't know, health-care system or a lot of other things. It's a lot about how teams work, about communication, about trying to define what a problem is such that designers can work on it – and designers cannot work if there is no design problem there. They may start with one problem and then revise it and remake and reinterpret it and reframe it and end up solving something that is a little bit different than the original problem, but there has to be some kind of problem announcement. Here, the question was what would that be? What would the design problem even be before we can hand it over to designers? It's not about design. It doesn't make it any less important or any less interesting also for designers, absolutely. But what I want to say is that when you use a shared database, you have to have very good reasons to do so. You have to make sure that the particular dataset that you've chosen is really the most suitable way to probe the kind of question that you're interested in, and that's not always what's happened here."

'Ewan', the design team leader: "I'd love to respond to it (laughter). I think you raise some valid points. I think, you know, I really hate it, and love it, when people are putting design next to market research. As it happens – and again excuse me for saying this – I think maybe for some of the sessions, or maybe for the untrained eye, it could look like market research. But remember that the people that are there in the room are *designers*. There are very few non-designers

in the room. And you could say, 'We are designers who have expanded into market research', but I don't want to call it that. We just say, you know, 'How can we get closer to the source for the people we are designing for?'. And I think, at least when I was studying graphic design and industrial design, which I originally came from, it was all about 'finding the problem' – and kind of 'what's the pain, what is the problem, what is the need?', where I felt that you know we need to also think about elevating *pleasure*. How can we find something that is not a problem, but something that feels really good, and how can we elevate that? And I think to be able to do that we need to bring a cross-functional team – a multi-disciplinary team – into the field and I think the design process starts there. First, framing the question, then understanding the users. And, you know, when you guys approach- same when you do any project, we don't go- we are rock solid in the process, in the structure of it, but we have no idea where it would end. We told you guys that we would come into conceptualizing and create things and stuff like that, but the process didn't allow us to do it, because we weren't mature enough. The data we got to get us into that mode [did not come] before after you guys were gone. [...] we weren't ready. So after you guys left, then you know – the designers kind of came out of us and we started creating and we started sketching, making prototypes, but we weren't ready before. I think that's- so in a way, I understand that you only had the data you had, but I would be really angry if it was labeled as 'market research' (laughter). [...] this time around, we needed a much longer time in kind of that 'Oh, what is sexy commitment? What is progress together?'. We needed to galvanize that before we could start designing with it."

[...]

Kees Dorst: "What I like about this particular dataset is that if you look at the development of design – industrial design – over the years, you can almost tell that story based on it being an increase in *complexity* of the area that you work in, which means that you include more research into sort of finding out what you need to be designing etc., etc. So, I just find it-, you also see that design struggles with that, in the sense that okay, you can actually gather all that data, but how do you encapsulate that, or bring that into your design process? And when does your design process actually not work anymore, because it's too complex and you don't know how to navigate it anymore? And what do you do then – does it become engineering? You split up problems? Or how do you work with it? So, what I like about this dataset is that it is actually, it's capturing something that is a movement around design that design has to deal with and it hasn't quite found the solution for. So, looking at the dataset, you guys [addressing Ewan] go around in circles every now and then. I think part of that is that complexity – and how do you deal with it? And it's very interesting to see what that actually means for design, or what that means for new methods, or whatever we need to be developing, as designers, for designers. To deal with these kind of things. So, for me it's very contemporary, and it is how design is expanding, and how- broader than concentrating on okay, this is the core design activity of say, concept design, or something. So, I didn't mind that so much, but, yeah."

Robin Adams: "Sure. So I'm appreciating the conversation because I think, for me, part of the strengths of a shared dataset, in the form of the shared dataset, is that it brings us together and it forces us to ask tough questions. It forces us to engage, and rethink, and revisit the same lenses that we've been wearing and asking if those lenses still work. And I think it brings in new perspectives and new methodologies and new voices, and that's not something I see in other communities [...] And so, connected to your idea of complexity, I think one of the directions with the shared dataset is that they just keep getting more authentic and more complex and more in these pieces, and I think they'll keep creating spaces for us to revisit the assumptions that we have and the evolution of these ideas, and address them in new ways."

Peter Lloyd: "Having been involved in I think about four datasets now, and having organised one too, I've always had the feeling that we're somehow missing the essence of designing; that we've never quite got to the core of what designing is. I mean, the first dataset was protocol analysis and I remember thinking 'design cognition is all locked up in long-term memory, we can't get it out of there except as the small verbal fragments that appear in short-term memory'. It seemed like the actual thinking of designing was locked away out of reach of verbalisation. And then the one that I organised with Janet, with architecture meetings, I think a few people said 'this is isn't architectural design they are just having meetings' (laughter). Even Robin's dataset on design teaching, I ended up thinking, you know, is this really designing here I'm looking at or some kind of political process involving the exercise of authority. Of course designing is all these things and more, which is what makes the datasets interesting. They challenge our own expectations about what we think design is and the way it manifests in the world."

Later in the symposium, topics raised during the panel debate were picked up again by Rianne Valkenburg, who had been asked to give a closing talk on her impressions of what was learned during the symposium. In her talk Rianne drew connections between issues discussed at DTRS11 and those addressed in her new book on the design thinking practices of innovators (see Valkenberg, Sluijs, & Kleinsmann, 2016). Below is an edited down version of the main points covered in her talk:

Rianne Valkenburg: Research always starts with curiosity: having a sincere interest in a topic or situation. In the case of design research this curiosity involves designers. We really want to understand what moves them. What do they do when they try to make sense in the complexity of designing? What goes around in their head when they are creative? Twenty-five years ago I fell in love with this research field and its community because of this humbleness: design researchers were trying to learn from *Designers* at work (with a capital D, since we looked up to them). Because design researchers felt that the complex process description, mainly from the engineering field, did not really explain the creative process our designers were facing, the first empirical studies started. These entailed observations of designers at work, trying to grasp what designing really was, among others the classic study by Hales in 1987, or the study by Cross and Cross in 1996 trying to understand the success of racing car design by "stalking" Gordon Murray. For a long time, design researchers relied on this curiosity and intuition to find new insights (almost like designers do).

The field has evolved since then. Over the past decades a profound body of knowledge on design expertise has been established. This is visible at this conference too: the body of knowledge is applied onto the data of DTRS11. This shows the maturity of the design research field, and is visible in several presentations showing that applying this knowledge onto the data can raise new insights. There are also occasions where proven theory is applied to the data and it doesn't fit. Some parts do fit, but for other parts the design behaviour is too fuzzy and chaotic to fit theory. This resulted in statement during presentations saying, "The data don't fit the model", "The team isn't realising what they are doing" and "The designers should be doing otherwise". Now, of course, you may blame the data if they are not collected well, but I think we all agree that that is not the case here at all. The DTRS11 team did a very good job in capturing design reality. The use of reality is very deliberate: these data reflect real-life designing. These designers are not chosen to participate in an experiment. They are neither students, applying a design process, with no commitment whatsoever to any outcome. The people observed in the data are real designers, with years of experience and working for an international renowned brand. They earn their living by identifying and designing solutions to fit their company's business. They are being paid to come up with results. And most important of all: they are here in the room listening to what research professionals say about their work.

How normative or prescriptive can and should we get as design researchers? When a theoretical model does not fit real data, then an interesting question should arise on the validity of the model. To my mind, the promise of a single case-study such as this is not so much to use it as a way of testing whether the designers are "doing a good job" compared to existing models, but instead to use it to explore how real-world contemporary design actually takes place, and use that as a source of inspiration for how the models might be challenged or changed. Perhaps the "mis-behaving" of the designers when compared to existing models, have some reasonable and insightful explanations?

The dilemma we see here is design research versus design practice. We should redefine our ultimate goal in the end. Do we do research to understand theory and publish or do we actually want to help and support designers in the future? Design research and design practice should really go hand in hand and support each other, and that means respecting each other and having a sincere curiosity in each other.

A theme that came up quite unexpectedly was on the definition of the notion "design" itself. After 25 years of study you might expect that we kind of had defined our field. However, during the panel debate Gabi Goldschmidt introduced the need to "Study design at its core" and raised the question of whether the data related to design activity at all. So, where does design "end" and these other activities begin? And should or shouldn't we as a community embrace the complexity of design?

At DTRS11 a very broad variety of research topics have been presented, including communication, framing, problems and solutions, timeframes, insights and empathy, iterations, conflicts, collaboration, user-centredness, co-creation, linguistics, leadership, notes, and cross-cultural aspects. Are the research questions broader, or deeper? What is the focus and what is the scope of design?

In my innovation practice, I encounter that problems become more and more complex and that new expertise is needed to solve the issues at hand. More and more people are involved, stemming from different organizations, with different drivers and aims. More and more I work with non-designers in a project that is aimed to come up with breakthrough solutions. Is that not design? It is real, at least to me, dealing with that team. And I love the challenges every day again. For me it symbolizes just the kernel of design: providing a process to act in complex situations, where others should freeze. Isn't this just what Tim Brown (2009) means in his book "Change by design", where he argues that: "Design thinking takes the next step, which is to put these tools into the hands of people who may never have thought of themselves as designers and apply them to a vastly greater range of problems".

Is the DTRS11 data about designing? Designers see it as designing. As in my daily practice, people see it as designing. As I see it, the focus of design may shift, towards more stakeholders, more topics, more complexity, but the scope is still the same: creating a "wow" for people in the end. Are we as researchers to discuss the shift in focus of design in practice? Or could we learn here from the design practitioners and – as they do – include all stakeholders in the discussion, and elevate pleasure for that audience when presenting our results. I sincerely hope we embrace the new designing and the complexity it brings. The challenge we face is to grasp all the opportunities that lie within our reach. With the claim that Tim Brown makes on the potential of design thinking: let us – together practitioners and researchers in design, really make this world a little bit of a better place to live. Then what are the new research challenges that come from there?

5 THE FUTURE OF SHARED DESIGN DATA

The lively debate helped highlight key questions for future data-sharing events to attempt to answer in making decisions on where to orient the camera. First, to what extent is the dataset aimed at deductively addressing specific models or aspects of design versus to what extent is the dataset aimed at allowing for new angles to be inductively explored? While the past DTRS events have successfully proven that both purposes may be served simultaneously in the same dataset, it is also apparent that the dataset qualities making it suitable for one are different from the qualities that make it suitable for the other. Second, any domain will, over time, find itself revisiting the definition of its core constructs. Definitional debates are often focused on exploring so-called *boundary conditions*. As the domain changes over time, it becomes important to establish whether new areas and activities may be subsumed under the "design" heading. But while the boundary conditions may be debated, the domain-central constructs change at a slower pace. No one would question that a renowned designer performing sketching activity of what is later deemed world-class architecture is *design*. But is he or she designing when talking to potential users?

The DTRS11 dataset explicitly sought to extend and capture the expanding boundaries of design. Implicitly an assumption was made that when we followed and captured the process of a professional design team working for a large international design company doing a design task full-time, then what was caught on tape was *designing*. Is this a valid assumption? While we believe the assumption is warranted, the panel debate illuminated that the assumption is precisely debatable. It can be argued that one should stick to the core of design activity in order to ensure that only *design* is captured, or it is possible to argue that one should not stick to the core in order to learn from fringe or evolving design phenomena. A key learning point in capturing future shared data is that pointing the camera toward professional designers tackling a design problem does not in itself alleviate the researcher from considerations of whether design activity is actually taking place. In setting up the DTRS11 dataset, we, the organizers, made editorial decisions (see also Lloyd & Oak, this volume): we screened out off-task behavior (e.g., jokes and office gossip between the designers), we did not trace the designers in their activities outside of those involving team design dialogue, which excluded, for example, individual sketching activity, bathroom visits and bus rides. We purposefully intended to capture a wide range of design activities, including interactions with lead-users. Were these data collection decisions well-chosen? The panel debate highlight that any decision as to where to point (and not to point) the camera involved the implicit claim that designing is (versus is not) taking place in that direction. In the absence of one consensual and overarching definition of design, we maintain that the assumption that tracing professional designers doing a design task was focally about design can be defended, but that does not mean it is not debatable. As Rianne Valkenburg commented – it all starts with the *definition*. Should design be defined as a profession, a type of activity or process, or be judged *post hoc* from the qualities of the resulting design outcome?

As the panel debate richly illustrates, shared data do have a future in design research. A key point to take away from DTRS11 and the Open Science agenda seems to be the need for trying hard to ensure that future data-sharing projects become available to a wider research and educational community, beyond the timeline of the shared data process. The DTRS datasets, while shared across a large number of research teams, have not been preserved and shared openly beyond the timeframe of the symposia. This lack of preservation is unfortunate as it makes later data re-use, replication, and cross-dataset comparison studies difficult. There are many reasons for this lack of data preservation: DTRS data have been collected for a specific collaborative purpose and centrally involve behavioral data where data-use agreements have not typically foreseen the need for long-term data preservation at the collection stage. In DTRS11 the initial intention was to collect data preservable and shareable beyond the symposium duration for wider use in the design research community. However, since the data centrally involve videos of organizational design processes with sensitive intellectual-property information, it proved impossible to agree with the case organization on terms that allowed for wider sharing of the data beyond what was needed by the symposium participants for the purpose of researching for this book.

Design research is in need of data build-up, allowing for the accumulation of design cases, generalizability, and the pursuit of reproducible findings both within and across datasets. It is time that a research infrastructure is established to enable the continual preservation of design-research video-data, for wider use in the design research and education communities. Such an infrastructure will, though, have to struggle equally with the difficulties of getting access and documenting the design processes as well as the methodological choices associated with data capture, and the ethical, intellectual-property and personal restrictions in using and sharing the resulting data. Nonetheless, in the continual effort to move towards Open Science it is of pivotal importance that future efforts allow for long-term and wide-open sharing of data in order to reap the full benefits that shared design data can afford.

6 THEMES FROM DTRS11

The qualities of the DTRS11 dataset sparked a multitude of research questions, which, for the purpose of the present book have been categorized into a number of themes:

Several papers related to **team dynamics and conflicts (Part 1)** in multicultural design teams, focusing on aspects related to team conflict (Chapter 3, Paletz, Sumer, & Miron-Spektor), communication (Chapter 6, Awomolo, Jabbariarfaei, Singh, & Akin), leadership (Chapter 5, Chen, Neroni, Vasconcelos, & Crilly), relationships with stakeholders (Chapter 4, Ylirisku, Revsbæk, & Buur), and how the team changed their conversation topics over time (Chapter 7, Chan & Schunn).

The expanding design-space boundaries were explored in a set of papers covering **designing across cultures (Part 2)**, focusing on design thinking as a situated cultural practice (Chapter 9, Clemmensen, Ranjan, & Bødker), the articulation of cross-cultural assumptions (Chapter 11, Gray & Boling), and differences and values (Chapter 10, Dhadphale). Crossing cultural divides need not, however, involve crossing borders, as both within-team and within-organization cultural differences to different organizational stakeholders were also extensively explored (Chapter 3, Paletz, Sumer, & Miron-Spektor; Chapter 8, Smulders & Dunne; Chapter 22, Adams, Aleong, Goldstein, & Solis; Chapter 4, Ylirisku, Revsbæk, & Buur).

Aligning with the idea that design team protocols may reveal design cognition, a subset of papers explored the **cognitive and metacognitive aspects of design thinking (Part 3)**. Two papers explored metacognitive processes relating to fluctuating epistemic uncertainty (Chapter 14, Christensen & Ball) and process awareness (Chapter 12, Valgeirsdottir & Onarheim), along with papers covering the cognitive functions of the frequently used, but little studied, design material of Post-its (Chapter 13,

Table 1.1 Quantitative and qualitative methods applied across book chapters.

Chapter	Approach	
	Qualitative	Quantitative
3	Psychological Factors Surrounding Disagreement in Multicultural Design Team Meetings	Content analysis (pre-defined coding scheme)	Regression
4	Resourcing of Experience in Co-Design	Interaction analysis (Symbolic Interaction)	
5	The Importance of Leadership in Design Team Problem-Solving	Content analysis (pre-defined coding scheme), Qualitative observations (without pre-defined coding scheme)	Frequency analysis
6	Communication and Design Decisions in Cross-Functional Teams	Content analysis (pre-defined coding scheme), Sociometric data analysis	Frequency analysis
7	A Computational Linguistic Approach to Modelling the Dynamics of Design Processes		Topic modelling, Latent Dirichlet Allocation
8	Disciplina: A Missing Link for Cross Disciplinary Integration	Grounded Theory, Qualitative observations, abductive coding	
9	How Cultural Knowledge Shapes Design Thinking	Qualitative observations, Content analysis (pre-defined coding scheme)	Frequency analysis
10	Situated Cultural Differences: A Tool for Analysing Cross-Cultural Co-Creation	Typological analysis (pre-defined coding scheme), Qualitative observations (without pre-defined coding scheme)	
11	Designers' Articulation and Activation of Instrumental Design Judgments in Cross-Cultural User Research	Discourse analysis, Thematic analysis	Word frequency
12	Metacognition in Creativity: Process Awareness Used to Facilitate the Creative Process	Qualitative observations, Content analysis (pre-defined coding scheme)	Quantitative coding (word search), Linear analysis
13	Grouping Notes Through Nodes: The Functions of Post-it Notes in Design Team Cognition	Qualitative observations, Conversation analysis, Multimodal analysis	
14	Fluctuating Epistemic Uncertainty in a Design Team as a Metacognitive Driver for Creative Cognitive Processes	In vivo analysis (pre-defined coding scheme)	Regression
15	Temporal Static Visualisation of Transcripts for Pre-Analysis of Video Material: Identifying Modes of Information Sharing	Observational case studies	Temporal Static Visualisation, Dynamic Data Visualisation

#	Title		
16	Combining Computational and Human Analysis to Study Low Coherence in Design Conversations	Topic Markup Scheme	Latent Semantic Analysis
17	"Comfy" Cars for the "Awesomely Humble": Exploring Slangs and Jargons in a Cross-Cultural Design Process	Content analysis (pre-defined coding scheme)	Variety analysis
18	Design {Thinking | Communicating}: A Sociogenetic Approach to Reflective Practice in Collaborative Design	Interaction analysis, Sound-wave display	
19	Unpacking a Design Thinking Process With Discourse and Social Network Analysis		Discourse-linguistic analysis, Semantic analysis, Network analysis
20	Design Roulette: A Close Examination of Collaborative Decision-Making in Design From the Perspective of Framing	Qualitative observations, Content analysis	
21	Planning Spontaneity: A Case Study About Method	Qualitative observations, Open coding, Affinity diagraming	
22	Problem Structuring as Co-Inquiry	Qualitative observations, Collaborative inquiry	
23	Designing the Constraints: Creation Exercises for Framing the Design Context	Qualitative observations	
24	Cracking Open Co-Creation: Categorizations, Stories, Values	Frame Analysis, Membership Categorization Analysis	Word frequency analysis
25	From Observations to Insights: The Hilly Road to Value Creation		Semantic gravity and density
26	Empathy in Design: A Discourse Analysis of Industrial Co-Creation Practices	Discourse analysis, Thematic analysis	
27	Information-Triggered Co-Evolution: A Combined Process Perspective	Qualitative observations, Content analysis (pre-defined coding scheme)	Comparisons of information types to transition types
28	Team Idea Generation in the Wild: A View From Four Timescales	Qualitative observations, morphological analysis, Open coding,	
29	Structures of Time in Design Thinking	Qualitative observations	Timeframe analysis
30	Tracing Problem Evolution: Factors That Impact Design Problem Definition	Qualitative observations, Inductive coding	

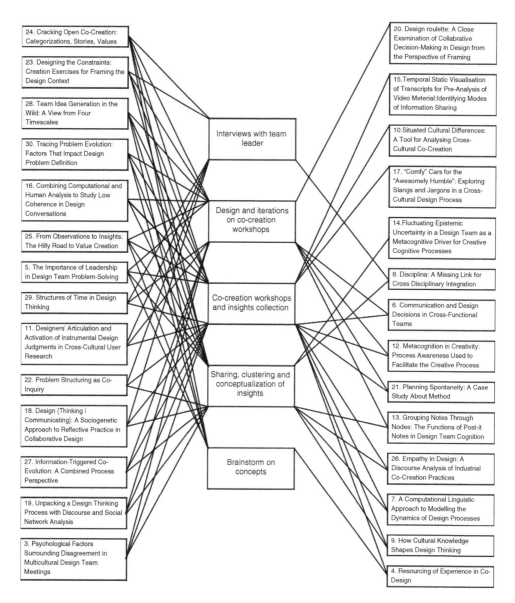

Figure 1.1 Overview of data usage across chapters.

Dove, Abildgaard, Biskjaer, Hansen, Christensen, & Halskov; Chapter 14, Christensen & Ball), and how to study modes of information sharing (Chapter 15, Wulvik, Jensen, & Steinart).

The topic of **design talk (Part 4)** identified the language (Chapter 19, Bedford, Arns, & Miller) and communication (Chapter 18, Jornet & Roth) of design as productive elements, with particular emphasis on design slang and jargon (Chapter 17,

D'souza & Dastmalchi), and conversational focus shifts in the form of low coherence turns (Chapter 16, Menning, Grasnick, Ewald, Dobrigeit, Schuessler, & Nicolai).

Framing in design (Part 5) emerged as a central concept across several studies, discussing framing as a skill (Chapter 20, McDonnell), problem structuring in collaborative inquiry processes (Chapter 22, Adams, Aleong, Goldstein, & Solis), the design of constraints (Chapter 23, Eckert & Stacey), and planned spontaneity (Chapter 21, Turnhout, Annema, Goor, Jacobs, & Bakker).

The DTRS11 dataset also centrally involved user interactions, and topics relating to **co-creating with users (Part 6)** included user empathy (Chapter 26, Hess & Fila), stories and values in co-creation (Chapter 24, Lloyd & Oak), and how to identify patterns of movement in designing for users (Chapter 25, Dong & McDonald).

The longitudinal nature of the dataset allowed for the studies of **design iterations across time (Part 7)**, including explorations into time structures (Chapter 29, Teixeira, Shafieyoun, Rosa, Cai, Li, Xu, & Chen) and different timescales (Chapter 28, Shroyer, Turns, Lovins, Cardella, & Atman), and the tracing of how themes (Chapter 25, Dong & McDonald; Chapter 7, Chan & Schunn) and problems (Chapter 30, Daly, McKilligan, Murphy, & Ostrowski) evolve or co-evolve (Chapter 27, Cash & Gonçalevs) over the duration of the dataset.

The individual chapters not only cross-link in terms of their central themes. On the previous pages we highlight how the chapters cross-link in terms of their chosen methods (see Table 1.1) and which parts of the data the researchers or research teams decided to utilize in their analysis efforts (see Figure 1.1).

REFERENCES

Adams, R. & Siddiqui, J. A. (eds.) (2016) *Analyzing Design Review Conversations*. West Lafayette, IN: Purdue University Press.

Adams, R. S., Cardella, M., & Purzer, S. (eds.) (2016). Analyzing design review conversations: Connecting design knowing, being and coaching. Special Issue of *Design Studies*, vol. 45, part A.

Adams, R.S., Radcliffe, D., and Fosmire, M. (2016). "Designing for Global Data Sharing, Designing for Educational Transformation." Advances of Engineering Education, Special issue on data sharing, 5(2), pp. 1–24.

Baker, A., van der Hoek, A., Ossher, H., & Petre, M. (2011). Studying Professional Software Design. Special Issue of *IEEE Software*, vol. 29, no. 1.

Baker, M. (2016). 1,500 scientists lift the lid on reproducibility. *Nature*, 533, 452–454.

Brown, T. (2009). *Change by design: How design thinking transforms organizations and inspires innovation*. New York: Harper Collins.

Christensen, B. T., & Abildgaard, S. J. J. (2017). Inside the DTRS11 Dataset: Background, Content, and Methodological Choices. In: Christensen, B. T., Ball, L. J. & Halskov, K. (eds.) *Analysing Design Thinking: Studies of Cross-Cultural Co-Creation*. Leiden: CRC Press/Taylor & Francis.

Christensen, B. T., & Ball, L. J. (2014). Studying Design Cognition in the Real World Using the 'In Vivo' Methodology. In P. Rodgers & J. Yee (eds.) *The Routledge Companion to Design Research*, pp. 317–328. Abingdon, UK: Routledge.

Christensen, B. T., & Schunn, C. D. (2007). The relationship of analogical distance to analogical function and pre-inventive structure: The case of engineering design. *Memory & Cognition*, 35(1), 29–38.

Cross, N., & Cross, A. C. (1996). Winning by design: The methods of Gordon Murray, racing car designer. *Design Studies*, 17(1), 91–107.

Cross, N., Christiaans, H., & Dorst, K. (eds.) (1996). *Analysing Design Activity*. Chichester, UK: John Wiley & Sons.

Davies, S. P. (1995). Effects of concurrent verbalization on design problem solving. *Design Studies*, 16(1), 102–116.

D'souza, N. (2016). Investigating design thinking of a complex multidisciplinary design team in a new media context. Special section of *Design Studies*, vol. 46.

Dorst, K. (ed.) (1995). Analysing Design Activity. Special issue of *Design Studies*, vol. 16, no. 2.

Dorst, K., & Cross, N. (2001). Creativity in the design process: Co-evolution of problem-solution. *Design Studies*, 22(5), 425–437.

Ericsson, K. A., & Simon, H. A. (1993). *Protocol Analysis: Verbal Reports as Data* (Revised edition). Cambridge, MA: MIT Press.

Finke, R. A., Ward, T. B., & Smith, S. M. (1992). *Creative cognition: Theory, research, and applications*. Cambridge, MA: MIT Press.

Hales, C. (1987). *Analysis of the engineering design process in an industrial context*. Doctoral dissertation, University of Cambridge, UK.

Lloyd, P., & McDonnell, J. (eds.) (2009). Values in the Design Process. Special issue of *Design Studies*, vol. 29, no. 2.

Lloyd, P., Lawson, B., & Scott, P. (1995). Can concurrent verbalization reveal design cognition? *Design Studies*, 16(2), 237–259.

McDonnell, J., & Lloyd, P. (eds.) (2009). *About: Designing – Analysing Design Meetings*. London, UK: Taylor & Francis.

McDonnell, J., & Lloyd, P. (eds.) (2009). Analysing Design Conversations. Special issue of *CoDesign*, vol. 5, no. 1

Munafó, M. R., Nosek, B. A., Bishop, D. V., Button, K. S., Chambers, C. D., du Sert, N. P., Sominsohn, U., Wagenmakers, E.-J., Ware, J. J., & Ioannidis, J. P. (2017). A manifesto for reproducible science. *Nature Human Behaviour*, 1, 0021.

Petre, M., & Van Der Hoek, R. (2013). *Software Designers in Action: A Human-Centric Look at Design Work*. Boca Raton, FL: Chapman and Hall/CRC.

Petre, M., van der Hoek, A., & Baker, A. (eds.) (2010). Studying Professional Software Design. Special issue of *Design Studies*, vol. 31, no. 6.

Schön, D. A., & Wiggins, G. (1992). Kinds of seeing and their functions in designing. *Design Studies*, 13(2), 135–156.

Schooler, J. W., Ohlsson, S., & Brooks, K. (1993). Thoughts beyond words: When language overshadows insight. *Journal of Experimental Psychology: General*, 122(2), 166–183.

Valkenberg, R., Sluijs, J., & Kleinsmann, M. (2016). *Images of Design Thinking: Framing the Design Thinking Practices of Innovators*. Amsterdam, Netherlands: Boom.

Chapter 2

Inside the DTRS11 Dataset: Background, Content, and Methodological Choices

Bo T. Christensen & Sille Julie J. Abildgaard

ABSTRACT

From October 2015 to January 2016 we followed and video recorded a professional Scandinavian design team in a user involvement department solving a design task for a worldwide manufacturer within the automotive industry. Central elements of the observed design processes included designing with co-creation along with user understanding and user experiences. We followed parts of the second phase of the design team's design project, observing the design team in their process of developing a concept package with both a soft delivery (e.g., strategies and sales channels) and more tangible delivery (e.g., mock-ups of accessories to their target users). Centrally, the dataset involved cross-cultural co-creation, in that the Scandinavian design team travelled to China to conduct user sessions with Chinese lead users.

In Part 1 of this chapter we outline the methodologies that we applied in getting access to the data, in collecting the data, in sampling the data, in processing the data and in distributing the data to the research teams to enable them to analyse and interpret the data. In Part 2 we provide a detailed description of the content of the dataset. We do this by summarising for each distributed video the design team that was observed and the design project that was being pursued.

PART 1: METHODOLOGY

1.1 Getting access

Using video in design research often raises concerns about how to gain access to and permission from participants and organisations alongside issues of ethics. A string of organizations were approached through both network-based contacts and general calls for expressions of interest in participation via intermediaries. Several organiza-tions expressed interest in participating, but issues relating to confidentiality as well as the anticipated extensive video recording of team processing proved to be difficult points for many organizations to agree on. The case company was eventually located through intermediary personal contacts directly to the design team leader. We estab-lished contact and set up a non-committal meeting to explain our intentions with the project and discuss any concerns and implications. The design team gave their permis-sion for us to follow their process and to undertake the video recordings. The nature of the company and the project made it necessary for everyone participating in the videos to sign informed consent forms, and for all research teams to sign both a Data Use Agreement and a Company Secrecy Agreement (described below).

1.1.1   Informed consent and classified information

The research set-up required written consent from all participants. This was obtained prior to the recordings by providing an information sheet (Informed Consent) to all members of the design team, along with the two stakeholders and the external consultants. All participants on tape signed the Informed Consent. All participants therefore remain anonymous in any publications and public material resulting from this project.

In order to protect all classified corporate information the following steps were taken. First, during data collection, a note was made by the DTRS11 participating observer (supported by the design team) on which (if any) parts of the video might contain company secret information. Second, prior to distribution of data to research teams the design team was able to view, read and screen all information that was to be distributed in order to clear the dataset for any company secret information. Third, before getting access to the videos and materials collected at the User Involvement Department, each DTRS11-enrolled researcher was required to sign a company 'Secrecy Agreement'. This needed to be signed by every person getting access to the videos through a password-secure server. Fourth, publications stemming from the DTRS11 data collection were required to omit any company secret information as well as any information that might identify company employees or other participants represented in the dataset (e.g. participants in the co-creation sessions). Such anonymity in publications was achieved through the use of pseudonyms and other identity-veiling actions.

As per the Secrecy Agreement, the researchers needed to consult the DTRS11 organizers if they were in doubt as to whether information contained in their publications might violate confidentiality. If company secret information was present, then this was primarily resolved by changing the information to something similar but non-classified in order for the reader to still understand the design process, while making the content non-confidential. Pictures or drawings were only used in the publications with the acceptance of The Company. Names were anonymized in all transcripts and written documents originating from the DTRS11 dataset, and a transcript and pseudonym key was provided for use.

1.2   In situ data collection

Video data were collected with the same design team over a four month period, tracing day-to-day the natural course of the design process. All data were collected *in situ*, in the design team's natural environments, rather than in a controlled environment or an experimental set-up.

We began the data collection for DTRS11 in September 2015 at the design team's Scandinavian office, and repeatedly attended meetings until the end of January 2016, with a return visit in March 2016, where a follow-up interview was conducted. During our observation period, we recorded more than 150 hours of *in situ* footage of the design team's daily routines.

The aim of the data collection was to provide high quality video recordings of interactions between the professional designers in their ongoing design process, and their interaction with stakeholders and lead users. To maximize the quality of the dataset an effort was made to follow the natural design process and not interfere with the normal work routines in the team. A DTRS11 student assistant worked from the design team's office on certain days and thereby was able to set up cameras at short notice when the design team engaged in meetings or dialogues concerning the design process. The student assistant, David, took on the role of a participating observer during the data collection period, which helped to develop a familiarity with the characteristics of the work routines of the team and different settings. He also followed the design team to China for their 10-day field trip to conduct video observations during two co-creation workshops with lead users, and to collect recordings of the design processes, ideations, and concept development in between the two co-creation workshops. David was present at the design team's office and throughout the field trip to China, which proved critical in order to decide when and how to record, where to position the equipment, and how to deal with practical problems such as securing a clear visual image and good quality sound.

Data were collected using multiple cameras with synchronized microphones recording to hard disks (see Table 2.1 and Figure 2.1 for specifications). Each recording set-up was unique in the way that no camera was fixed, but mounted on the go when the design team met for a meeting or when a spontaneous conversation occurred. We used power banks when possible to ensure an easy, flexible and cord-free set-up that could record for many hours without frequent battery changing.

Before commencing the video data collection we decided to complement the video recordings with an interview with the design team leader. The interview was framed as an introduction to the design project for the DTRS11 research teams, told by the design team leader himself. The interviewer brought a semi-structured interview guide and an audio recorder. The semi-structured interview format gave a loose structure to the interview, diving into certain elements of the project, that is, the project plan, the expected deliveries, the design approach and so on, but the semi-structuredness also allowed room for the team leader to explain his thoughts and reflections.

After the data collection period and after the dataset was distributed to the researching teams in January 2016, we conducted a follow-up interview with the team leader, following the same semi-structured format as the initial interview. This time we encouraged the research teams to contribute with questions which they felt could enrich their analyses and clarify potential uncertainties that might have occurred during the research process. The follow-up interview was conducted and shared 2 months into the research process (March 2016), with the intention to encourage interactions between the actual design team being studied and the researchers working with the data.

1.3 Sampling the collected data

Any collection of timewise-disjointed video requires some editorial decisions. We had much more video data than we could possibly transcribe or distribute for DTRS11 (150+ hours of video was collected – see also technical report), and based on the experiences of earlier DTRS data-sharing events (see Christensen *et al.*, 2017; Chapter 1 of this volume), we intended to select approximately 10–15 hours of video for

Table 2.1 Recording specifications.

Video specifications
Cameras in total: 4 (GoPro HERO3+ black edition)
Recording quality: 1080 p/30 fps/wide-angle
Time-lapse: Pictures were taken every 10 sec.
Data amount: 9 gb/hour (approx.)

Audio specifications
Audio-recorders in total: 4 (Zoom h1)
Recording quality: 24 kbit/48 kHz .wav-file
Data amount: 1 gb/hour (approx.)

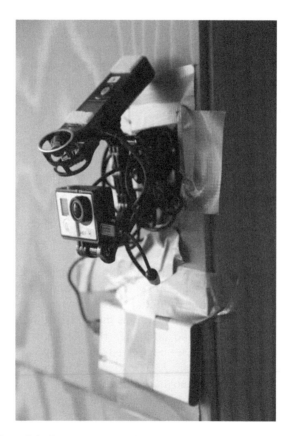

Figure 2.1   Camera set-up with power bank and audio recorder.

distribution. A number of criteria guided our selection of the final videos: we wanted to display a range of different types of design activity performed by the same team over time; we wanted to follow a design development process over an extensive period of time; we wanted to follow the design team across contexts 'in the wild'; we wanted video/audio of good quality, we did not want lengthy engagement on camera with company secret information. These criteria were guided by an effort to end up with a dataset that reflected original features that we had not previously seen in shared design datasets derived from field observations. In particular, we've never seen fully longitudinal studies of design teams and we've never seen a team being followed carrying out multiple types of design activity. So, achieving these criteria was an ambition from the outset.

As we followed the design team over time, it became apparent to us that the data we were collecting could centrally be characterized as involving *co-creation* with lead-users and involved multiple *cross-cultural* design interactions (both within the organization and across continents in communicating with lead-users and moderators). This led us to coin the theme of 'cross-cultural co-creation' for the DTRS11 workshop, which further guided which videos we would eventually select (e.g., we included videos involving activities before, during and after specific co-creation sessions). We evaluated all materials, going over field-notes and preliminary observations in order to review data, but also undertaking detailed assessments of selected fragments in meetings and discussing which fields of inquiry might be pursued and which analytical approaches could form a focus for further research. We sampled sessions from different stages in the design process and from different meeting set-ups. The DTRS11 dataset was designed to provide multiple entry points of analysis, allowing researchers a wide range of analytic options. The videos included collaborative design activities at various stages of the team's design process, including planning, ideation, designing and executing two co-creation workshops with lead users. The recorded sessions and meetings included variations in structure and stages in the design process, such as stakeholder meetings, meetings with external consultants, core-team meetings, workshops, sprint sessions, brainstorming sessions, spontaneous idea generation sessions and briefing sessions. The design team employed many different modalities when working; at a minimum speech, gestures, written comments, post-its, drawings and pictures were visible in most of the recorded sessions. To constitute the final DTRS11 dataset we selected 15 hours and 24 minutes of video recordings and 1 hour and 56 minutes of audio recordings of two interviews with the design team leader, along with additional pictures and material. This final dataset included 20 sessions from different stages in the design process and from different meeting set-ups, supplemented by the background interview and follow-up interview with the design team leader, for a total of 22 data sessions. A few sessions with extensive engagement with company secret information were intentionally excluded from the final dataset. All sessions came with full transcripts, supplementary material and English translations where necessary. Data pertaining to the Chinese lead-users were only distributed in the form of time-lapse videos along with the step-by-step moderation guides as additional material. The time-lapse videos were included with the intention that the viewer would quickly be able to get a general idea about what took place in the co-creation sessions. Originally we wanted to share these recordings in full length, but since the sessions ended up being in Chinese in a noisy environment, with multiple simultaneous co-occurring conversations, it proved to be impossible to transcribe, translate, and anonymize the data.

1.4 Packaging and distributing the data

Following data sampling we transcribed the selected data and generated illustrations of the set-up for each session (cameras in use, people present etc.) as well as data descriptions to help provide a quick overview.

Table 2.2 Transcription notation.

.	Indicates a natural end of a sentence (falling intonation).
,	Indicates a natural short break in the spoken sentence.
!	Indicates animated intonation.
?	Indicates questioning intonation.
aa'	Indicates a word whose ending is cut off.
: or ::	Prolonging of sound. Two colons indicate a strong prolonging.
aa-	Interrupted speech. Incomplete or cut off utterance.
>aa<	Rapid speech.
<aa>	Slow speech.
AA	Pronounced with high-volume.
(.)	Short break in utterance (less than 0.5 second).
(..)	Longer break in utterance (between 0.5 and 2 second).
[aa]	Hard brackets indicate when overlapping speech begins and ends.
(aa?)	Unclear utterance or word. Transcriber's best guess within parentheses.
(aa)	Paralinguistic or non-verbal activity. Used for transcriber to describe laughs, mimics, gestures, movement, gaze or attention. Longer breaks are also noted within parentheses.
XXX	Confidential material.

1.4.1 Transcription of video

For researchers to undertake a close, detailed inspection of the data we transcribed all talk in the 20 selected videos along with the two interviews. Three students and research assistants transcribed the selected data using both video and audio files as sources of content. When transcribing videos a range of details emerge from repeated viewings of the data, and the video reveals a complexity in the participants' actions that an audio recorder, or just a separate camera, cannot capture. To prevent the transcriptions from becoming too detailed we used a simple transcription notation (Jefferson, 1984) inspired by Conversation Analysis (CA) to transcribe talk and in some cases activity (see Table 2.2).

The separate videos along with spilt-screen views and the transcriptions allowed for a detailed analysis, whether the researching teams were interested in only talk or in additional modalities such as gestures, body language, materiality and group interaction. The lengths of pauses or silence were not measured in milliseconds, but were indicated. A few segments in the transcriptions were altered in some instances where the content threatened company secret information, or contained content involving ethical concerns. Confidential material was marked with XXX and all names, company-specific products and places, and other confidential material was concealed using pseudonyms. In order to translate these names all researching teams used a separate "Anonymization Key". This makes it possible to locate the same person/company across publications and chapters in the present volume.

The transcripts were entered into Excel files, and segmentation took place according to turn-taking in talking. In a few instances, overlapping speech (marked with the hard brackets [ ]) was moved to the subsequent segment in order for the reader to follow the conversation. See Table 2.3 for an illustrative transcript fragment. All segments were numbered and all research teams were asked to use the same reference style, making it possible to locate the current segment across chapters. The reference

Table 2.3 Illustrative transcript fragment.

Segment	Initial of the person talking	
89	K	Yeah, why why why, and then we have them (.) "now you've defined health, or "good life", within our interpretation" (.) and then we ask them "you feel there's something in our interpretations missing or is wrong?".
90	A	"Now spend five minutes, eh to: eh write down what, what else: is missing".
91	K	"In "good life" ", yeah.
92	A	In this direction.
93	E	[So >why why why< −]
94	K	[What is missing or what is wrong, or]
95	E	Yeah, so maybe we do some clustering here and maybe we add some, some theme headers or something like that.
96	A	Yeah. And remember that (.) we talked about that we: are actually not a part of this because, we don't want our [eh: exactly, perception of health]
97	E	[Western: (.) to pollute it]
98	A	So in one group there's actually just three people (.) doing this exercise, ehm: (..) so I think that's (..) it's also putting eh some kind of pressure on these people (.) that they need to- thre'- three people needs to: (.) share back (.) to six people.
99	K	Yeah (..)

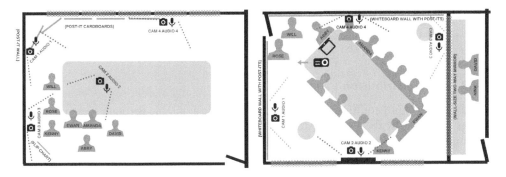

Figure 2.2 Plan examples from two different *in situ* recording set-ups.

style is (vX, Y), with v for video number, and Y number referring to the segment number in the transcript. For example (v03, 23) refers to video 03, segment 23.

1.4.2 Data session set-up descriptions

Set-ups from the selected 20 videos were illustrated and made available within the DTRS11 database. The numbers of cameras in use and available angles were illustrated for each recording (see Figure 2.2) along with people present in the room, materials, artefacts and the technology in use. The respective set-ups for all recordings also appear in the initial seconds of each distributed movie – immediately after the data disclaimer. Thus camera angles and audio sources vary for each recorded session.

Table 2.4 Overview of data in the DTRS11 dataset.

Session no.	Title	Date (yyyy.mm.dd)	Length (minutes)
01	Background interview with Ewan	2015.10.09	32
02	Designing co-creation workshops	2015.10.28	21
03	Iterations on workshop design	2015.10.28	55
04	Iterations on co-creation workshops	2015.11.09	77
05	Designing co-creation workshop day 2 (CC2)	2015.11.11	100
06	Co-creation workshop day 1 (CC1)	2015.12.01	Time-lapse 10
07	Debrief CC1	2015.12.01	18
08	Sharing insights from CC1	2015.12.02	56
09	Clustering insights from CC1	2015.12.03	52
10	Iterations on CC2 design	2015.12.03	34
11	Linking insights from co-creation to project	2015.12.04	39
12	Briefing stakeholders on CC1	2015.12.04	16
13	Co-creation workshop day 2	2015.12.07	Time-lapse 10
14	De-brief CC2	2015.12.07	78
15	Sharing insights from co-creation workshops part 1	2015.12.08	54
16	Sharing insights from co-creation workshops part 2	2015.12.08	34
17	Sharing insights from co-creation workshops part 3	2015.12.08	21
18	Clustering insights from CC1+CC2	2015.12.09	86
19	Recap with consultants	2015.12.09	38
20	Recap with stakeholders	2015.12.10	37
21	Brainstorm on concept and products	2016.01.26	88
22	Follow-up Interview with team leader	2016.05.03	84

1.4.3 Data distribution

All data (video and additional materials) were made available through a secure online server provided by the Danish e-Infrastructure Cooperation (DeIC), a service specifically aimed at scientific collaboration. The service is primarily intended for working with and sharing research data as well as for the protection of large datasets.

To allow for the longitudinal study of design development, the data were presented in chronological order on the server along with user manuals, guidelines and additional supplementary materials for download. Table 2.4 depicts a chorological overview of all sessions supplied, with number, title, date and length in minutes.

1.4.4 Additional material

Pictures and written material from the sessions had been collected with permission from the team and were added to the dataset. During meetings and workshops high-quality pictures were taken of whiteboards, post-its, materials and notes, and were added to the corresponding session. In addition, each GoPro camera was set to take pictures every 10 seconds, making it possible to add before, during and after pictures of the work that the participants conducted in each meeting. We were also able to obtain copies of most of the materials used and discussed in the recorded design process such as slideshows, the design team's own schedules and notes, moderation guides, and drafts

Table 2.5 Additional materials.

Project brief: "project at a glance"

A short pitch presentation, aimed at external persons. The project brief is a short overview of the design challenge along with the project plan, the involved stakeholders and the expected soft and tangible deliveries.

Internal newsletter (30th of October 2015)

A news item from an internal email sent from the design team, and the two stakeholders, to the different departments in The Company. In the newsletter the design team gives a brief description of the background for the project (Phase I), they explain what they have done so far and where they are heading.

Moderation guide from CC1 and CC2

Two moderation guides used by the external consultants, Rose and Will, during the two co-creation workshops. The documents are the consultant's scripts and contain both the expected program along with a rough manuscript, keywords and practical reminders.

Field plan: "from insights to concept – field plan"

A preliminary day-to-day program for the two co-creation sessions, preparation plans, the preliminary program for the expected workload in the entire team (the design team, the two stakeholders and the consultants) along with programs for the different design workshops such as "insight workshop" "iteration workshop", "create workshop", and "wrap-up workshop" during the field-trip. The Field Plan was created before the field trip to China and was adapted many times during the trip.

Participant list for co-creation workshops

A list of the nine lead user participants in the co-creation workshop. The list states descriptive data such as income, status, car, products, habits etc.

Project deliveries (selected slides from Phase II delivery report)

Selected slides from the design team's delivery report were added as extra material to the dataset (in May 2016). The slides included an overview of the Phase II process, where the design team describes each part of the process from learnings from Phase I, preparations for Phase II, the field trip with the two co-creation sessions, the analysis afterwards to the final deliveries (concepts). Included are slides showing mock-ups of the deliveries manifested in the sub-brand and main concepts.

---

of presentations for stakeholders. This allowed researchers to read the participant's notes, slideshows, and drafts while the design process evolved during the observed months (see Table 2.5 for an overview). All materials went through an anonymization and depersonalization process before being shared for research, as per the Secrecy Agreement. These supplementary materials cannot be shared in publications to ensure confidentiality and Intellectual Property protection.

PART 2: BACKGROUND AND CONTENT

2.1 Participants

The design team/user involvement department

Ewan, Abby and Kenny are the core of the design team within the User Involvement Department and were also part of the Phase I of the project. The three core members were working full-time on the project when we followed them from September 2015 to March 2016. In the User Involvement Department they work with research, design and prototyping within user experiences. Ewan, Abby and Kenny have all been working in the same department the last four years and also worked together in their previous employment, joining the Company as a team. They have educational backgrounds within Communication Design, Graphic Design, Multimedia Design, Humanistic Informatics and Engineering, and have 8–10 years of professional design experience after completing their degrees. The three normally work together on different design projects and know each other well. Nina is an intern during the period of our observation and has a minor role in the design team. She participates in some of the meetings in Scandinavia and she travels with the design team on the field-trip to China. She is involved in a sub-project of Phase II concerning the stakeholder relations, thus working closely with the stakeholders Tiffany and Hans during the field-trip.

The two stakeholders/Accessories Department

Tiffany and Hans are from an Accessories Department in Scandinavia specialized in car accessories and products which can be purchased after car production, and cradle-to-grave products. They collaborate with the design team and are responsible for the project "navigation" or "priming" in relation to other stakeholders and different departments, and the overall implementation of the finished project. They participate in some meetings with the design team in Scandinavia to stay updated on the progression in the project. They travel with the design team to China and participate in some meetings and one of the co-creation workshops, while also conducting research on the Chinese market and they attend meetings with stakeholders while in China.

The external consultants

Rose, Amanda and Will are external consultants with expertise in Asian markets. Two of them are Design Thinking experts, and they all have backgrounds within user and market research. Rose and Amanda were external consultants on Phase I in the project and thus know the design team well. Rose has as also worked with Kenny, Abby and Ewan previously. Will is working with the design team for the first time and is thus new to the project partners and the team dynamics. The three consultants are only full-time on the project during the field-trip to China, where they participate in all meetings on equal terms as the design team members. They also assist in translating Chinese to English, and in translating cultural diversities and traditions. Will and Rose facilitate and moderate the two co-creation workshops. Amanda participates in the workshops on same terms as the design team.

2.2   Project description: The design task

Below we describe the project background (Phase I, which was not observed for DTRS11), and the project description for Phase II (which is the phase covered in the present dataset).

Phase I: Need for a user-centric approach

The first part of the project (Phase I) begins in early 2015 and involves a research phase during the summer of 2015. The design team conducts field research in an Asian city with premium car users. The project aims at answering questions like "*We sell lots of cars in Asia, but the accessory take-rate is very low, why is that?"* and *"In which direction should we aim in order to reach the consumers in a "post-modern world"?* These questions are carried on through Phase II and remain a goal for the project.

In Phase I the design team investigates "the general life" of the Chinese premium car users in terms of transportation needs, habits, pains, pleasures and aspirations. The design team presents a list of recommendations to The Company based on the insights from the research in Phase I. They suggest changes in the storytelling about the company, new portfolio items/lines, and updated sales channels, in order to increase take-rates in China. The design team suggests a more "user-centric" approach and investment in gathering user insights, which becomes the topic of Phase II.

Phase II: Co-creating and redefining car accessories concepts

We follow the design team in the process of planning, designing and executing two co-creation workshops with lead users. Most of the planning and designing phases take place in Scandinavia, while the team executes the co-creation workshops in China. The project is a collaboration between the Accessories Department and User Involvement Department. The two departments (Accessories and User involvement) collaborate in order to *"explore and understand the users in an Accessories scope through qualitative research."* (Unpublished project description).

2.2.1   The design task

The design team sets out to investigate, develop and create, in dialogue with their lead users, a concept package, which redefines what, how and where The Company offers accessories to lead users (referred to as *urbanites*) in China. The design task for Phase II is formulated in these terms:

> *"How might we evoke and capture the attention of the urbanite so that we secure their emotional engagement and establish long-term Company brand/product/service commitments?"*

On the next page (Figure 2.3) is a visualization of the project plan divided into seven parts (the last part, Part 7, is left out of the picture, but is described below). We met and observed the design team during Part 1 "Scope & Plan" and Part 2 "Stakeholder Involvement" of Phase II of the design project, and followed them in Part 3 "Concepting & Field Prep", where the design team designed the co-creation workshops, which took place in China. We travelled with them to capture Part 4 "In-Field Concept Co-Creation" of the design process. In China two co-creation workshops with Chinese lead users took place. In between and after the two co-creation workshops the design team started Part 5 "Concept Refinement", where the design team, three Asian consultants, and in some sessions the two stakeholders, analysed and discussed their observations from the co-creation workshops. During the field-trip the design team turned to Part 6 "Concept Package Assembly" with regard to the softer part of the delivery (e.g., story and strategy on Figure 2.4). In January we met the design team back in Scandinavia, when the Part 6 "Concept Package Assembly" was underway. We observed them while they were conducting a brainstorm session on ideas for the final delivery with four colleagues from their User Involvement Department. The rest of Part 6 "Concept Package Assembly" and the last part of the design project, Part 7

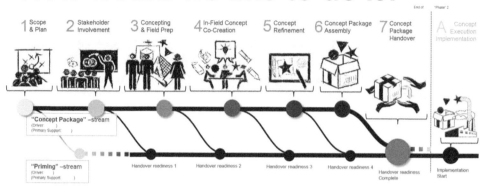

Figure 2.3 The Project plan (Project Brief, 2015).

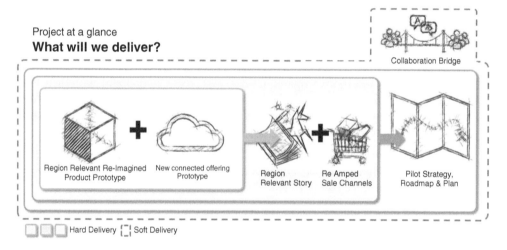

Figure 2.4 Hard and Soft Deliveries (project brief, 2015).

"Concept Package Handover", took place in February and March 2016, and were not observed.

The deliveries (see Figure 2.4) were planned to be "interconnected" and stated in the project brief as a "*merger of product, story and sales channels*", which the team based on the key findings from Phase I, from meetings with the Special Products Unit, the Chinese department, and especially from the co-creation with their core users.

2.3   Content summaries of the 22 individual sessions

In the following, we provide a brief summary to the 22 sessions that appear in the DTRS11 shared dataset.

Background interview with Ewan (v01)

Prior to the video observation at the company, we conducted a qualitative interview with the design team leader, Ewan, at the office of the User Involvement Department. During the interview Ewan gives an overview of the design team's plan for the next months and what they expect to deliver at the end of the project.

Keywords: Design process, Project brief, User experiences, Organization, Project plan

Designing co-creation workshops (v02)

The first recording to appear in the dataset is an all-day meeting, where the design team begins designing the two co-creation workshop sessions, which they are to execute in a month's time. The team begins with an in-plenum sharing of previous experiences with co-creation and expectations for the current project. They discuss their understanding of co-creation and try to define a shared understanding of what co-creation is and what it involves. They go on to discuss different co-creation workshop elements and how to involve the users with different exercises and themes to discuss. The discussion also concerns the importance of the physical environment for the co-creation workshops. A lot of notes are taken on post-it notes with regard to various elements and tools to use and remember during a co-creation session.

Keywords: Co-creation, User involvement, Shared understanding, Physical environment, Post-it notes

Iterations on workshop design (v03)

Following the previous session on this day, the design team continues to take the first steps in designing the co-creation workshop sessions. Having shared previous co-creation experiences and future expectations for the project, Abby, Kenny, David and Nina, led by Ewan, do a 5 minute individual brainstorm with post-it notes on how to screen the right participants for the co-creation sessions. This is followed by a joint participatory session, where everybody presents his or her post-it notes with ideas. Spin-off occurs during these presentations, and minor clustering of the brainstorm takes part. Finally, the team has a detailed discussion with regard to translation challenges, and they are in doubt as to whether it will be possible to carry out the co-creation sessions in English.

Keywords: Brainstorm, Idea generation, Clustering, User involvement, Cultural differences

Iterations on co-creation workshops (v04)

In this session the design team iterates further on the design of the co-creation workshops. Prior to the recording the team has done a sprint planning session, and held a Skype-meeting with Amanda, one of the external consultants, who gave inputs from her expertise in Chinese culture, which the design team then takes with them into the iteration. They discuss and plan the first co-creation workshop in detail using a whiteboard and post-it notes as material. They also discuss expectations and considerations of the element of translation and how the workshop moderator will conduct the session. Besides deciding the program of the day in detail, they also focus on how to ask questions that will help "open up" for the workshop participants' thoughts in relation to certain "pillars" of different themes (e.g., "Health").

Keywords: Co-creation, Workshop design, Moderation, Framing, Cultural understanding

Designing co-creation workshop day 2 (v05)

Following a weekly update and detailed walkthrough of the plans for the first co-creation workshop over Skype with the consultants (Amanda, Rose, and Will), the design team go on to have their first iteration on the set-up of the second co-creation workshop. In this session they plan and design the second co-creation workshop in detail, and many themes and elements are discussed. They furthermore discuss and plan different concrete workshop exercises in detail, explaining their ideas behind these exercises. An exercise where the workshop participants are to vote for and invest in certain themes of interest is especially discussed and formed in detail. During all of the above, they use post-it notes from previous sessions as input in the planning process and continuously use whiteboards for outlining and noting their plans and thoughts.

Keywords: Co-creation, Workshop design, User involvement methods, Design iterations, Post-it notes

Co-creation workshop I (v06)

This session is the first recoding in China, where the first co-creation workshop is conducted with seven Chinese lead users. Ewan, Abby, and Kenny are present, alongside Rose and Will who facilitate and moderate the sessions (in Chinese). Two translators translate simultaneously from Chinese to English so that Ewan, Abby and Kenny can follow what is going on, ask questions and bid into the conversation with help from the translators. David and Nina take notes behind a one-way mirror. The workshop begins with a greeting session followed by a presentation of the plan led by Rose and Will. They move on to a "Health" & "Good life" themed session where the participants explain and discuss, vote and cluster their interpretation of the themes using post-it notes as design material.

Keywords: Co-creation, User involvement, Brainstorming, Clustering, Value definitions

Debrief co-creation workshop day I (v07)

Immediately after the first co-creation workshop, the consultants and the design team gather in a debriefing session, sharing observations and notes with each other. The debriefing is done in two groups: David, Will, Amanda, and Abby in one; Ewan, Rose, Kenny, and Nina in the other. In this session we follow the group with Will, who starts by translating and explaining the different post-it clusters that were written and put on one of the walls by the workshop participants. Supplemented by Amanda, Will also explains some of the participants' characteristics and statements, and attempts to summarize the apparent relevant findings. These include among other things how the participants conceive of leisure time, family relations, and general ideas about the theme of "Health" & "Good life".

Keywords: Debriefing, Sharing experiences, Cultural translation, User insights, Evaluation

Sharing insights from co-creation workshop day 1 (v08)

The design team and the consultants spend a whole day sharing and working with the findings and observations from the first workshop. In this debriefing session the first workshop is discussed and noticeable observations are shared. The external consultants and moderators, Will and Rose, relay their insights and findings from their respective co-creation workshop groups in PowerPoint presentations. Both Will and Rose describe their impressions of some of the different workshop participants' personalities, characteristics, and their personal stories and values. The focus is on the participants' statements in relation to different themes, such as health, safety, family, and social status.

Keywords: User insights, Value definitions, Conceptualisation, Cultural translation, Ideation

Clustering insights from the first co-creation workshop (v09)

Having spent the previous day sharing insights from the first co-creation workshop session, the design team now meets with two of the external consultants, Amanda and Rose, for a post-it clustering session with the aforementioned insights. The team start analysing these insights in relation to their Phase I findings, and begin to adjust and design the plans for co-creation workshop day 2, which they continue to do in the following video.

Keywords: Workshop design, Concept development, Clustering, User insights, Co-creation

Iterations on the design of the second co-creation workshop (v10)

Following the previous session held earlier the same day the design team and the external consultants now move on to iterate on the design of the second co-creation workshop day. Using post-it notes to organize their plans and ideas, they adjust their previous plan for the second workshop based on insights and experiences from the first workshop. Their focus is especially on optimizing the "Company profile exercise" where the participants are to act as investors, choosing what product or company to invest in. They also discuss how to utilize the introduction of two new co-creation workshop participants to the original group of participants.

Keywords: Workshop design, Brainstorm, Iterations, User involvement methods, Framing

Linking insights from co-creation to project (v11)

In this session the design team meets with the consultants once more to iterate on the set-up of the second co-creation workshop day. Prior to this meeting, they have worked in separate groups, updating and adjusting exercises and slideshows. Kenny starts by describing their findings from the first co-creation workshop, and the team continues to discuss further adjustment. They go on to adjust the moderation guides. Furthermore, they continuously work towards having a shared understanding of the co-creation workshops' connections to the overall project, also serving as preparation for a briefing session with the stakeholders later the same day.

Keywords: Workshop design, Shared understanding, Collaboration, User involvement methods, Design process

Briefing stakeholders on the first co-creation workshop (v12)

Following the previous session, where the design team and the consultants worked on linking insights from the co-creation workshops to the project, we follow a stakeholder briefing with Hans and Tiffany. Here Ewan brings them up to date on both the insights from the first co-creation workshop day, as well as the plan for the second co-creation workshop day. The stakeholders comment on the workshop design and discuss the knowledge they have gained about the users.

Keywords: Collaboration, Stakeholder relations, Workshop design, Conceptualisation, User insights

Co-creation workshop 2 (v13)

The second co-creation workshop with the Chinese lead users follows the same set-up as the first workshop. Ewan, Abby, and Kenny from the design team are present together with the external consultants, who also facilitate and moderate the sessions (in Chinese). Two translators translate simultaneously from Chinese to English. David and Nina from the design team, and this time also the two stakeholders, Tiffany and Hans, take notes behind the two-way mirror. Themes and concepts from the first workshop are revisited in the first part of the workshop in a "Freedom/Enjoy Life Concept Testing" session, a theme which has been derived from a previous "Health/Good Life" theme. The participants move on to the "Company Profile exercise", followed by a concept/service development exercise linked with the company profile they created earlier. Afterwards the participants pitch their company materials and give feedback to each other.

Keywords: Co-creation, User involvement, Concept development, Brainstorming, Service design

De-brief co-creation workshop 2 (v14)

Immediately after the co-creation workshop the design team, the consultants, and the stakeholders have a debriefing session. They share and discuss observations and notes with each other from the workshop, and begin to connect some of the insights to the overall project themes and concepts.

Keywords: Debriefing, Sharing experiences, Cultural translation, User insights, Evaluation

Sharing insights from co-creation workshops (v15, v16 and v17)

The design team, the stakeholders, and the three consultants are all attending this all-day insights and analysis workshop based on the second co-creation workshop. Kenny,

Nina, Tiffany, Rose and Abby are working in one group. David, Hans, Ewan, Will and Amanda are working in another group. In this session we are following the first group while they are sharing some of their observations from the second workshop. We follow the group led by Rose, one of the external consultants. Rose initiates the session with a translation and summary of her observations from the two co-creation workshops. Kenny and Abby generate ideas based on Rose's inputs and create post-it notes, which are being placed in central view for commenting. The five group members go on to discuss products, concepts, and story in broad terms based on observation notes and Rose's inputs and translations. The group generates new ideas for products, services and concepts on post-its and places them on the wall, next to post-its and notes from previous sessions. The group discuss different product features and concepts (e.g., monitoring, personalized, environmental, and non-profit).

Keywords: Cultural translation, Concept development, Service design, Storytelling, Idea generation

Clustering insights from the two co-creation workshops (v18)

The team engages in a scheduled "Concept Iteration Workshop". The design team and Amanda, one of the external consultants, cluster insights from the two co-creation workshops according to certain opportunity areas. They use these insights to cluster ideas for accessory products (e.g., related to air purification), marketing, and storytelling, as well as how to communicate these ideas to the stakeholders. Amanda presents an "innovation-matrix" and an "opportunity area template" developed on the basis of conclusions from Phase I and the previous days. After the joint session, they split into two groups, Kenny and Abby in one, Ewan and Amanda in the other, where they continue to cluster insights and ideas regarding to "story", "product", and "sales".

Keywords: Concept development, Ideation, Storytelling, Product development, Innovation matrix

Recap with consultants (v19)

After the previous insight-clustering earlier the same day, the design team meets up with all of the external consultants, Amanda, Rose and Will. Ewan starts by presenting different findings and ideas, followed by Kenny, who sums up the idea clustering he and Abby made prior to this meeting. The group discusses how to design services and concepts in relation to environmental awareness in an Asian versus Western culture, also addressing issues of status and value connected to the design of certain services and products. They all discuss how to further relate their ideas to the overall project, and continue clustering.

Keywords: Idea generation, Clustering, Recap, Cultural differences, Service design

Recap with stakeholders (v20)

On the final day in China, this video captures the final recap session, "Wrap up", where the design team, the stakeholders, Tiffany and Hans, and the consultants are present. Led by Ewan, the stakeholders are brought up to date on the outcome from the field-trip, the findings and the current concept designs, based on the two co-creation workshops. Ewan explains the concepts to the stakeholders, Tiffany and Hans, in a pitch-like format. The stakeholders bring to the design team up to date on their stakeholder relations in China such as current product lines, car dealers, and sales challenges. The group discuss their overall impressions from the trip and how the new insights contribute to the design project in relation to products, services and concepts for new accessories.

Keywords: Recap, Stakeholder relations, Conceptualisation, Service design, Asian market

Brainstorm on concept and products (v21)

Back in Scandinavia the core of the design team Ewan, Abby and Kenny arrange two brainstorming sessions with two colleagues in each round (four colleagues in total) to receive new inputs on the concepts and ideas they have been developing since returning from China. The sessions begin with Ewan giving a short brief on Phase I + Phase II and their main theme/concept "Good life". Ewan explains about the user insights from the field trip to China and from Phase I. The colleagues are given the task to create as many and "wild" ideas as possible for accessories based on the brief Ewan have just given them. Each round the colleagues are encouraged to go "wild" and also think of accessories beyond products, as stories, communication, sales channels, accessories "kits" etc. The design team mentions the emotional angle in the Good Life theme as being central, and ask for the two colleagues to think in "enablers" with reference to what *"enables Good Life for the Chinese users"*.

Keywords: Brainstorming, Inspiration, Concept development, Service design, Divergence

Follow-up interview with team leader (v22)

In May 2016 we meet the design team leader, Ewan, for a follow-up interview and a talk about the final parts of the project. In the interview Ewan reflects on the different parts of the project based on the project brief. He describes the process after the last observation in January, and explains how the delivery process took place and was received in the organisation. He reflects on his own satisfaction with the process and on which obstacles the design team met during the process. He talks about co-creation as a design approach and how he uses post-it notes as design material.

Keywords: Evaluation, Design process, Co-creation, Design Thinking, Design materials

2.4   Outcomes: Deliveries from Phase II

The deliveries from the project are, as planned, soft deliveries, that is, mock-ups and wireframes exemplifying a new line of accessories, events and digital products. The deliveries are handed over to relevant stakeholders in a Phase II delivery report, where the design team presents their reflections on Phase I & Phase II. They explain how Phase I was about understanding the reasons for low takes rates on accessories in China, and that Phase II was about finding the "key" to increase the take rates. The design team suggest a new strategy with a user-centred focus, which goes beyond the products and accessories themselves, in order to create relevance and offerings that are desirable to the Chinese lead users. The way to achieve this is through a new sub-brand line and two-main concepts, designed upon their insights gained from Phase I and the co-creation sessions in Phase II. The sub-brand and concepts are delivered in the report as mock-ups and wireframes, illustrating aesthetics, functionality and content. The sub-brand and the two concepts are presented visually in the delivery report as a complete "kit". The "kit" involves mock-ups of new car accessories, illustrations of pop-up events, wireframes of a sub-brand website, wireframes of an app with description of different user scenarios, along with mock-ups of the physical environment and presence at the dealership stores. The two main concepts are designed according to certain themes, which was conceptualised during the design project such as "Health" and "Good life" involving design values such as sustainability and social responsibility. The design team's deliveries were described in the shared, unpublished DTRS11 technical report with excerpts and mock-ups from the original delivery report, but cannot be replicated here for reasons of confidentiality.

REFERENCE

Jefferson, G. (1984). Transcription Notation. In J. Atkinson and J. Heritage (eds), *Structures of Social Action*, New York: Cambridge University Press.

Part I

Team Dynamics and Conflicts

Psychological Factors Surrounding Disagreement in Multicultural Design Team Meetings

Susannah B. F. Paletz, Arlouwe Sumer & Ella Miron-Spektor

ABSTRACT

As multicultural teams become the norm, research is increasingly directed to understanding their challenges and opportunities. Our research contrasts two theories of creativity in multicultural teams. The dual-process model focuses on the effects of level of diversity, whereas cross-cultural psychology focuses on specific cultural compositions. In individualistic cultures, team members see conflict as an opportunity, tend to express more conflicts, and benefit from conflict compared to collectivistic cultures that emphasize harmony and see conflict as threatening. The relative representation of members from these cultures may affect team dynamics, conflict, and creativity. This study tested these theories using the DTRS11 dataset. We coded over 3100 speaker turns for the presence of disagreements, or micro-conflicts, and examined the effects of conflict phase and team diversity on creativity in the moment, promotion, and prevention approaches to conflict using Linguistic Inquiry Word Count (LIWC) measures. We found that micro-conflicts increased insight words in the moment of the conflict. Individuals in more diverse team meetings of Scandinavians and Southeast Asians expressed fewer conflicts than teams dominated by Scandinavians and were less likely to focus on potential gains and opportunities when experiencing micro-conflicts. Interestingly, regardless of conflict, the more culturally diverse teams were more likely to use insight words and promotion words overall. There were no effects for prevention. These findings support extant theory, and extend it to different types of heterogeneous teams in a real-world design setting. This study is novel in combining theory on team cultural diversity with a micro-process method.

1 INTRODUCTION

To compete in a worldwide market, many organizations are increasingly drawing on a multinational and multicultural workforce. Past diversity research suggests that multicultural teams are more likely to have conflict and difficulties, but be more creative (Stahl, Maznevski, Voigt, & Jonsen, 2010). Researchers are divided as to whether interpersonal conflict can benefit team creativity. On the one hand, minority opinion dissent and disagreement can increase creativity (e.g., Nemeth, 1986). Yet, meta-analyses show that self-reported conflict is generally negatively related to team performance and creativity (de Dreu & Weingart, 2003; de Wit, Greer, & Jehn, 2012). Other studies suggest that mild task conflicts engender more information acquisition, interest, and excitement (Todorova, Bear, & Weingart, 2013) and can foster creativity (de Dreu, 2006). Recently, Paletz, Miron-Spektor, and Lin (2014) suggested that cultural values and meanings shape the way team members approach and express conflict, which then determines whether conflict can improve creativity. This project ties together two separate paradigms to explore the role of disagreement in multicultural team meetings: (1) research on team micro-conflicts, or brief disagreements; and (2) research on multicultural teams, conflict, and creativity. Prior studies used retrospective, self-reported accounts of long-term team processes, and thus may suffer from perceptual and recall biases (Paletz, Schunn, & Kim, 2011). Instead, we analyze micro-conflicts, which are brief, expressed, interpersonal (intrateam) disagreements (Paletz et al., 2011). Similar to other micro-process analyses, this approach examines real team behavior. We address three important questions: Does cultural composition affect the expression of micro-conflicts in multicultural team meetings? Are culturally diverse teams more creative, measured via verbal expressions of insight? And, does the cultural composition of the team affect the tendency to react to conflict as an opportunity or as a threat? We analyzed micro-conflicts in meetings of a design team in a large Scandinavian company.

1.1 Micro-conflicts and creativity

We define conflict as interpersonal and intrateam disagreement, as when "a divergence of values, needs, interests, opinions, goals, or objectives exists" (Barki & Hartwick, 2004, p. 232). We focus on brief, interpersonal, expressed disagreements. Similar to more macro-conflicts, studies on micro-conflicts suggest that these conflicts are more prevalent in teams with difficulties (Paletz et al., 2011), and precede creativity and analogy (Chiu, 2008; Paletz, Schunn, & Kim, 2013). Yet, unlike macro-conflicts, micro-conflicts are not associated with particularly high levels of negative affect (Paletz et al., 2011). The micro-conflict approach is a fruitful way of conceptualizing conflict when seeking to investigate fine-grained team processes, particularly the differential effects of disagreement on psychological reactions. Our study responds to calls to extend micro-process research to multicultural teams, conflict, and creativity (Srikanth, Harvey, & Peterson, 2016).

Creativity involves many cognitive processes, such as mental simulation, analogy, incubation, and insight (e.g., Ball, Onarheim, & Christensen, 2010; Ward, Smith, & Vaid, 1997). In the creative cognition literature, insight refers to the flash of suddenly realizing a correct answer (Metcalfe & Wiebe, 1987), which entails a significant leap in the understanding of a problem and a move from implicit to explicit knowledge (Hélie & Sun, 2010). By contrast, the lay definition of insight includes "the act or result of apprehending the inner nature of things..." (Merriam-Webster, n.d.). For this study, we use a measure that reflects the explicit expression and discussion of creativity and problem solving, involving mental search, information restructuring, deep thinking, and discernment, which is more in line with the broad definition of creative insight used by Friedman and Förster (2000).

1.2   Multicultural teams, conflict and creativity

Despite its importance to increased workplace diversity, research on conflict and creativity in multicultural environments is limited (e.g., Zhou & Su, 2010). Existing research has suggested positive but moderated effects of team cultural diversity on both conflict and creativity, resulting in a dual-process model (Stahl *et al.*, 2010; Srikanth *et al.*, 2016). This model suggests both positive and negative effects of diversity: Such teams have poorer group cohesion and more conflict due to social identity processes, but their increased informational diversity leads to greater creativity. Paletz and her colleagues (2014) incorporated cross-cultural psychology and social cognition to modify this theory, proposing that team members' approach to conflict could determine whether disagreements in multicultural teams foster or hinder creativity. If team members see conflict as an opportunity, they systematically explore and process information (i.e., use a promotion orientation), and gain new knowledge and insights. However, if they experience conflict as risky and threatening, they may react by restricting information processing and relying on their dominant uncreative responses (engage a prevention orientation, Higgins, 1997; Friedman & Förster, 2001, 2005).

Whether or not a disagreement is perceived as an opportunity or a threat could depend, in part, on the cultural composition of the team (Paletz *et al.*, 2014). Culture consists of imperfectly shared, learned meanings and unspoken assumptions (Rohner, 1984; see also Daly *et al.*, 2017). Culture can occur at many levels simultaneously, such as within nations, organizations and groups, and culture can be endorsed by individuals (Erez & Gati, 2004). Individuals from collectivistic cultures focus on maintaining harmony, and thus generally avoid conflict or respond to it by trying to compromise (Friedman, Chi, & Liu, 2006; Oetzel & Ting-Toomey, 2003). In contrast, people from individualistic cultures value uniqueness and are more likely to respond to team conflict by scrutinizing task issues, engaging in deliberate processing of information and greater creativity (de Dreu, 2006; Jehn, 1995). Thus, culture can drive the interpretation and expression of and tolerance for team conflict.

1.3   Research questions and predictions

This study combines these theories to test the relationship between disagreements (micro-conflicts), creativity, and prevention and promotion approaches, depending on the cultural composition of a multicultural team. Specifically, this study goes beyond the usual homogenous versus diverse operationalization of diversity to incorporate theory about underlying cultural composition. This study uses the 11th Design Thinking Research Symposium (DTRS11) dataset, which includes audio-video recordings of creative meetings that involved either individuals of Western nationality (predominantly Scandinavians) or these same individuals when they were later collaborating with Southeast Asian teammates (Christensen & Abildgaard, 2017). Relative to Southeast Asians, Scandinavians are more individualistic, but score similarly on the tightness/looseness dimension of culture: Tight cultures have stronger situational constraints and norms, regardless of what those norms are (Gelfand *et al.*, 2011). Tight collectivistic cultures emphasize harmony and conflict avoidance that may then hinder the ability to leverage conflict into greater creativity. In contrast, in tight individualistic

Table 3.1 Predictions from diversity theory versus cross-cultural psychology.

Factor	Dual-process model (diversity theory)	Cross-cultural psychology
Expressed conflict	Diverse teams > homogeneous teams	Individualistic teams > collectivistic teams
Level of creativity	Diverse teams > homogeneous teams	Individualistic teams > collectivistic teams
Promotion and prevention approaches	Homogeneous teams more promotion, less prevention vs. diverse teams	Individualistic teams more promotion, less prevention vs. collectivistic teams
Impact of conflict on creativity	Mainly negative but can be positive, depending of type of conflict (e.g., relationship vs. task conflict)	Positive if conflict sparks promotion, negative if conflict sparks prevention.

cultures, team members conform to norms and values that encourage the expression of unique and diverse perspectives (Goncalo & Duguid, 2012; Goncalo & Staw, 2006). The cross-cultural literature would thus suggest that compared to more diverse groups, low diversity groups from individualistic cultures (Scandinavians) would approach disagreements as opportunities rather than as threats, resulting in more of a promotion focus and less of a prevention focus, and leading to more creativity. The cross-cultural literature would predict that multicultural teams of predominantly individualists would also have more conflict than teams mixed with or dominated by collectivists. In terms of creativity, cross-cultural psychology suggests that teams with mostly individualistic members should have more creativity, while the dual-process model suggests greater creativity in more diverse teams. We contrast these approaches and compare multicultural teams dominated by individualistic members (mainly Scandinavians) with multicultural teams that include collectivistic members (Southeast Asians; see Table 3.1 for predictions).

Our fine-grained process analysis also enables us to ascertain whether the increased creativity, promotion, and prevention occur before, during, or after micro-conflicts, providing an exploratory analysis of the timed relationship between these variables. Thus, we test the effects of multicultural team composition on (1) expressed conflict and (2) creativity; the effects of conflict phase on (3) creativity, (4) promotion, and (5) prevention; and the interaction effects of conflict phase and team composition on (6) creativity, (7) promotion, and (8) prevention. For our measurement of creativity, we focus on the relevant cognitive process of deeper information processing as characterized by the use of particular language in team conversations. We then follow up with exploratory tests to determine whether the findings might be due to differences between people or changes in behavior depending on team context.

2   METHOD

The DTRS11 dataset includes over 13 audio-video recordings of a professional design team working on a specific design task for a multinational automobile company. The meetings, which included planning, brainstorming, and insight extraction, were mainly in English and transcribed by DTRS11 staff into speaker turns (Christensen & Abildgaard, 2017).

2.1   Participant-designers and research context

The core design team consisted of three Scandinavians, one Eastern European, and a bicultural Scandinavian-East Asian, who were two women and three men. They started their work in a Scandinavian city, and then traveled to an East Asian city to conduct an in-depth co-creation workshop with local participants. There, the core design team was joined by three Southeast Asian colleagues (two women, one man) who assisted with both running the workshop and extracting insights from it. They were joined by two additional Scandinavians (one male, one female), who were stakeholders for the project.

2.2   Meeting sampling and characteristics

Of the 22 DTRS11 sessions, we chose seven meetings/video clips to analyze, of which three were highly diverse (v10, v11, and v18) and four were predominantly the Scandinavian core design team (v02, v03, v04, v05, although only 18% of v05 was coded due to time constraints). These seven were chosen based on having transcribed interactions between team members (i.e., not be an interview) and clear audio. They had to be inherently creative meetings (i.e., not be recap or debrief meetings). The first meetings had four or five predominantly Scandinavian designers. The three highly multicultural team meetings were in the middle of or later in the team process and had five or seven designers. In all, we analyzed 179 minutes from Western team meetings (1768 turns) and 142 minutes from highly diverse meetings (1340 turns), for a total of 321 minutes (5.35 hours) and 3108 turns.

2.3   Measures

2.3.1   Micro-conflicts and conflict segment types

We coded for micro-conflicts at the turn level as to whether the turn included conflict or not (Paletz *et al.*, 2011). Micro-conflicts were identified when a speaker disagreed with something said previously, whether via tone, gestures, body language, or words. The first and second authors double-coded all of the turns, both reading the transcript and listening/watching to the audio-video recording simultaneously and replaying as many times as necessary. After training, which was necessarily on the DTRS transcripts, reliability was adequate (average kappa for the conflict code across each video clip = 0.44, most 0.59–0.61 for clips after training; raw agreement 90–97%).[1] All disagreements were resolved via consensus discussions to reduce noise (Smith, 2000), resulting in a valid final coding (see below).

Micro-conflict events were clusters of turns relating to a specific conflict topic. For example, in planning for the second day of co-creation, one of the designers suggested that having the participants from the first day pitch their companies would be motivational, but a team member disagreed. This second individual believed that the pitch approach would encourage competition, which would shift the focus away from

Table 3.2 An example micro-conflict from v03 (381–390).

Turn	Speaker	Transcript	Conflict
381	A	But I think we need a sec'- I mean, also for the first session. We'll be seven people, seven respondents and then we will be mingled in there as well, so we'll be: almost too many to just be in one group the whole day, so we'll- we will be- we will be split out, I foresee. And then of course Amanda and the other guy will be the moderators of each group, so: they can't be the translators as well, so I'm thinking-	1
382	K	No no, but what we talked about before was then the translator must somehow mingle back and forth between the two groups.	1
383	A	Yeah I don't think that's realistic.	1
384	K	No no, but that's what we talked about before.	1
385	A	Yeah.	0
386	K	But I agree, it's not optimal in any way. But if we have two translators, then we have the technical issue, how do we: make sure then- then eh: (.) the system supports two translators at the same time.	1
387	E	I guess, yeah. But that I guess- that is just the- they need to choose, with the channel. Channel one, channel two, they can't listen to both at the same time. Or, if it's several people in the back, someone can listen to channel one-	1
388	K	Yeah, but then the one translator kind of loses effect, and then you don't get any value out of having two translators.	1
389	E	Except for us that are in the room.	1
390	A	Yeah because I'm also thinking that, I mean, most of us will be in the room, anyway. So people will- I mean the people sitting out- outside could be the China-org for example, ehm.	1
391	E	Yeah, so it will be- oh sorry. Yeah so okay, this is the scenario for break out, right? Here's the behind the scenes.	0
392	N	Mhm.	0

Note. 1 = presence of micro-conflict, 0 = absence.

the goals of the morning. Three other team members then argued with the second team member and tried to understand his position. Conflict micro-events ranged from this long event to minor single-turn events, such as when one designer disagreed with another's idea to have the workshop participants wear color-coded clothes when they arrived on the first day. A separate turn-by-turn example is in Table 3.2.

In order to examine differences for our dependent variables based on conflict, we took a similar strategy to Chan, Paletz, and Schunn's (2012) examination of the effects of analogy on uncertainty. We segmented most of the coded transcript into five types: (1) pre-conflict segments, which were 5 turns prior to a conflict micro-event; (2) during micro-conflict events, which were turns that were identified specifically as conflict from beginning to end of a particular micro-conflict event; (3) immediate post-conflict segments, which were five non-conflict turns immediately after a micro-conflict event; (4) delayed post-conflict segments, five turns immediately after the post-conflict segments; and (5) baseline segments, which were segments of 5 turns, total, at least 10 turns away from the other segment types. Five turns were chosen for the segments because there was an average of 10 turns a minute for the clips chosen. Past research suggests that windows of 30–60 seconds are fruitful for examining micro-processes (e.g., Chan *et al.*, 2012; Paletz *et al.*, 2013).

We identified 52 micro-conflict events out of 475 total segments (11%). Although the other types of segments were five turns long, the micro-conflict segments ranged from one to 24 turns long. If micro-conflict events occurred with fewer than 15 turns between them, no pre- or post-conflict segments were determined in order to maintain stability and clarity of the different phases. This strategy enabled a clean comparison baseline, pre-, during-, and post-conflict segments (Chan *et al.*, 2012). There were 317 baseline segments, providing ample variability of word usage for our dependent variables.

Qualitatively, the designers were very positive and used a great deal of mitigating language around their micro-conflicts, as compared to predominantly American groups we have coded (e.g., Paletz *et al.*, 2011). To validate our coding scheme, we tested the effects of conflict phase on relevant words using the same methods described below. Conflict phase had a significant negative effect on assent words and positive effects on negation and differentiation words. Assent (e.g., OK, yes) words were 59% less frequent during conflict events and 28% less frequent during delayed post-conflict compared to baseline.[2] Differentiation words (e.g., hasn't, but, else) were 38% more common during and 24% more common just after a micro-conflict compared to baseline.[3] Similarly, negation words were 84% more likely during micro-conflict events versus baseline.[4] These findings provide evidence for the validity of the coding.

2.3.2 Multicultural diversity

The segmentation strategy distilled the unit of analysis to 475 block segments. Of these segments, 56% (266) were of low diversity teams (mainly Scandinavians; dummy coded as 0) and 44% (209) were highly diverse teams (highly multicultural; Scandinavians, Southeast Asians, and others, dummy coded as 1).

2.3.3 Dependent variables: Psychological factors of prevention, promotion, and creativity

The dependent variables were operationalized using Linguistic Inquiry Word Count (LIWC) 2015 word lists. LIWC is a software package that automatically assesses text for a variety of variables via counting words from a validated dictionary of word sets (Pennebaker, Boyd, Jordan, & Blackburn, 2015). LIWC has been widely used to assess psychologically meaningful categories (Tausczik & Pennebaker, 2010). LIWC2015 software was run on the transcripts, resulting in a word count for each category for each speaker turn. Then, the word counts for each category were summed across the segments. We operationalized a prevention approach as the frequency of *risk* words (e.g., danger, doubt), and a promotion approach as the frequency of *reward* words (e.g., prize, benefit, take). We used the number of *insight* words (e.g., think, know, consider) as a measure for information processing and creativity. Pennebaker and colleagues (2015) report the corrected internal reliability of these word lists as adequate (insight alpha = .84, reward alpha = .69, risk alpha = .68). Because the LIWC is proprietary software that replaces human coding, no more than these example words are publically available (see Pennebaker *et al.*, 2015). These measures provide a summed number of words per predetermined slice of text (see Table 3.3 for examples).

Table 3.3 Examples of Insight, Reward, and Risk LIWC word count results per turn.

Turn	Speaker	Transcript	LIWC
Number of Insight Words from v18			
379	AM	So I think we should do it together, 'cause we are in a different groups and then we need to cross, but, we can write it on our own, before we put it out we just share:, to see if anyone- anything spark other thoughts, [and we'll let it].	2
380	E	[Yeah, that was-], yeah.	0
381	AM	(..) So, what are the others, around there you think?	1
382	K	I guess we can:- you already kind of, eh:, structured it.	0
Number of Reward Words from v08			
48	A	But I think also it was also just kind of in the small daily routines that something just had to be changed a little bit, all the time, so that you don't, I don't know eat the same food every day, or, so it just eh in: (.) which is not about status but is about (.) something else.	0
49	AM	Access, yeah. Access to many different experiences.	2
50	A	Yeah.	0
Number of Risk Words from v07			
85	W	Yeah. So, it's personal safety and security, but at the same time it's also safety and protection for the family, indirectly.	4
Number of Risk Words from v04			
23	E	putting them in to a situation, where they: they fail, but they see that it's okay to fail	2

Prior research suggests the LIWC insight word list serves as a proxy for complex information processing associated with creative cognition. In a review by Tausczik and Pennebaker (2010), LIWC insight words were indicative of active processing of information, reconstrual, and verbal complexity. In addition, insight words become more common as writers age and gain experience. One question is whether the insight word list essentially measures epistemic and psychological uncertainty (Ball *et al.*, 2010; Christensen & Ball, 2017; Trickett *et al.*, 2005). This possibility was tested against other LIWC word category lists: tentativeness, which includes hedge words (e.g., maybe, perhaps) and certainty (e.g., always, never).[5] Proportion of insight words were uncorrelated with either the proportion of tentativeness words ($r_s = .02, p > .60$) or certainty words ($r_s = .075, p > .10$). These findings suggest that insight measured via LIWC is generally unrelated to either uncertainty or certainty, and is not conceptually or empirically the same as epistemic uncertainty.

2.3.4 Analyses and covariates

Given that the LIWC dependent variables were discrete count data with a skewed distribution (many zeroes, then ones, etc.), and that the variance of these variables was greater than the mean, we analyzed the data with negative binomial regression with maximum likelihood estimation using SPSS 21. As appropriate, we used an "offset variable" of the total number of words for each segment,[6] ensuring that this substantial

Table 3.4 Descriptive statistics.

Variable	M	Median	SD	Maximum
Insight (Creativity)	2.83	2.00	3.35	38
Reward (Promotion)	0.96	0.00	1.69	17
Risk (Prevention)	0.23	0.00	0.62	6
Total number of words per segment	84.06	71.00	71.73	799

Note. $N = 475$. The minimum for each factor is 0; the minimum total number of words is 6.

underlying variability (6 to 799 words) was controlled for in each analysis. To best interpret the results, we provide the $Exp(B)$, an incidence rate ratio, such that the percent difference between categorical variables is $100*(Exp(B) - 1)$ (see below).

Two covariates were tested: gender of speakers, or percent of turns spoken by females in the segments ($M = 36\%$, Median $= 40\%$, $SD = 24\%$), and team size. For team size, the videos were from 4-person teams (159 segments, 33.5%, all low-diversity), 5-person teams (174, 36.6%, of which 39% were very diverse), and 7-person teams (142, 29.9%, all very diverse). Each of these covariates could potentially have an effect on word usage. For example, men may talk more than women in mixed-gender groups and use more assertive speech, whereas women use more affiliative speech (Leaper & Ayres, 2007).

For each dependent variable, we first tested a full model that included the covariates, the two main effects (team composition and conflict segment type), and the interaction between team composition and conflict segment type. If any of the covariates were not significant, they were excluded from the final model. We tested the planned contrasts for our dependent variables for each micro-conflict phase as compared to baseline (e.g., pre-conflict versus baseline, during micro-conflict vs. baseline) for both the main effect of conflict phase and the interaction effect with team composition. In these analyses, each effect is controlled for by all the other variables in the model.

3 RESULTS

We address the research questions after displaying the descriptive statistics for the dependent variables (Table 3.4).

3.1 Cultural diversity and micro-conflicts

We compared the proportion of micro-conflicts experienced by the highly diverse team relative to those experienced by the less diverse team. We created a segment-level variable (1 = conflict block, 0 = not conflict block) and examined the frequency of the conflict blocks for the different types of teams. Using logistic regression controlling for team size and total number of words, team diversity was negatively related to conflict, with highly multicultural teams expressing significantly less conflict, $B = -1.80$ (.02, 1.37), $SE = .77$, Wald $\chi^2 = 5.47$, $Exp(B) = 0.17$ (0.04, 0.75), $p = .019$.

Table 3.5 Specific contrasts of conflict phase, team composition, interaction, and covariates on insight words.

Specific contrasts	B (95% confidence interval)	SE B	Wald χ^2	Exp(B) (95% confidence interval)	p
(Intercept)	−3.37 (−3.51, −3.24)	.07	2443.55	0.03 (0.03, 0.04)	<.001
Covariates					
Proportion female turns	−0.32 (−0.58, −0.06)	.13	5.85	0.73 (0.56, 0.94)	.016
7- vs. 4-person teams	−0.36 (−0.59, −0.12)	0.12	9.44	0.70 (0.56, 0.88)	.002
5- vs. 4-person teams	−0.09 (−0.25, 0.07)	0.08	1.28	0.91 (0.78, 1.07)	.257
Conflict phase contrasts vs. baseline					
Pre-conflict	0.09 (−0.13, 0.32)	.11	0.68	1.10 (0.88, 1.37)	.410
During conflict	0.26 (0.09, 0.43)	.09	9.33	1.30 (1.10, 1.54)	0.002
Immediate post-conflict	0.05 (−0.19, 0.28)	0.12	0.15	1.05 (0.83, 1.32)	0.701
Delayed post-conflict	0.11 (−0.09, 0.31)	.10	1.08	1.11 (0.91, 1.36)	0.300
Multicultural Diversity	0.38 (0.20, 0.56)	.09	16.84	1.46 (1.22, 1.74)	<.001

3.2   Conflict, cultural diversity and creativity

For a fine-grained understanding of the relationship between conflict and creativity, we tested the effects of conflict and multicultural diversity on the frequency of insight words. None of the interaction contrasts between conflict phase and team diversity were significant ($ps > .15$). The final model without the interactions was significant, $\chi^2(8, N = 475) = 24.52$, $p = .002$ (Table 3.5). Conflict phase had a positive effect on the usage of insight words. There were 30% more insight words used during conflict compared to baseline. The increased frequency of insight words occurred only during the conflict and not before or after it (see Figure 3.1). In addition, culturally diverse teams were 46% more likely to use insight words than less diverse teams. These findings suggest that the increase in creativity (as insight words) occurs during – but not prior or after – the conflict, and that on average, culturally diverse teams use more insight related words regardless of conflict.

For example, in one micro-conflict (v04, 458–476), a designer questioned the methods planned for the moderator during the co-creation workshop to generate discussion among participants, and the novelty of the resulting ideas. The designer used the word "think" to express both his/her own opinion ("I think") and participant insights in the discussion (e.g., "… maybe they're not thinking about it…").

3.3   Promotion and prevention approaches to conflict

In terms of a promotion-focused approach to conflict (measured using reward words), team size was not a significant covariate, and so not included in the final model. The significant final model included the two main effects (conflict phase and team composition), their interaction, and the covariate of percent of turns spoken by females, $\chi^2(10, N = 475) = 36.41$, $p < .001$ (Table 3.6). We found main effects for conflict phase and team diversity. In terms of conflict phase and controlling for all other effects

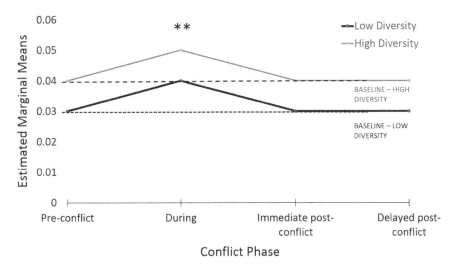

Figure 3.1 Insight words by conflict phase. These estimated marginal means control for team size and team diversity.

Table 3.6 Specific contrasts of conflict phase, team composition, interaction, and covariates on reward words.

Specific contrasts	B (95% confidence interval)	SE B	Wald χ^2	Exp(B) (95% confidence interval)	p
(Intercept)	−4.65 (−4.90, −4.39)	.13	1288.12	0.01 (0.01, 0.01)	<.001
Covariates					
Proportion female turns	−0.88 (−1.39, −0.37)	.26	11.48	0.42 (0.25, 0.69)	.001
Conflict phase contrasts vs. baseline					
Pre-conflict	−0.17 (−0.84, 0.50)	.34	0.25	0.84 (0.43, 1.65)	.619
During conflict	0.46 (0.03, 0.89)	.22	4.45	1.59 (1.03, 2.45)	.035
Immediate post-conflict	0.33 (−0.22, 0.89)	.28	1.39	1.40 (0.80, 2.43)	.238
Delayed post-conflict	0.72 (0.25, 1.18)	.24	9.12	2.05 (1.29, 3.26)	.003
Multicultural Diversity	0.77 (0.48, 1.05)	.15	27.67	2.15 (1.62, 2.86)	<.001
Interaction contrasts					
Pre-conflict* Diversity	−0.18 (−1.18, 0.83)	.51	0.12	0.84 (0.31, 2.28)	.725
During conflict* Diversity	−0.69 (−1.44, 0.06)	.38	3.24	0.50 (0.24, 1.06)	.072
Immediate post-conflict *Diversity	−0.54 (−1.49, 0.41)	.48	1.24	0.58 (0.23, 1.50)	.265
Delayed post-conflict *Diversity	−0.91 (−1.68, −0.14)	.39	5.34	0.40 (0.19, 0.87)	.021

in the model, there was a significant increase in the use of promotion-related words during the micro-conflict events (59% more reward words compared to baseline) and during the delayed post-conflict phase (105% more reward words vs. baseline). When the team was more multicultural, it used 115% more promotion words compared to when it included mainly Scandinavians. Finally, there was a significant interaction between conflict phase and team diversity on promotion words. When the teams included Southeast Asians, they were 60% *less* likely to use promotion-related words 30–60 seconds after the conflict event compared to when the teams were less diverse.

One instance of conflict within a low-diversity group involved debating the value of having the co-creation workshop moderator also serve as an additional translator. One designer referred to his/her past experience, describing the benefits and drawbacks for idea exchange when moderation and translation were performed by the same person within a multi-language group. Two other designers discussed drawbacks and benefits from including an extra translator in the co-creation group. This discussion uses the word "gain" frequently in the context of describing possible benefits in various scenarios involving multiple translators (v03, 487–498).

We tested similar models with prevention words as the dependent variable. We found no significant relationship with any of the variables of interest – not for conflict phases, diversity, or their interaction ($p > .16$).

Together, these findings suggest three separate effects for promotion, even when controlling for each independent variable: The teams tended to use promotion-oriented words during conflicts, and more diverse teams used more promotion words generally. There was also an increase in promotion-oriented words in a delay after conflict, but mainly when the team was less diverse. Interestingly, conflict and team diversity did not affect the expression of prevention-oriented words.

3.4   Follow-up tests of Scandinavians across context

One question is whether these findings are due to the introduction of new, Southeast Asian team members who have different linguistic styles, or due to changes in behavior by the core Scandinavian design team. We created versions of our variables only summed for the Scandinavian participants from the original, core team: number of conflict turns per segment and number of insight, reward, and risk words from turns spoken by these individuals (A, D, E, K, and N, or ADEKN). Choosing only segments where at least one word was spoken by these individuals ($N = 465$), we ran similar negative binomial regressions with the log of the non-zero total of the ADEKN words as the offset variable and the social context – diverse or less so – as the main predictor variable. Percent female was not significant with this limited sample, and team size almost entirely confounded the multicultural variable, so these analyses were run without any covariates. For these individuals, there was no difference between the multicultural team meetings and the Scandinavian team meetings for number of conflict turns overall, insight words, or risk words.[7] However, there was a significant difference in reward words, such that these same Scandinavians used more promotion words when in the presence of their Southeast Asian colleagues and later in their processes (multicultural $M = .02, SE = .001$; Scandinavian-only $M = .01, SE = .001$), $B = 0.56$ ($-0.32, 0.80$), $SE = .12$, Wald $\chi^2 = 21.17$, $\text{Exp}(B) = 1.75$ ($1.38, 2.22$), $p < .001$. Given the sparsity of data, these findings are simply suggestive; but they imply that some, though not all of the study's overall findings may be due to differences in the styles of individuals and cultures present in the team. In other cases, the team members themselves may be changing their behavior because of the social context.

4   DISCUSSION

This is the first project, to our knowledge, to test the relationship between conflict, prevention and promotion approaches, and creativity in culturally diverse teams using a micro-process method. In aiming to reconcile competing theories in the literature, this project deliberately took a team-centric view of culture and creativity. By taking this perspective, this study presumed individuals within a team meeting start to speak similarly to each other (Brennan & Clark, 1996), and their conversations were influenced by the dynamics and cultural combination within the group. Using communication analyses, we showed that micro-conflicts increased creativity in the moment. More diverse, mixed individualistic-collectivistic teams were less likely to have conflicts in our sample, and were less likely to focus on potential gains *when* experiencing micro-conflicts relative to multicultural teams with mainly individualistic members (despite overall using more promotion words in multicultural teams). The tendency to focus on prevention (risks and losses) was not affected by micro-conflicts nor team diversity. The differences in the presence of conflict may be due to differences in conflict expression between the Scandinavian core design team versus their Southeast Asian colleagues and the Scandinavian stakeholders, but the greater focus on promotion words overall in multicultural teams was, at least in part, due to changes in behavior on the part of the Scandinavian core team. Together, these findings support cross-cultural theories for multicultural teams. Yet, in line with the dual-process theory, highly diverse teams were more creative, regardless of conflict. These findings advance knowledge on micro-conflicts and creativity in the context of the development of a multicultural team.

4.1   Implications and explanations

This study suggests that teams vary in their expression and approach to micro-conflicts, and that team cultural composition is key. Importantly, this study found some support for both the dual-process model and cross-cultural psychology. With regards to the dual-process model, the highly diverse teams (which were less individualistic) were more creative. This finding may have been due to added creativity and deep processing of the Southeast Asian team members who joined the team later, but it is not possible to tell if their level of insight words changed compared to if they had been in a homogenous team. In line with cross-cultural theory, the highly diverse teams had *less* conflict, suggesting that diversity can promote creativity without requiring conflict. Also in support of cross-cultural theory, individualistic teams had a delayed boost for promotion-oriented words from conflict, but the more diverse teams did not, potentially due to cultural (or individual) differences in responses to conflict. The delay could be due to the time it takes to express, listen, and encode a conflict after it has stopped. Interestingly, the teams experienced greater creativity during but not before or after the conflict, suggesting that conflict and creativity may co-occur, rather than have a simple causal relationship as has been previously argued.

Two findings still beg explanation: the lack of finding for prevention, and the higher levels of promotion words overall for the very diverse teams. Both findings could be due to the overall positivity of this team, such that they simply were not prevention-oriented at all: As most of the individuals had worked successfully together in the past, the team was likely high on trust, which can provide a buffer to negative group processes (Srikanth *et al.*, 2016). Another explanation is that we compared teams that mainly include individualists to more balanced teams. Multicultural teams dominated by collectivistic members are more likely to avoid conflicts (Oetzel & Ting-Toomey, 2003). As an alternative explanation, the lower conflict and greater promotion for the more multicultural groups could be due to temporal dynamics in group processes, in addition (or exception) to the cultural composition. Indeed, the Scandinavians increased their promotion words in the presence of their Southeast Asian colleagues. Because the more diverse teams all worked later in the team's lifecycle, they could have benefitted from the less diverse teams having already argued over group processes, lessening the amount of arguing they had to do (Paletz *et al.*, 2011; Tuckman & Jensen, 1977).

This study has implications for both practice and theory. It provides empirical evidence that very diverse teams can thrive, and that small disagreements (micro-conflicts) need not be avoided. Indeed, micro-conflict events were beneficial for in-the-moment creativity for both types of groups. This study also suggests a blending of theories is in order, such that the dual-process model should incorporate cross-cultural (and temporal) theories.

4.2   Limitations and strengths

As with any study, the unique data set and methods offer strengths but also some limitations. First, the structure of the designers' project meant that the multicultural teams were all later in their work processes, larger in size, and in the East Asian location, such that some of the team composition effects could be the result of the group development or the salience of the location rather than team diversity. Essentially, these were not independent teams, but the same team with added (and subtracted) members over time. The more multicultural teams were also, therefore, different in consistent ways in terms of individual personalities, roles in the project, and knowledge of Chinese culture and language. These differences in composition could be causing our findings directly, such that the effects are caused by differences in team composition separate from culture, or indirectly, in that particular personalities, motivations, and roles helped create a team culture that then drove these effects. For instance, if the Scandinavian team members just happened to be cooperative and individualistic, their personalities could drive the creation of a team culture that valued those attitudes and behaviors. Relatedly, this study focused on a single case from a specific company, limiting its generalizability. This particular team's dynamics were roughly aligned with the literature, but may be more descriptive of longer-lasting, productive, positive teams. Given the small number of individuals involved, we cannot truly separate the personal, social, and cultural. That noted, this limitation was also a strength: The DTRS11 project enabled a deep dive into a rich dataset of a real-world, creative team in action.

Second, the kappa for conflict was modest. In this particular dataset, the coders initially struggled to distinguish between a constraint and a disagreement. By double-coding all of the text and discussing each coder disagreement, these nuanced determinations could be made. Indeed, the associations between the conflict coding and relevant LIWC word lists (e.g., negation) suggest that the final coding was a valid measure of disagreement.

Third, the operationalization of creativity as deep processing insight words necessarily limits the study to one type of creative cognitive process, and as expressed aloud. Future research could test other types of creativity, such as analogy and mental simulation, as well as the relationship between LIWC insight words to these other types of creativity and to when teams have demonstrable breakthroughs (see Shroyer *et al.*, 2017). In addition, the LIWC was designed to analyze a larger number of base words, with under 50 being undesirable (Pennebaker Conglomerates, 2016). The segments analyzed ranged from 6 to 799 words ($M = 84$), implying that many, but not all, would be sufficient.

Ideally, these findings would be extended and replicated with more teams at different phases in the development process. Future studies could also examine teams numerically dominated by Southeast Asians and test whether those teams avoid conflicts and use prevention related words during those conflicts *early* in the life of the teams. Computational social science methods, such as the LIWC, have the potential to swiftly analyze large datasets: By collaborating with computational linguists, it may be possible to speed up conflict coding and apply it to more datasets. Combined with human annotation, team processes can be more readily unpacked and examined.

4.3 Conclusion

Multicultural teamwork is becoming more prevalent in organizations, posing new challenges and opportunities for creativity management. Our research suggests that although conflict may stimulate creativity, cultural diversity contributes to creativity beyond conflict. This study is one of the first to examine micro-processes in diverse teams (Srikanth *et al.*, 2016). In addition, this study's findings highlight the importance of studying diversity via examining the detailed cultural composition of team meetings.

ACKNOWLEDGEMENTS

This project was supported in part by a Research Activities Request grant to the first author by the University of Maryland Center for Advanced Study of Language. The authors are grateful to Shauna Sweet for statistical advice.

NOTES

1 Kappas from 0.40 to 0.59 are considered moderate, 0.60 to 0.79 substantial, and 0.80 and above outstanding (Landis & Koch, 1977). In addition, when rating categories are used relatively frequently or infrequently, intercoder reliability may be lower (Smith, 2000).

2 The final model for assent words, which included effects for team size, team composition, conflict, and the interaction between conflict and team composition, was significant, χ^2 (11, N = 475) = 87.16, $p < .001$. The results for during conflict vs. baseline were $B = -0.88$ (-1.19, -0.58), $SE = .15$, Wald $\chi^2 = 32.93$, Exp(B) = 0.41 (0.31, 0.56), $p < .001$; for delayed conflict versus baseline contrast, $B = -0.32$ (-0.63, -0.02), $SE = .15$, Wald $\chi^2 = 4.43$, Exp(B) = 0.72 (0.54, 0.98), $p = .035$. Multicultural teams had 47% fewer assent words, ($p < .001$), seven person teams had 54% more assent words compared to four person teams, $p = .002$), and multicultural teams had 85% more assent words just after conflict (postcon1) compared to Western-only teams ($p = .011$).

3 For differentiation words, the results for during conflict vs. baseline were $B = 0.32$ (0.17, 0.48), $SE = .08$, Wald $\chi^2 = 16.09$, Exp(B) $= 1.38$ (1.18, 1.62), $p < .001$; for postcon1 vs. baseline, $B = 0.22$ (0.02, 0.41), $SE = .10$, Wald $\chi^2 = 4.75$, Exp(B) $= 1.24$ (1.02, 1.51), $p = .03$. Only conflict and team size were included in the final model of differentiation words, $\chi^2(6, N = 475) = 27.00, p < .001$.

4 The final model for negation words, which included main effects for team composition, conflict, and the interaction between the two, was significant, χ^2 (9, N $= 475) = 38.56$, $p < .001$ (although team composition itself was not significant, $p > .80$). In addition to the effects for during conflict versus baseline, $B = 0.61$ (0.27, 0.95), $SE = .17$, Wald $\chi^2 = 12.54$, Exp(B) $= 1.84$ (1.31, 2.58), $p < .001$, during the post1 segment, multicultural teams were twice as likely to use negation words compared to Scandinavian-only team meetings during the same phase, $B = .69$ (.02, 1.37), $SE = .34$, Wald $\chi^2 = 4.04$, Exp(B) $= 2.00$ (1.02, 3.93), $p = .044$.

5 Correlations tested using Spearman rho due to the lack of normality of the variables ($n = 475$).

6 Specifically, the natural log of the total words per segment.

7 For number of conflict turns overall, $B = -0.66$ (-1.54, 0.23), $SE = .45$, Wald $\chi^2 = 2.11$, Exp(B) $= 0.52$ (0.21, 1.26), $p = .146$; insight words, $B = 0.10$ (-0.04, 0.23), $SE = .07$, Wald $\chi^2 = 2.01$, Exp(B) $= 1.10$ (0.96, 1.26), $p = .156$; or risk words, $B = -0.14$ (-0.64, 0.36), $SE = .26$, Wald $\chi^2 = 0.31$, Exp(B) $= 0.87$ (0.53, 1.43), $p = .577$.

REFERENCES

Ball, L. J., Onarheim, B., & Christensen, B. T. (2010). Design requirements, epistemic uncertainty, and solution development strategies in software design. *Design Studies, 31*, 567–589. Doi: 10.1016/j.destud.2010.09.003

Barki, H., & Hartwick, J. (2004). Conceptualizing the construct of interpersonal conflict. *International Journal of Conflict Management, 15*, 216–244. Doi: 10.1108/eb022913

Brennan, S. E., & Clark, H. H. (1996). Conceptual pacts and lexical choice in conversation. *Journal of Experimental Psychology: Learning, Memory, and Cognition, 22*, 1482–1493.

Chan, J., Paletz, S. B. F., & Schunn, C. D. (2012). Analogy as a strategy for supporting complex problem solving under uncertainty. *Memory and Cognition, 40*, 1352–1365.

Chiu, M. M. (2008). Effects of argumentation on group micro-creativity: Statistical discourse analyses of algebra students' collaborative problem-solving. *Contemporary Educational Psychology, 33*, 382–402. Doi:10.1016/j.cedpsych.2008.05.001.

Christensen, B. T., & Abildgaard, S. J. J. (2017). Inside the DTRS11 Dataset: Background, Content, and Methodological Choices. In: Christensen, B. T., Ball, L. J. & Halskov, K. (eds.). *Analysing Design Thinking: Studies of Cross-Cultural Co-Creation*. Leiden: CRC Press/Taylor & Francis.

Christensen, B. T., & Ball, L. J. (2017) Fluctuating Epistemic Uncertainty in a Design Team as a Metacognitive Driver for Creative Cognitive Processes. In: Christensen, B. T., Ball, L. J. & Halskov, K. (eds.). *Analysing Design Thinking: Studies of Cross-Cultural Co-Creation*. Leiden: CRC Press/Taylor & Francis.

Daly, S., McKilligan, S., Murphy, L. & Ostrowski, A. (2017) Tracing Problem Evolution: Factors That Impact Design Problem Definition. In: Christensen, B. T., Ball, L. J. & Halskov, K. (eds.). *Analysing Design Thinking: Studies of Cross-Cultural Co-Creation*. Leiden: CRC Press/Taylor & Francis.

de Dreu, C. K. W. (2006). When too little or too much hurts: Evidence for a curvilinear relationship between task conflict and innovation in teams. *Journal of Management, 32,* 83–107.

de Dreu, C. K. W., & Weingart, L. R. (2003). Task versus relationship conflict, team performance, and team member satisfaction: A meta-analysis. *Journal of Applied Psychology, 88,* 741–749.

de Wit, F. R. C., Greer, L. L., & Jehn, K. A. (2012). The paradox of intragroup conflict: A meta-analysis. *Journal of Applied Psychology, 97,* 360–390.

Erez, M., & Gati, E. (2004). A dynamic, multi-level model of culture: From the micro level of the individual to the macro level of a global culture. *Applied Psychology: An International Review, 53,* 583–598.

Friedman, R., Chi, S.-C., & Liu, L. A. (2006). An expectancy model of Chinese-American differences in conflict-avoiding. *Journal of International Business Studies, 37,* 76–91. Doi:10.1057/palgrave.jibs.8400172

Friedman, R. S., & Förster, J. (2000). The effects of approach and avoidance motor actions on the elements of creative insight. *Journal of Personality and Social Psychology, 79,* 477–492.

Friedman, R. S., & Förster, J. (2001). The effects of promotion and prevention cues on creativity. *Journal of Personality and Social Psychology, 81,* 1001–1013.

Friedman, R. S., & Forster, J. (2005). Effects of motivational cues on perceptual asymmetry: Implications for creativity and analytical problem solving. *Journal of Personality and Social Psychology, 88,* 263–275.

Gelfand, M. J., Raver, J. L., Nishii, L., Leslie, L. M., Lun, J., et al. (2011). Differences between tight and loose cultures: A 33-nation study. *Science, 332,* 1100–1104.

Goncalo, J. A., & Duguid, M. M. (2012). Follow the crowd in a new direction: When conformity pressure facilitates group creativity (and when it does not). *Organizational Behavior and Human Decision Processes, 118,* 14–23.

Goncalo, J. A., & Staw, B. M. (2006). Individualism-collectivism and group creativity. *Organizational Behavior and Human Decision Processes, 100,* 96–109.

Hélie, S., & Sun, R. (2010). Incubation, insight, and creative problem solving: A unified theory and a connectionist model. *Psychological Review, 117,* 994–1024.

Higgins, E. T. (1997). Beyond pleasure and pain. *American Psychologist, 52,* 1280–1300.

Jehn, K. A. (1995). A mulitmethod examination of the benefits and detriments of intragroup conflict. *Administrative Science Quarterly, 40,* 256–282.

Landis, J. R., & Koch, G. G. (1977). The measurement of observer agreement for categorical data. *Biometrics, 33,* 159–174.

Leaper, C., & Ayres, M. M. (2007). A meta-analytic review of gender variations in adults' language use: Talkativeness, affiliative speech, and assertive speech. *Personality and Social Psychological Review, 11,* 328–363.

Merriam-Webster. (n.d.). *Insight.* Merriam-Webster.com. [Online] Available from: http://www.merriam-webster.com/dictionary/insight. [Accessed: September 27, 2016]

Metcalfe, J., & Wiebe, D. (1987). Intuition in insight and noninsight problem solving. *Memory & Cognition, 15,* 238–246.

Nemeth, C. J. (1986). Differential contributions of majority and minority influence. *Psychological Review, 93,* 23–32.

Oetzel, J. G., & Ting-Toomey, S. (2003). Face concerns in interpersonal conflict: A cross-cultural empirical test of the face-negotiation theory. *Communication Research, 30,* 599–624. Doi:10.1177/0093650203257841

Paletz, S. B. F., Miron-Spektor, E., & Lin, C.-C. (2014). A cultural lens on interpersonal conflict and creativity in multicultural environments. *Psychology of Aesthetics, Creativity, and the Arts, 8*(2), 237–252.

Paletz, S. B. F., Schunn, C. D., & Kim, K. (2011). Intragroup conflict under the microscope: Micro-conflicts in naturalistic team discussions. *Negotiation and Conflict Management Research, 4*, 314–351.

Paletz, S. B. F., Schunn, C. D., & Kim, K. H. (2013). The interplay of conflict and analogy in multidisciplinary teams. *Cognition, 126*(1), 1–19.

Pennebaker Conglomerates. (2016). *How it works.* [Online] Available from: http://liwc. wpengine.com/how-it-works/. [Accessed June 15, 2016]

Pennebaker, J.W., Boyd, R.L., Jordan, K., & Blackburn, K. (2015). *The development and psychometric properties of LIWC2015.* Austin, TX: University of Texas at Austin. [Online] Available from http://liwc.wpengine.com/wp-content/uploads/2015/11/LIWC2015_ LanguageManual.pdf. Doi: 10.15781/T29G6Z

Rohner, R. P. (1984). Toward a conception of culture for cross-cultural psychology. *Journal of Cross-Cultural Psychology, 15*, 111–138. Doi: 10.1177/0022002184015002002

Shroyer, K., Turns, J., Lovins, T., Cardella, M. & Atman, C. J. (2017) Team Idea Generation in the Wild: A View from Four Timescales. In: Christensen, B. T., Ball, L. J. & Halskov, K. (eds.) *Analysing Design Thinking: Studies of Cross-Cultural Co-Creation.* Leiden: CRC Press/Taylor & Francis.

Smith, C. P. (2000). Content analysis and narrative analysis. In H. T. Reis & C. M. Judd (eds.), *Handbook of research methods in social and personality psychology* Cambridge, U.K.: Cambridge University Press. pp. 313–335.

Srikanth, K., Harvey, S., & Peterson, R. (2016). A dynamic perspective on diverse teams: Moving from the dual-process model to a dynamic coordination-based model of diverse team performance. *Academy of Management Annals, 10*, 453–493.

Stahl, G. K., Maznevski, M. L., Voigt, A., & Jonsen, K. (2010). Unraveling the effects of cultural diversity in teams: A meta-analysis of research on multicultural work groups. *Journal of International Business Studies, 41*, 690–709.

Tausczik, Y. R. & Pennebaker, J. W. (2010). The psychological meaning of words: LIWC and computerized text analysis methods. *Journal of Language and Social Psychology, 29*, 24–54.

Todorova, G., Bear, J. B., & Weingart, L. R. (2014). Can conflict be energizing? A study of task conflict, positive emotions, and job satisfaction. *Journal of Applied Psychology, 99*, 451–467. Doi: 10.1037/a0035134

Trickett, S. B., Trafton, J. G., Saner, L., & Schunn, C. D. (2005). 'I don't know what's going on there': the use of spatial transformations to deal with and resolve uncertainty in complex visualizations. In: M. C. Lovett, P. Shah (eds.) *Thinking with data* Mahwah, NJ: Erlbaum. pp. 65–86.

Tuckman, B. W., & Jensen, M. A. C. (1977). Stages of small-group development revisited. *Group Organization Management, 2*, 419–427.

Ward, T. B., Smith, S. M., & Vaid, J. (Eds.) (1997). *Creative thought: An investigation of conceptual sructures and processes.* Washington, DC: American Psychological Association.

Zhou, J., & Su, Y. (2010). A missing piece of the puzzle: The organizational context in cultural patterns of creativity. *Management and Organization Review, 6*(3), 391–41.

Chapter 4

Resourcing of Experience in Co-Design

Salu Ylirisku, Line Revsbæk & Jacob Buur

ABSTRACT

The process through which people's experiences are resourced in co-creation has gained little attention. Involving different people is fundamental in today's multi-stakeholder endeavours, and knowledge of the process of resourcing experience is relevant for developing innovation practices in organisations. We develop a framework for the study of resourcing of experience that builds on G. H. Mead's pragmatist theory. The resourcing of experience is a social undertaking, where experiences are made available to co-designers through articulation. We identify the responsive, conceptual and habitual characteristics of the resourcing of experience and investigate how these characteristics are observable in the situated interactions in the DTRS11 dataset. Through the analysis we pinpoint specific ways that the design team fosters the resourcing in the collaboration with co-designers. The paper suggests ways to resource experience that design teams can make use of in the rich involvement of co-designers.

1 INTRODUCTION

The first Design Thinking Research Symposium was organised in the early 1990s (Cross, Dorst, & Roozenburg, 1992), and since then the meaning of design thinking has broadened to consider complex, open and dynamic social processes (Dorst, 2011). Today design projects are increasingly collaborative and involve multiple stakeholders, such as users, product managers, consultants and representatives of various organisations. The involvement of different stakeholders, and not only users, is now integral to the organisation and conduct of design projects. Inviting various stakeholders to participate requires careful attention to how the process is introduced to them and how their contribution unfolds in interactions and over time (Buur & Larsen, 2010; Heape, Larsen, & Revsbæk, 2015).

Our paper takes a starting point in the theoretical construct of *resourcing*, which we have coined to help explain the complexities of design workshop facilitation. 'Resourcing' refers to the *negotiated use of what is available* for co-designing. The concept contrasts the objectivist view of 'resources' often found in management literature. This view assumes 'a resource' to have an essential identity independent of its use, whereas the concept of resour*cing* emphasises that co-designers (including the designers and facilitators) engage in negotiated social interaction with what is available to them. User-experience is often considered available as a fixed 'resource' in the form of user representations, similar to objectivist resources (e.g., Fleury, 2012). By attending to the resourcing of 'experience' we mean to acknowledge 'experience' as continuously shaped, re-contextualised, and re-prioritised in design projects over their duration. In this paper we explore how participants' *lived experience* is *made available through articulating* and *responded to by others* in the co-design events.

We study the DTRS11 dataset recorded in a concept design process by a European car manufacturer. The observed design events take place in China and Scandinavia.

2   RESOURCING OF EXPERIENCE

We approach the resourcing of experience through a situated lens, where the co-designers negotiate and signify what they perceive relevant for the project. The present work builds on studies of how designing happens (i.e., how the embodied design interactions unfold and are embedded within particular settings; e.g., Bucciarelli, 1994; Heinemann, Landgrebe, & Matthews, 2012; Matthews & Heinemann, 2012), and on studies of how designing is contextualised (i.e., how the process is made part of the surrounding reality), such as Hyysalo's (2010, 2012) studies of how users are represented in the process.

We focus on situations described by the pragmatist philosopher G. H. Mead (1932) as 'emergent events'. In such events novelty occurs for the participants in a way that calls for the restructuring of involved participants' understanding of the situation. According to Mead, the surprising event, which obstructs or troubles ongoing action and challenges the 'accepted structures of relations,' stands out as 'data' to the participants involved. The novel occurrence becomes the nodal point from which a new meaning and understanding arises as a new structure of relations of past experience emerges.

The organisational theorist Ralph Stacey (2011), drawing on Mead in his complexity theory perspective on organizational life, uses the term 'abstracting' about all forms of thinking about and reflecting upon experience. He writes: "*Articulations (…) in narrative form involve selecting and simplifying and, in that sense, abstracting from experience*" (Stacey, 2011, p. 415). According to him, the selection involved in a narrative account (of some experience) serves not only a matter of simplification but also one of elaboration in that it gestures to other people present to respond to it. Similarly, building on Mead's work, Herbert Blumer investigated the role verbally expressed 'concepts' have for addressing such emergent events (Blumer, 1998/1930, p. 158): "*The concept permits one to catch and hold some content of experience and make common property of it.*" Thus, the articulation of experience in spoken language mediates the re-organising of past experience occurring in emergent events. As such Blumer (1998/1930) argued that through concepts people can "*isolate and arrest a certain experience which would never have emerged in mere perception.*"

In Mead's process philosophy the notion of 'experience' is an ontological term to account for the temporal relationship between the individual and their environment (Mead, 1934). Much of what we in this paper mean by 'experience,' when we analyse the 'resourcing of experience,' is in Mead's vocabulary better equivalent to what Mead (1932) calls 'the past,' that is, past experience or previous experience. When talking about resourcing of experience we aim to understand how and to what effect participants' previous experience and knowledge is accentuated and articulated in specific co-design events. Symbolic concepts play an important role in this, as a symbolic concept may enable people to orient to a situation in a similar way (Blumer, 1998/1930). In his related theorising, Donald Schön underlined this role of concepts, in terms of the 'displacement of concepts' (Schon, 1963, p. ix):

> *"The displacement of concepts is central to the development of all new concepts and theories, whether they have to do with science, invention, or philosophy. The process is nothing less than our way of bringing the familiar to bear on the unfamiliar, in such a way as to yield new concepts while at the same time retaining as much as possible of the past."*

In our vocabulary of *resourcing* the narrative and conceptual articulations play two roles: (1) in themselves they are experience resourced and made available to others; and (2) they invite further elaboration and resourcing of additional experience. The identified and articulated 'themes' used by the design team in the DTRS11 dataset (one example is the theme of 'conscious commitment') can be expected to sum up impressions, observations, and reflections about a generalised Chinese consumer, and yet, to trigger associations to related experiences and invite new interpretations. The re-articulation of and re-attending to the themes can also be expected to serve the continuing elaboration of experience by people engaged in the particular design events, as Stacey (2011, p. 415) writes:

> *"Narrative articulations of experience require interpretation in particular contingent situations. Their aim is not simplicity (…), but rather their aim is the opening up of accounts of experience for greater exploration in order to develop deeper understanding."*

Based on these considerations of experience, we develop an analytical framework with three characteristics of 'articulating experience,' which are relevant for concept design:

1 *Articulating of experience is responsive.* What and how people talk about their past experience depends on the local interactions with the involved participants. Immediacy matters.
2 *Articulating of experience is conceptual.* People use particular words to address and explain experiences, to suggest associations, and to invite others to think about themes.
3 *Articulating of experience is habitual.* People with different professional backgrounds have different response-sensitivities and they take notice of, promote, and talk about things in differing ways. Habits can be learned, and hence, articulating of experience can be 'trained'.

Our research questions are:

– What can we learn about how the participants resource articulated experiences in the co-design events in the DTRS11 dataset?

– What practical means does the design team use to resource the experience of the participants in the co-design events in the DTRS11 dataset?

3 APPROACH TO ANALYSIS

We approach the analysis especially through the process philosophy of Mead (1932, 1934), which emphasises a paradoxical notion of time. It foregrounds the 'living present' as the seat of reality from which previously lived experience is reinterpreted for the purpose of continuous meaningful conduct in interplays of intention among a multitude of stakeholders. We focus on the role that the resourced experience has for the unfolding action in the project. Resourcing of experience partly occurs through stories and explanations and partly materialises as posters and sticky notes in the workshop space.

We start our analysis by focusing on two situations at different stages of the design process. The first event is a session where the extended team shares insights from the first co-design workshop in China on the 2nd of Dec 2015. The second event is the last event available in the DTRS11 data (on the 26th of Jan 2016). We assume that these two events will be different regarding how the participants resource their articulated experience due to the learning that happens. We expect to find differences in how the core team responds to particular terms. We focus on the emergence of the key term of 'conscious commitment,' which was presented as a key result of the project. It was used as the tagline for the key design concept in the project's deliverable (dated 3rd of May 2016), which underlines the centrality of this term for the project.

We utilised the initial transcripts by the DTRS11 committee, went back to review the original data to verify the content, and then re-did the transcripts to highlight the essential elements for the inquiry into resourcing of experience.

4 ANALYSIS

The notion of 'commitment' appears for the first time in the data in a session where the extended team shares insights from the first Co-Creation workshop (CC1) in China on the 2nd of December 2015. We begin our analysis here, when the term 'commitment' was first articulated while the entire design team was present.

4.1 Emergence of 'commitment'

Present in the situation is the core design team (Ewan, Abby and Kenny), external consultants (Amanda, Rose and Will), intern (Nina), and participant observer (David). They have all been participating in the CC1 with Chinese consumers the day before, which Will and Rose were facilitating. The space is staged with note-boards with sticky notes from the workshop, Figure 4.1. There is a screen displaying the view from Amanda's computer. In the session Will explains his observations first, and then Rose goes through her observations. Both of them follow a pattern where they first introduce the Chinese participants of their group in the workshop, then summarise what they consider key findings. Next, the following exchange takes place.

Figure 4.1 Rose (2nd from left) points at a board with a sticky-note with the text 'COMMITMENT'. The board rests on the floor behind Abby (4th from left).

Excerpt 1. (v08/280-290)

01: Amanda: So, sorry. When that person talk about recycling for the sake of recycling, that one lady.

02: Amanda: Did the other agree and do the same or not?

03: Rose: So I felt that they didn't. They're definitely not at to her level. They, because that's why.. When we talk about our theme *(Points at a white plate with sticky notes behind Abby. One of the notes has a text ''COMMITMENT'' on it.)*

06: Rose: Brian, he called it ''commitment''. He said the English word first and then he called it more like, ''taking up the role or promise''. So that is a little bit softer than how she is. But I think there was a general sense that they saw the importance of it. And, I think, just maybe not so much like.

07: Amanda: Not in action, yeah.

08: Nina: But she said that she doesn't care if other people know. But if her friends find out, that's good.

09: Rose: Yes she wasn't gonna go out for others.

10: Amanda: She didn't do it for others.

11: Rose: She really was doing it for herself.

12: Nina: For, yeah, for herself.

13: Ewan: But it didn't hurt, it didn't hurt if someone saw it.

14: Rose: Of course.

15: Nina: Yeah.

4.1.1   *Notes on the responsive character of resourcing articulated experience*

In terms of resourcing experience, Rose is here articulating her experience of what happened the day before. She brings up the term 'commitment' as introduced by one of the Chinese workshop participants. In her mentioning of "*Brian, he called it "commitment". He said the English word first and then he called it more like taking up the role or promise,*" Rose responds to a question posed by Amanda. The question inquires into 'recycling' and into being radical about 'recycling for the sake of recycling'. In an earlier remark in the session (v08/161, before Excerpt 1) Rose presented the workshop participant Heidi: "*So, she was the one who, amongst the four, who cared the most about the environment and recycling, especially towards the end. She really believed in being eco-friendly for the sake of like, you just need to play a part in doing that.*" Amanda's later question "*did the other agree and do the same or not?*" serves for Rose to continue elaborating on her story about the workshop participants' accounts. She begins her response by saying, "*So I felt that they didn't. They're definitely not at to her level.*" She stops her sentence abruptly, looks at a board on the floor behind Abby and points. It has the sticky note with the label "commitment" on it. She points in the precise moment when she says "*theme*" in the sentence "*When we talk about our theme*"; referring to the talk in the workshop group of Chinese consumers. Her gesture provides visual support for the others concerning the name of the theme and it is then that Rose articulates the theme name "commitment". In their accounts, Rose and Nina thus link the issues of 'eco-friendly' and 're-cycling' to the 'commitment' theme.

The resourcing of experience changes from articulation of what is remembered (about the particular workshop participants Heidi and Brian) into a discussion of what to conclude from what Rose and Nina report these participants said. The above discussion gains interest in the light of how prominent the 'commitment' theme becomes later in the project. Nina's comment, "*She doesn't care if other people know. But if her friends find out, that's good,*" is first followed by Amanda's response, returning to her experience of this workshop participant, as if correcting the picture that is being outlined in the discussion. "*She didn't do it for others*", Rose responds. As if following up on her own earlier remark about this workshop participant's preoccupation with recycling 'for the sake of recycling', Rose concludes: "*She really was doing it for herself*". Nina responds by dwelling for a while on this remark "*for, yeah, for herself.*" Ewan, however, departs from the account of this workshop participant by re-articulating the point initially expressed by Nina: "*But it didn't hurt, it didn't hurt if someone saw it.*"

4.1.2   *Notes on the conceptual character of resourcing articulated experience*

As stated in Section 2, people use particular words to address and explain experiences, to suggest associations, and to thus invite others to elaborate on their experience. The participants' sharing of insights from the Chinese consumer workshop draws on numerous discretely identifiable parts, ideas, or thoughts. In earlier work we have named these 'conceptual entities' (Ylirisku, 2013). Amanda's question exemplifies this: "*So, sorry. When that person talk about recycling for the sake of recycling, that one lady. Did the other agree and do the same or not?*" Building up her sentence, Amanda refers to the separate entities of 'that person,' 'the person's talk about recycling,' 'that one lady,' 'the other(s),' and 'do the same'. Conceptualising in terms of these specific entities that separate, for example, 'that one lady,' 'the other(s),' and 'do the same' is followed by Rose's response that teases out the difference between two separate and, in Amanda's question, opposed entities. Rose's response revolves around this introduced conceptual split between 'that one lady' and 'the other(s)', bringing her to elaborate on how the standpoint of 'the other(s)' (in particular Brian) diverts from the standpoint perceived and described to be 'that one lady's standpoint – a response which eventually introduces the theme of 'commitment' taking the conversation on 'recycling' into a different direction.

4.1.3 Notes on the habitual character of resourcing articulated experience

The theme of 'commitment' is not yet habitual to the design team members this early in the project. The articulation of the theme appears coincidental rather than planned and intended, even if the theme is already materialised in the posters present in the session. Reference to the theme is not yet characterised by a certain *order of things presented* (e.g., hierarchically starting from a label or by enumerating concepts that relate to a whole), *highlighting of the theme* (e.g., repeating and coming back to the theme name), nor *fluency in its articulation* (e.g., long turns of explaining the contents with little repair of the talk). Instead, the appearance of the term 'commitment' is more associative than decisive. It starts out as a fragment in a fragmented answer that builds into a small elaboration that one particular participant reportedly stated. Hence, talking about 'commitment' in the context of this project is not yet habitual for Rose, nor to the other participants.

4.1.4 Notes on the strategic ways in which the team resources experiences

- The design team has asked the involved Chinese consultants to prepare for a review of CC1, that is, the consultants provide a version of their experience in terms of the most relevant-for-the-project issues emerging from the workshop.
- The review session takes place in China soon after the workshop, that is, within a day. The memory of the workshop is fresh, and the Chinese surroundings and the physical takeaways from the workshop (i.e., posters, etc.) serve as a specific enabler of workshop recollections in the resourcing of the participants' experiences.
- The design team uses materials (sticky notes on the boards) to carry textual traces of discussions forward. As visuals these enable the stories to be cued, elaborated and associated.
- All the involved people are co-located in one shared physical space. Hence people use their embodied means of communicating, including hand gestures, pointing, etc., to elaborate and to relate to each other and to the articulated themes in the conveyed experience.
- The design team listens with an attitude of using what they hear. They take notes for the next day's meeting.

4.2   Condensing the theme of 'commitment'

The following interaction takes place in a session two months after the previous transcript, towards the end of the DTRS11 dataset. Participating in the "Brainstorming on
Concept and Products" session (video 21) are the core design team members (Ewan,
Kenny and Abby) and two internal clients (Paul and Steven). The design team has
introduced the activity to the clients as a 'brainstorming' session and asked them to
make notes of the (product) ideas they may think of on the sticky notes provided. A set
of posters with headlines, some of which are 'Conscious Commitment,' 'Progressing
Together' and 'Evolving Status Symbol', hang on the wall. The posters also contain
smaller text and some have figures and illustrations in them, Figure 4.2. Ewan explains
the theme of 'conscious commitment':

```
Excerpt 2. (v21, 075)
Ewan:    (Stands in front of the poster with the text
         'conscious commitment') Conscious commitment is about
         this dual effect of being really dedicated to a more
         collectivism's values, that you are reinvesting in
         society and doing the right thing. Not only for you,
         your family, but also other people. (…) As a return,
         as you do this, you get social recognition that would
         elevate your status.
         This comes to life in the sense that you are now,
         suddenly you're having questions around ''Where does
         the >you know< the food that I eat, where does that
         come from?'' ''What do they feed the chicken and the
         pig that I eat'' and then even further ''What about
         the people that take care of the, the pig that
         I eat?'' And so on.
         So probably, it suddenly becomes much more kind of
         triple bottom line thinking that it has ever been
         before. And your investment there, and the values
         that you are suddenly advertised to that or admit,
         will come back to you in the form of social
         recognition and social elevation.
```

4.2.1   The articulation of experience becomes habitual

At this point in the design project the design team has made several decisions on what to
focus on in the ideation. The themes re-articulated in the brainstorm session have been
discussed recurrently throughout meetings. Articulating the theme 'conscious commitment' has become habitual to the design team, which is visible in how Ewan explains
it. Ewan starts hierarchically, presenting an overarching label 'conscious commitment,'
before moving to enumerating concepts that relate to this overall whole. Ewan's notion
of 'dual effect' is an abstraction of the two conceptual entities respectively articulated
as "*reinvesting in society*" and "*elevate your status*". That Ewan mentions the 'dual
effect' before explaining what the two parts of the dual effect are, along with his fluency

Figure 4.2 Ewan explains the theme 'conscious commitment' while standing in front of a poster with the theme name.

in articulating it, serves to evidence the habitual articulation of the theme of 'conscious commitment' (e.g., the long turn of explaining the content without reading the small print in the poster and with little repair of talk).

4.3 Ideating on 'commitment'

Once the design team has explained the themes on the posters in the room (see Figure 4.2) Ewan asks Steven about his thoughts on the project presentation, which leads into a set of responses on the theme that Steven suggests. Excerpt 3 shows how the core design team members (Ewan, Abby and Kenny) resource each other's project-specific experience in the elaboration of the new meaning for 'recycling' in the project context. Facing a visiting client becomes as much an effort of resourcing the project-specific experience of the design team members in new ways as it is about exploring and integrating the experience that a new participant may articulate.

Excerpt 3. (v21, 329–356)
```
01:  Ewan:     And Steven, what did you think when you heard
                all this? (…)
02:  Steven:   (…) Ehm, okay, I wrote some stuff here
                (looking at his stack of sticky notes) So, I
                wrote down. I wrote these all when you (points
                at the wall with posters) were presenting the
```

different slides on the wall. **Durability**, eh,
with the tagline **for me and other people**. So if
we make sure that the customers know that our
accessories are made of sustainable >maybe not
sustainable< but durable materials. Then they
can pass them on to, eh, generations and
generations.

03:	Ewan:	Ah okay! So sustainable in **that sense**. That you are keeping it so you don't have to produce it again. It's not like going to waste when you get rid of it.
04:	Steven:	Yea', Yea'
05:	Ewan:	Can you elaborate a little on that?
06:	Steven:	Yes. (Stamps a note to the board with the text: ''DURABILITY FOR ME AND OTHER PEOPL'') (Ewan walks at the note)
07:	Abby:	What was the tagline?
08:	Kenny:	Durability
09:	Ewan:	''For me and other people'' (*Follows with finger the text on the note*) meaning kind of in a timeline?
10:	Steven:	Yea'
11:	Ewan:	Not in the moment, but over time.
12:	Steven:	Yes (Stamps a note from his hand to the board near the previous one. Note text: ''SPARE PARTS -- FOR ME AND OTHER PEOPLE'')
13:	Kenny:	So I guess it's very much about how you explain or communicate your (INAUDIBLE). So you communicate it in terms of for example a family: how it lives on for generations to generation. Rather than telling that ''this thing has been made out of eh SCANDINAVIAN steel'' and stuff like that.
14:	Ewan:	Mhm. (Moves the two notes, which are attached to each other, to the top of the board)
15:	Kenny:	That they might not be able to relate to and they don't know if that's actually true. But they can relate to the family member's something else.
16:	Ewan:	Yeah.
17:	Abby:	Could it also be about.. If, I mean, if it's recycled? So it's not the same product that lives on but that's actually recycled.

18:   Ewan: Well, that's interesting, yeah.
19:   Steven: Yeah, yeah, it could.
20:   Abby: And that's why it lives on.
21:   Ewan: (...) That's it 'cause it has the two paths that
                you can **re-cycle** it or **down-cycle** it into other
                parts. Or is it the same object, that just be
                inherited, but is so fucking strong that it
                actually. You don't even see that people have
                been using it. In fact it is literally the same
                thing. Yeah, that's two interesting approaches.
                (...) It gets value through the, the action of
                passing it on.
22:   Abby: Mmm
23:   Kenny: Mmm
24:   Ewan: Some of the most powerful eh Rolex advertising
                or Omega advertisement I've seen is like. It's
                about the father who gives the watch to his
                son. And it's like.. It's not.. They say
                something that is not right about the watch. It
                is about the tradition. Yeah time, the
                tradition, new starting. You are starting
                something. He would give it to his son, and so
                on and so on. And it's, yeah. I think that
                really appealed to me, like wow. I remember
                when I got something from my father that used
                to be my grandfather's. And stuff like that.
                It's like wow. Suddenly you are not only a
                person in the moment. You are a person that
                stretches out towards time.
25:   Steven   Yeah, so **that's what I wanted to say.**
26:   Others   (laughter)
27:   Ewan   Yeah, yeah (laughs). I think that's the
                interesting (touches the two notes on the board
                from Steven) that actually **handing stuff down**
                adds value instead of taking value away.
                (Moves at a pile of sticky notes, and writes
                a new note, which he attached to the two notes
                from Steven. The note has the text: ''PASSING
                THE ITEM ON ADDS VALUE TO THE ITEM (OMEGA ADD)''

4.3.1   Notes on the responsive character of resourcing experience

When Steven links his notion of 'durability' to 'sustainability' Ewan responds with
an energised remark: "*Ah okay! So sustainable in that sense*". The linking of 'durability' and 'sustainability' becomes a novel nodal point that makes the design team
challenge their idea of sustainability that has grown habitual in the project by now.
Eventually, linking 'sustainability' and 'durability' in this way leads to the resourcing of

supplementary experiences on behalf of the design team members. The most prominent incident of this is Ewan's articulated recollection of the Rolex/Omega advertisement and his childhood memory of being handed down a token from his grandfather.

In one respect Excerpt 5 illustrates the resourcing of Steven's impressions in response to Ewan's presentation. Experience-probing questions from Ewan such as *"Steven, what did you think when you heard all this?"* and *"Can you elaborate a little on that?"* lead Steven to share his notes. Had the design team aimed for a further resourcing of Steven's experience, they could have enquired into why Steven signified these particular aspects when listening to Ewan's presentation in the session. Instead, the more elaborate resourcing of experience in this incident happens on behalf, not of the external clients, but of the design team members – estranged towards their own habitual conception by the new appraisal of 'durability' as a form of 'sustainability'.

Eventually Steven's account leads into a set of turns, where the core design team begins resourcing their experience in ways that responds to what Steven has just proposed. Abby's question, *"What was the tagline?"* is answered by Kenny *"Durability"* and then Ewan *"For me and other people"* in word-to-word repetitions of what Steven has written on his note. In this, Steven's contribution is being resourced, signified by the design team members, and in return resources new aspects of the project-specific and otherwise lived experience on behalf of the design team member. The text on the sticky note anchors the conceptual aspect of the resourcing in this situation.

4.3.2 Notes on the conceptual character of resourcing experience

The restructuring of relations, and the reorganising of understanding, elicited by Steven's linking of 'sustainability' and 'durability' in the brainstorm session in Excerpt 5 is not only detectable in the participant articulations, but also in the arranging of materials in the setting. Related sticky notes are attached to each other, collected on the board, and moved into a particular location of the board, and thus, put into relation with what is already there. Ewan's closing gesture of writing *"passing the item on adds value to the item (Omega add)"* on a sticky note, and putting it on the board by the notes on 'durability' and 'spare parts – for me and for other people' written by Steven closes off this particular line of ideation (before the design team and the client visitors move on to additional explorations).

4.3.3 Notes on the strategic ways in which the team resources experiences

– The design team has invited participants external to the project, but who work at the case company in a project-relevant department. Hence, the invited participants can expectedly respond in relevant ways by drawing on their company-relevant experiences.
– The design team have prepared posters and use them in presenting the themes. The physical material makes permanent traces in the ideas discussed and serves as a lever in the reorganising of understanding as emergent events challenge established structures of meaning. Written notes are likely to assist the team in simplifying to either further elaborate or to condense into future concept descriptions. Some of the texts and ideas on the sticky notes in the case study design project thus reappear in the project concept description.

–   The team is actively elaborating on what the participants contribute, and while doing so they articulate their own past experiences for others to resource in the process. The team also effectively elaborates on the experienced surprises and works towards assembling new materials for the project.

5 KEY FINDINGS

We developed an analytical framework of three characteristics of resourcing experience. The framework is based on Mead's pragmatist philosophy and related insights from Schön, Blumer and complexity theory from Stacey. Below we conclude on *the academic exploration into the process of resourcing experience*, which aims at conceptual development of the term resourcing experience, and on the *strategies for resourcing experience in practice*, which highlight such ways of resourcing experience that may be useful for designers in exploratory projects.

5.1 Exploration into the process of resourcing experience

We analysed the DTRS11 dataset through the lens of resourcing experience with the focus on its three characteristics as a *responsive*, *conceptual*, and *habitual* process. The analysis of moment-by-moment interactions through this conceptual lens enabled us to show how the experience of both the design team members and the various stakeholders in the process (specifically consultants and internal clients) was resourced. The analysis is based on the central assumption that experience is made available for others in the situation through articulating, and that it is possible to discern how experience was 'used' in co-design by attending to the responses of the others in the situation. By 'use' we refer to how articulated experiences become 'significant symbols' (Mead, 1934) in the conversation, taken up by others or otherwise responded to and collaboratively reflected on, thus over time made into 'design material' for and by the participants of the co-design event.

5.1.1 Findings about the responsive character

Human social interaction is said to be both context-shaped and context-renewing (Heritage, 1984). Complexity theory-informed accounts of human interaction show that this is simultaneous: In local interaction individuals respond to their sense of the context and any interaction is immediately co-constitutive of the context in which it is made to make sense (Stacey, 2011). Especially interesting for the present study are such moments when something unexpected occurs to which a participant responds by sharing new aspects of (their lived) experience. In such moments a new order of happenings and understandings takes place, even if on a small scale. An example in the data is from the Excerpt 1 of Amanda's enquiry into the 'recycling for the sake of recycling' in response to a an incomplete narrative by Rose. Being asked to elaborate Rose makes a reference to the theme of 'commitment' and then develops this into 'taking up the role or promise'. Further articulating her experience from the consumer workshop spurs the dialogue between team members. Another prominent example of prior experience being articulated in response to what is discussed is the articulation of Ewan's story about the 'Rolex/Omega add'. Eventually, this storytelling materialises in a note on the whiteboard.

5.1.2   Conceptual

Verbal concepts are used throughout the process to articulate experiences, to probe experience, and to facilitate co-design work. According to Blumer (1998/1930) concepts allow people to establish shared points of view and aid a group of people to orient themselves to a situation in similar ways. In the illustrated events of resourcing experience, any response of the participants closely relates to the overall project or the topic discussed. For example, when Rose is asked about 'recycling' she related this to the theme of 'commitment' which, as she explained it, was part of what the participants discussed when talking about recycling. The continuous and repetitive work around the theme of 'commitment' witnesses the cultivation of the concept of 'commitment' in collaborative work over time. Ewan's presentation of the theme of 'conscious commitment' in Excerpt 2 displays a clear three-part structure from broad concept to concrete details, and then back to a broader theoretical concept of 'triple bottom-line' thinking. To borrow Blumer's terms (1998/1930), such conceptual work is what enables the team to *"isolate and arrest a certain experience which would never have emerged in mere perception"*.

5.1.3   Habitual

The habitual character of resourcing of experience is displayed in the studied interactions in two main ways: first, as the 'training' or socialisation as members of a certain societal culture; and second, as the 'training' and learning that takes place in particular, longitudinal project-specific work. Due to their life-long membership of the Chinese culture, the external consultants, who were present in the co-design workshops and in the 'sharing of insights' session, have certain response-sensitivities and are therefore likely to notice and select out incidents and experiences from the co-creation workshops that their Scandinavian colleagues in the design team might not make a point of. We observe such a process in Rose's report on her observations and memories from the workshop in the 'sharing of insights' session. Related to project-specific learning and coordinating, we observed the design team to develop an increasingly structured way of articulating the topic of 'commitment'. Over time an order, structure, and flow of articulating the theme emerges: In the early stages articulations were unstructured, longwinded, and involved multiple people, not expressing one central and clear idea. Later, the notion of 'commitment' was cultivated to the point of one team representative conveying this to external consultants in an orderly narrative structure.

   The habitual character of resourcing experience can also be understood in terms of expectations. Once the team becomes (more) familiar with a theme, they form particular expectations on what the theme 'is' and this, in return, conditions and constrains the resourcing of 'new' aspects of experience in specific ways. The incident with Steven in the brainstorming session, who ties together previously unrelated aspects of the material, thus challenging the habitual articulations of the design team, was shown to resource the renewed and recollected memories of the team members and project manager Ewan. In terms of Schön's (1963) displacement of concepts, this *"bringing the familiar to bear on the unfamiliar"* in new situations enables the resourcing of previously not articulated team member experience. As mentioned, we have previously conceptualised the development of expectations in terms of 'response-sensitivities' to emphasize the role of expectations on what people take notice of, promote, and talk about, and how they do so in differing ways.

5.2 Strategies for the resourcing of experience in practice

In the following we discuss in what way the design team can be said to strategically organise their practice of resourcing experience: (1) the team *attended to the articulated experience with a resourcing attitude,* that is, in order to use what was brought about; (2) they *captured experiences in materialised simplifications*; and (3) they *relied on cultural, professional and project-specific cultivation of habits* (or response-sensitivities).

5.2.1 Attending to articulated experience with a resourcing attitude

The team actively took notice of what their co-designers articulated. They asked questions, wrote notes, and responded by articulating how they understood what the co-designers stated. Examples of this are found in Excerpt 1, where Ewan concludes on the basis of the discussion on recycling "*But it didn't hurt, it didn't hurt if someone saw it*" and on the next day he presents a note he made stating "*I wrote down (...) other people discover your commitment and sacrifice, and you get elevated to a higher status level.*" They took what they heard others articulate as ingredients into their own subsequent articulation. And they also resourced some of the articulated experiences of their own, as displayed in Ewan's writing of the note on his own Rolex/Omega story in session 21.

5.2.2 Capturing experiences in materialised simplifications

Writing notes appeared to be an assumed assurance that the things attended would be carried on. Some stories and articulations were not referred to in a note, possibly because they were not listened to in the overt dialogue, and they might have been forgotten or would emerge again at a later point in the work. The acts of writing appeared as a sign of perception of meaning, or value in what was articulated. Interestingly, the core team members urged the co-designers also to make such notes. For example, the Chinese consultants had prepared slide sets that were used in the review of the CC1 experience. The consultants needed to simplify their experience into shareable notes on the slides that they used when articulating their insights about what they learned from the participants of the co-design workshop. Even though it is only implicit in the data, it is quite likely that the team had explicitly asked the consultants to prepare these presentations.

The theme 'commitment' was initially stated in a single remark, and the term was not returned to again in the session where the name was initially coined. The theme, however, was captured on a sticky note from the workshop. It was placed above a group of notes, and it appeared as a theme label. The next day the team went through what was on the boards, they discussed the topic, and re-articulated the notion of 'commitment' multiple times during that session. In the late session 21 the theme, which was originally brought up as a characteristic of the Chinese user experience, was materialised as a set of posters with texts and figures.

5.2.3   Relying on cultural, professional and project-specific training of habits

The project was about creating new offering for the Chinese market. The team organised the action in a way that allowed them to gain access to Chinese consumers' experience of this market, and they involved a set of people, which they appeared to have carefully chosen to serve this purpose. They invited the Chinese consumers and worked with Chinese consultants. The Consumers had experiences living in the targeted culture, and the Chinese consultants acted as translators and re-articulators of what was stated in the workshops. The team also invited experts from their own company (stakeholders as well as internal clients), who contributed not only by commenting on the evolving contents of the process and suggesting new ideas, but also by serving as partners to whom the core team had to articulate what they considered relevant.

As a pointer for potential future work, one could analyse resourcing in terms of 'interplay of intentions' (Stacey, 2011) among the participants throughout the process. That which is carried on in the emergent themes necessarily reflects also the figuration of power between the people involved. This might be especially useful for understanding collaborations across culturally diverse groups.

6   CONCLUSIONS

In this paper we set out to investigate how the resourcing of experience happened in an international and multi-cultural design project with the aim of developing appropriate analytic methodology, to deepen understanding of the value of involving different people in co-design. We intended to portray the apparently well-working strategic means that designers seem to apply in their practice while resourcing the lived experience of people involved in the co-design. We conducted the analysis by building on Mead's process philosophy and on our earlier work with the theoretical concept of *resourcing*. We use the term resourcing to refer to the *negotiated use of what is available* for co-designing. Other people's experience is not directly observable, but people make it available through articulating. Based on pragmatist philosophy, symbolic interactionism, and process theory, we outlined three characteristics of articulating experience (*responsive, conceptual,* and *habitual*) for the purpose of analysis.

We analysed the DTRS11 dataset that covers an exploratory design project with a European Car Manufacturer conducting a concept design project for a Chinese market. From the full project data we focussed on a handful of key incidents, where the resourcing of participants' experiences appeared to have a significant influence on what is by the design team articulated as outcomes of the project. These incidents were investigated regarding how experience was articulated and responded to by the participants in the studied project.

As a result we conclude that by attending to three characteristics of articulating experience, we can make discoveries into how experience is *used* as a resource in designing. The investigation accentuates the role of learning within a project, thus showing emergent themes and concepts to appear in an increasingly ordered, structured, and flowing fashion in later stages of the design process. The close scrutiny of how experience is being resourced in design enables us to explicate how the design team members approach the resourcing of experience strategically. We found that: (1) they *attend to articulated experience with a resourcing attitude*, that is, in order to use what is being brought about; (2) they *capture experiences in materialised simplifications*; and (3) they *rely on cultural, professional and project-specific training of habits* (or response-sensitivities).

REFERENCES

Blumer, H. (1998). Science Without Concepts (1930). In: *Symbolic Interactionism: Perspective and Method* University of Chicago: University of California Press. pp. 153–170.

Bucciarelli, L. L. (1994). *Designing engineers*. Cambridge, Mass: MIT Press.

Buur, J., & Larsen, H. (2010). The quality of conversations in participatory innovation. *CoDesign*, 6(3), 121–138. https://doi.org/10.1080/15710882.2010.533185

Cross, N., Dorst, K., & Roozenburg, N. (eds.). (1992). *Research in design thinking*. Delft, The Netherlands: Delft Univesity Press.

Dorst, K. (2011). The core of "design thinking" and its application. *Design Studies*, 32(6), 521–532.

Fleury, A. (2012). *Drawing and acting as user experience research tools*. ACM Press. p. 269. https://doi.org/10.1145/2350046.2350101

Heinemann, T., Landgrebe, J., & Matthews, B. (2012). Collaborating to restrict: a conversation analytic perspective on collaboration in design. *CoDesign*, 8(4), 200–214. https://doi.org/10.1080/15710882.2012.734827

Heritage, J. (1984). *Garfinkel and ethnomethodology*. Cambridge [Cambridgeshire]; New York, NY: Polity Press.

Hyysalo, S. (2010). *Health Technology Development and Use: From Practice-Bound Imagination to Evolving Impacts*. New York: Routledge.

Hyysalo, S. (2012). Accumulation and Erosion of User Representations or How is Situated Design Interaction Situated. In: Viscusi, G.; Campagnolo, G. M. & Curzi, Y. (eds.) *Phenomenology, Organizational Politics, and IT Design: The Social Study of Information Systems*. pp. 196–220. IGI Global.

Jornet, A. & Roth, W. (2017). Design {Thinking | Communicating}: A Sociogenetic Approach to Reflective Practice in Collaborative Design. In: Christensen, B. T., Ball, L. J. & Halskov, K. (eds.) *Analysing Design Thinking: Studies of Cross-Cultural Co-Creation*. Leiden: CRC Press/ Taylor & Francis.

Matthews, B., & Heinemann, T. (2012). Analysing conversation: Studying design as social action. *Design Studies*, 33(6), 649–672. https://doi.org/10.1016/j.destud.2012.06.008

Mead, G. H. (1932). *The Philosophy of the Present* (2002 ed.). Amherts, NY, USA: Prometheus Books.

Mead, G. H. (1934). *Mind, self, and society: From the standpoint of a social behaviorist*. Morris, C. W. (ed.). University of Chicago Press.

Schon, D. A. (1963). *Displacement of Concepts*. London: Tavistock Publications.

Stacey, R. (2011). *Strategic management and organisational dynamics: the challenge of complexity to ways of thinking about organisations* (6th edition). New York: Prentice Hall.

Yliristku, S. (2013). *Frame it Simple! Towards a theory of conceptual designing* (Doctoral dissertation). Aalto University, Helsinki, Finland.

Chapter 5

The Importance of Leadership in Design Team Problem-Solving

*Chih-Chun Chen, Maria A. Neroni, Luis A. Vasconcelos &
Nathan Crilly*

ABSTRACT

Design team problem-solving is challenging to study because the strategies and
behaviours exhibited by teams can vary considerably with team composition, design
environment and task demands. As a consequence, the tools and methods developed to
improve team strategies and behaviours are not always empirically informed. Existing
work on team problem-solving tends to adopt one of two perspectives. The first relates
to design cognition and focuses on how the team represents and tackles the problem.
The second relates to teamwork and focuses on how the composition, structure and
dynamics of the team affect the way it works together. In this study, we adopt both
these perspectives when analysing the dialogue and behaviours of a professional design
team in meetings over the course of a project. Content analysis of team members' verbal
communications (with pre-defined codes) is used to characterise the team's problem-
solving strategy while qualitative observations (without pre-defined codes) are used
to characterise team dynamics, integration and effectiveness during different problem-
solving episodes. Our findings suggest that leadership can play an important role in
design problem-solving. The team leader shapes the team's problem-solving strategy by
coordinating team discussions and helps the team to build up shared representations
by facilitating closed-loop communication. These findings suggest that good leadership practices – and the team processes that they facilitate – are important in small
task-driven teams and not only in larger team units such as organisations.

I INTRODUCTION

The size and complexity of real world problems very often means that designers work in
groups to develop a solution. For this reason, interactions between people are an integral part of the design process, with communications ranging from short queries within
the design team to long meetings with other organisations collaborating on the project
(Perry & Sanderson, 1998). Therefore, in order to support designers in their work, it is
necessary to understand what happens when groups of people work together to solve
a problem. Such understanding could be used to guide design practice and to inform
the development of design tools and methods that work effectively in a team context.

Communication between members of the design team reflects both how the team
approaches the problem and how they relate to each other as a team (see also Bedford

et al., 2017). These represent two distinct but related perspectives. The "collaborative problem-solving" perspective (hereafter referred to simply as the "problem-solving" perspective), focuses on how the team represents and tackles the problem. This might be in terms cognitive operations (Ball *et al.,* 1994; Stempfle & Badke-Schaub, 2002; Ward *et al.,* 1999) or in terms of "shared mental models" held by the design team (Bierhals *et al.,* 2007; Cannon-Bowers *et al.,* 1993; Peterson *et al.,* 2000). This perspective addresses questions such as how the problem is approached and reasoned through, how the problem is represented by the team, or which information exchanges are involved. Another perspective, the "teamwork", focuses on how the composition, structure and dynamics of the team affect the way the group works together on the problem (e.g., Moenaert *et al.,* 2000; Sonnenwald, 1996). This perspective addresses questions such as which roles team members adopt, how team members respond to each other, and how the patterns of interaction change or persist during problem-solving.

Whilst the role of a leader in collaborative design can be vital to team success (Wiltschnig *et al.,* 2013) and can exert considerable influence on the internal dynamics and outcomes of a team (Yukl, 1994), empirical studies of the design process often neglect this aspect, leaving much to be learnt about how the figure of a leader impacts on the overall performance of the design team (Sarin & McDermott, 2003). It has been hypothesised that the behaviour of the team leader is important in determining how the team operates with respect to both problem-solving and teamwork perspectives (Zaccaro *et al.,* 2001). In the current study, we show how analysing empirical data from both perspectives can provide insights into the relationship between team leadership and team processes, and between these processes and team effectiveness. We also relate our findings to earlier work reported in Stempfle & Badke-Schaub (2002), which analyses team members' communications in ephemeral student design teams from the problem-solving perspective. Contrasting with this previous work with novice designers, in the current study we have access to the design process of professionals (expert designers). By identifying commonalities and discrepancies in the results of this earlier study and ours, we are able to further elaborate on our findings and identify further opportunities for investigation.

2   BACKGROUND AND RELATED WORK

In collaborative problem-solving, different cognitive operations have been shown to occur to different degrees depending on the complexity of the design problem (Grogan and de Weck, 2016; Weingart, 1992), the stage of the design process (French, 1985; Howard, Culley, & Dekoninck, 2007), and on designers' levels of experience (Ahmed *et al.,* 2003; Waldron & Waldron, 1996). It has also been shown that different design teams might adopt distinct problem-solving strategies in terms of transitions between different design stages (e.g., goal clarification, solution generation, analysis, evaluation) and activity focus (e.g., content, process). These different strategies can have implications for the team's performance. For example, it has been observed that if the team does not devote enough time and effort to clarifying the goal or analysing the problem then they can prematurely commit to ideas without proper consideration (Stempfle & Badke-Schaub, 2002).

In real-world design team problem-solving, in order to ensure that discussion and problem-solving are directed, it is usually the case that one of the team members has a more dominant role in driving the discussion (Dieter & Schmidt, 2013). This leadership role might be assigned or it might emerge through the course of team interactions. In both organisations and work groups, leader behaviour has been shown to have a significant impact on team performance (outcome with respect to a given task) and effectiveness (efficiency with which the team delivers the outcome). This effect has been explained in terms of the (often reciprocal) influence of leader behaviour on team processes, which then gives rise to different performance outcomes (Zaccaro et al., 2001), both by bringing out (or failing to bring out) the best in individual team members (Reiter-Palmon & Illies, 2004) and by influencing team dynamics (Williams et al., 2010). These team processes include:

- Cognitive effects, whereby the team leader helps the team to make sense of the task, to better search for and process information, and to establish accurate "shared mental models".
- Motivational effects, whereby the team leader sets high performance goals and provides various strategies to attain these goals.
- Emotive effects, whereby the team leader encourages emotional control norms, thus minimising outright conflict, even when opinions differ and conditions are stressful.
- Coordinating effects, whereby the team leader matches members' capabilities to roles and activities, provides feedback, and aggregates the contributions.

Due to the mutual influence between team leadership and team processes, and to the fact that the team leader is also a team member, it is often difficult to consider these two factors separately with respect to team performance and effectiveness. Furthermore, some studies suggest that although quality of leadership has a role to play in design team performance and effectiveness, other factors such as "group climate" and "group organisation" can be more important (Badke-Schaub & Frankenberger, 1999). In addition to models of leadership effects, numerous models of the more general relationship between team processes (also referred to as "team dynamics" or "teamwork") and team effectiveness have been proposed. A meta-analysis of studies addressing such models (Salas et al., 2005) identified five important aspects of teamwork (the "big five") that collectively cover the factors deemed to be important in the teamwork related literature:

- Team leadership, which refers to the ability to direct and coordinate the activities of other team members, assess team performance, assign tasks appropriately, build team capacity, motivate team members, plan and organise, and establish a positive atmosphere (see also Marks et al., 2000);
- Mutual performance modelling, which refers to team members' understanding of the team environment and their adoption of appropriate strategies to monitor each other's performance (see also Bierhals et al., 2007; Cannon-Bowers et al., 1993; Peterson et al., 2000);
- Backup behaviour, which refers to team members' anticipation of each other's needs from an accurate understanding of their responsibilities;

–   Adaptability, which refers to the team's ability to adjust strategies through the use of backup behaviour and reallocation of resources; and

–   Team orientation, which refers to team members' propensity to take the behaviour of others into account during group interaction and their belief in the importance of the team's goals over those of individual members. Additionally, three mechanisms were hypothesised to underlie these "big five" aspects:

–   Shared mental models, which refers to knowledge structures of the task and of how the team will operate to solve it (see also Stout *et al.*, 1999);

–   Mutual trust, which refers to the shared belief that team members will perform their roles and protect the interests of their teammates (see also Webber, 2002); and

–   Closed-loop communication, which refers to the exchange of information between team members and the assurance that the information has been transferred successfully, usually through explicit confirmation that the information has been received (see also McIntyre & Salas, 1999).

There are several reasons why aspects of teamwork such as the "big five" and their underlying coordinating mechanisms are difficult to study empirically. Firstly, team behavior may not always be consistent and may depend on context or attributes of the team. For example, it has been found that teams tend to become more effective over time as members learn to work together (see also Bowers *et al.*, 1998; Dyer, 1984). These effects may also have different underlying mechanisms in different teams, such as in establishing procedures for working together (Dyer, 1984), developing shared knowledge and beliefs (Edmondson, 2003), effective communications (Bowers *et al.*, 1998), developing interpersonal relationships (Tuckman, 1965), and establishing roles and performance expectations (Jordan *et al.*, 1963; Morgan *et al.*, 1994). Secondly, various biases can arise when employing only survey and observational methods, including attributional bias, recollection bias and self-presentation bias (Cobb and Mathieu, 2000; Donaldson & Grant-Vallone, 2002). One way of overcoming these issues is by using behavioural markers in the verbal and non-verbal communications between team members. These can provide empirical clues as to which mechanisms are operating. Just as the communications between team members help establish which problem-solving strategies are being adopted by the team, the behaviours and interactions between team members can help establish which team structures and dynamics are at work. In the current study, we take an integrated approach to design team problem-solving across different design episodes with both quantitative analysis of team discussion content and qualitative observations of team member interactions.

3   METHODS AND DATA

We studied the communicative acts and interactions between three designers during seven collaborative sessions in order to understand which strategies the team adopted and how team members interacted. The three designers, Ewan, Abby and Kenny, made up the core team, and Ewan was recognised as being the team leader. The dataset analysed came from the second part of the project; hence prior to all the sessions analysed, the team had already had around three months' experience working together in this

Table 5.1 Details of the sessions included in the analyses, indicating the additional participants present (other than the core design team).

Session #	Title	Length (mins)	Additional participants
2	Designing co-creation workshops	21	1 intern; 1 participating observer
5	Designing co-creation workshop day 2	100	1 participating observer
9	Clustering insights from CC1	52	1 intern; 1 participating observer; 1 consultant
10	Iterations on CC2 design	34	1 intern; 1 participating observer; 2 consultants
11	Linking insights from co-creation to project	39	1 participating observer; 2 consultants
18	Clustering insights from CC1 and CC2	86	1 intern; 1 participating observer
21a	Brainstorm on concepts and products	60	2 colleagues (asked to generate ideas)
21b	Brainstorm on concepts and products	28	2 colleagues (asked to generate ideas)

particular configuration. In each of the sessions analysed, they were working together to accomplish a problem-solving task (see Table 5.1). We selected the sessions based on the presence of all the core team members and where communications between them made up the majority of utterances in the session (i.e., we excluded co-design sessions involving many users and interviews with only the team leader).

In some of the sessions analysed, there were other people present (consultants, interns, observers) who also contributed to the discussion; their utterances were not included in the analysis as it was less clear that they constituted part of the team (e.g., an external consultant might be seen more as an advisor than a team member). However, their impact on team dynamics was included in the qualitative observations to explore the potential for external advisors and observers to influence team behaviour.

The high level characteristics of the sessions analysed are shown in Table 5.1. Note that Session 21 was counted as two separate Sessions (21a and 21b) because it was split into two videos. As such, we analysed a total of eight design sessions. We combined two different methods: *content analysis* of the transcripts (with a pre-defined coding scheme) and *qualitative observations* of the videos (with no pre-defined coding scheme).

3.1 Content analysis of team dialogue

Content analysis was used to identify the problem-solving strategies adopted by the team during the design process. We applied the same coding scheme and mappings as those used in by Stempfle and Badke-Schaub (2002), as summarized in Table 5.2 with example code assignments. Although this coding scheme was developed primarily to be applied to a product development scenario, we adopted a generalised interpretation of the design stages and activity focuses (e.g., solution ideas might be options for how to solve a given problem rather than candidate product ideas). We chose to adopt this coding scheme as it explicitly links communicative acts (continuous utterance by a single speaker) to design steps and activity focuses, and the earlier findings (from Stempfle & Badke-Schaub, 2002) permitted us to draw further conclusions by providing a frame of reference for our analyses (see Section 5).

Code	Design step	Activity focus
Goal-related statement or question E.g. "... we need to create a problem around whatever we have, and then they need to solve the problem...." Requirement-related statement or question E.g. "... what I'm thinking is that we need to make it as flexible as possible."	Goal clarification	Content
Solution idea E.g. "So, we will prime them, yeah. They will know it's a creative session, they will know that and so on and we- we need to put that into the screener, the reactions around that and so on and so on."	Solution generation	
Question E.g. "... how do you prototype communication and marketing?" Answer, piece of information E.g. "... he didn't even know he was going to have this interview ..." Hypotheses, implication E.g. "... it's depending on which people we get in, and we can control it somewhat, but then again it's really difficult to filter it ... then we have a screening session with them or something, we can't do that."	Analysis	
Question for opinion or evaluation E.g. "But isn't it hard to ..."?" Positive evaluation/agreement E.g. "... it's very correct what you said around this ..." Negative evaluation/agreement E.g. "My gut feeling is that we shouldn't." Expression of uncertainty E.g. "... Yeah. I'm a little afraid that enjoy life is a little too big."	Evaluation	
Decision E.g. "... so: All people introduce themselves on a basic level, except the two new ones, which are doing the more in depth."	Decision	
Planning, discussion of task completion order "... maybe we can go one by one and just spend half a minute to talk about what we think it should be." Assigning tasks to group members	Planning	Process
Question E.g. "... should we move it over to an actual whiteboard?" Answer, piece of information E.g. [responding to question above] "yeah" Hypotheses, implication E.g. "maybe other stuff will come up ..."	Analysis	
Question for opinion or evaluation E.g. "How about keeping this open and continue tomorrow?" Positive evaluation/agreement E.g. [responding to question above] "yeah, sure" Negative evaluation/disagreement E.g. "So that's (.) the result of eh: one and a half hour work, and I'm sorry it looks like so few words-"	Evaluation	
Decision, arrangement E.g. "... let's continue it tomorrow."	Decision	
Control of group members "... Write it down." Summary, consensus E.g. "... I think we have been iterating and discussing at the same time, so I think we're on the same."	Control	
Residual E.g. "Mhm."	Residual	

The coding scheme was applied to transcripts of the eight sessions. Excluding the residual utterances (those which did not map to any of the codes) and utterances by individuals other than the core design team, a total of 2259 communicative acts were analysed over the eight sessions. We also allowed the application of more than one code to a cell (either because the utterance could be assigned more than one code or because different parts of the utterance needed to be assigned different codes). We felt this method of coding would serve as the best proxy for the prevalence of the different themes.

4   QUALITATIVE OBSERVATIONS OF TEAM DYNAMICS

Qualitative observations were used mainly to characterise teamwork processes. Two researchers, hereafter referred to as 'R1' and 'R2' (the second and third authors, with a background in Cognitive Psychology and Design Research, respectively) provided qualitative observations of team behaviour and effectiveness from watching videos of the eight sessions. Neither of these researchers were involved with the transcript content analysis described above (which was conducted by the first author, with a background in Complexity Science). The two researchers worked independently of each other to generate 1–2-page descriptive reports on each of the sessions with respect to the interactions between the group members and the operation of the group as a whole. These descriptive reports provided information about the group dynamics underlying the patterns of communication identified in the content analysis. Based on these descriptions, the two researchers also gave qualitative ratings (high, low) in terms of team "integration" (how integrated the team was during the session, i.e., high integration, low integration; this is sometimes also referred to as team "cohesiveness") and team "effectiveness" (how well the team worked together to achieve the session outcome(s), i.e., high effectiveness, low effectiveness). Note that "effectiveness" should be seen as distinct from "performance" in that the two researchers did not consider the quality of the outcome(s) itself but only evaluated the team processes and interactions involved in achieving the outcome. This distinction is also noted in (Salas *et al.*, 2005). As a final methodological step, all three researchers involved in the analyses compared their observations to identify and discuss discrepancies, and to propose potential explanations for particular observations based on their analyses.

5   FINDINGS

As outlined in Section 3, we conducted both content analysis and qualitative observations with the objective of better understanding how the team's collaborative problem-solving strategy (the "problem-solving" perspective) relates to the composition, structure and dynamics of the team (the "teamwork" perspective). Table 5.3 summarises how the different analysis outputs of our findings relate to the different perspectives, data and analyses described above (we also cross reference this to the corresponding section of this chapter for ease of reference).

5.1   Team communications and problem-solving

From our content analysis of the transcripts, we considered both the frequencies of communicative acts associated with different design steps and different focuses of activity (see Table 5.4), and the frequencies of the transitions between them.

Table 5.3 Analysis outputs in terms of perspectives, data and method of analysis.

Perspective	Data	Analysis	Analysis outputs	Section
Problem-solving	Transcripts	Content analysis	Frequencies of communications of different design steps and activity focuses	4.1
Problem-solving	Transcripts	Content analysis	Frequencies of transitions between communications of different design steps and activity focuses	4.1
Problem-solving	Videos	Qualitative observations	Freeform observations of the team in operation	4.1
Teamwork	Transcripts	(Counts only)	Frequencies of communications from different individuals	4.2
Teamwork	Transcripts	(Counts only)	Frequencies of transitions between communications from different individuals	4.2
Teamwork	Transcripts	Content analysis	Frequencies of communications of different design steps and activity focuses from different individuals	4.2
Teamwork	Transcripts	Content analysis Frequencies of transitions between communications of different design steps and activity focuses from individuals		4.2
Teamwork	Videos	Qualitative observations	Freeform observations of team in operation	4.2 4.3
Teamwork	Videos	Qualitative observations	Qualitative ratings of team cohesiveness and effectiveness	4.3

With respect to the various stages of the design process, we found the highest proportion of communication to be concerned with analysis (42%), followed by solution generation (23%) and then evaluation (14%). However, there was also considerable variation between the sessions, indicating that contextual factors may be at play and no single "one size fits all" team behaviour. For example, a far smaller proportion of communication in Session 2 is centred on solution generation compared to Session 11: 7% and 32%, respectively (see Table 5.4).

When moving between design stages in their discussion, the team seem to go through periods of solution generation, followed by analysis (see Table 5.4). This is indicated by the majority of communications on solution generation being followed by further communications on solution generation (51%) or by analysis (24%). By contrast, evaluation and goal clarification rarely follow solution generation (both < 20%), suggesting that the selection and rejection of solution(s) tends to be achieved through analysis rather than through simple evaluative judgements. This interpretation is further supported by the qualitative observations; both R1 and R2 observed that solutions

Table 5.4 Communications concerned with different design steps and activity focuses.

Session	2	5	9	10	11	18	21a	21b	Overall
Total utterances	105	795	285	186	195	326	305	62	2259
By design step (content focused only, but proportions are with respect to all utterances)									
Goal clarification	16 (15%)	92 (12%)	36 (13%)	26 (14%)	20 (10%)	31 (10%)	18 (6%)	8 (13%)	247 (11%)
Solution generation	7 (7%)	181 (23%)	52 (18%)	48 (26%)	63 (32%)	76 (23%)	70 (23%)	21 (345)	518 (23%)
Analysis	45 (43%)	375 (47%)	132 (46%)	63 (34%)	59 (30%)	134 (41%)	120 (39%)	18 (29%)	946 (42%)
Evaluation	31 (30%)	81 (10%)	39 (14%)	37 (20%)	35 (18%)	50 (15%)	39 (13%)	4 (6%)	316 (14%)
Decision	0 (0%)	19 (2%)	0 (0%)	4 (2%)	3 (2%)	3 (1%)	0 (0%)	0 (0%)	29 (1%)
Control	2 (2%)	2 (0%)	0 (0%)	0 (0%)	0 (0%)	1 (0%)	0 (0%)	0 (0%)	5 (<0.5%)
By activity focus (this includes all non-residual utterances)									
Content	101 (96%)	750 (94%)	259 (91%)	178 (96%)	180 (92%)	295 (90%)	247 (81%)	51 (82%)	2061 (91%)
Process	4 (4%)	45 (6%)	26 (9%)	8 (4%)	15 (8%)	31 (10%)	58 (19%)	11 (18%)	198 (9%)

were seldom explicitly evaluated negatively. Instead, members of the team tended to propose further solutions as alternatives or use analysis to show that a solution might not be adequate (where analysis includes exploring questions, hypotheses and implications). Additionally, the fact that a significant proportion of communications on evaluation are followed by communications on analysis suggests that evaluative judgements are themselves analysed.

With respect to activity focus, the vast majority of communication centred around "content" (91% across all sessions), that is, the design task itself, rather than "process" (9% across all sessions), that is, how the task should be executed. It is interesting to note that the proportion of communication dedicated to process was higher in Sessions 21a and 21b, where two colleagues joined the team (see Table 5.4). Explanations for this are explored in Section 5.

5.2 Member participation and leadership

In order to better understand the team dynamics underlying the patterns of communication and shifts in discussion focus described above in Section 4.1, we analysed team communications with respect to individual team members. In Table 5.5, the first value in the parentheses is the proportion of each individual's utterances which concern the design step or activity focus, that is, how prevalent the design step or activity focus is for the individual; the second value in the parentheses is the proportion of utterances (across all core team members) concerning the design step or activity focus that the

Table 5.5 Summary of the coded communicative acts for the three members of the core design team (see text for explanation of the different percentages).

	Ewan (team leader)	Abby	Kenny	Total
Total utterances	1064 (47%)	683 (30%)	512 (23%)	2259
By activity focus (including all non-residual utterances)				
Content	954 (90%;46%)	639 (94%;33%)	468 (91%;25%)	2061 (91%)
Process	110 (10%;56%)	44 (6%;22%)	44 (9%;22%)	198 (9%)
By design step (content focused only, but proportions are with respect to all utterances)				
Goal clarification	128 (12%;52%)	70 (10%;28%)	49 (10%;20%)	247 (11%)
Solution generation	231 (22%;45%)	159 (23%;31%)	125 (24%;24%)	515 (23%)
Analysis	454 (43%;48%)	285 (41%;30%)	207 (40%;22%)	946 (42%)
Evaluation	123 (12%;39%)	113 (17%;36%)	80 (16%;25%)	316 (14%)
Decision	12 (1%;41%)	11 (2%;38%)	6 (1%;21%)	29 (1%)
Control	3 (0%;60%)	1 (0%;20%)	1 (0%;20%)	5 (<1%)

individual is responsible for, that is, their contribution to the set of communications relating to that step or focus.

The values in Table 5.5 indicate that the relative proportions of communication dedicated to the different design steps and activity focuses were similar for the different designers (the first values in the parentheses are similar along the rows representing each focus of discussion). At the same time, individuals' participation remained fairly consistent across different focuses of discussion (the second values in the parentheses are similar within the columns representing each designer). For example, the proportion of communications concerning analysis was 43%, 41% and 40% for each of the three core team members, while the proportion of communications concerning solution generation were 22%, 23% and 24%, despite the differences in overall participation (1064, 783 and 512 utterances, respectively). This suggests that there was no "specialisation" of roles in the team with respect to different design activities and contrasts with findings from some previous studies (e.g., Goldschmidt, 1995). Consistent with the team leader speaking more than the others across all sessions, most of the shifts in discussion focus were made by him. To illustrate this, Table 5.6 shows designers' transitions between different activity focuses. As in Table 5.5, the first value in the parentheses indicates the designer's contribution to all the transitions of that type while the second indicates the prevalence of the particular type of transition with respect to all of the designer's transitions.

As indicated by the values in Table 5.6, the majority of content-to-process and process-to-content transitions came from Ewan. However, relative to his overall contribution, the proportion of his utterances that shifted the focus was not significantly

Table 5.6 Shifts in activity focus for different team members (see text for explanation of the different percentages).

Activity focus transitions	Ewan (team leader)	Abby	Kenny	Total
Content-to-Content	888 (93%;46%)	605 (95%;31%)	439 (94%;23%)	1932 (86%)
Content-to-Process	65 (7%;52%)	33 (5%;26%)	28 (6%;22%)	126 (5%)
Process-to-Content	66 (60%;52%)	33 (75%;26%)	28 (64%;21%)	127 (6%)
Process-to-Process	44 (40%;62%)	11 (25%;15%)	16 (36%;23%)	71 (3%)
Total	1063 (47%)	682 (30%)	511 (23%)	2256

Table 5.7 Designers' transitions to each other (see text for explanation of the different percentages).

Speaker transitions	From Ewan	From Abby	From Kenny	Total
To Ewan	366 (35%;34%)	426 (62%;40%)	271 (53%;26%)	1063 (47%)
To Abby	414 (39%;61%)	94 (14%;14%)	175 (34%;26%)	683 (30%)
To Kenny	284 (26%;55%)	163 (24%;32%)	65 (13%;13%)	512 (23%)
Total	1064 (47%)	683 (30%)	511 (23%)	2258

greater than for Abby or Kenny. (We have not reported the transitions between design process focus due to constraints on space, but a similar trend was observed).

Probing more deeply into the dynamics of the team, we also consider the interactions between each pair of designers. In Table 5.7, the first value in the parentheses indicates the proportion of the "source" individual's communications that were followed by the "target" member's communication, while the second value indicates the proportion of communications to the "target" individual originating from the "source".

As shown in Table 5.7 and consistent with a leadership role (see Marks *et al.*, 2000), there were more interactions between Ewan and Abby and Ewan and Kenny than between Abby and Kenny. In other words, Ewan served as a sort of hub for the design team, with communications from each of the other two designers being relayed through him. Ewan also spoke continuously more than the other two designers, as indicated by the higher rate of transition to another communication of his own (35% compared to 14% in the case of Abby and 13% in the case of Kenny).

5.3 Team effectiveness and integration

In addition to considering how a team's problem-solving strategy manifests itself in interactions between team members, we sought to better understand the relationship between team dynamics and the team's effectiveness. Consistent with existing literature (e.g., Guzzo & Dickson, 1996), which positively associates team cohesiveness and team effectiveness, we found that sessions that were rated high in team integration also tended to be rated high in team effectiveness (see Table 5.8). The exceptions to this were Session 2, which was extremely short (only 21 minutes), where group integration was low despite high effectiveness, and Sessions 9 and 18, where group integration was high but effectiveness was low. Alongside the team integration and team effectiveness

Table 5.8 Balance between designers' participation in different sessions. In the final column, the distributions of the designers' contributions are given in square brackets: [Ewan; Abby; Kenny].

Session	Group integration		Group effectiveness		Team balance (Gini coefficient)	
	R1	R2	R1	R2	By overall volume	By # of communications
2	Low	Low	High	High	0.37	0.25 [0.46;0.33;0.21]
5	Low	Low	Low	Low	0.27	0.21 [0.43;0.35;0.22]
9	High	High	Low	Low	0.43	0.26 [0.49;0.27;0.24]
10	Low	High	High	High	0.34	0.34 [0.50;0.34;0.16]
11	High	High	High	High	0.55	0.45 [0.62;0.17;0.21]
18	High	High	Low	Low	0.06	0.05 [0.36;0.31;0.33]
21a	High	High	High	High	0.48	0.35 [0.55;0.26;0.19]
21b	Low	Low	Low	Low	0.81	0.50 [0.66;0.16;0.18]

ratings, Table 5.8 shows the degree of "team imbalance" in both overall volume of communication and number of communications, as indicated by the Gini coefficient (a normalized measure of inequality in a frequency distribution ranging from 0 to 1; in this case, the higher the Gini coefficient, the greater the imbalance in number of utterances from different team members). From the values in Table 5.8, we can see that sessions with greater imbalance (Sessions 10, 11, 21a, 21b) also tended to be rated more highly in terms of both team integration and effectiveness (the exception being 21b) while the sessions with more balanced participation by each of the designers (Sessions 2, 5, 9, 18) were rated low in either integration and effectiveness (or both).

At first, the association between team imbalance and integration and effectiveness might seem counter-intuitive, since group integration is often characterised in terms that would suggest equal participation by members. However, considering the qualitative observations allow us to identify a possible mechanism behind this association (see Table 5.9). Sessions rated low in integration (I) and/or effectiveness (E) tend to also be those where contributions are uncoordinated, or where there is no clear leadership. By contrast, sessions that were rated positively for both integration and effectiveness were those where there is clear leadership and the Team Leader is engaged with the discussion.

6   DISCUSSION, IMPLICATIONS AND CONCLUSIONS

In Section 4, we identified several relationships between the designers' approaches to problem-solving (the "collaborative problem-solving" perspective) and the interactions between the designers (the "teamwork" perspective). These are summarised below:

- The majority of discussion centres around the content of the design task rather than the process for addressing it, although when people outside the core design team participate in the task, there is more communication focusing on process (Section 4.1).
- The largest portion of the team's discussions centre around analysis, followed by solution generation. Explicit evaluation of solutions is rare, with proposed

Table 5.9 Representative extracts from the descriptive reports generated by the two Researchers. Under the Ratings columns, 'I' stands for Integration and 'E' stands for effectiveness.

Session	R1 From descriptive report	Ratings	R2 From descriptive report	Ratings
2	"They start with a round of individual inputs and then things get a bit more chaotic."	I: Low E: High	"Ewan and Abby have a more central role in the discussion."	I: Low E: High
5	"The session is marked with constant clashes, which means issues that could be solved quickly take much longer to solve."	I: Low E: Low	"Abby and Kenny often disagree with Ewan and propose alternative solutions."	I: Low E: Low
9	"Designers agree or converge most of the time ... questioning or judging others' ideas was more common than outright clashes."	I: High E: Low	"There are not very many instances of disagreement amongst the designers ... The atmosphere is more relaxed ..."	I: High E: Low
10	"There are parallel activities going on ... There are many moments throughout the session when designers are either away from the room or keep themselves deliberately away from the discussion."	I: Low E: High	"Ewan has a very central role in the discussion ... The group shares ideas and mostly agrees."	I: High E: High
11	"Everyone is interacting well and seems engaged."	I: High E: High	"All the designers participate actively in the discussion ... The group work in a very productive way."	I: High E: High
18	"Everyone is participating in the discussion in a smooth way."	I: High E: Low	"In the first part of the session, all the designers participate in the discussion ... In the second part of the session, the group seems to be less integrated ... Ewan does not lead much and Abby is far more active than in previous sessions."	I: High E: Low
21a	"Abby takes control more in this session and directs people to do certain things."	I: High E: High	"Abby frequently intervenes during Ewan's summary to clarify things."	I: High E: High
21b	"The discussion phase is disparate and not very effective."	I: Low E: Low	"Ewan does not participate much in brainstorming."	I: Low E: Low

solutions being eliminated mostly through analysis and generation of better solutions (Section 4.1).

- The participation of different team members remains largely consistent across different sessions and focuses of discussion (Section 4.2).
- The team leader dominates the discussion in terms of both number of communications and shifts in discussion topic (Section 4.2).
- The team leader is more likely to speak before and after each of the two other designers than they are likely to speak before and after each other, serving as a "hub" for communication (Section 4.2).

–   Team integration and team effectiveness are positively associated with each other (Section 4.3).
–   Sessions with high team integration and effectiveness also tend to be those where the team leader is more actively engaged in the discussion (Section 4.3).

These findings complement the results reported by Awomolo *et al.* (2017) where it was observed that Ewan was responsible for a large portion of the design decisions made, but also accepted decisions made by other individuals when they had more experience or knowledge in a particular area. Also, Wulvik *et al.* (2017) observed that Ewan acted as an "Intergroup Star" (Sonnenwald, 1996) and facilitated integration between the core team and other individuals present in the meeting. These findings support the importance of team leader behaviour with respect to team effectiveness through facilitating more team integration and in determining the problem-solving strategy adopted by the team.

In the section that follows, we seek to place our findings in the wider context of design research by comparing them with those of the study reported by Stempfle and Badke-Schaub (2002). Based on these comparisons and the findings already reported, we then propose an account of design team problem-solving which integrates both "problem-solving" and "teamwork" perspectives.

6.1   Comparison with ephemeral student design teams

The findings obtained by Stempfle and Badke-Schaub (hereafter referred to as the "2002 study") serve as a frame of reference for our findings. Dieter and Schmidt (2013) outline two important characteristics that distinguish student design teams from experienced design teams. The first is that student design team members are all close to the same age and level of formal education. The second is that team members are peers and no member has "official" designated authority over the other team members. As a consequence, student design teams often prefer to work without a designated leader and instead operate in a "shared leadership" environment. In comparing our results with those obtained in the 2002 study, we also compare the behaviour of experienced designers in an established team (the current study) with the behaviour of novices in ephemeral teams. While we avoid making strong assertions based on this comparison, two key observations can be made.

Firstly, the team in our study devoted far less time to process-focused team communication: only 9% compared to 32%, 23% and 34% in the three groups in the 2002. It may be that there is less need for explicit process-related communication in the presence of a team leader who drives the discussion and helps the other team members to understand each others' roles. Consistent with this, there is more process-related communication when two colleagues join the team in Sessions 21a and 21b and have to actively participate (see Section 4.1), but this remains anecdotal in the absence of other data.

Secondly, the portion of team communication that was concerned with analysis in the current study (42%), was comparable to that observed across the three teams in the 2002 study (46%). The portion concerned with evaluation was also similar, 14% and 13% respectively. However, the portion of team communication concerned with solution generation was far higher in the current study than it was in the 2002

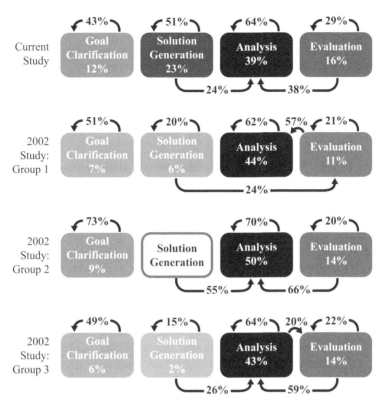

Figure 5.1 Graphs of transitions between Goal Clarification, Analysis, Solution Generation and Evaluation. For clarity, only transition proportions exceeding 20% are shown.

study: 23% across sessions in the current study and 4% across teams in the 2002 study. It is unclear whether this results from differences in the nature of the tasks or whether it results from differences in the team. Indeed, as with the communication focuses reported above, we observed considerable variation in the proportions of communication concerned with each of the different design steps across the different sessions. It is notable that compared to the teams in the 2002 study, the team in the current study tended to use analysis (often questions) rather than evaluation to select and discard potential solutions. This might lead to team members feeling more comfortable to freely share their solution ideas. Furthermore, from the qualitative observations, we found that solutions were rarely explicitly evaluated negatively, with team members proposing alternative solutions or using analysis to show that a solution might not be adequate. This was also reflected in the probabilities of transitioning between different design steps (see Figure 5.1):

–   The proportion of transitions from solution generation to evaluation in the current study was 14%, which was slightly lower than in two of the groups in the 2002 study.

- The proportion of transitions from solution generation to (more) solution generation in the current study was 51% (across all sessions), which was much higher than in any of the groups in the 2002 study.
- The proportion of transitions from solution generation to analysis in the current study was 24%, which was much higher than in any of the groups in the 2002 study.

Further work should seek to determine whether other established professional design teams also show similar patterns.

6.2 Team leadership, team effectiveness and team integration

This study allows us to sketch out an account of design team problem-solving that integrates the problem-solving and teamwork perspectives where team leadership plays a crucial role. Specifically, team leadership shapes collaborative problem-solving through the following two mechanisms.

- *Guiding problem-solving strategy.* The team leader guides the design discussion (as indicated by shifts in focus and design step) and sets the goals and requirements of a particular session. Variation in discussion trajectories can be attributed to the different demands of different tasks. Through guiding problem-solving discussions, team leadership exerts "cognitive" effects on team processes (see Section 2).
- *Closed-loop communication and shared mental models.* The team leader often serves as a "communication hub" who receives ideas and opinions from the other team members, and relays them back to the speaker (closed-loop communication) and the rest of the team. This helps the team to build a "shared mental model" of both the content of the task and the problem-solving process. Through this closed-loop communication mechanism, team leadership exerts both "cognitive" and "coordinating" effects on team processes (see Section 2).

It should also be emphasised that acceptance of an individual as the team's leader is itself an important part of team members' shared mental model of the team. It is from this acceptance of leadership that the team allows the problem-solving process to be shaped by the team leader.

6.3 Further work

While our findings point to the account given above, further work would need to be conducted to validate and further elaborate on the underlying problem-solving and teamwork mechanisms. Firstly, the current study focused on a single design team working on a specific design task. However, it has been observed that the respective importance of different teamwork factors can vary depending on the type of task (see McGrath, 1984; Stewart & Barrick, 2000; Strauss, 1999). For example, in (Stewart & Barrick, 2000) it has been shown that in conceptual tasks, the relationship between team integration (which the authors call "interdependence") and performance differs from that observed in behavioural tasks, both in terms of and direction. Further work should seek to systematically investigate these differences and identify potential moderators of teamwork effects. Secondly, the design team in the current study consisted of only three team members with similar levels of experience. It would be worth conducting further studies with different numbers of team members and different team compositions to determine how far the current findings generalise.

7 CONCLUSION

Given the importance of collaboration in solving many real world design problems, it is important to understand how people work together in a team when tackling a design task. In this study, we studied design problem-solving through the application of two separate methods. This allowed us to synthesise problem-solving and teamwork perspectives, showing that team leadership plays a crucial role in collaborative problem-solving through shaping and structuring discussions, and by facilitating shared mental models through closed-loop communication. With respect to design practice, design tools and collaboration methods, it would be worth investigating how the dynamics associated by effective team leadership might be facilitated or monitored, even in teams with no "official" leader. However, just as important as the specific findings reported is the recognition that richer empirically-grounded models of team problem-solving can be formulated by combining perspectives and methods. Such models would provide stronger empirical foundations for developing design tools and methods that are effective in a team context.

ACKNOWLEDGEMENTS

This work was funded by the UK's Engineering and Physical Sciences Research Council (EP/K008196/1) and by the CAPES Foundation, Ministry of Education of Brazil (BEX 11468/13-0).

REFERENCES

Adams, R. S. & Siddiqui, J. A. (eds.) (2016). *Analyzing design review conversations*. West Lafayette, Indiana: Purdue University Press.

Ahmed, S. Wallace, K. M. & Blessing, L. T. M. (2003). Understanding the differences between how novice and experienced designers approach design tasks. *Research in Engineering Design*, 14, 1–11.

Awomolo, O., Jabbariarfaei, J., Singh, N. & Akin, Ö. (2017) Communication and Design Decisions in Cross-Functional Teams. In: Christensen, B. T., Ball, L. J. & Halskov, K. (eds.). *Analysing Design Thinking: Studies of Cross-Cultural Co-Creation*. Leiden: CRC Press/ Taylor & Francis.

Badke-Schaub, P., & Frankenberger, E. (1999). Analysis of design projects. *Design Studies*, 20(5), 465–480.

Badke-Schaub, P., Neumann, A., Lauche, K. & Mohammed, S. (2007). Mental models in design teams: a valid approach to performance in design collaboration? *CoDesign*, 3(1), 5–20.

Ball, L. J., Evans, J. St. B. T. & Dennis, I. (1994). Cognitive processes in engineering design: A longitudinal study. *Ergonomics*, 37, 1753–1786.

Bedford, D. A. D., Arns, J. W. & Miller, K. (2017) Unpacking a Design Thinking Process with Discourse and Social Network Analysis. In: Christensen, B. T., Ball, L. J. & Halskov, K. (eds.). *Analysing Design Thinking: Studies of Cross-Cultural Co-Creation.* Leiden: CRC Press/Taylor & Francis.

Bierhals, R., Schuster, I., Kohler, P. & Badke-Schaub. (2007). Shared mental models – linking team cognition and performance. *CoDesign*, 3(1), 75–94.

Bowers, C. A., Weaver, J., Barnett, J. & Stout, R. (1998). Empirical validation of the SALIANT methodology. Paper presented at the *RFO HFM Symposium of Collaborative Crew Performance in Operational Systems*, Edinburgh, Scotland.

Cannon-Bowers, J. A., Salas, E. & Converse, S. (1993). Shared mental models in expert team decision making. In: N. J. Castellan (ed.). *Individual and Group Decision Making: Current issues*, 221–246. Lawrence Erlbaum Associates, Hillsdale

Cohen, M. S., Freeman, J. T. & Thompson, Bl. (1998). *Critical thinking skills in tactical decision making: A model and a training strategy.* Cannon-Bowers, J. A., Salas, E. and Converse, S. (eds.) American Psychological Association, Washington. 155–159.

Dieter, G. E., & Schmidt, L. C. (2013). *Engineering design* (5th ed). New York: McGraw-Hill.

Dyer, J. L. (1984). Team research and team training: A state-of-the-art review. In Muckler, F. A., Neal, A. S. and Strother, L. (eds). *Human factors review*, Santa Monica, CA, Human Factors Society. 285–323.

Edmondson, A. (2003). *Framing for learning: Lessons in successful technology implementation.* California Management Review, 45(2), 34–54.

Edwards, B. D., Day, E. A., Arthur, W. & Bell, S. T. (2006). Relationships among team ability composition, team mental models, and team performance. *Journal of Applied Psychology*, 91(3), 727–736.

Ellis, A. P. J. (2006). System breakdown: the role of mental models and transactive memory in the relationship between acute stress and team performance. *Academy of Management Journal*, 49(3), 576–589.

French, M. (1985*). Conceptual Design for Engineers*. London: The Design Council, The Pitman Press.

Goldschmidt, G. (1995). The designer as a team of one. *Design Studies*, 16(2), 189–209.

Grogan, P. T. & de Weck, O. (2016). Collaboration and complexity: an experiment on the effect of multi-actor coupled design. *Research in Engineering Design*, February, 2016.

Guzzo, R. A. & Dickson, M. W. (1996). Teams in Organizations: Recent Research on Performance and Effectiveness. *Annual Review of Psychology*, 47, 307–338.

Howard, T., Culley, S. & Dekoninck, E. (2007). Creativity in the Engineering Design Process. *Proceedings of the International Conference on Engineering Design*, ICED'07. 329–330.

Jordan, N., Jensen, B. T. & Terebinsky, S. J. (1963). The development of cooperation among three-man crews in a simulated man-machine information system. *Journal of Social Psychology*, 59, 175–184.

Lim, B-C. & Klein, K. J. (2006). Team mental models and team performance: a field study of the effects of team mental model similarity and accuracy. *Journal of Organizational Behavior*, 27(4), 403–418.

Marks, M. A., Zaccaro, S. J. & Mathieu, J. E. (2000). Performance implications of leader briefings and team-interaction training for team adaptation to novel environments. *Journal of Applied Psychology*, 85(6), 971–986.

Mathieu, J. E., Heffner, T. S., Goodwin, G. F., Salas, E. & Cannon-Bowers, J. A. (2000). The influence of shared mental models on team process and performance. *Journal of Applied Psychology*, 85(2), 273–283.

Mathieu, J. E., Maynard, M. T., Rapp, T. L. & Mangos, P. M. (2006). Interactive effects of team and task shared mental models as related to air traffic controllers' team efficacy and effectiveness. Annual meeting of the Academy of Management Conference, 2006.

McIntyre, R. M. & Salas, E. (1995). Measuring and managing for team performance: Emerging principles from complex environments. In: R. A. Guzzo & E. Salas (eds.) Team effectiveness and decision making in organizations, San Francisco: Jossey-Bass. pp. 9–45.

Michaelsen, L. K., Watson, W. E., & Black, R. H. (1989). A realistic test of individual versus group consensus decision making. Journal of Applied Psychology, 74, 834–839.

Moenaert, R. K., Caeldries, F., Lievens, A. & Wauters, E. (2000). Communication Flows in International Product Innovation Teams. Journal of Product Innovation Management, 17, 360–377.

Morgan, B. B., Salas, E. & Glickman, A. S. (1994). An analysis of team evolution and maturation. The Journal of General Psychology, 120, 277–291.

Peterson, E., Mitchell, T. R., Thompson, L. & Burr, R. (2000). Collective efficacy and aspects of shared mental models as predictors of performance over time in work groups. Group Processes and Intergroup Relations, 3(3), 296–316.

Reiter-Palmon, R. & Illies, J. J. (2004). Leadership and creativity: Understanding leadership from a creative problem-solving perspective. The Leadership Quarterly, 15, 55–77.

Salas, E., Cannon-Bowers, J. A. & Johnston, J. H. (1997). How can you turn a team of experts into an expert team? Emerging training strategies. In: Zsambok and Klein, G. (eds.) Naturalistic decision-making, Hillsdale, NJ, Lawrence Erlbaum. pp. 359–370.

Salas, E., Sims, D. E. & Shawn Burke, C. (2005). Is there a "big five" in teamwork? Small Group Research, 36(5), 555–599.

Sarin, S. & McDermott, C. (2003). The Effect of Team Leader Characteristics on Learning, Knowledge Application, and Performance of Cross-Functional New Product Development Teams. Decision Sciences, 34(4), 707–739.

Sonnenwald, D. H. (1996). Communication roles that support collaboration during the design process. Design Studies, 17, 277–301.

Stempfle, J. & Badke-Schaub, P. (2002). Thinking in design teams – an analysis of team communication. Design Studies, 23, 473–496.

Stout, R. J., Cannon-Bowers, J. A., Salas, E. & Milanovich, D. M. (1999). Planning, shared mental models, and coordinated performance: An empirical link is established. Human Factors, 41, 61–71.

Tuckman, B. W. (1965). Developmental sequence in small groups. Psychological Bulletin, 63, 384–399.

Waldron, M. B. & Waldron, K. J. (1996). The influence of the designer's expertise on the design process. In: Waldron, M. B. and Waldron, K. J. (eds.) Mechanical design: theory and methodology. Berlin: Springer. pp. 5–20.

Ward, T. B., Smith, S. M. & Finke, R. A. (1999). Creative Cognition. In: R. J. Sternberg (ed.). Handbook of creativity. Cambridge: Cambridge University Press. 190–212.

Webber, S. S. (2002). Leadership and trust facilitating cross-functional team success. Journal of Management Development, 21, 201–214.

Weingart, L. R. (1992). Impact of group goals, task component complexity, effort, and planning on group performance. Journal of Applied Psychology, 77(5), 682–693.

Williams, H. M., Parker, S. K. & Turner, N. (2010). Proactively performing teams: The role of work design, transformational leadership, and team composition. Journal of Occupational and Organizational Psychology, 83, 301–324.

Wiltschnig, S., Christensen, B. T. & Ball, L. J. (2013). Collaborative problem-solution co-evolution in creative design. Design Studies, 34(5), 515–542.

Wulvik, A., Jensen, M. B. & Steinert, M. (2017) Temporal Static Visualisation of Transcripts for Pre-Analysis of Video Material: Identifying Modes of Information Sharing. In: Christensen, B. T., Ball, L. J. & Halskov, K. (eds.) *Analysing Design Thinking: Studies of Cross-Cultural Co-Creation*. Leiden: CRC Press/Taylor & Francis.

Yukl, G. (1994). *Leadership in organizations* (3rd ed.). Englewood Cliffs, NJ: Prentice Hall.

Zaccaro, S. J., Rittman, A. L. & Marks, M. A. (2001). Team leadership. *The Leadership Quarterly*, 12, 451–483.

Chapter 6

Communication and Design Decisions in Cross-Functional Teams

Olaitan Awomolo, Javaneh Jabbariarfaei,
Nairiti Singh & Ömer Akin

ABSTRACT

Solving complex problems requires input from stakeholders with different disciplinary expertise. Each stakeholder contributes to solution finding and generation by bringing a different set of skills to approaching the problem. Teamwork is crucial for performing most design tasks as these different roles and perspectives can complement each other, far beyond the scope of a single individual. Proper communication between team members is required to reap the benefits of teamwork. Communication in teams has been shown to be influenced by the relationship between team members, hierarchies present within teams and team diversity. Communicative acts have also been shown to represent team processes. However, research on the relationship between team member communication and its impact on team outcomes, team performance and team decision making is limited. Even more limited are studies that identify when communication is most important during design and the impact of functional boundaries within the team. Using the DTRS11 data, we studied the influence of communication on team design decisions made and the communication patterns that emerge to create different decisions, focusing on the influence of the presence of functional boundaries. We found that roles and positions influence design decisions made and in a well-integrated cross-functional team, the presence of multiple functional groups can reduce this effect. Having a cross-functional team led to increased input on decisions made, following an integrative approach, than in a team with a single functional group.

1 INTRODUCTION

Complex problems are often ill-defined. There is a lack of clearly defined solution requirements and ambiguous criteria to determine when a solution is found (Fischer, Greiff & Funke, 2011; Simon, 1977). It is rare to find a complex problem composed of just one component problem. Usually, they are made of multiple component problems that must be resolved and integrated before a solution can be accepted. These component problems are often diverse and require different sets of skills and knowledge to solve. Therefore, complex problem solving requires input from multiple stakeholders with different functional backgrounds. Each stakeholder contributes to solution finding and generation by bringing a different set of skills to approaching the problem. Design tasks have been described as ill-defined, complex problems (Akin, 1986; Goel,

1995; Goldschmidt, 1997; Seitamaa-Hakkarainen, 2000; Simon, 1969), making the use of design teams necessary for most tasks and projects.

Teamwork is crucial for performing design tasks as these different roles and perspectives can complement each other, far beyond the scope of a single individual (MacWilliam, 2016). Numerous benefits can be reaped from teamwork in problem solving such as: improved efficiency, information processing and time management; an increased sense of belonging among team members (Lean Enterprise Institute, 2014); and improved innovation and productivity. Team processes such as dialogue and negotiation are used to establish effective working relationships amongst the team members and "allow the team to effectively communicate, manage conflict, make decisions, and problem solve" (Stephoudt & Mariotta, 2011, p. 1). These processes need to be effectively managed to avoid the adverse effects of working in teams such as: conflict; poor coordination and time management; and incompatibility among team members (Salvendy, 2012). To obtain the best possible performance from teams, all members must work together to ensure that all individual efforts contribute towards achieving a similar, shared overall goal (Bater, 2014).

Design teams are diverse and cross-functional in response to the complexity of the problems they solve. Cross-functional teams are made up of members with different disciplinary and functional knowledge. Rather than belonging to the same work group, team members perform different roles and can attend to different aspects of the problem to be solved. When working effectively, cross functional teams can show "improved team integration and co-ordination by spanning organizational boundaries, leading to improved problem solving and decision making" (Inc, 2006). Team members' interaction influences how design problems are framed (Chan & Schunn, 2017). The different perspectives and roles present can lead to improved outcomes. However, cross-functional teams particularly need to balance multiple demands. They are expected to manage the integration of component solutions developed as well as the integration of project teams themselves.

Design has been described as a social process involving practices such as "participation, reflection and action". Communication, defined as human behavior that facilitates meaning sharing through the unidirectional or bi-directional transmission and exchange of information (Lane, 2000) is a fundamental component of the design process. Effective communication during the design process, between design team members improves team efficiency and allows the benefits of teamwork to be reaped. "Without proper communication between the design team members, time and effort will be wasted which will lead to losses for all stakeholders involved. Streamlining and improving the communication process will improve team member morale, leading to improved design outcomes with less wasted efforts and resources" (Varner, 2012). Communication is a critical issue in the working of cross functional teams as integration is required not only between functionally similar team members but also across functional boundaries. Functional boundaries exist either due to hierarches that define roles within organizations or along professional divides responsible for specific tasks (Litt, 2000). If improperly managed, these boundaries limit effective communication between members of cross-functional teams. Frequent and effective communication promotes coordination, mutual support and cooperation between team members, allowing for an open process that leverages individual skills and abilities to maximize problem-solving.

Effective problem-solving depends on the right decisions being made by design team members. This makes decision-making as well as the decisions made important in design activity. Badke-Schaub and Gehrlicher (2003) define decisions as a step in the problem-solving process that involves making choices between alternatives, where the problem-solving process is defined by six 'critical situations': (1) goal-clarification; (2) solution generation; (3) analysis; (4) evaluation; (5) decision; and (6) control (Stempfle & Badke-Schaub, 2002). Hansen and Andreasen (2004) found "different descriptions of decision making in design literature focusing on (1) finding the best solution from given or generated alternatives, (2) validating the acceptability of a proposed solution, (3) determining the process or what to do next, and (4) identifying the elements of a decision". Proper communication between team members leads to a rich problem solving and decision making process. In cross-functional teams where functional boundaries can influence communication, the implications on design decisions made particularly needs to be explored.

Research on design activity has been carried out to understand creativity in the design process and the cognitive processes of designers, with an emphasis on solution finding during design tasks (Akin & Akin, 1998; Badke-Schaub & Gehrlicher, 2003; Dorst & Cross, 2001). Very few of these have explicitly studied the relationship between team processes and design outcomes (Stempfle & Badke-Schaub, 2002). Extensive work has been carried out on team design activity focused on how teamwork influences idea generation and creativity during the design process (Badke-Schaub et al., 2010; Ensici et al., 2013; McDonnell, 2012; Perttula et al., 2006) and problem solving (Carroll et al., 1980; Goldschmidt, 1995; Olson et al., 1992; Wiltschnig et al., 2013). Several papers have also looked at interaction and communication between design team members, but only a small number explored the relationship between team member communication and design outcomes, specifically idea generation.

There is existing literature on the factors that influence team communication such as the relationship between team members (Pinto & Pinto, 1990; Smith et al., 1994), diversity within the team (Joshi & Roh, 2009; Knight et al., 1999; Stahl et al., 2010) and hierarchies, both internally within the team and externally, with the larger organization the team forms a part of (Perretti & Negro, 2006). N. Cross and A. C. Cross (1995) studied roles and relationships between team members and their relationship with successful task completion, while Sosa et al. (2004) focused on the misalignment between communication patterns and design interface patterns across boundaries in cross functional teams. Stempfle and Badke-Schaub (2002) identified that communicative acts are representative of team processes and that team diversity significantly influences communication. In addition, informal social networks behind existing traditional hierarchies are the drivers of performance in organizations (R. Cross & Prusak, 2002; Krackhardt & Hanson, 1993).

However, few studies have explicitly focused on the relationship between team member communication and its impact on team outcomes, team performance and team decision making. Fewer yet have identified when communication is most important (Hassall, 2009) in team design contexts and how this varies with increased functional boundaries within the team. We conducted an analysis of transcriptions and video recordings of design team meetings to determine when communication is most critical during design activity by looking at the interaction between team members and the design decisions made. We were interested in the influence of team interaction on

Table 6.1 Research questions.

Question	Description
Question 1	a. To what extent are team decisions influenced by communication roles and hierarchy in design teams?
	b. How does the presence of functional boundaries affect this relationship?
Question 2	a. Are there different communication patterns that emerge to create different types of design decisions?
	b. How does the presence of functional boundaries affect this relationship?

design decisions made and if interaction varied by the type of design decision made. We were also interested in identifying the effects of the presence of functional boundaries on this relationship to provide new information on the relationship between team communication and team outcomes in the design context. The research questions posed are outlined in Table 6.1.

2 METHODS

2.1 Methods used

The DTRS11 dataset comprised of three teams: (1) the design team; (2) the external consultants; and (3) the accessories department. To address the research questions posed in the previous section, we set up a comparative analysis of two different team configurations: (1) the design team; and (2) the design team and the external consultants. We performed a content analysis of relevant data in the DTRS11 dataset, specifically those related to the planning of and preparation for the co-creation workshops. We also focused on the teams present in the datasets to allow comparisons of different team configurations, which, as noted previously with the emphasis on cross-functional teams, was an important aspect of the study. We used phrases present in the transcription provided as the unit of analysis. From these phrases, categories and themes were identified and organized into codes through the thorough examination of the transcribed data. To tackle the research questions guiding this study, we followed up the content analysis with methods of sociometric data analysis. The analysis of frequencies together with the use of sociograms and data visualization techniques allowed us to map decisions made as well as relationships within and between members of the teams studied. This provided insight into the roles and hierarchies existing within the groups while allowing us to identify the communication patterns and differences due to the presence of functional boundaries.

2.2 Data selection

From the DTRS11 data (Christensen & Abildgaard, 2017), we selected three data folders containing videos and transcripts of the meetings that best captured the information we required for our analysis. First, to ensure some level of consistency and standardization in our comparison of teams, we selected data folders that described

Table 6.2 Taxonomy of design decisions (Akin & Lin, 1995).

Decisions	Description	Example
Design decision	"Design decisions are any all intentional declarations of information as valid for the design problem at hand, directly concerning the product being developed" (Akin and Lin, 1995), in this case, the Co-Creation Workshops	And then, number two is of course a willingness to co-create, so they are- they want to create, they wanna:- they wanna participate, but they also want to push something forward. And the last one, the most luxury one, is that- their ability: to communicate.
Process decision	Process decisions, while also intentional declarations of information do not directly refer to the product but to the process of creating the product.	Then maybe get a little inspiration, so a share back session and I brought some of the stuff I had from before and maybe you have some experience from before that you wanna share.

the planning stages of the co-creation workshops. This inclusion criterion left us with folders that either involved the designing of the co-creation workshops or the iterations on the workshop design (Data folders: 03 Designing co-creation workshops; 03 Iterations on workshop design; 02 Iteration on co-creation workshops; 05 Designing co-creation workshop day 2; 10 Iterations on CC2 design). Then we went through the data to identify what teams were present at the meetings we originally selected. Due to our focus on cross-functional teams, we decided on having at least one data folder with the design team and one data folder with the design team and either the accessories department or the external consultants. The objective of this approach was to have two data folders of approximately equal length and similar content that would allow us study the effect of functional boundaries on team communication and decision making. The final data folders selected for analysis were Dataset 02: Designing co-creation workshops which was primarily used as training data, Dataset 03: Iterations on workshop design, and Dataset 10: Iterations on co-creation workshop day 2.

2.3 Codification

Three researchers with prior exposure to content analysis (OA, JJ and NS) developed the preliminary content categorization matrix guided by previous research in the areas of design decisions and decision making. We defined decisions as a series of steps beginning with a problem or opportunity and ending in actions taken, like findings by Hansen & Andreasen (2004). Based on this and using the definitions and taxonomy of design decisions developed by Akin and Lin (1995), where design decisions were differentiated from process decisions as those directly involved in the design process, phrases were identified as containing design specific decisions, that is, those directly related to the planning for the co-creation workshop.

Focusing on these design decision phrases identified, two of the researchers (JJ and NS) independently searched for themes and content categories related to the co-creation activity that was being planned. They met with an additional researcher (OA) to combine similar categorization for themes and address questions, while

Table 6.3 Definitions of content categories.

Codes	Definitions	Example
Approach	A general idea of how things will be done during the Co-Creation workshops. The main theme of the co creation workshops. The form of the workshops	Eh: *but there is so many different approaches, and some are super effective one place, but not effective another place and* so on, so. So the purpose of this meeting, which maybe will take an hour, two hours, let's see: how long we have steam for.
Tools	A device equipment or tangible technique to be used during the co-creation workshop to facilitate proceedings.	And embrace it and I think it's:- it's very correct what you said around this, *you know we need to create these tools and this session to match with the people we get in,* but it's really difficult to know what people we get in.
Users	The target demographic for the Co-Creation workshops. People who will provide information during the workshop.	You know when you go in to the interviews it's one one one one, you can control the situation pretty well, *but here we are- you have seven people in a room. If they don't play well together it's really really difficult-* it's like hurting- hurting cats.
Constraints	Challenges, potential problems and difficulties that may occur during the workshop and have to be accounted for.	That's very nice expectations and I think you are:- we're touching up on all: *the stuff that is cool, but all this enormous amount of stuff that is difficult as well*
Exercises	Tasks, activities and applications to be used by the users or on the users during the workshops.	Are you- are you going to prime them or are you allowed to prime them? Or what can-
Moderation	The flow and scheduling of the co-creation workshop. Planning for the control and organization of the two co-creation sessions.	It's no good that someone just- someone's wife signed then up- them up and, *I mean, we really need to make sure that Rose or whoever is doing the recruitment* is talking to the [persons who is actually coming].
Cultural factors	Context specific information directly related to the location and setting of the co creation workshops (in China).	Yeah, and just a comment to what you said, *I mean when it is China it's even more eh: (.) something that we really need to consider,* if- if they're going to present themselves with a title and everything.
Content	Specific information regarding the co-creation workshop. Things that the workshop will contain, things that will be focused on.	*But maybe we need to be a little bit more spe'- clear with it,* because last time we also thought that everybody was primed and knew what [they were going into].

refining the codes and content categories identified. After clarifying the definition of each code, the researchers developed a final categorization matrix that included the inclusion and exclusion criteria for each code. The two researchers (JJ and NS) worked independently to review and code the selected data as required. Inter-coder reliability ranged from 65% originally to 90% after review and clarification of the code categories. The content categories are shown in the Table 6.3.

2.4   Measures – team member characteristics

We used two codes to describe team member characteristics from information provided in the data synopsis and data overview documents provided by the DTRS organizers

Table 6.4 Team member characteristics.

Name	Functional background	Role
Ewan	The Design Team	Team Leader/Designer
Abby	The Design Team	Design researcher
Kenny	The Design Team	Design researcher
Nina	The Design Team	Intern
David	The Design Team	DTRSII participant observer
Will	External Consultants	Market researcher
Rose	External Consultants	Design researcher/Design Thinking expert
Amanda	External Consultants	Design researcher/Design Thinking expert

in the DTRS database. The functional background of team members was defined by the team the members belonged to, that is, either the design team in 02 Designing co-creation workshops and 03 Iterations on workshop design or the design team and the external consultants in 10 Iterations on co-creation workshop day 2. Team member roles in the team were also obtained from the data synopsis and used to describe team member characteristics.

2.5 Sociometric and frequency analysis

To identify decisions made and how communication roles within teams affect that (RQ 1), frequency counts were initially used to summarize team member characteristics with respect to the content categories obtained. This provided insight on which team members were responsible for specific decisions made. To understand the emergent communication patterns and their impact on design decisions made (RQ 2), network analysis approaches were used. Network analysis utilizes tools to organize and visualize the links between nodes in a network group while reveals the spatial centralization of the network (Alexander, 2009). Network perspectives provide a glimpse into relational interaction between individuals in a group, providing explanations of social structures and communication within the group (Cheliotis, 2010). Graphical methods and diagrams were used to map communication and interaction between team members as captured in the transcribed data. Sociograms were developed using Discursis, "a computer-based, textual analysis software package developed by researchers at the University of Queensland to understand the content and relationships present within textual data" (Discursis, 2016). The sociograms allowed us visualize communication patterns between team members and identify how they related to the design decisions made.

3 RESULTS

3.1 Frequency analysis of content categories

A frequency analysis was performed on the results of the content analysis and coded data. This provided a summary of all three datasets while allowing for comparisons

Table 6.5 Frequency analysis of content categories.

Content categories	Dataset 02: Designing co-creation workshops		Dataset 03: Iterations on workshop design		Dataset 10: Iterations on co-workshop day 2	
	Frequency	Percentage	Frequency	Percentage	Frequency	Percentage
Content	16	15%	135	33%	66	29%
Cultural factors	2	2%	3	1%	2	1%
Moderation	4	4%	45	11%	45	20%
Exercises	8	8%	44	11%	20	9%
Constraints	12	12%	14	3%	6	3%
Users	36	34%	90	22%	49	22%
Tools	14	13%	27	7%	8	4%
Approach	13	12%	56	14%	31	14%

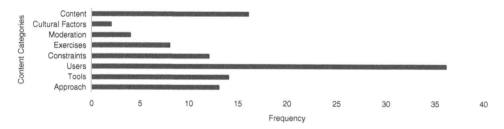

Figure 6.1 Frequency of content categories in Dataset 02: designing co-creation workshop.

to be made as required. The results of the frequency analysis of the coded data in the three datasets analyzed are shown in Table 6.5.

In Dataset 02: Designing co-creation workshops, the majority of the design decisions made by design team members were about the users (34%), that is, the target population to be used in the co-creation workshop. Decisions about users were followed by decisions on the content of the co-creation workshops (15%) along with those on the tools to be used during the workshop (13%), the approach to the workshop (12%) and possible constraints posed during the workshop (12%). This is shown in Figure 6.1.

In Dataset 03: Iterations on workshop design, the majority of the design decisions made by the design team members were about the content of the co-creation workshop (33%), that is specific information regarding how the co-creation workshops would occur, on the design of the workshop. Decisions about content were followed by decisions on the users (22%), the approach to the workshop (14%) and the moderation of (11%) and exercises to be performed during the workshop (11%). This is shown in Figure 6.2.

In Dataset 10: Iterations on co-creation workshop day 2, the majority of the design decisions made by the design team members were about the content of the co-creation workshop (29%). These decisions on the content of the workshop were followed by decisions on the users of the workshop (22%), the moderation and organization of

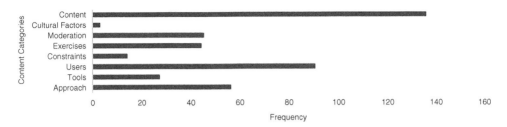

Figure 6.2 Frequency of content categories in Dataset 03: iterations on workshop design.

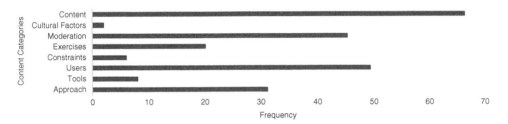

Figure 6.3 Frequency of content categories in Dataset 10: Iterations on co-creation workshop day 2.

the workshop (20%) and the approach to the workshop (14%). This is shown in Figure 6.3.

Comparing all three datasets analyzed, decisions on the content of the workshops and the users of the workshops were consistently high throughout all three datasets. However, there was more similarity in the frequency of decisions on the approach to the workshop, the users of the workshop and the content of the workshop between dataset 03: iterations on workshop design and dataset 10: iteration on co-creation workshop day 2. Decisions on cultural factors, though present in all three datasets was consistently low through all three datasets. Decisions on the moderation of the workshop increased in frequency (4% to 20%) from the first to the third dataset while decisions on tools decreased (14% to 7%) from the first to the third dataset.

3.2   Frequency analysis of content categories with respect to team members

The frequency analysis on the content categories gave us a sense of what was occurring in the analyzed datasets. To understand how decisions were influenced or affected by the team measures, we also performed an analysis of frequencies of the two team measures, the role of the team members and the functional background of the teams, that is in Dataset 10 both the Design team and the External consultants are present.

In Dataset 02, results of the analysis show that the design team leader, Ewan, was consistently responsible for the design decisions made across all categories. When dealing with the Approach, Users and Content of the co-creation workshop content categories, the researcher, Abby, followed the Team leader, Ewan, and occasionally

Table 6.6 Frequency analysis of content categories with respect to team member roles in Dataset 02.

Content categories	Team leader (Ewan)		Design researcher (Abby)		Design researcher (Kenny)		Intern (Nina)		Observer (David)	
	Freq.	Percent	Freq.	Percent	Freq.	Percent	Freq.	Percent	Freq.	Percent
Content	7	39%	3	17%	7	39%	1	6%	0	0%
Cultural factors	0	0%	2	100%	0	0%	0	0%	0	0%
Moderation	1	25%	2	50%	1	25%	0	0%	0	0%
Exercises	4	50%	1	13%	1	13%	1	13%	1	13%
Constraints	6	50%	2	17%	4	33%	0	0%	0	0%
Users	15	42%	9	25%	8	22%	2	6%	2	6%
Tools	7	50%	2	21%	3	21%	0	0%	2	14%
Approach	5	38%	1	38%	5	38%	1	8%	1	8%
TOTAL	45	42%	22	21%	29	27%	5	5%	6	6%

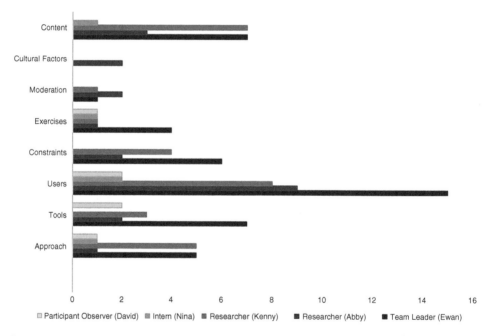

Figure 6.4 Frequency of content categories with respect to team member roles in Dataset 02: Designing co-creation workshop.

matched the design decisions made. Ewan, Abby and Kenny made most of the design decisions in this dataset, which was expected as Nina, being an intern, would be less likely to make decisions while David, as the participant observer, had limited but insightful contributions to the decisions made. The flow of decisions made in Dataset 02 follows the traditional hierarchy expected in a team, where although input was obtained from all team members, the bulk of the design decisions came from the team

Table 6.7 Frequency analysis of content categories with respect to team member roles in Dataset 03.

Content categories	Team leader (Ewan)		Design researcher (Abby)		Design researcher (Kenny)		Observer (David)	
	Freq.	Percent	Freq.	Percent	Freq.	Percent	Freq.	Percent
Content	56	41%	38	28%	39	29%	2	1%
Cultural factors	2	67%	0	0%	1	33%	0	0%
Moderation	18	40%	16	36%	11	24%	0	0%
Exercises	14	33%	16	37%	12	28%	1	2%
Constraints	8	57%	5	36%	1	7%	0	0%
Users	29	32%	37	41%	22	24%	2	2%
Tools	7	26%	12	44%	7	26%	1	4%
Approach	26	46%	19	34%	10	18%	1	2%
TOTAL	160	39%	143	35%	103	25%	7	2%

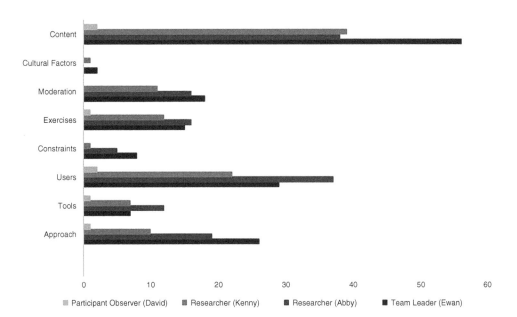

Figure 6.5 Frequency of content categories with respect to team member roles in Dataset 03: Iterations on workshop design.

leader. The results of the analysis of Dataset 03: Iterations on Workshop design are shown in Table 6.7 and Figure 6.5.

In Dataset 03, results of the analysis show that, once again, through most of the content categories, the design team leader, Ewan, was consistently responsible for the design decisions made. When making decisions involving the content of the workshop, moderation of the workshop, the constraints faced and the approach to the workshop, Ewan made the bulk of the decisions, however, unlike the previous dataset, researcher Abby, made most of the design decisions regarding the users, the tools and the exercises

Table 6.8 Frequency analysis of content categories with respect to team member roles in Dataset 10.

Content categories	Team leader (Ewan) Freq.	Percent	Design researcher (Abby) Freq.	Percent	Design researcher (Kenny) Freq.	Percent	Observer (David) Freq.	Percent
Content	17	26%	11	17%	14	21%	1	2%
Cultural factors	0	0%	0	0%	0	0%	0	0%
Moderation	10	22%	6	13%	0	0%	1	2%
Exercises	6	30%	4	20%	0	0%	0	0%
Constraints	2	33%	1	17%	1	17%	0	0%
Users	9	18%	11	22%	6	12%	1	2%
Tools	4	50%	1	13%	0	0%	0	0%
Approach	11	35%	1	3%	3	10%	0	0%
TOTAL	59	26%	35	15%	24	11%	3	1%

Content categories	Researcher (Will) Freq.	Percent	Researcher (Rose) Freq.	Percent	Researcher (Amanda) Freq.	Percent
Content	12	18%	8	12%	3	5%
Cultural factors	0	0%	2	100%	0	0%
Moderation	8	18%	16	36%	4	9%
Exercises	7	35%	2	10%	1	5%
Constraints	1	17%	1	17%	0	0%
Users	12	24%	7	14%	3	6%
Tools	1	13%	1	13%	1	13%
Approach	7	23%	7	23%	2	6%
TOTAL	48	21%	44	19%	14	6%

for the co-creation workshops. This is a deviation from the first analyzed dataset where decisions followed a traditional team leader hierarchy. In this dataset, there was more input and contribution from team members regarding decisions made.

The results of the analysis of Dataset 10: Iterations on Co-Creation Workshop day 2 are shown in Table 6.8 and Figure 6.6.

A comparison of the team member hierarchies based on their total contribution to making design decisions through each analyzed dataset is shown in Figure 6.7.

Having two teams present in dataset 10 allows for an analysis of the effects of functional boundaries on the decisions made. On observing the decisions made by the design team and the external consultants, across most categories the presence of the functional boundaries within the group did not affect the decisions made, since both teams focused on the same decisions showing similar frequencies of categories. The two teams show a high level of integration in the decisions made. However, there were a few differences worth noting. While the design team placed more emphasis on the Content of the workshop and users of the workshop and made more design decisions in these categories, the external consultants were more concerned with the

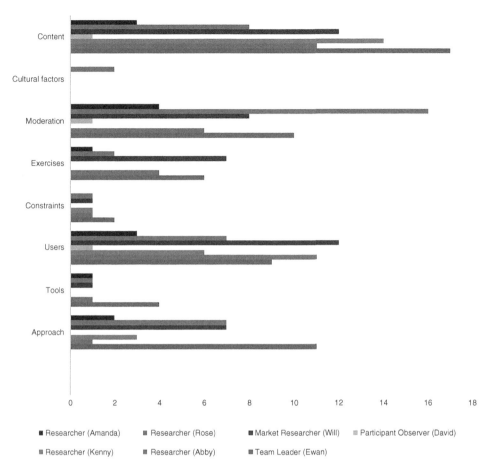

Figure 6.6 Frequency of content categories with respect to team member roles in Dataset 10: Iterations on co-creation workshop day 2.

moderation of the workshop and the cultural factors that could influence the outcomes of the workshop. This result is expected as the external consultants oversaw the actual workshop and served as moderators for it.

However, in looking at the individual contributions to Dataset 10, it is interesting to note that unlike the earlier datasets, different team members contributed to the different categories in different ways. While the design team leader, Ewan, made the most design decisions in the Approach and Content categories, researcher Rose contributed the most to decisions on cultural factors and the moderation of the co-creation workshops. Like the second analyzed dataset, the team followed a more democratic process of making decisions rather than one dominated by the design team leader. Although he contributed the most overall to design decisions made, the presence of more team members did not seem to significantly influence the decisions made during the meeting.

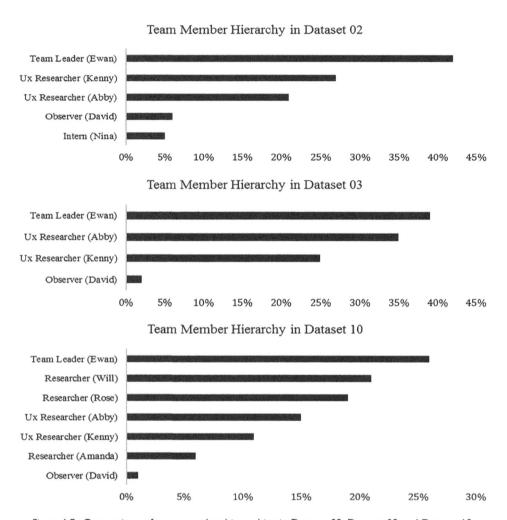

Figure 6.7 Comparison of team member hierarchies in Dataset 02, Dataset 03 and Dataset 10.

3.3 Analysis of team member interactions

To address RQ 2, we used "the Discursis approach, an information visualization technique that uses recurrence plots to identify trends within time series data" (Watson, Angus, Gore, & Farmer, 2015). We limited this analysis to the two datasets that would be compared, Dataset 03: Iterations on workshop design, and Dataset 10: Iterations on co-creation workshop day 2. The Discursis plots for the initial sections of the analysis of the two datasets, comprising of the first 100 utterances with design decisions, are shown in Figure 6.9. The conversation plots obtained allowed us to examine the "turn-by-turn similarities in concepts between team members present in both datasets as well as the engagement and interaction of team members" (Watson *et al.*, 2015) with respect to the decisions made. We examined each dataset first, then, in smaller chunks

Table 6.9 Frequency analysis of content categories with respect to the functional boundaries.

Content categories	The design team (Ewan, Abby, Kenny, David)		The external consultants (Will, Rose, Amanda)	
	Frequency	Percentage	Frequency	Percentage
Content	43	65%	23	35%
Cultural factors	0	0%	2	100%
Moderation	17	38%	28	62%
Exercises	10	50%	10	50%
Constraints	4	67%	2	33%
Users	27	55%	22	45%
Tools	5	63%	3	37%
Approach	15	48%	16	52%
TOTAL	121	53%	106	47%

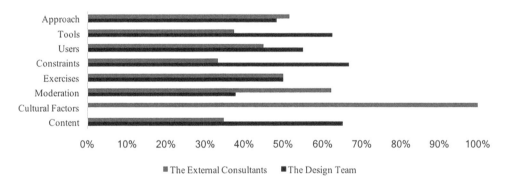

Figure 6.8 Frequency of content categories with respect to team member roles in Dataset 10: Iterations on co-creation workshop day.

of 25 to 50 utterances. Overall, a major difference observed between the two datasets was the density of concept repetition in Dataset 10 compared with that in Dataset 03, as can be observed from Figure 6.8. This result is consistent with the findings from the frequency analysis where Dataset 10 had more team members contributing to the design decisions made than in Dataset 03. The higher number of individuals present in Dataset 10 would also indicate a denser communication plot than in Dataset 03 with fewer individuals present.

Studying the Discursis plots in more detail (50 utterances), we observed that the communication plots were denser, signaling more interaction, as shown in Figure 6.10. Also, by studying the sharing of concepts between team members, which we call inter-member similarity in content and concepts, as indicated by the doubly-shaded rectangles, we find that concepts are shared between team members in Dataset 10 than in Dataset 03. This may also be a result of there being more individuals present

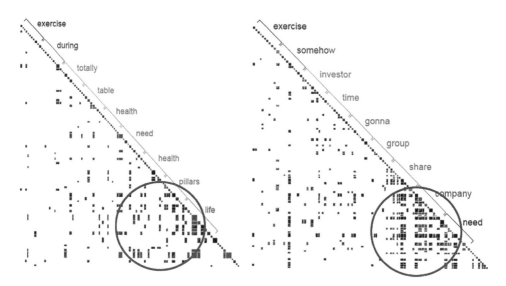

Figure 6.9 Discursis plots comparing Dataset 03 (left) and Dataset 10 (right), highlighting differences in concept repetition.

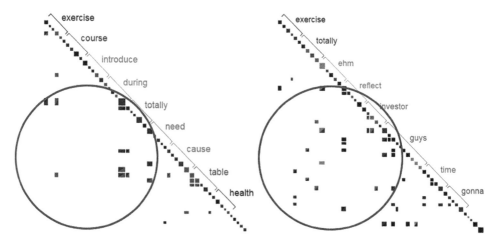

Figure 6.10 Discursis plots of 50 utterances comparing Dataset 03 (left) and Dataset 10 (right), highlighting differences in inter-member similarity.

in Dataset 10 than in Dataset 03. This indicates that the interaction and integration between team members in Dataset 10 was better than in Dataset 03 due to this difference in inter-member similarity.

Dataset 03 shows more instances of intra-member similarity than Dataset 10, suggesting that individuals were more likely to hold on to discuss and possibly resolve concepts on their own rather than having input from other team members, as seen in Dataset 10. This indicates that the team members in Dataset 10 adopted a more integrative approach to making design decisions than they did in Dataset 03.

The turn-by-turn analysis showed that there was a high level of engagement and concept repetition in both datasets, as can be seen by the connected recurrence of design decisions. Significant sections of individuals repeating their decisions or modifying their decisions were observed in both datasets, although this was more pronounced in Dataset 10 than in Dataset 03. It could be said that Dataset 10 provided more interesting conversation turns than Dataset 03, and we attribute this to the presence of two functional teams rather than one, that is Dataset 10 had members from both the design team and the external consultants.

4 DISCUSSION

The aim of this study was to identify how team member roles and hierarchies influenced communication and design decisions made while accounting for the influence of functional boundaries. This was done to identify and visualize the communication patterns created by cross-functional teams and how this influenced the outcomes and decisions made by cross-functional team members, when compared against a team without functional boundaries. Through the analysis of selected datasets provided in the DTRS11 dataset, specifically Dataset 03 and Dataset 10, with Dataset 02 used as training data to develop the content categories and test the methods of analysis, we could identify the impact of team roles on design decisions made while visually presenting the interaction and integration patterns between team members.

Identifying the design decisions made allowed us to focus specifically on information regarding the co-creation workshops rather than the planning and coordination of team meetings in the datasets. By coding and sorting the design decisions into content categories, we organized the flow of the workshop design decisions. This content categorization allowed us to identify the contributions of individuals and their roles. The content categories were developed using content analysis procedures. The three analyzed datasets, Dataset 02, Dataset 03 and Dataset 10 were used to come up with the different iterations on the content categories. By adopting a systematic approach to the content analysis we limited the effect of subjectivity inherent in the content analysis process.

We used analyses of frequencies of coded data to observe, on aggregate, which individuals and specific roles were responsible for certain decisions, as well as providing an overview of the different patterns that emerged. From the results of the frequency analyses of the coded datasets we found that different team roles and team members contribute in different ways to the design decisions made. Results showed that the team leader (Ewan) consistently provided and contributed to majority of the design decisions made in the datasets with just the design team present. The intern (Nina) contributed minimally to the design decisions made, as can be seen from the frequency analysis in Table 6.6: Frequency Analysis of Content Categories with respect to Team Member Roles in Dataset 02.

The analysis of frequencies carried out on Dataset 10 showed that having more than one functional group changes the dynamics of interaction between team members. We observed that having more people present in Dataset 10 reduced the overall dominance of the design team leader (Ewan) over the design decisions made, as can be seen in Table 6.8: Frequency Analysis of Content Categories with respect to Team Member Roles in Dataset 10 and Figure 6.6: Frequency of Content Categories with respect to Team Member Roles in Dataset 10. We attribute this to the high level of integration observed between the two functional teams present in Dataset 10, the design team and the external consultants. We also observed that the different teams and team members contributed to the different content categories with the design team placing emphasis on the content, tool and users of the co-creation workshop and the external consultants placing more emphasis on the moderation of and cultural factors impacting the workshop. We predict that more information on the background and expertise of the team members will allow more detailed conclusions to be made about the individual contributions to the content categories of the design decisions.

The analysis of interactions carried out with Discursis validated the results obtained in the frequency analyses. Discursis also allowed us to observe the turn by turn interaction between team members and visualize patterns that emerged for different design decisions. Results showed that different patterns and densities of content occur throughout the datasets analyzed. The most interesting result obtained was that having more functional groups present led to a more integrative and democratic approach in making design decisions, where more team members contributed to the decisions made and were involved in the resolution of issues.

We identify similarities in these results and in the analysis performed by Paletz, Sumer, and Miron-Spektor on diversity and conflict in multicultural teams (Paletz *et al.,* 2017). They found that more diverse teams were less likely to have conflicts, although this was mediated by several psychological factors. We suggest that the integrative approach to decisions could be a factor that contributed to the lack of conflict. This brings the topic of factors that influence diversity in teams into question. Further work in this area could study different team characteristics and individual characteristics to identify which are more likely to influence design team outcomes. Also, from the work of Jornet and Roth (2017), we are interested in the sociogenetic perspective and how it ties the social, communicative process with the individual, mental process. As Discursis allowed us to model the communication process of the design team, we are particularly interested in how a mental model would map onto a communication model and the insights this mapping could provide.

Although we cannot generalize these results to all cross-functional teams due to the absence of specific information on the team members, the nature of prior engagements or interactions and the small size of the dataset used, we believe the analyses, comparison and results presented here will provide insight into the workings of cross-functional teams in design contexts.

ACKNOWLEDGEMENTS

The analysis for this DTRS11 submission was based on previous work done for previous DTRS meetings – especially those organized by Robin Adams (DTRS10). This analysis was made possible with support from David Angus of the Discursis team who provided access to the software.

REFERENCES

Adams, R.S. and Krautkramer, C. (2016). Harnessing the Elusive Expertise of Big Picture Thinkers. *Proceedings of the University-Industry Interaction Conference, Amsterdam, June.*

Adams, R. S., Aleong, R., Goldstein, M. & Solis, F. (2017). Problem Structuring as Co-Inquiry. In: Christensen, B. T., Ball, L. J. & Halskov, K. (eds.) *Analysing Design Thinking: Studies of Cross-Cultural Co-Creation.* Leiden: CRC Press/Taylor & Francis.

Akin, Ö., & Akin, C. (1998). On the process of creativity in puzzles, inventions, and designs. *Automation in Construction.* [Online] 7(2), 123–138. Available from: https://www.researchgate.net/publication/223595254_On_the_process_of_creativity_in_puzzles_inventions_and_designs [Accessed February 15, 2017]

Akin, Ö., & Lin, C. (1995). Design protocol data and novel design decisions. *Design Studies.* [Online] 16(2), 211–236. Available from: http://www.sciencedirect.com/science/article/pii/0142694X9400010B [Accessed February 15, 2017]

Alexander, M. L. (2009). Qualitative social network research for relational sociology. In *Proceedings of The Australian Sociological Association Conference. Citeseer.* [Online] 1–15. Available from: http://citeseerx.ist.psu.edu/viewdoc/download?doi=10.1.1.458.1779&rep=rep1&type=pdf [February 15, 2017]

Badke-Schaub, P., Gehrlicher, A. (2003). Patterns of Decisions in Design: Leaps, Loops, Cycles, Sequences and Meta-processes. *DS 31: Proceedings of ICED 03, the 14th International Conference on Engineering Design, Stockholm.* [Online] 313–314. Available from: https://www.designsociety.org/publication/24012/patterns_of_decisions_in_design_leaps_loops_cycles_sequences_and_meta-processes [Accessed February 15, 2017]

Badke-Schaub, P., Goldschmidt, G., & Meijer, M. (2010). How does cognitive conflict in design teams support the development of creative ideas? *Creativity and Innovation Management.* [Online] 19(2), 119–133. Available from: http://onlinelibrary.wiley.com/doi/10.1111/j.1467-8691.2010.00553.x/abstract [Accessed February 15, 2017]

Bater, R. (2014, March 5). *IBM Rational Design blog: The importance of a team in design.* [Online] Available from: https://www.ibm.com/developerworks/community/blogs/b62e9130-7f69-4b25-b248-40beaee24d55/entry/the_importance_of_a_team_in_design?lang=en [Accessed May 10, 2016]

Carroll, J. M., Thomas, J. C., Miller, L. A., & Friedman, H. P. (1980). Aspects of solution structure in design problem solving. *The American Journal of Psychology* [Online] 269–284. Available from: https://www.jstor.org/stable/1422232?seq=1#page_scan_tab_contents [Accessed February 15, 2017]

Chan, J. & Schunn, C. D. (2017). A Computational Linguistic Approach to Modelling the Dynamics of Design Processes. In: Christensen, B. T., Ball, L. J. & Halskov, K. (eds.) *Analysing Design Thinking: Studies of Cross-Cultural Co-Creation.* Leiden: CRC Press/Taylor & Francis.

Cheliotis, G. (2010). *Social Network Analysis.* [Online] Available from http://www.slideshare.net/gcheliotis/social-network-analysis-3273045 [Accessed February 15, 2017]

Christensen, B. T. & Abildgaard, S. J. J. (2017). Inside the DTRS11 Dataset: Background, Content, and Methodological Choices. In: Christensen, B. T., Ball, L. J. & Halskov, K. (eds.) *Analysing Design Thinking: Studies of Cross-Cultural Co-Creation.* Leiden: CRC Press/Taylor & Francis.

Cross, N., & Cross, A. C. (1995). Observations of teamwork and social processes in design. *Design Studies.* [Online] 16(2), 143–170. Available from: https://www.researchgate.net/publication/229248689_Observations_of_Teamwork_and_Social_Processes_in_Design [Accessed February 15, 2017]

Cross, R., Borgatti, S. P., & Parker, A. (2002). Making invisible work visible: Using social network analysis to support strategic collaboration. *California Management Review.* [Online]

44(2), 25–46. Available from: http://cmr.ucpress.edu/content/44/2/25 [Accessed February 15, 2017]

Discursis. (2016). *Discurris :: Home.* [Online] Available from: http://www.discursis.com/ [Accessed May 25, 2016]

Dorst, K., & Cross, N. (2001). Creativity in the design process: co-evolution of problem-solution. *Design Studies.* [Online] 22(5), 425–437. Available from: http://www.sciencedirect.com/science/article/pii/S0142694X01000096 [Accessed February 15, 2017]

Elo, S., & Kyngäs, H. (2008). The qualitative content analysis process. *Journal of Advanced Nursing.* [Online] 62(1), 107–115. Available from: https://www.ncbi.nlm.nih.gov/pubmed/18352969 [Accessed February 15, 2017]

Ensici, A., Badke-Schaub, P., Bayazit, N., & Lauche, K. (2013). Used and rejected decisions in design teamwork. *CoDesign.* [Online] 9(2), 113–131. Available from: http://www.tandfonline.com/doi/full/10.1080/15710882.2013.782411?scroll=top&needAccess=true [Accessed February 15, 2017]

Fischer, A., Greiff, S., & Funke, J. (2011). The process of solving complex problems. *Journal of Problem Solving.* [Online] 4(1), 19–42. Available from: http://docs.lib.purdue.edu/jps/vol4/iss1/3/ [Accessed February 15, 2017]

Goldschmidt, G. (1995). The designer as a team of one. *Design Studies.* [Online] 16(2), 189–209. Available from: http://dx.doi.org/10.1016/0142-694X(94)00009-3 [Accessed February 15, 2017]

Hansen, C. T. & Andreasen, M. M. (2004). A mapping of design decision-making. *DS 32: Proceedings of DESIGN 2004, the 8th International Design Conference, Dubrovnik, Croatia.* [Online] 1409–1418. Available from: https://www.designsociety.org/publication/19931/a_mapping_of_design_decision-making [Accessed February 15, 2017]

Hassall, S. L. (2009). *The relationship between communication and team performance: testing moderators and identifying communication profiles in established work teams.* [Online] 1. Available from: http://eprints.qut.edu.au/30311/ [Accessed February 15, 2017]

Inc. (2006). *Cross-Functional Teams.* Available from: http://www.inc.com/encyclopedia/cross-functional-teams.html [Accessed March 1, 2016]

Jornet, A. & Roth, W. (2017). Design {Thinking | Communicating}: A Sociogenetic Approach to Reflective Practice in Collaborative Design. In: Christensen, B. T., Ball, L. J. & Halskov, K. (eds.) *Analysing Design Thinking: Studies of Cross-Cultural Co-Creation.* Leiden: CRC Press/Taylor & Francis.

Joshi, A., & Roh, H. (2009). The role of context in work team diversity research: A meta-analytic review. *Academy of Management Journal.* [Online] 52(3), 599–627. Available from https://www.researchgate.net/publication/275714198_The_Role_of_Context_in_Work_Team_Diversity_Research_A_Meta-Analytic_Review [Accessed February 15, 2017]

Knight, D., Pearce, C. L., Smith, K. G., Olian, J. D., Sims, H. P., Smith, K. A., & Flood, P. (1999). Top management team diversity, group process, and strategic consensus. *Strategic Management Journal.* [Online] 20(5), 445–465. Available from: http://onlinelibrary.wiley.com/doi/10.1002/(SICI)1097-0266(199905)20:5%3C445::AID-SMJ27%3E3.0.CO;2-V/abstract [Accessed February 15, 2017]

Krackhardt, D., & Hanson, J. R. (1993, July 1). Informal Networks: The Company Behind the Chart. *Harvard Business Review.* [Online] Available from: https://hbr.org/1993/07/informal-networks-the-company-behind-the-chart [Accessed March 14, 2016]

Lane, D. (2000). *Communication Competence Perspective.* [Online] Available from: http://www.uky.edu/~drlane/capstone/commcomp.htm [Accessed March 16, 2016]

Lean Enterprise Institute. (2014). *Jidoka in the Lean Lexicon ©.* [Online] Available from: http://www.lean.org/lexicon/jidoka [Accessed March 16, 2016]

Litt, B. (2000). *Crossing Functional Boundaries*. [Online] Available from: http://www.project management.com/articles/11548/Crossing-Functional-Boundaries [Accessed March 16, 2016]

MacWilliam, I. (2016). *The importance of teams – Knowhow Nonprofit* [Page]. [Online] Available from: https://knowhownonprofit.org/people/teams/about-teams-and-types-of-team/importance [Accessed February 29, 2016]

McDonnell, J. (2017). Design Roulette: A Close Examination of Collaborative Decision-Making in Design From the Perspective of Framing. In: Christensen, B. T., Ball, L. J. & Halskov, K. (eds.) *Analysing Design Thinking: Studies of Cross-Cultural Co-Creation*. Leiden: CRC Press/Taylor & Francis.

Olson, G. M., Olson, J. S., Carter, M. R., & Storrøsten, M. (1992). Small group design meetings: An analysis of collaboration. *Human-Computer Interaction*. [Online] 7(4), 347–374. Available from: http://dl.acm.org/citation.cfm?id=1461843 [Accessed February 15, 2017]

Watson, B. M., Angus, D., Gore, L., & Farmer, J. (2015). Communication in open disclosure conversations about adverse events in hospitals. *Language & Communication*. [Online] 41, 57–70. Available from: http://www.sciencedirect.com/science/article/pii/S0271530914000755 [Accessed February 15, 2017]

Paletz, Susannah B. F., Sumer, A., & Miron-Spektor, E. (2017). Psychological Factors Surrounding Disagreement in Multicultural Design Team Meetings. In: Christensen, B. T., Ball, L. J. & Halskov, K. (eds.) *Analysing Design Thinking: Studies of Cross-Cultural Co-Creation*. Leiden: CRC Press/Taylor & Francis.

Perretti, F., & Negro, G. (2006). Filling empty seats: How status and organizational hierarchies affect exploration versus exploitation in team design. *Academy of Management Journal*. [Online] 49(4), 759–777. Available from: http://amj.aom.org/content/49/4/759.abstract [Accessed February 15, 2017]

Perttula, M. K., Krause, C. M., & Sipilä, P. (2006). Does idea exchange promote productivity in design idea generation? *CoDesign*. [Online] 2(3), 125–138. Available from: http://www.tandfonline.com/doi/pdf/10.1080/15710880600797942 [Accessed February 15, 2017]

Pinto, M. B., & Pinto, J. K. (1990). Project team communication and cross-functional cooperation in new program development. *Journal of Product Innovation Management*. [Online] 7(3), 200–212. Available from: http://dx.doi.org/10.1016/0737-6782(90)90004-X [Accessed February 15, 2017]

Salvendy, G. (2012). *Handbook of human factors and ergonomics*. [Online] John Wiley & Sons. Available from: https://books.google.com/books?hl=en&lr=&id=WxJVNLzvRVUC&oi=fnd&pg=PA3&dq=Handbook+of+Human+Factors+and+Ergonomics&ots=pXppKTSBq7&sig=oMrou2HpPTTqhn-YEb7gKdKMh08 [Accessed February 15, 2017]

Seitamaa-Hakkarainen, P. (2000). *Design tasks as ill-defined complex problems*. [Online] Available from: http://www.mlab.uiah.fi/polut/Yhteiskunnalliset/teoria_ill_defined_complex.html [Accessed February 15, 2017]

Simon, H. A. (1977). The structure of ill-structured problems. *Models of discovery*. [Online] 304–325. Available from http://link.springer.com/chapter/10.1007/978-94-010-9521-1_17 [Accessed February 15, 2017]

Sosa, M. E., Eppinger, S. D., & Rowles, C. M. (2004). The misalignment of product architecture and organizational structure in complex product development. *Management Science*. [Online] 50(12), 1674–1689. Available from: http://pubsonline.informs.org/doi/abs/10.1287/mnsc.1040.0289 [Accessed February 15, 2017]

Sonnenwald, D. H. (1996). Communication roles that support collaboration during the design process. *Design Studies*. [Online] 17(3), 277–301. Available from: http://www.sciencedirect.com/science/article/pii/0142694X96000026 [Accessed February 15, 2017]

Stahl, G. K., Mäkelä, K., Zander, L., & Maznevski, M. L. (2010). A look at the bright side of multicultural team diversity. *Scandinavian Journal of Management*. [Online] 26(4), 439–447. Available from: http://dx.doi.org/10.1016/j.scaman.2010.09.009 [Accessed February 15, 2017]

Stempfle, J., & Badke-Schaub, P. (2002). Thinking in design teams-an analysis of team communication. *Design Studies*. [Online] 23(5), 473–496. Available from: http://www.sciencedirect.com/science/article/pii/S0142694X02000042 [Accessed February 15, 2017]

Stephoudt, B. B., & Mariotta, A. B. (2011). Definition and differences between team and task process. *MIT Collaboration Box*. [Online] 1–4. Available from: http://web.mit.edu/collaborationtbox/module3/team-task.pdf [Accessed February 15, 2017]

Varner, E. (August 1st, 2012). *The Importance of Communication in Design*. Available from: http://skyrocketgroup.com/the-importance-of-communication-in-design/ [Accessed February 15, 2017]

Wiltschnig, S., Christensen, B. T., & Ball, L. J. (2013). Collaborative problem-solution co-evolution in creative design. *Design Studies*. [Online] 34(5), 515–542. Available from: http://www.sciencedirect.com/science/article/pii/S0142694X13000033 [Accessed February 15, 2017]

Chapter 7

A Computational Linguistic Approach to Modelling the Dynamics of Design Processes

Joel Chan & Christian D. Schunn

ABSTRACT

Cross-cultural interactions within the design team (and/or with external participants, as in co-design processes) can impact the team in many ways. In modeling changes in design processes over time, it is important to capture changes in key aspects of design processes, such as understanding of the problem and emergence of crucial new ideas. However, modeling such foundational change can be very effort-intensive for researchers. In this short paper, we explore the use of computational linguistic methods – such as topic modeling – as a cost-effective supplement to traditional methods in providing a "quick and dirty" birds-eye view of what designers are talking about, and how those topics change over time. Specifically, we use Latent Dirichlet Allocation – a foundational topic model approach – to explore changes in the diversity and prominence of topics over time in the DTRS11 dataset. We test the impacts of variations in key modeling assumptions and parameters on the quality of the resulting model. Overall, our topic models identify a robust shift in diversity of topics following the second co-creation session. We also find significant changes (both decreases and increases) in the prominence of several topics. These high-level patterns provide a quantitative complement to qualitative intuitions, and raise interesting new research questions (e.g., did the design team learn more from the second co-creation session compared to the first? If so, what changed?). In summary, our analysis demonstrates the benefits and potential limitations of computational linguistic methods as a supplement to traditional in-depth qualitative analysis of design processes.

1 INTRODUCTION

Recent work on sources of creativity in design (Fischer, 2005) has drawn attention to how cross-cultural interactions – where designers and/or users interact and/or collaborate across cultural boundaries – can be another instance of environment-influenced design. Theoretically, cross-cultural interactions can have at least two important kinds of impacts on design processes. First, cross-cultural interactions could facilitate (re)framing of the design problem (e.g., which constraints get more attention, what is the end goal of the design problem; Dorst, 2011; Dubberly & Evenson, 2008; Schön, 1983). Design reasoning involves not simply working out a model of *how* to solve a problem (how to achieve some desirable state), but working out what that desirable state is in the first place. This joint specification results in a *frame* – "general implication that by applying a certain working principle we will create a specific value" (Dorst, 2011, p. 524) – which can then be used to do problem solving or engineering specification of *what* artifacts or designs would achieve the desirable state. Cross-cultural interactions could facilitate the development of good frames in a number of ways, such as forcing disclosure of (often implicit) assumptions and constraints designers and/or users strive to achieve shared mental models. Second, cross-cultural interactions could facilitate broader exploration of the design space. One possible mechanism is by recombination of diverse knowledge elements, which has been both hypothesized (Paletz & Schunn, 2010) and shown to lead to more novel, and sometimes more creative (i.e., both more novel and more useful) design ideas (Chan & Schunn, 2015; Uzzi, Mukherjee, Stringer, & Jones, 2013).

Nevertheless, many open questions remain about the precise processes by which cross-cultural interactions exert influence on these key components of design thinking. For example, what are the relevant processes by which cross-cultural interactions actually influence design? It is possible cross-cultural interactions change which ideas are considered by adding new constraints. Alternatively, it could be that cross-cultural interactions produce more total ideas under consideration by having team members with greater variety in prior experiences or assumptions (i.e., less knowledge/belief redundancy within the team). Similarly, there are open questions about how these processes actually unfold over time. For example, if cross-cultural interaction adds new ideas, does it do so by adding new ideas continuously or do new ideas only emerge later in the process? For example, perhaps one cultural perspective is initially dominant and it takes time for the other culture's views to emerge. Alternatively, it may take time for new hybrid ideas to be developed.

Protocol analyses of actual design cognition (e.g., design conversations), including over long time scales (e.g., the lifespan of a project), can provide significant insight into these issues. A crucial component of such analyses is having a formal model of changes in design processes over time, for characterizing changes in understanding of the problem or emergence of crucial new ideas. However, traditional methods of protocol analysis for modeling change (e.g., linkography; Kan & Gero, 2006) can be extremely effort-intensive. It is not uncommon to spend weeks to months coding and analyzing just a few hours of design conversations. The high cost of traditional methods makes it prohibitively difficult to analyze larger corpora of design activities, limiting researchers' ability to observe processes operating at longer time scale than a few hours of design meetings (the DTRS11 dataset is one such example of such larger corpora).

Our working hypothesis is that computational linguistics can provide a cost-effective supplement to traditional methods of design protocol analysis, opening up new avenues of design research. Consider two motivating examples: one from our prior work, and one from a different set of design researchers. In a previous project, we showed through analysis of a large corpus (thousands of design ideas and inspiration sources) that the benefits of diverse inspiration sources depended on the degree of iteration that followed it (Chan & Schunn, 2015). This nuanced analysis was possible because we were able to use Latent Dirichlet Allocation (LDA; Blei, Ng, Jordan, & Lafferty, 2003) – a computational method for automatically discovering semantic "topics" in large unstructured collections of text – to analyze the semantic diversity of the inspiration sources over time. In this effort, traditional methods of coding diversity (e.g., with human pairwise similarity judgments) would have been prohibitively costly. Consider also Song, Dong, and Agogino (2003), who analyzed design activities of student design teams over an entire semester. They showed that cycles of divergence between team members in terms of their shared understanding of the design space, but constrained within an overall increasing trend of coherence, was a pattern of successful design. Their analysis was made possible by applying Latent Semantic Analysis (LSA; Landauer, Foltz, & Laham, 1998) – another computational linguistic technique for identifying semantic similarity between texts – to measure semantic coherence between design documents (e.g., reports, reflections, concept descriptions) produced by each team member.

In this short paper, we extend the application of computational linguistic methods from analysis of design documents (as in the previous two examples) to analysis of design *conversations*. Specifically, we explore whether topic modeling – a specific class of computational linguistic methods that automatically discover semantic structure in collections of text – could be applied to model change over time in design conversations (e.g., in revealing whether the number of topics being considered changed over time and which topics change in prominence). Our primary goal is a proof-of-concept: can a well-formed topic model adequately describe semantic activities and changes in design conversations, and generate interesting hypotheses to test? A secondary goal is to contribute a set of methodological reflections, including potential limitations of computational linguistics approaches to design conversations, and recommendations for how to apply those approaches.

2 METHODS

2.1 Data selection

Our motivating interest is analyzing changes in understanding of the problem and/or emergence of new ideas over time. In our initial read-through of the corpus, we noted that the sessions prior to the first co-creation session (CC1) were primarily focused on planning the co-creation session, rather than discussing substantive topics related to the problem. Therefore, we focused our analysis on all sessions following CC1 for which transcripts were available. These selected sessions are shown in Table 7.1. Our selection criteria yielded a dataset comprised of a total of 4,286 transcribed utterances for almost 9 hours of conversations across 13 sessions.

2.2 Latent Dirichlet allocation

Probabilistic topic modeling (Blei, 2012) is a major computational approach for understanding large collections of unstructured text. Topic models belong to the family of unsupervised machine learning methods (e.g., K-means clustering, and LSA) which do not require training sets (i.e., partial data that has already been hand-coded) to build a model. Within this family, topic models are distinct in that they are better suited for human understanding of not just the relationship between documents in a collection, but also human understanding of the "reasons" for the hypothesized relationships, e.g., the "meaning" of particular dimensions of variation (Stevens, Kegelmeyer, Andrzejewski, & Buttler, 2012). Topic models have this property largely because the

Table 7.1 Session ID and description of the sessions selected for analysis.

Session id	Session description
7	Debrief CC1
8	Sharing insights from CC1
9	Clustering insights from CC1
11	Linking insights from CC to project
12	Briefing stakeholders on CC1
14	Debrief after CC2
15	Sharing insights from CC2 1
16	Sharing insights from CC2 2
17	Sharing insights from CC2 3
18	Clustering insights from CC1 and CC2
19	Recap with consultants
201	Recap with stakeholders
202	Recap with stakeholders

algorithms underlying these models produce dimensions in terms of clusters of tightly co-occurring words. Thus, they have been used most prominently in applications where understanding of a corpus is a high priority goal, rather than just information retrieval performance like in a Google search. Example prior applications benefitting from human understanding of results include knowledge discovery and information retrieval in repositories of scientific papers (Griffiths & Steyvers, 2004), describing the structure and evolution of scientific fields (Blei & Lafferty, 2006a, 2006b), and discovering topical dynamics in social media use (Schwartz *et al.*, 2013).

In this paper, we use Latent Dirichlet Allocation (LDA), a foundational approach to topic modeling. LDA assumes that documents are composed of a mixture of latent "topics" (occurring with different "weights" in the mixture) that generate the words in the documents. Formally, LDA defines topics as (Dirichlet) probability distributions over words. As a fictitious example, a "genetics" topic can be thought of as a probability distribution over the words {phenotype, population, transcription, cameras, quarterbacks}[1], such that words closely related to the topic {phenotype, population, transcription} have a high probability in that topic, and words not closely related to the topic {cameras, quarterbacks} have a very low probability. Using Bayesian statistical learning algorithms, LDA infers the latent topical structure of the corpus from the co-occurrence patterns of words across documents. This topical structure includes: (1) the topics in the corpus, that is, the sets of probability distributions over words; and (2) the topic mixtures for each document, that is, a vector of weights for each of the corpus topics for that document.

2.3 Modeling approach

2.3.1 Preprocessing transcript text

To prepare the data for LDA, several preprocessing steps were necessary. Some were standard to LDA, and some were specific to the conversational nature of the data. For clarity, we describe each of the preprocessing steps in sequence, noting whether each step is standard practice or novel and specific to conversational data.

First, we removed all special characters and phrases from the transcription conventions (e.g., >< marks that were used to denote rapid speech, gesture descriptions). The details of this step are specific to the transcription conventions of the particular dataset being analyzed. Next, we removed conventional "stop-words" (i.e., function words that do not carry much semantic meaning by themselves, e.g., "the", "by", "this"). This step is standard practice in computational linguistics. We also created a filler-word stop-word list to filter out common conversational "tics" in this domain that do not reflect content (e.g., *"I think"*, *"basically"*, *"actually"*, *"ehm"*, *"mmm"*). This filler-word stop-list was iteratively created by: (1) running and inspecting the output of the models; and (2) analyzing the frequency distribution of words. Finally, we also removed words that were above a certain frequency threshold. The insight behind removing filler stop-words and highly frequent words is that topic models can be dominated by these words, thereby hiding the unique content in each separate topic. High frequency word dominance is found by inspecting the topics that result, which will be dominated by these more frequent words that occur across topics. What we want is for there to be frequent and infrequent words that are distributed unevenly across documents (defined below), so that we can actually distinguish between documents, and ultimately, between topics. This processing step therefore yields two free parameters: (1) **composition of the filler-word stopword list**; and (2) **frequency threshold for filtering out frequent words**. In a later section, we describe in more detail the process we used to select the final settings for these parameters.

2.3.2 Segmenting lines into documents

The final step before estimating the topic models is to define the "documents" that will serve as input to the model. LDA tries to "guess" the latent topics that most likely produce the words in a given document. It assumes that documents are composed of one or more topics, and this assumption is what enables LDA to learn useful topics. Because of this core assumption, LDA works best when documents are the "right" size: documents that are too small do not provide enough context for words to co-occur (e.g., short conversational turns), making it hard for the model to learn global co-occurrence patterns, while documents that are too large (e.g., whole sessions) may mix in too many topics, making it difficult for the model to distinguish between words. In many datasets, such as scientific papers or patent descriptions, the "document" is already fixed in the dataset structure. But in a transcript, the most commonly used segmentation of the corpus into documents (i.e., individual lines) may not be useful. Therefore, this processing step yielded our third free parameter: **number of transcript lines that compose a document**. We experimented with different segmentations, from treating a single line as a document, to treating multiple subsequent lines as documents. We initially expected grouping consecutive lines (i.e., documents of 2 lines each) to yield the best performance, reasoning that back-and-forth turns are likely to be more suitable semantic "chunks" than single lines, while larger chunks may combine semantic turns that should be separated. However, this assumption turned out to be incorrect (see below), possibly because arbitrary grouping of consecutive turns captures both topic maintenance and topic shift.

As a side note, because of the nature of conversational data, the combination of stopword removal and document segmentation already results in the removal of very short, no-content turns (e.g., *"um"*, *"yeah"*), so no additional work is required to deal with the problem of finding content in such short turns.

2.3.3 Building the topic model

The final step is to build the topic model. LDA requires the user to specify the number of topics to estimate upfront. Thus, this final step yields our fourth and final free parameter to select: **how many topics should the model learn?** There is no hard-and-fast rule for selecting the number of topics, except to try out different settings to see which setting provides useful views of the data. In general, however, fewer topics provide "coarser" or more general views of the topics, while more topics provide more fine-grained views of the topics.

We used the gensim library (Řehůřek & Sojka, 2010) to learn the topic model. This library uses the online LDA algorithm by Hoffman, Bach, and Blei (2010). Since the model iterates on simultaneous estimation of topic content and topic distribution across documents, we ran the algorithm for each parameter setting combination for 5,000 iterations.

2.3.4 Refining analysis parameters

Table 7.2 summarizes the four free parameters that needed to be refined through careful analysis. Although this refinement process is often hidden to readers of research applications of LDA, it is useful in when developing new applications (e.g., to this new context of understanding design representations) to show other researchers how parameters influence results.

The refinement had an intuitive, iterative stage and a systematic stage. In the first, iterative stage, we constructed the filler-word stop-words and phrases by running a model, checking the output topics for "contamination" (e.g., the same word/phrase littered across many topics), removing that word/phrase, re-running the model, and then checking if the topics improve (retaining the change if they did). The final set of filler-word stop-words are shown in the Appendix. In the second stage, we systematically searched in the parameter space, creating topic models for each combination of parameters. Table 7.2 shows the parameter space we searched. In total, we estimated and evaluated 252 models.

Our evaluation criterion was the real world coherence of the topics that resulted. We used C_V, an automated coherence measure developed by Röder, Both, and Hinneburg (2015). This measure formalizes the intuition that the set of words that compose a topic (i.e., the words that have high probability given a particular topic) should tend to co-occur more often in the "real world": for example, the topic {football, pitch, boots, jersey} is more coherent than the topic {football, kitchen, computer, dress} because the words in the first topic tend to co-occur more often than the words in the second topic. In the case of C_V, the "real world" is defined as co-occurrence patterns of words in the English Wikipedia. Although the mathematical details of the measure are out of the scope of this paper, the interested reader is referred to Röder *et al.* (2015) for more details. Early use of LDA used perplexity to evaluate topic models. Perplexity measures the model's ability to predict what words

Table 7.2 Parameters refined during model building.

Step	Parameter	Range considered	Final setting
Preprocessing transcript text	Filler stop-words/phrases: which?	n/a	See appendix
	Frequency filter for frequent words: how frequent?	[150, 200, 250, 300]	$<= 300$
Segmenting lines into documents	Number of lines in document: how many?	[1, 2, 3]	1
Building the LDA model	Number of topics: how many?	[10, 15, 20, 30, 40, 50, 60]	10

would show up in a new set of documents that the model has not yet seen. However, topic coherence is the new standard for evaluating topic models, since Chang and colleagues (2009) showed that the perplexity measure is very often uncorrelated with human-judged topic interpretability.

2.3.5 Applying the topic model to the data

Once we converged on a final model, we applied it to the data. We decided the most interesting first-pass analysis would be to analyze the topic composition of each session and how that changed over time. This step is called inference in topic modeling, where new documents are "projected" into an existing model space: basically, the model tries to infer which topics were most likely to have produced the words in that document, given the words in the document.

2.4 Final parameter settings

There was considerable variation in topic coherence across the parameter space. To explore how our parameters influenced the topic model quality, we fitted a regression model predicting coherence with the three parameters of number of topics, bin size, and frequency filter threshold. All three parameters were statistically significant predictors of coherence at $p < .001$. In general, increasing the number of topics resulted in a curvilinear (decelerating) decrease in topic coherence, controlling for bin size and frequency filter threshold. Similarly, increasing bin size tended to decrease topic coherence in a linear fashion, controlling for number of topics and frequency threshold. Finally, increasing the frequency threshold tended to *increase* topic coherence in a linear fashion, controlling for bin size and number of topics. The overall regression model was statistically significant, $F(4, 247) = 585.7$, $p < .001$, but also accounted for approximately 90% of the variance ($R^2 = 0.90$), suggesting that variations in coherence were well-accounted for by variations in our parameters.

Figure 7.1 visualizes how coherence changes with each of the 3 parameters. Descriptively, we can see a strong curvilinear trend for number of topics, and linear trends for both frequency threshold and bin size. We note that these results were surprising to us, underscoring the importance of systematically refining model parameters, since prior intuitions may lead the modeler astray.

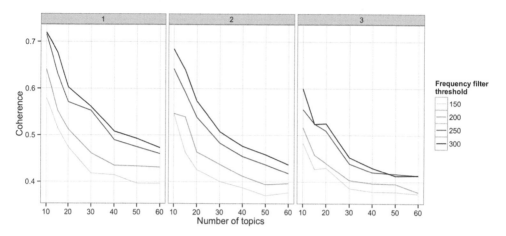

Figure 7.1 Variations in topic coherence by frequency filter threshold (line color), bin size of transcript lines to form a document (panel), and number of topics (x-axis).

We used topic coherence to select our final parameter settings (see Table 7.2), choosing the topic model with the highest coherence. This point in the parameter space defines a dataset with 1,743 documents. Note that we selected a frequency threshold of 250 since, at 10 topics and document size of 1 utterance, coherence was equivalent to a frequency threshold of 300, and, qualitatively, the topics seemed more "meaningful" (although equally coherent) at the threshold of 250.

Since we used single utterances as the bin for document definition, this means we had a nearly 60% reduction in the number of utterances from removing stop words and high frequency utterances (i.e., there were many utterances that our approach defines as content-free, at least in the sense of defining a topic – utterances of agreement or disagreement have meaning, but the underlying "topic" exists only by extension of the topic introduced in the prior context). We also note that this number of documents is relatively small by conventional topic modeling standards, where datasets tend to run in the tens of thousands to millions of documents, and it was therefore an open question whether this approach would successfully capture meaningful topics with a relatively small overall dataset.

2.5 Analyses

Our analysis focused on what we thought was the most natural first question: **are there changes in the topic composition of sessions over time?** For example, are there fluctuations in the diversity of topics under discussion across sessions? To explore this, we calculated Shannon entropy of the topic-weight distribution for each session. Shannon entropy is commonly used to measure diversity of information by quantifying the inequality in distribution over discrete categories (Harrison & Klein, 2007). Topic entropy for each session is given by the following equation:

$$-\sum_{k=1}^{K} p_k \times \ln(p_k) \tag{7.1}$$

where p_k is the proportion of probability mass in the session that is allocated to topic k. Entropy increases with the *evenness* of spread of probability mass across the topics. For example, consider three hypothetical sessions: (a) designers talk mostly about one topic; (b) designers talk mostly about three topics; and (c) designers talk relatively evenly about 10 different topics. Entropy would be *lowest* in Session a (focus only on one topic), and highest in Session c (discussing many different topics), with the three-topic Session b being somewhere in between. As a second example, consider two hypothetical sessions: (d) designers talk evenly about two topics; and (e) designers talk 90% about one topic and only 10% about a second topic. Here entropy is higher in Session d than Session e, matching our intuitions of entropy capturing the evenness of topic probability mass.

3   RESULTS

3.1   Preliminaries

For visualization purposes, we represent topics with word clouds of the top 30 words associated with that topic with high probability (see Figure 7.2). The size of the word is proportionate to its probability given the topic: in other words, the words that are more closely associated with a given topic appear larger in that topic's word cloud. Word color is not meaningful, but rather is just used to make nearby words of equal size more distinguishable.

The topic model is able to produce a number of sensible topics that map intuitively to the conversations in the transcripts. For example, Topic 9 seems to map to a prominent "tea analogy" developed by the designers to explain levels of expertise and identification with a particular class of purchases. Topic 1 seems to map to a notion of value associated with purchases, another prominent theme in the design meetings. Topic 6 describes a notion of achieving status by looking different. Topic 3 describes the importance of branding. However, there are also topics that are difficult to comprehend, such as Topics 2, 4, 7, and 10. From our informal analysis, approximately half of the ten topics appear to be easily understood. As our subsequent analysis demonstrates, this mixed quality in topics does not prevent the model from yielding potentially informative and interesting hypotheses.

3.2   Changes in topic entropy over time

We now turn to our central analysis of the paper: are there observable shifts in topic patterns over time in the data? Figure 7.3 shows topic entropy for each session, in chronological order.

Descriptively, we see a marked upward shift in entropy between Session 12 and Session 14, which then remains high for the remaining sessions. That is, the designers shifted to discussing a broader range of topics more evenly; intuitively, we may think of this as an expansion of the idea space being considered. This is an interesting pattern, because that shift coincides with the second co-creation session (CC2), suggesting that topic entropy increases following co-creation. Quantitatively, a t-test comparing mean entropy of pre-CC2 ($M = 0.905, SE = 0.010$) and post-CC2 sessions ($M = 0.971, SE = 0.004$) is statistically significant, $t(5.7) = 6.2, p = .001$.

Figure 7.2 Top 30 most probable words for each of the 10 topics. Words are sized according to their probability given the topic.

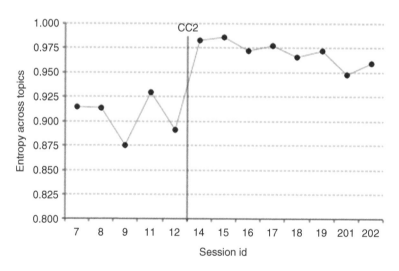

Figure 7.3 Changes in topic entropy over sessions.

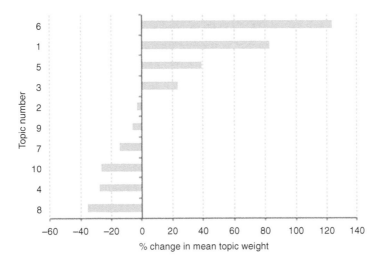

Figure 7.4 Percent change in mean topic weight before and after CC2.

3.3 Emergence of new topics

To explore this pattern further, we compared the mean topic weights for each topic across pre- and post-CC2 sessions: were there particular new topics that emerged in prominence? Figure 7.4 shows the percent change in mean topic weights from pre-to post-CC2. This analysis highlights two topics that approximately double in prominence, and three topics that slightly decline in prominence by 20–30%, after CC2.

An in-depth analysis of all aspects of this pattern is out of the scope of this paper, which simply seeks to refine the approach and establish its basic utility. Thus, we explore just one topic in depth (Topic 1) that nearly doubles in prominence, as a proof of concept of the hypothesis-generating potential of topic model-based analysis. Topic 1 is about the importance of values in purchasing decisions. This topic in particular is intriguing, because it was highlighted as an interesting major opportunity area for the company, as the following snippet illustrates:

"...what we come to realize all the time is that it's so cool that we are within the accessory bubble, because accessory makes people able to sample the THE COMPANY values, without making the full commitment of having a car for example. Eh and that's kind of cool, like buying a car is a huge deal, and in China for sure because it's not only about the practicalities it's also about "what does this communicate about me?" So: through: good and relevant accessories we can start building the brand without- we can get people on board (.) kind of if we can get people from here to here faster. And we used- and this is kind of a template we have used, and probably we will take this farth'- further, which have different-like, this is the insight, this is the story and quote, value position, it's a little bit like this, but like more in a package. And one of the things that we talked about here is like a scenario, eh, for doing this, could for example be okay, so there's a

Mercedes guy, or like this is an example for a conquest, a conquest within acces-
sory. There's a Mercedes guy, who has a friend over in his car, and he has both a
THE COMPANY or a THE COMPANY approved eh air purifier. And the friend
kind of asks "wow, I thought you were like a Mercedes guy, you always had a
Mercedes?" Eh and then the guy says "yes, I am a Mercedes guy, but above that
I am a: person that is committed to: eh to really get, you know the right quality:
and this air purifier is made in eh- by like sustainable material and it is really really
good". So they are then using THE COMPANY as a way to brand themself. And
it's even a little controversial. And you- we might not have made this air purifier,
but we have approved it, or we have done this kind of cross over, where we have
associated us with another brand and we are using their ways- eh their kind of
trust, in a way, and they are using our trust. And that would be amazing, like,
having a THE COMPANY accessory in another brand as, not necessarily because
it is the best: air purifier, but because it's- it is really good and it states something
about them. (Ewan, in v20: Recap with stakeholders, line 29).

This notion of accessory as an ambassador of the company's values, which forms an important component of the target users' purchasing decisions, revealed an opportunity to build trust with the users via communicating particular values of the company in the form of the accessory. It would be interesting to further trace precisely where in the meeting this idea arose, and to what extent cross-cultural interactions were responsible for synthesizing this insight from CC2.

4   DISCUSSION

Our primary goal in this short paper was to provide a proof-of-concept of a computational linguistic approach to modeling change in design processes. We show that topic modeling can yield quick overviews of potential semantic patterns in the data, and uncover quantitative trends that may be fruitful to investigate further. We identified two such patterns: (1) a sudden and sustained increase in diversity of topics being discussed from before to after CC2, which seemed partially attributable to (2) significant rise in prominence of two specific (and interpretable) topics after CC2. These patterns raise interesting follow-on areas for investigation in this dataset about the efficacy of CC2 for uncovering new insights (finding 1 suggests that CC2 was successful), and provide pointers for precisely *what* those new insights might be (e.g., an understanding of the connection between values and accessory purchases).

Our secondary goal was to illustrate the benefits and limitations of the method, and offer some reflections on how to best use the method. In terms of benefits, our proof-of-concept illustrates two unique potential benefits of a topic modeling approach. First, analysis of changes in topic proportions and diversity over time can **help researchers quickly identify which parts of datasets (especially large ones) to sample for qualitative screening.** From start to finish, a first-pass run through the topic modeling approach takes about 20–30 minutes. The first-pass topic model can provide a quick sense of which parts of a dataset might contain interesting shifts in *semantic* patterns, which are subtle and more difficult to quickly identify manually compared to shifts in the amount of activity in a given moment (e.g., number of words spoken, or turns taken). Second, topic modeling can **provide formal/quantitative patterns that complement qualitative observations** (e.g., shifts in topic diversity). These quantitative patterns can yield hypotheses that are testable across datasets where designers are working on different problems.

Beyond these benefits, however, we argue that having access to a computational semantic model of the dataset can support many other rich research questions in their own right. For example, the semantic model can quantify the degree of coherence between team members' mental models (e.g., see Menning *et al.*, 2017), allowing researchers to characterize how levels of coherence might vary as a function of the degree of cultural diversity present in a given conversation, and use those variations to predict the emergence of significant shifts in the team's understanding of the design problem. Analyzing topic proportions over time could also allow researchers to trace the emergence of new concepts, similar to how Howard Gruber documented the evolution of Darwin's insights over time from analysis of his diaries (Gruber & Wallace, 2001), or how Kevin Dunbar documented the emergence of key insights in microbiology labs (Dunbar, 1997). Computational approaches could make it feasible to trace these insights over many different design contexts, improving our ability to make principled generalizations about how design processes actually unfold. We believe there is much potential in adopting a computational approach to asking these questions, especially as large records of design activity and conversations become available online, such as design conversations in open-source software communities, or on online design platforms like OpenIDEO (Chan, Dow, & Schunn, 2015; Chan & Schunn, 2015) and Climate CoLab. Because computational approaches greatly decrease the degree of effort required to analyze a single dataset, it may become possible for researchers to bridge the gap between detailed protocol analyses and quantitative comparisons across different design conditions to discover relationships between design strategies and outcomes, and general patterns of design strategies across many different contexts.

In terms of reflections, our analysis demonstrates how systematic refinement of model parameters can be a crucial complement to adjustments based on intuitive understanding of the data. This is usually hidden from view in topic modeling applications, but is likely to be especially important for applying topic modeling to design conversations. We also found that even the best topic model we could learn from the data was not perfect in terms of producing topics that were easily comprehensible and interesting. We suspect that this ceiling in performance may be partially due to the relatively small (effective) size of the corpus (after removing non-content words); that is, we expect greater analytic precision with increased dataset size. The need for a sufficient number of documents may be especially acute for modeling conversational text, since it tends to be much noisier than the written texts typically used for topic modeling. Nevertheless, the mixed quality of topics did not hinder identification of interesting high-level quantitative patterns in the data.

In conclusion, we believe this short paper demonstrates that computational linguistic methods indeed can provide a valuable cost-effective method for modeling change over time, and hope that methods like topic modeling find wider use in the design research literature, especially as the community takes on larger corpora in consortium approaches like DTRS or through automated transcription methods.

NOTE

1 We use this notation {word1, word2, word3} to describe probability distributions over words.

REFERENCES

Blei, D. M. (2012). Probabilistic topic models. *Communications of the ACM, 55*(4), 77–84. https://doi.org/10.1145/2133806.2133826

Blei, D. M., & Lafferty, J. (2006a). Correlated topic models. *Neural Information Processing Systems.*

Blei, D. M., & Lafferty, J. D. (2006b). Dynamic topic models. *Proceedings of the 23rd international conference on Machine learning.* ACM. 113–120.

Blei, D. M., Ng, A. Y., Jordan, M. I., & Lafferty, J. (2003). Latent Dirichlet Allocation. *Journal of Machine Learning Research.* 993–1022.

Chan, J., Dow, S. P., & Schunn, C. D. (2015). Do The Best Design Ideas (Really) Come From Conceptually Distant Sources Of Inspiration? *Design Studies.* [Online] 36, 31–58. Available from: https://doi.org/10.1016/j.destud.2014.08.001

Chan, J., & Schunn, C. D. (2015). The importance of iteration in creative conceptual combination. *Cognition* [Online] 145, 104–115. Available from: https://doi.org/10.1016/j.cognition. 2015.08.008

Chang, J., Gerrish, S., Wang, C., Boyd-graber, J. L., & Blei, D. M. (2009). Reading tea leaves: How humans interpret topic models. *Advances in neural information processing systems.* 288–296.

Dorst, K. (2011). The core of "design thinking" and its application. *Design Studies* [Online] 32(6). 521–532. Available from: https://doi.org/10.1016/j.destud.2011.07.006

Dubberly, H., & Evenson, S. (2008). On Modeling: The Analysis-synthesis Bridge Model. *Interactions.* [Online] 15(2), 57–61. Available from: https://doi.org/10.1145/1340961. 1340976

Dunbar, K. N. (1997). How scientists think: On-line creativity and conceptual change in science. In T. B. Ward, S. M. Smith, & J. Vaid (Eds.) *Creative thought: An investigation of conceptual structures and processes.* Washington D.C. 461–493.

Fischer, G. (2005). Distances and diversity: sources for social creativity. *Proceedings of the 5th conference on Creativity & cognition* ACM. 128–136.

Griffiths, T. L., & Steyvers, M. (2004). Finding scientific topics. *Proc Natl Acad Sci U S A.* [Online] 101 Suppl 1, 5228–35. Available from: https://doi.org/10.1073/pnas.0307752101

Gruber, H. E., & Wallace, D. B. (2001). Creative work: The case of Charles Darwin. *American Psychologist, 56*(4), 346–349.

Harrison, D. A., & Klein, K. J. (2007). What's the difference? Diversity constructs as separation, variety, or disparity in organizations. *Academy of Management Review, 32*(4), 1199–1228.

Hoffman, M., Bach, F. R., & Blei, D. M. (2010). Online Learning for Latent Dirichlet Allocation. In J. D. Lafferty, C. K. I. Williams, J. Shawe-Taylor, R. S. Zemel, & A. Culotta (eds.) *Advances in Neural Information Processing Systems* [Online] 36, 856-864. Curran Associates, Inc. Accessed from: http://papers.nips.cc/paper/3902-online-learning-for-latent-dirichlet-allocation.pdf

Kan, J. W. T., & Gero, J. S. (2006). Acquiring Information from Linkography in Protocol Studies of Designing. *Design Studies.* 1–26.

Landauer, T. K., Foltz, P. W., & Laham, D. (1998). An introduction to latent semantic analysis. *Discourse Processes, 25*(2), 259–284.

Menning, A., Grasnick, B. M., Ewald, B., Dobrigkeit, F., Schuessler, M. & Nicolai, C. (2017) Combining Computational and Human Analysis to Study Low Coherence in Design Conversations. In: Christensen, B. T., Ball, L. J. & Halskov, K. (eds.) *Analysing Design Thinking: Studies of Cross-Cultural Co-Creation.* Leiden: CRC Press/Taylor & Francis.

Paletz, S. B. F., & Schunn, C. D. (2010). A Social-Cognitive Framework of Multidisciplinary Team Innovation. *Topics in Cognitive Science*, 2(1), 73–95.

Řehůřek, R., & Sojka, P. (2010). Software Framework for Topic Modelling with Large Corpora. *Proceedings of the LREC 2010 Workshop on New Challenges for NLP Frameworks.* 45–50. Valletta, Malta: ELRA.

Röder, M., Both, A., & Hinneburg, A. (2015). Exploring the Space of Topic Coherence Measures. *Proceedings of the Eighth ACM International Conference on Web Search and Data Mining.* [Online] 399–408. New York, NY, USA: ACM. Available from: https://doi.org/10.1145/2684822.2685324

Schön, D. A. (1983). *The Reflective Practitioner. How Professionals think in Action.* New York, NY: Basic Books.

Schwartz, A. H., Eichstaedt, J. C., Kern, M. L., Dziurzynski, L., Ramones, S. M., Agrawal, M., Shah, A., Kosinski, M., Stillweil, D., Seligman, M. E. P. & Ungar, L. H. (2013). Personality, Gender, and Age in the Language of Social Media: The Open-Vocabulary Approach. *PLOS ONE.* [Online] 8(9), e73791. Available from: https://doi.org/10.1371/journal.pone.0073791

Song, S., Dong, A., & Agogino, A. M. (2003). Time variation of design "story telling" in engineering design teams. *Proceedings of the International Conference on Engineering Design '03*, Vol. 3.

Stevens, K., Kegelmeyer, P., Andrzejewski, D., & Buttler, D. (2012). Exploring Topic Coherence over Many Models and Many Topics. *Proceedings of the 2012 Joint Conference on Empirical Methods in Natural Language Processing and Computational Natural Language Learning.* [Online] 952–961. Stroudsburg, PA, USA: Association for Computational Linguistics. Available from http://dl.acm.org/citation.cfm?id=2390948.2391052

Uzzi, B., Mukherjee, S., Stringer, M., & Jones, B. (2013). Atypical Combinations and Scientific Impact. *Science.* [Online] 342(6157), 468–472. Available from https://doi.org/10.1126/science.1240474

APPENDIX: FINAL SET OF FILLER AND HIGH-FREQUENCY STOP-WORDS AND PHRASES

actually	for example	literally
ahm	gonna	maybe
also	guess	mean
amanda	heidi	mhm
basically	I guess	mmh
believe	I mean	mmm
brian	I see	mostly
definitely	I think	much
didn't	i'll	necessarily
doesn't	indicate	okay
eh	kind	possibly
ehm	kinda	probably
example	know	really
feel	like	right

say

shouldn't

something

sorry

stuff

susan

THE COMPANY

THE COMPANY-PRODUCT

they're

thing

think

totally

wanna

wasn't

we

we guess

we mean

we see

we think

we're

won't

wouldn't

yeah

yen

yes

you're

Part II

Designing Across Cultures

Chapter 8

Disciplina: A Missing Link for Cross Disciplinary Integration

Frido Smulders & David Dunne

ABSTRACT

Innovation processes are the work of many disciplines that each contribute and add value to the object of innovation. Research addressing the handover between sequential dependent users within the innovation process is scarce. Based on the DTRS11 dataset this paper investigates such transfer and handover of results, consisting of user insights, concepts and branding strategies as developed by a team of designers to sequentially dependent intermediate users from another department within the same organization. We seek to understand whether user experience-methodologies are applicable to intermediate users. A lens consisting of theoretical notions from design and organization sciences was used to analyse the dataset. The lens consisted of two optics: an integration optic to investigate how the handover package became an integrated whole; the second related to the creation of a transition model that supports the boundary spanning interactions including the design of the handover package. Like the notion of persona representing the end user, we identify a specific element of the transition model as a 'disciplina': a representative understanding of the receiving actors from the other department. The lens allows us to investigate how the design team actually apply these disciplina-insights for purposes of the content and package integration as well as for the design of their interactions with intermediate users, including the design of the hand-over meeting. We find that the designers did not see the actors from the receiving department as users that can be subjected to user experience-methodology and therefore made only limited use of their own user-centred repertoire to build such a disciplina. In failing to build a rich and fully representative disciplina, the designers did not optimize the handover package for the intermediate users.

I INTRODUCTION

User-centred design involves the end user in the development process. This paper introduces the results of a research project that considers the in-process user, that is, users of the partly-finished design located downstream of the actual designers, as another valid and important stakeholder. These users, in this paper referred to as intermediate users, need to complement and add to the work being handed over to them by the designers and are essential to the progress of the innovation process. They need to insert further life into the designers' conceptual ideas and are sequentially related to the upstream designers (Thompson, 1967).

Typical design literature addressing downstream actors is found under various DfX-labels such as 'assembly', 'manufacturing', 'logistics', etc. Most of the literature in this area focuses on lowering costs in manufacturing and assembly and preventing costly iterations in late stages of the innovation process. There is, however, very limited literature that addresses the human and social constraints of downstream actors that will form the subject of the present paper.

This paper deals with a sequential situation where designers prepare a design 'package' that needs to be further developed by downstream departments. Our goal is to analyse the design process with a view to describing how integration happens, focusing particularly on how the design 'content' – the results of the design department's work – is integrated with the design 'package', elements that make the design usable to intermediate users. These latter elements aim to set in motion a next set of actors that will need to develop and deliver their contribution to the innovation process. We are particularly interested in understanding to what extent and in what form these design actors see the downstream actors as users of the package delivered to them, and, if so, how UX methodology is applied.

The literature on Design-Manufacturing Integration (DMI) aims to bring manufacturing knowledge into the NPD process by way of integration mechanisms, like cross-functional teams, co-location, individual integrators, group-based design reviews and formalization, in order to achieve a better design and a smooth transition to manufacturing (e.g., Rusinko, 1999; Vandervelde & Van Dierdonck, 2003). These integration mechanisms approximate participatory design approaches in organizing for the participation in development activities of downstream actors, be it in-house users and/or end users.

From an organizational science perspective, Van de Ven (1986) describes the bringing of new ideas into 'good currency' within the organization as one of the central problems of innovation. A more recent paper by Baer (2012) argues that creativity and implementation are two different things, underscoring Van de Ven's observation. Baer shows that 'creativity and idea implementation are two different activities within the innovation process' and that the nature of the idea, its degree of creativity, is not necessarily positively related to implementation (Baer, 2012, p. 1114). Implementation instrumentality and network abilities (strong ties) possessed by the creative actors are believed to have positive influence on implementation of creative ideas. The author concludes that to understand implementation the influence of the personal and relational contingencies should be taken into consideration. This is exactly what we are interested in. We want to identify the mechanism by which design actors may influence communication and acceptance of the new idea by intermediate actors in downstream departments.

We make use of a dataset that reports on the design activities performed by a professional design team working on a specific design task for a company ('the Company') within the automotive industry. We integrate design methodology, organization science and communications in describing the design and interaction activities related to the handover package.

The paper is structured as follows. First, we describe the dataset and our methodology. We then describe the development of our lens, consisting of variables identified within the design and organizational sciences. We then describe the application of the lens and its subsequent results. We introduce additional literature to discuss these findings and draw conclusions regarding the perspective on user-centred design where the user is located within the actual innovation process. This paper ends by indicating some challenging directions for further research.

2 DATASET

This paper is part of the 11th Design Thinking Research Symposium (DTRS11) organized by the Copenhagen Business School. The organizers of DTRS11 have collected a dataset of a design team working for and within the automobile industry, consisting of video, audio, documents, pictures and additional information.

The data follow a design team in Phase II of a design project comprising team activities related to planning, workshop design, workshop execution and results analysis. The design is commissioned by the Accessory Department of a large international automaker. The actual ideation workshops are co-creation sessions in which future users from the Asian context participate.

3 METHODOLOGY

The empirical study concentrates on the background interview with the project leader, supporting documentation and transcripts of meetings related to the two end-user co-creation sessions that are part of the DTRS11 dataset.

The background interview discusses the project in its contextual setting in terms of its position within the overall innovation process, and the relationships between the design team and other actors within the company. Supporting documents include an overview of the innovation process and the responsibilities of the key participants. The meeting transcripts either discuss the design of the co-creation sessions or elaborate on the results from these sessions.

For our study, an important element of the design process is the 'innstyrings'-stream, the infusion of the results from the first design process into the implementation 'machine' of the organization, which will be referred to in this paper as the 'package'-design process. This process aims at smooth hand-over of the design 'content' by the design department to the actors belonging to the Accessory Department, here referred to as intermediate users.

Our purpose is to describe how the design team's proposals and perspective are integrated with the rest of the organization. We pursue this in two ways: first, by gaining insight into how design actors handle this situation with different, sometimes conflicting, requirements attached to the end users and the intermediate users; and second, by understanding what actually happens during these meetings related to the stitching of these stakeholder requirements into a whole, as well as how plans are made for handover of the total package to the intermediate actors.

Specifically, the following research questions guide our investigation:

Whether integration occurred:

1   How did the team get from the initial brief to the solution that integrated the results of the 'content-design' and the 'package-design' process? What was the evidence of integration in the process?

By this first research question we want to find out if content and package are actually integrated or remained distinct from each other and the organization's innovation process.

How integration occurred:

2   How were the needs of different 'users' identified? What were differences in methodology for these purposes? Did one dominate over the other? If so, why and how? Did they seem to empathize with one more than the other? How was a communication strategy for the intermediate users developed?
3   What, if any, difficulties were caused by the changing emphasis on the intermediate user versus the end user?

With these second and third research questions, we hope to shed light on the processes and activities that aim at integration regarding package design and handover.

The unit of analysis is the utterances of the actors in the project-related meetings and in interviews. Our analysis was conducted in the spirit of Grounded Theory (Glaser & Strauss 1967), building theory iteratively and progressively in several stages. Because of limited time and singularity/one-sidedness of the dataset (one project, one set of actors), we decided to start with an initial lens based on UX methodology as we expected the designers to apply this at least for the end users and possibly also for the intermediate users. Subsequently, we alternated between data and literature while sharpening our lens until we reached a form of saturation:

1   Initial examination of the dataset to frame the empirical situation using input from UX-literature and to increase our theoretical sensitivity;
2   Informed by our first observations, literature review integrating theory from the design and organization sciences to develop a second version of our lens, or theoretical perspective, on the problem related to intermediate users;
3   Analysis of the DTRS11 dataset with this lens in mind, but open to the possible occurrence of events that fall outside it;
4   Revisiting the literature and theory in light of findings from the dataset, and again updating of the lens;
5   Establishing validity of the initial findings through member checking in the form of a follow up interview with the project leader.

Each cycle of data analysis was independently performed and coded by two researchers from differing backgrounds (industrial design and business) to ensure reliability. The respective cycles of sharpening the lens and maturing theoretical perspective was done jointly.

4   FROM LENS OF ANALYSIS TO THEORETICAL PERSPECTIVE

As mentioned above, we have been developing our theoretical perspective in a grounded manner by alternating between two sets of activities: analysing the empirical data and sharpening our theoretical lens through literature reviews. We thus developed the lens iteratively until we reached theoretical saturation and the lens became the resulting theoretical perspective on design for intermediate users as presented in this paper. We provide below an overview of the building blocks that informed this perspective.

4.1 Developing the lens from a UCD perspective

The initial theoretical sensitivity for analysing the data, that is the first lens, was informed by User-Centred Design (UCD) methodology. Based on an overview by Gulliksen *et al.* (2003) we identified the following elements of UCD:

1 User focus, which includes taking into account the context of use and the goals of the users and development of personas;
2 Active user involvement during development;
3 Continuous involvement of usability experts;
4 Prototyping early and frequently with involvement of users;
5 Evaluating use in context;
6 Simple design representations for interactions.

We applied these during our initial investigations of the data and looked for utterances that addressed our focus on intermediate users.

If design actors adopt a user-centred approach towards the intermediate users, then they need to build up a representative user model. This is similar to the notion of a persona, but not identical. A persona has more archetypical characteristics representing *unknown* future users, whereas intermediate are *known individuals* working in a known department downstream of the designers, which opens up the possibility for including much more situational knowledge in the user model. We therefore introduce the notion of a 'disciplina' as representative for the intermediate users in the context of their own departmental habitat. The disciplina will be further elaborated below.

Gulliksen *et al.*'s second and third UCD elements relate to the involvement of users and usability experts. From the data it is clear that two of the intermediate users were actively involved in some parts of the design process. In this case, it is not clear who the usability experts are or may be. Because of the proximity of some actual intermediate users it is questionable if these are necessary at all. However, the purpose of a package design that accompanies the design content is to add usability elements directed at facilitating the use of that content by the intermediate users.

The fourth and fifth element focus on prototyping and actual use evaluation in context respectively. In case of intermediate users that need to work with unfinished designs, the prototyping and evaluation elements are of different nature as we will see later. This is similarly the case for element six on simple and understandable design representations.

4.2 Input from design and organizational science literature

As we alternated between data and theory it became clear that the situation of the intermediate users is different in other respects. In the case of consumers as users there is a user need that operates as a market pull mechanism. Intermediate users, on the other hand, may not have an actual need waiting to be fulfilled by the design. Of course, the company might be in need of new products and services, but intermediate users may not represent this need. Hence the motivation of intermediate users to accept the design and bring it into good currency within their department could be an important factor in determining whether they are adopted.

In line with this observation, Smulders (2007) suggested that innovating actors in a sequential setting should have a 'transition (mental) model' as the flipside of the design assignment. The transition model 'contains knowledge, abilities and attitudes that are prerequisite to boundary spanning team activities' (Smulders, 2007, p. 51) and thus should include the disciplina.

There seem to be three dimensions of the transition model, one that has a generic character and contains knowledge and past experiences about change, interventions and innovation processes. The second, the disciplina, being the specific and to some extent detailed knowledge about the intermediate users in their own departmental context, being their thought (Dougherty, 1992) and object worlds (Bucciarelli, 2002). The third relates to shared experience between the design team and the intermediate users. Each of these dimensions is discussed below.

The generic dimension: As previously noted, the past experiences, generic knowledge and abilities inform the plan of interactions with the intermediate users and addresses issues such as how to involve intermediate users; when to involve them; how to keep them informed on progress, etc. An additional concern is the design of the handover situation in which the design results are delivered to the intermediate users.

Following the literature on generic knowledge on change and interventions (Smulders *et al.*, 2003), we also include the use of intermediary objects (Boujut & Blanco, 2003) and boundary objects in generic knowledge. The purpose of these objects is to connect the two different social worlds (Star & Griesemer, 1989; Carlile, 2002; Smulders, 2006).

Other elements of generic knowledge include understanding the constituents of the departmental absorptive capacity (e.g., Cohen & Levinthal, 1990), that is, the capability of managing new knowledge to create (new) value for the organization (Zahra & George, 2002). Also relevant is generic knowledge on design and innovation processes and finally capabilities to assess these processes in the particular circumstances of the intermediate users' departmental context.

The specific dimension (the disciplina): The disciplina should include the specifics of intermediate users and their departmental object and thought worlds. Just as personas include end users' behaviour, goals and motives, the disciplina should inform the designers about intermediate users' likely response to the content and package, and suggest strategies for package design and communication.

Based on the disciplina, the designers should be able to create sufficient empathic understanding of the actual intermediate users as well as on the disciplinary processes these actors are involved in (Smulders *et al.*, 2003). Assessing the quality (complexity, capabilities, etc.) of these processes and the constraints (time, budget, workload, etc.) will provide them with a feeling of what we term their 'Obtainable Absorptive Capacity', which sits between the potential and realized absorptive capacity introduced by Zahra and George (2002).

Shared experience: A final element of the transition model is related to shared experience and strong ties as enablers for acceptance. The intermediate users interpret meaning in the context of their own experience, some of which is shared with the

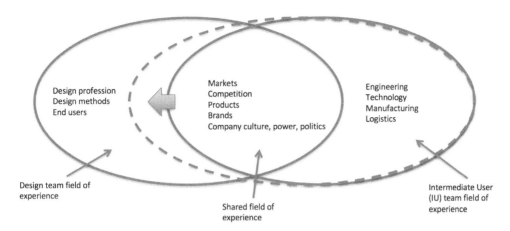

Design profession
Design methods
End users

Markets
Competition
Products
Brands
Company culture, power, politics

Engineering
Technology
Manufacturing
Logistics

Design team field of
experience

Shared field of
experience

Intermediate User
(IU) team field of
experience

Figure 8.1   Shared experiences as influencer of receiving stakeholder's belief in the design.

design team, while some is independent. We assume that the intermediate team's field of experience comprises its knowledge of engineering, technology, manufacturing and logistics. Its shared field of experience with the design team includes political and organizational factors, such as knowledge of the organization's products, its brands, the corporate culture and power structure; and economic and industrial factors such as market dynamics, competition, and so on (Figure 8.1).

Both teams are interested in how the design meets end user needs while meeting organizational goals. The intermediate user team's understanding of the economic and industrial context will influence its belief that the design ('content') uniquely meets the needs of end users, while its understanding of political and organizational issues will influence its belief that it fits the company's goals and capabilities.

By expanding the area of shared experience (the broken line in Figure 8.1), the design team can improve its understanding of the intermediate user's frame of reference and adapt both the design and the method, or channel, of interacting with them.

4.3 The occurrence of integration

Lastly, we return to the first research question, how integration of design content and deliverable package occurred. From a design sciences perspective, Dunne and Martin (2006) define integrative thinking as 'creative resolution of the tension' between competing points of view. According to Lawson and Dorst (2009), design involves balancing and integrating stakeholder needs; based on both perspectives we expect the designers to arrive at an integrated whole that takes into account the end user needs as well as the intermediate users' capabilities. Alavi and Tiwana (2002) suggest that such a synthesis is not only a re-arrangement of parts, but also the creation of new meaning. Based on this we examined the DTRS11 dataset to see if creative resolution of these competing views has occurred and the content and package became one integrated whole with new meaning.

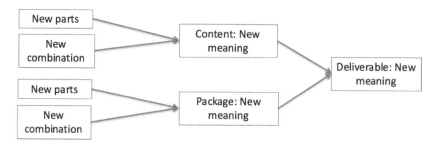

Figure 8.2 The first optic of analysing lens: The occurrence of integration.

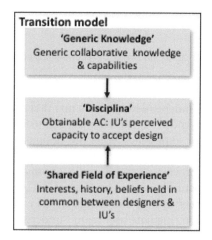

Figure 8.3 Second optic of our analysing lens: Transition model including the disciplina.

4.4 Summarizing our theoretical perspective

In summary, the perspective we have been able to construct by our research activities contains two optics: integration and transition model.

In Figure 8.2 we show how we have modeled the question of integration. For design integration to occur, content, package and deliverable should combine to provide new meaning for the receiver, either as the result of new parts or recombination of existing parts.

The second optic focuses on the transition model that could be held by the designers. The transition model has three building blocks: generic knowledge and capabilities on collaborative innovation, situational knowledge resting in the disciplina and shared field of experience. Generic knowledge and shared field of experience each contribute to the formation of the disciplina.

From this second optic it is possible to create a model of successful acceptance by the intermediate users. According to our analysis of the data and elaborations of the theoretical perspective, there are three constituting factors influencing this success. The first is related to the designers' ability to apply their generic knowledge and insights in

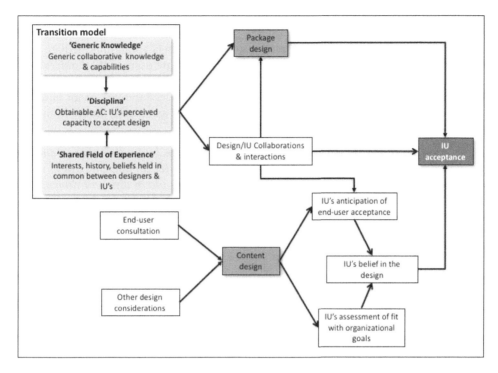

Figure 8.4 Theoretical model depicting the constituting factors of IU acceptance.

the disciplina for having fruitful and collaborative interactions with the intermediate users. The second relates to how the disciplina informed the design of the package encapsulating the actual design content. The third factor is attached to this design content and the intermediate users' anticipation of acceptance by end users and its assessment of organizational fit, which increases the likelihood that they will believe in the design. Our model is informed by the theoretical discussion above and displayed in Figure 8.4.

5 FINDINGS

In our analysis, we reviewed the transcripts of all design team meetings, with a particular focus on situations in which: (i) intermediate users were mentioned by the design team; (ii) one or more representatives of the intermediate users (IU's) were present and involved in the discussion; and (iii) the design team appeared to be implicitly considering intermediate users' interests or preferences in their absence.

We will discuss our findings related to the ingredients of our theoretical perspective and its two optics, followed by comments on the strategy adopted by the design team for developing the package.

5.1   Development of a transition model

5.1.1   Generic collaborative knowledge and capabilities

Fit with Process. As the project progresses, the team's beliefs about intermediate users become evident. At one point, Ewan compares intermediate users to 'flatlanders' who can only see in one dimension, referring to a 19th-century book by that title (Abbott, 1884). They are considered to have 'just a simplified understanding', or an outdated one. The perspective of the rest of the company is seen as severely limited, and the team's job to convince it to see things as it sees them.

> *[Intermediate Users] have zero idea, they don't even know that there's another world. It's kind of like, that book: called 'Flatland'? It's an amazing book, wrote in the 1800s, it's a journey, where you will meet people in different dimensions, so first you encounter one image. It's just one guy, a dot, and he doesn't [know anything about anything].*(v18,122)

Intervention. Ewan's remark also provides insight into the disciplina he has in mind. He realises that his results cannot just be handed over by sending a report to the intermediate users, but that they must be 'educated':

> *[... that's also part of the education] ... kind of the Flatland-education, they need to be taken into this three dimensional world. And this is how they think: they need to realize it themselves, have that epiphany: 'wow, maybe there are other areas that we didn't even see'.* (v18, 297)

In the end, this understanding of the absorptive qualities of intermediate users led to a week-long handover process.

5.1.2   Disciplina

Disciplina and Fit with Process. At other moments, the way of working by the intermediate user is referred to as 'machine-like', implicitly assuming that these intermediate users prefer automated and routinized ways of working. Ewan suggests in the final interview and after the results have been put in the 'freezer' that:

> *[...] the gap between our team here in [Scandinavian city] and the team in [Scandinavian city] is maybe too big, was too wide, for [...] – being able to bridge in the best possible way [...]* (v22, 046)

Ewan and his team do not seem to have a great deal of trust in the Potential or Obtainable Absorptive Capacity of intermediate users, since it is frequently mentioned that they have high workloads and work constantly on routine activities, '*... if you come there they are not there ...*' and '*The Monday after they were tasked with other stuff*' (v22, 031).

Disciplina and Absorptive Capacity. Tiffany and Hans are from the Accessories Department and are treated as representatives of the rest of the company. In the briefing on the first co-creation workshop they say little but appear to express approval for the design team's work to date. During the debrief and insight sharing meetings, they provide context to the design team about history and some of the implications of their ideas. In one exchange the design team is interested in whether the Company has provided customized accessories in the past, and Tiffany explains that customers may choose customized accessories as they choose colour and finish. Hans, however, comments that this may complicate production, indicating some of the constraints and limitations of the downstream processes.

Intermediate User Beliefs. For the design team, Tiffany and Hans need to become ambassadors that enhance and facilitate the belief by other stakeholders in the company in these new ideas and broader direction:

> *... we also need to give them some tools to actually say that to the organization, 'no, but this is actually new', because it's not just to convince them. They need to feel it so that they can convince the rest of the organization.* (v18, 327)

Intervention Design. Following the debriefs on the second co-creation workshop, the design team has two meetings with the consultants to share insights and receive advice on how to interact with intermediate users. The consultants propose an 'innovation matrix' as an organizing framework, to position the present offerings and visualize the gap as a *'conversation starter'*. The design team and consultants agree that they want Tiffany and Hans to 'take ownership' and represent their ideas to the rest of the company. Part of this means that they have to answer the questions asked by their peers from the Accessory Department instead of the designers providing the answers. Such intervention strategies could be based on generic knowledge the consultants have on change and interventions, and as such would be part of the transition model of the designers.

Moving away from a strict focus on the physical product, we consider the 'values' behind the product/service mix. During the 'recap' meeting, Tiffany and Hans indicate some of the challenges in this, both from an internal perspective and from a dealer perspective:

EWAN: *What do you guys think are the biggest challenges for pushing something like this? For this is a way to try to move away, or at least add much more intangible value to our products.*

TIFFANY: *I think the biggest challenge is that China-org is really into lowering the price, period.* (v20, 063)

Tiffany and Hans eventually agree to 'explain it ... in a structured way'. The meeting concludes with an appeal from Ewan to continue two-way communication.

5.1.3 Shared field of experience

Several interactions between the design team and intermediate users indicate that there is considerable shared experience related to the brand and its dominant culture on safety.

While the Phase II design task is to build the Company's brand by engaging Active Urbanites, the design team seeks to change the Company more fundamentally. In the Background Interview, team leader Ewan expresses this as follows:

> *We have a higher goal in what we are doing, it's not ONLY about using the best methods for the projects that we are doing individually, but it's also to showcase a whole mindset or way of thinking to the organization.* (v01, 97)_

This aspiration towards a 'higher' goal is repeated at several points during team meetings by various members of the design team. However, since it is not spontaneously expressed by the two members of the stakeholder team, Tiffany and Hans, shared space between the two teams appears limited by a fundamental difference in goals.

While there is an apparently genuine desire to build mutual understanding with the stakeholder team, the design team approaches this aspect of the process with a degree of suspicion, even cynicism. Ewan recognizes the challenge as a political one, since

> *most of the people in our organization ... are engineers and don't necessarily subscribe [to our way of thinking].* (v01, 014)

Yet he also recognizes the need to engage them, since he frequently mentions that the intermediate users needed to be educated in the ways of working and new insights from the design team.

5.1.4   Package development

Throughout the process of content development and consultation, the team is anticipating how its ideas will resonate in the rest of the company. There are some instances of self-censoring of ideas as a result. The team concerns itself with developing a 'story' but it is not always clear whether the story is being developed for end users or for internal stakeholders. With the goal of imparting its vision, the team uses several means of communication: boundary objects, analogies and persuasion.

Disciplina and Intervention. As noted earlier, the team's goal is more ambitious than designing accessories: it wants to change the company's approach to be more like its own. Initially, Tiffany and Hans do not appear convinced:

EWAN: *I think we need to be allowed to dream like this. And I actually think it's possible.*
TIFFANY: *You can, if you have-*
HANS: *But, still if you can spread that, we have something that is so unique and really makes my life so much better, of course that will spread to your friends and so on, but-* (v20, 039)

It appears that at this point the intervention aimed at guiding Tiffany and Hans into their role as ambassadors has not yet succeeded.

Intermediate User Beliefs and Boundary Objects. In the follow-up interview, Ewan is asked whether the team has come up with a prototype; he replies that the sketches and mock-ups the team has developed can be considered prototypes. The innovation matrix is the critical piece given to Tiffany and Hans as the vehicle for disseminating the team's way of thinking to the rest of the company. These items can all be considered boundary objects (e.g., Carlile, 2002) that act as communication vehicles with other stakeholders.

Intermediate User Beliefs and Boundary Spanning. The team uses analogies both to further its own thinking and to facilitate communication, which seem to work as boundary spanning tools. At one point, Apple is referred to for comparison. In discussion with the consultants, and later with Tiffany and Hans, tea drinking is used to illustrate how users can progress from simple to more sophisticated levels of product engagement.

> *'I drink tea because I'm thirsty' … It doesn't really matter which tea. Then … ah, this tea … it's like strong, it wakes me up. In the evenings I like camomile tea. So they're starting to become a little aware of what's around … and then … communicate outwards: 'My friends know more about me and who I am because they have seen my specialized tea collection; I have some tea equipment'; and this is a role model, so 'people come to me 'cause I know so much about tea, and how to use it: and the subtle differences between this tea and this tea', and they can even hold a little tea ceremony in their house.* (v20, 027)

Such analogical stories in coherence with the stories supporting the handover package are used frequently. In their efforts to play a 'change agent' the designers look continuously for the right story that could serve as package design at the handover.

Intermediate User Beliefs and Persuasion. In the 'recap' meeting with Tiffany and Hans, Ewan and the design team become more didactic. At this stage, the design team knows what it wants and Tiffany and Hans' role is to accept its vision and convince the rest of the company.

6 DISCUSSION

In the follow-up interview, team leader Ewan expresses a sense of failure to engage the intermediate users Tiffany and Hans. His acknowledged reasons for this failure include a lack of time and priority, and a lack of social connection.

In terms of integration (Research Question 1), one could question whether the needs of all stakeholders were adequately taken into account to arrive at an integrated whole that contained new meaning for the intermediate users. The potential new meaning of the results, presently in the 'freezer', is awaiting realization. On the other hand, the sessions seemed to have resulted in new meaning for the designers, albeit in their own three-dimensional world. One could argue that insufficient integration of package and content occurred, since the usability for intermediate users was not sufficiently developed and integrated with the design content stemming from the end-user sessions.

Looking at our model (Research Questions 2 and 3), the design team did not explicitly develop a 'disciplina'. However, the team did aim to expand its shared field of experience with the intermediate user team by having two representatives from the latter team being partly present during the generative sessions. Such collaborative actions, like the week planned for the handover, typically are based on generic knowledge and understanding of collaborative innovation. On the other hand, it might be questioned whether this was sufficient to create adequate social connection.

This is not the whole story: apart from a few exchanges, little effort was made to understand Tiffany and Hans' field of experience to inform the disciplina. Through these limited exchanges, the design team learned about the importance of dealers and the likelihood that senior management would accept its ideas. However, this was a limited effort and there was no explicit attempt to understand the disciplina in a complete sense: the goals, constraints or mindset of the intermediate users, or the broader context within which they were operating. It seemed that the design team did not develop an adequate disciplina to develop a package that increases the usability of the content for the intermediate users.

Such efforts as were made to understand intermediate users' mindset and to inform a possible disciplina appeared to be hampered by simplistic assumptions. The 'flat-landers' analogy stands out as an example of such an assumption: in this way of thinking, the rest of the organization is considered to have a 'limited' perspective, as opposed to merely a 'different' one, from the design team, and is written off as being myopic.

The design team appeared to view communication as a one-way affair. Tiffany and Hans were brought into the process after the completion of the first workshop, and their only input was to deliberations on the second workshop, by which time the design team was relatively committed to a direction. Following this meeting the design team developed a plan for imparting its vision to Tiffany and Hans, through the boundary object of the innovation matrix and the package containing a mapped-out pilot strategy. Such efforts point to the generic knowledge on collaborative innovation as possessed by Ewan and his design team.

Tiffany and Hans were considered a 'bridge' to the rest of the organization, whose role was to pass the design team's vision on, not to help shape it. They ultimately agreed to comply, but it is no surprise that, as Ewan acknowledged in the follow-up interview, they felt excluded.

The inclusion of Tiffany and Hans in the process might not, however, have made a great deal of difference on its own. As two individuals in the Accessories Department, they could not be expected to have a complete view of the Company or its dealers. Broader consultation would most likely have been necessary to come up with a credible and adequate disciplina. The extensive effort made, in Phases I and II, to understand end users stands in contrast to the lack of effort to understand the mental models of the intermediate users.

One caveat to this interpretation should be stated: it is possible that both the design team and Tiffany and Hans had a good understanding of internal issues, obviating the need for an explicit disciplina. This argument, however, is unconvincing, as is demonstrated by the simplistic 'flatlanders' analogy.

7 CONCLUSIONS

We have seen in this paper that the intermediate user of unfinished (design) results deserves much more attention in academic research. In our first attempt to make sense of how design actors, or in general, innovation actors, take this group of intermediate users into account as serious stakeholders with their own needs, limitations and bound-aries we have come to formulate the new notion of 'disciplina', an analogy with the conceptual understanding of the persona as frequently applied by user-centred design-ers. The disciplina, as part of a transition (mental) model, should play a similar role in terms of providing a representative and adequate understanding of the sequentially dependent actors further downstream the internal value chain. Such understanding then should inform the development of the package in such a way that the actual (unfinished) design content forms an integrated whole and becomes usable for the intermediate users. Unlike the persona, the disciplina could (and should) include the specifics of intermediate users in coherence with their departmental context. Routines, processes, disciplinary logics, and the like, could and should all be included.

One might have expected that user-centred designers, by default and perhaps even unconsciously, apply their UX-methodologies to intermediate users. It became clear, however, that the designers in our dataset made only limited use of their own user-centred repertoire to build such a disciplina. It seems that they saw the intermediate users not as 'true' users, and therefore not really invested in the development of such disciplina.

On the other hand, the team included, if selectively, two of the intermediate users in their activities and consciously thought about the 'innstyrings'-stream, which clearly indicates their awareness of the challenges related to the transfer of new ideas to other departments (Van de Ven, 1986) and the availability of generic knowledge as one of the three dimensions of the transition model.

The design team's failure to build a rich and fully representative disciplina turned the designers into 'flatlanders' in their own right, incapable of recognizing the three-dimensional world of the receiving department. Seen from the perspective of joint responsibility of designers and intermediate users, the package design may not be the design team's task alone, but could be considered a shared task between the design team and the intermediate user team. In a truly collaborative situation it might be expected that both collaborative teams feel responsible for the quality of results, and feel the joint obligation to bring these into good currency within the company. This could have been caused by discrepancies between the designers' goal and that of the Accessory Department and maybe other stakeholders. Whereas the designers seemed to have aimed at showcasing 'a hole mindset or way of thinking to the organization' (v01, 097) and not just handing over the results of a design task.

In conclusion, an innovating actor such as a design team needs to develop and update an explicit disciplina, as part of a transition mental model, in order to provide an integrated deliverable that falls within the obtainable absorptive capacity of the intermediate users and is viable to the organization.

REFERENCES

Abbott, E.A. (1992[1884]). *Flatland: A Romance in Many Dimensions*. New York: Dover Thrift Edition.

Alavi, M., & Tiwana, A. (2002). Knowledge Integration in Virtual Teams: The Potential Role of KMS. *Journal of the American Society for Information Science and Technology*, 53(12), 1029–1037.

Baer, M. (2012). Putting creativity to work: the implementation of creative ideas in organizations. *Academy of Management Journal*, 55(5), 1102–1119.

Boujut, J.-F., & Blanco, E. (2003). Intermediary objects as means to foster co-operation in engineering design. *Computer Supported Cooperative Work*, 12, 205–219.

Bucciarelli, L.L. (2002). Between thought and object in engineering design. *Design Studies*, 23(3), 219–231.

Carlile, P.R. (2002). A pragmatic view of knowledge and boundaries: boundary objects in new product development. *Organization Science*, 13(4), 442–455.

Clark, K.B., Chew, W.B., & Fujimoto, T. (1992). Manufacturing for design: beyond the production/R&D dichotomy. In: G.I. Susman (ed.), *Integrating design and manufacturing for competitive advantage* New York: Oxford University Press. pp. 178–204.

Cohen, W.M., & Levinthal, D.A. (1990). Absorptive Capacity: A New Perspective on Learning and Innovation. *Administrative Science Quarterly, 35*(1), 128–152.

De Caluwé, L., & Vermaak, H. (2003) *Learning to Change: A Guide for Organizational Change Agents.* Thousand Oaks, CA: Sage.

Dougherty, D. (1992). Interpretative barriers to successful product innovation in large firms. *Organization Science, 3*(2), 192–202.

Dunne, D., & Martin, R. (2006). Design Thinking and How It Will Change Management Education: An Interview and Discussion. *Academy of Management Learning & Education, 5*(4), 512–523.

Glaser, B., & Strauss, A. (1967) *The Discovery of Grounded Theory – Strategies for Qualitative Research.* London: Weiderfeld and Nicolson.

Gulliksen, J., Göransson, B., Boivie, I., Blomkvist, S., Persson, J., & Cajander, Å. (2003). Key principles for user-centred systems design. *Behaviour & Information Technology, 22*(6), 397–409.

Lawson, B., & Dorst, K. (2003). *Design Expertise.* New York, NY: Taylor & Francis.

Rusinko, C.A. (1999). Exploring the use of design-manufacturing integration (DMI) to facilitate product development: A test of some practices. *IEEE Transactions on Engineering Management, 46*(1), 56–71.

Smulders, F.E. (2006). *Get synchronized! Bridging the Gap between Design & Volume Production.* (PhD thesis). Delft: Delft University of Technology.

Smulders, F.E. (2007). Team mental models: Means and Ends. *CoDesign, 3*(1), 51–58.

Smulders, F.E., De Caluwé, L., & Van Nieuwenhuizen, O. (2003). Last stage of Product Development: Interventions in existing processes. *Creativity and Innovation Management, 12*(2), 109–120.

Star, S.L., & Griesemer, J.R. (1989). Institutional Ecology, "Translations" and Boundary Objects: Amateurs and Professionals in Berkeley's Museum of Vertebrate Zoology, 1907–39." In: *Social Studies of Science* 19: 387–420.

Thompson, J.D. (1967). *Organizations in action.* New York: McGraw-Hill.

Vandervelde, A., & Van Dierdonck, R. (2003). Managing the design-manufacturing interface. *International Journal of Operations & Production Management, 23*(11), 1326–1348.

Van de Ven, A.H. (1986). Central problems in the management of innovation. *Management Science, 32*(5), 590–607.

Zahra, S.A., & George, G. (2002). Absorptive capacity: a review, reconceptualization, and extension. *Academy of Management Review, 27*(2), 185–203.

Chapter 9

How Cultural Knowledge Shapes Design Thinking – a Situation Specific Analysis of Availability, Accessibility and Applicability of Cultural Knowledge in Inductive, Deductive and Abductive Reasoning in two Design Debriefing Sessions

Torkil Clemmensen, Apara Ranjan & Mads Bødker

ABSTRACT

This paper challenges the 'core design thinking and its application' as outlined by Dorst (2011) and uses a dynamic constructivist notion of cultural-cognitive performance to analyse aspects of a design thinking process. Based on a qualitative analysis of some of the events in the DTRS11 dataset and using the theory of Dorst on design thinking as well as Hong and Mallorie's socio-cognitive theory of cultural knowledge networks, the paper shows how it is possible and useful to analyse design thinking from a cultural perspective. The results show that cultural knowledge, either as shared knowledge by the cross-cultural team or group specific knowledge, influences the Dorst design thinking equations across all the 16 episodes analysed in the DTRS11 dataset. Furthermore, most of the design discussions were approached by the designers as problem situations and were approached in a backwards manner, where the value to create was known in advance; however, the designers were using available cultural knowledge to figure out the unknown what (products/services) and how (working principles of why something would work or not work). In conclusion, the paper demonstrates a novel approach to understand how design thinking can be efficiently understood as a culturally situated practice.

1 INTRODUCTION

In this paper we explore assumptions about 'core design thinking and its application' as outlined by Dorst (2011). We do this with a focus on how cultural contexts shape design thinking (Clemmensen, 2009; Hong & Mallorie, 2004). Our contribution is towards a dynamic and situation specific analysis and a model of how cultural contexts shape the unfolding of design thinking. This is exemplified with data from the DTRS11 dataset.

Dorst's core design thinking 'equations' help formulating a clear and easy to follow analytical scheme, and they provide an overview of how thought processes can lead to innovation and 'outside the box' thinking. However, as Dorst himself acknowledges, his approach is problematic, as design thinking cherishes multiple perspectives and rich articulations over simplification. Like Kimbell (2011), we argue that a significant flaw in much thinking about design thinking is the oversimplification of the creative thought processes to be unaffected by cultural contexts. Kimbell (2011) on "Rethinking Design Thinking" emphasizes cultural components and external factors, which are hard to simplify without losing their meaning and therefore credibility.

Our preliminary findings on how cultural context shapes design thinking may hold significant promise to inform the DTRS11 community. An appropriate way of analysing this is by going beyond the lens of our initial predefined codebook based on Dorst's core design thinking equations, and focus instead on how culture play parts in the present data. This include analysis of cultural stereotypes of Chinese users, use, and products, and of the interaction between Western designers and Eastern facilitators. Compared to other relevant approaches, for example, *the team-centric view of culture and creativity approach* that assumes that design conversations are influenced by the dynamics and cultural combination within the group (Paletz *et al.*, 2017), or *the culturally situated difference approach* that aims to identify relatively stable local embodied practices that expresses group values (Dhadphale *et al.*, 2017), our focus is instead strictly on the dynamics within and across situations.

Our choice of theory for analysing how the cultural contexts shape design thinking is the dynamic and situational social-cognitive theory of culture (Hong & Mallorie, 2004), which has been adapted to IT design (Clemmensen, 2009; Pineda, 2014). This theory suggests that people can have more than one (and sometimes conflicting) loose networks of domain-specific cognitive structures (implicit theories, beliefs) at a time. Moreover, which of these networks is activated depends on what situational constraints are salient. Thus, the situation will determine which cultural cognitive system is *accessible, available and applicable* in the given situation. A Chinese-European will tend to think like a Chinese person when in a 'Chinese situation', and conversely, think like a European when primed by European icons, text, etc. A dynamic and situational social cognitive theory of culture is highly relevant in an analysis of the DTRS11 data, as it may be used to understand, for example, the group of Western designers' available cultural knowledge (including their stereotypes of customers/users and Eastern facilitators), the priming of their thinking by the design artefacts and materials, and what is socially appropriate to say and do in the different design situations.

In the rest of the paper, we present the theory, method, findings and discussion of our analysis of two of the design videos from the DTRS11 dataset. We use qualitative analysis informed by our reading of Dorst (2011) for core design thinking, and Clemmensen (2009) and Hong and Mallorie, (2004) for exploring the dynamic and situation specific application of knowledge in design by various participants. We analyse examples of core design thinking in relation to specific cultural aspects of the situation and the designers' cultural background. We propose an initial framework for exploring how culture shapes design thinking in dynamic and situation specific ways.

2 THEORY

2.1 Dorst's core design thinking

Dorst (2011) suggests that in order to describe and understand design thinking in its many variations, it may be useful to attempt a high level simple or 'sparse' description of design thinking. Though rich descriptions are important, as design unfolds in a dense context, we may learn something from thinking about basic reasoning patterns that humans use in problem solving in design, or what Dorst calls core design reasoning patterns. In particular, we may learn something from comparing different settings of the knowns and unknowns in design patterns and the ways designers reason about these. Dorst refers to these patterns as design thinking 'equations'.

Dorst argues that the most important or 'core' design reasoning pattern is *abduction*. This is what signifies design thinking or productive thinking. It is different from classic problem solving in that the outcome is value (e.g., 'good health') rather than results (e.g., the outcome of a calculation) of a mainly deductive or inductive analytical process. Dorst (2011) suggests two equations for abduction. The first kind of abduction is closed problem solving, *abduction 1*, where the designers do not know what thing or design artefact or service they are discussing, but they do know the working principle that will help achieve the aspired value. According to Dorst, *abduction 1* is a common way of working for professional designers.

The second kind of abduction is open problem solving, *abduction 2*, where the designer neither knows the thing to be designed nor the working principle, but only the aspired end value. Dorst argues that this is what designers do when they do conceptual design, for example, when there is no familiar working principle or design method that can guide the design. In such situations, the designer has two unknowns in the equation, which is a different situation from the everyday routine closed problem solving in abduction 1. In this second type of abduction, 'framing' can be used. Framing involves applying analogies from other design thinking scenarios with similar aspired end value to the problem at hand, which helps to identify the working principle and thing to design in current design scenario. McDonnell (2017) describes how *"within a reflective practice paradigm for designing, situations are 'framed' and the things that will be attended to within that framing are 'named'"*.

Dorst emphasizes that even though design thinking may be described as abductive reasoning, it is a mix of different ways of thinking, as designers use a lot of inductive and deductive reasoning to come up with ideas and rigorously test and evaluate them to assess whether a proposed design solution will work.

2.2 The dynamic constructivist theory of culture

The dynamic constructivist theory of culture is different from an essentialist view of culture. It does not see culture as a holistic entity, but rather as a set of loose and developing knowledge networks or domain-specific cognitive structures including theories and beliefs. Hong and Mallorie (2004) argue that domains and situations interact with more essentialist aspects of culture. People may hold more than one cultural meaning system, even if such systems may contain conflicting cultural knowledge, for example, conflicting cultural models of how to use design products (Clemmensen, 2009). In any given situation, the individual uses the knowledge that is most accessible. Hence, the accessibility, availability and applicability of particular cultural models of technology use will determine the experience of using the product. The concepts of accessibility, availability, and applicability are taken from the theory of knowledge activation, which underscores that cultural knowledge must be activated by something in order to be used. Hong, Benet-Martinez, Chiu, and Morris (2003) argue that acculturation can make specific cultural knowledge systems become available. Furthermore, prolonged exposure to a culture may increase the chronic accessibility of the shared knowledge in that culture (Hong *et al.*, 2003). The accessibility of each cultural knowledge system may thus vary as a function of situation. Applicability, sometimes referred to as appropriateness, has to do with the feasibility of acting out or focusing on specific culture-related behaviours. In a given social situation, this depends on whom you are with, what they know, and what norms for behaviour are present.

3   METHOD

We analysed two videos from the DTRS11 dataset (Christensen & Abildgaard, 2017), using qualitative analysis informed by our reading of Dorst (2011) for core design thinking, and Clemmensen (2009) and (Hong & Mallorie, 2004) for exploring the dynamic and situation specific application of knowledge in design by various participants. We could not analyse the videos of actual co-creation workshops that took place in China with Chinese users because of our lack of understanding of the Chinese language. Hence, we investigated the debriefing sessions that took place immediately after each of the co-creation workshops, depicted in Figure 9.1. These were V07: Debrief of co-creation workshop day 1 (CC1), and V14: Debrief of co-creation workshop day 2 (CC2).

The debriefing phases in V07 and V14 are critical to the whole design process because this is where the designers empathize with the actual Chinese user. Based on the insights generated in the collaboration with the users, the designer defines the problem statement, and a guiding statement that focuses on insights and needs of a particular user, or composite character developed in interaction with the user. In this

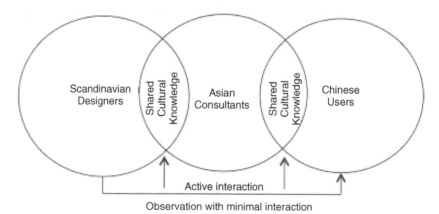

Figure 9.1   Shared cultural knowledge in the co-creation and debriefing workshops.

phase all the varied findings about individual users are put together and evaluated in light of the design themes (i.e., health, environment, self-reliance etc.) defined in Phase one of the design process. Teixera *et al.* (2017) provide further insights into the notions of timeframes for the often lengthy and fragmented design process.

Figure 9.1 demonstrates the interaction among the three main groups that participated in the co-creation workshops, Scandinavian designers, Asian consultants (who worked as facilitators) and the Chinese users. The overlapping areas in the Venn diagram depict the cross-cultural interactions between the designers and the consultants, and the Chinese users and the consultants. When the Asian consultants were moderating the co-creation workshops and actively interacting with the Chinese users, the Scandinavian designers were only observers, due to their lack of understanding of the Chinese language. After each workshop there was a debriefing session, where the Asian consultants debriefed the Scandinavian designers about the workshops in English.

3.1 Participants

Two cultural groups were present in the videos we analysed: Asian consultants and Scandinavian design team members. Out of the five Scandinavian design team members, two were external stakeholders who were not as actively involved in the above videos as the three core designers. The core design team consisted of three designers who from years of collaboration on design projects knew each other well.

The three Asian consultants had expertise in Chinese markets, and became part of the project during the field trip to China. They participated in the meetings on equal terms as the design team members. They were familiar with design thinking approaches and they aided in the translation of Chinese to English, as well the translation of cultural diversities and traditions. Two of the consultants were also facilitators for the two co-creation workshops.

3.2 Material and procedure

The analysis focused on the discussions among the designers, stakeholders and consultants, in the debriefing meetings CC1 and CC2, held after each of the co-creation workshops.

In these two videos the Asian consultants, Scandinavian designers and Scandinavian stakeholders shared observations and notes from the co-creation workshops, by doing activities such as brainstorming, problem solving, re-interpreting the personas of the Chinese users, and evaluating user responses to the questions based on overall project themes and concepts. Given the collaborative nature of these meetings, the designers and consultants were constantly talking, thereby providing a rich, ongoing, external record of their thinking and reasoning. We selected 16 episodes each of 2–10 minutes for our analysis, Table 9.1. The criterion for selecting the episodes was to have the two cultural groups actively participating in the discussion.

3.3 Coding/analysis

All the selected episodes were coded for Dorst equations. The coding was based on Dorst's model of design thinking: WHAT (thing) + HOW (working principles) leads to RESULT (observed). Table 9.2 shows the equations proposed by Dorst (2011), which we used in our analysis of the design thinking process.

Table 9.1 The 16 episodes analysed.

Videos	Duration	Content	The 16 episodes
V7: CC1 debrief co-creation	18 minutes	Sharing observations, translating and explaining the different post-it clusters that were put on one of the walls by workshop participants and facilitators. Explaining some of the participants' characteristics and statements, trying to draw insights about how the participants conceive of leisure time, family relations, and general ideas about the theme of "Health" and "Good life".	1. V7, 009–015 2. V7, 021–035 3. V7, 038–051 4. V7, 055–080 5. V7, 096–119 6. V7, 140–158
V14: CC2 debrief co-creation	78 minutes	Sharing and discussing observations and notes with each other from the workshop, and slowly beginning to connect some of the insights to the overall design project themes and concepts.	7. V14, 056–071 8. V14, 072–085 9. V14, 093–112 10. V14, 128–140 11. V14, 143–151 12. V14, 232–246 13. V14, 273–304 14. V14, 314–328 15. V14, 420–440 16. V14, 683–692

Table 9.2 Dorst (2011) equations.

Type of design reasoning	Dorst equation
Induction	WHAT + ??? leads to RESULTS
Deduction	WHAT + HOW leads to ???
Abduction 1 (Closed problem solving)	??? + HOW leads to VALUE
Abduction 2 (Open problem solving)	??? + ??? leads to VALUE
Framing	WHAT + HOW leads to VALUE FRAME

Table 9.3 The triple A of a dynamic, situation specific concept of culture. Adapted from Hong and Mallorie, 2004; Clemmensen, 2009.

Triple A	Definition
Availability	Existence of cultural knowledge structures (i.e., including stereotypes of customers/users and Eastern facilitators)
Accessibility	Getting primed to access the cultural knowledge structures
Applicability/Appropriateness	Appropriateness and/or feasibility of culture-related behaviours in situational context (i.e., in context of setting design goals)

In a second part of the analysis we applied a framework from the dynamic constructivist approach to the study of culture (Hong & Mallorie, 2004). For each identified design thinking method (i.e., induction, deduction, abduction) we further analysed whether the participants were having cultural knowledge available, whether they were making it accessible, and if it was appropriate in present cultural and design context. We call this the triple A: Availability, Accessibility, and Applicability, Table 9.3.

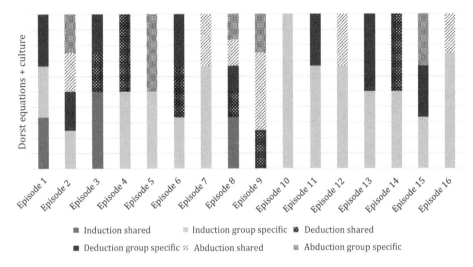

Figure 9.2 Distribution of group specific and shared cultural knowledge across 16 episodes of design thinking.

4 RESULTS

4.1 Overall results

Overall, our qualitative analysis of the 16 episodes indicated that abduction characterized the design thinking process. A combination of inductive and deductive thought processes was incorporated as part of the abductive thinking process. Abductive thinking was heavily shaped by cultural knowledge. In Figure 9.2, we show how culture, either as cultural knowledge shared by the whole cross-cultural team or group specific cultural knowledge, shaped the Dorst problem solving methods in design thinking across the 16 episodes.

As illustrated by Figure 9.2, cultural differences appeared and disappeared during the debriefing sessions, depending on the accessibility, availability and applicability of cultural knowledge of the team members in the concrete episodes. In all the 16 episodes there was interaction between the two cultural groups (Asian and Scandinavian). Below we present results from the in-depth analysis of two of these episodes (no 1 and 9) to illustrate how cultural knowledge shapes induction, deduction and abduction.

4.2 How cultural knowledge shapes deduction and induction

To illustrate how deductive and inductive design thinking is shaped by cultural knowledge, we analysed episode 1 from CC1 (V7, 09–15), see appendix 7.1 for the full transcript.

4.2.1 How deduction is shaped by available cultural knowledge

In episode 1, the Asian consultant W uses his *available* cultural knowledge to deduce that the Chinese user's behavioural data suggests that he lives a healthy life in the traditional Chinese way. This kind of thinking process is deductive as W is drawing conclusion about the user's personality based on the marketing data and the behavioural characteristics of a typical Chinese user:

> *So, eh, I think there was one guy who, the younger guy, who I think leads a slightly more disciplined life, I mean like, he's not married, he's not, you know, has his own family and whatever. He talks about things like sleeping early, going to bed by ten, waking up really early by six, you know, because your body starts to detox at eleven a clock. [V07, 09]*

The Dorst equation for deduction is WHAT + HOW = ??? (Dorst, 2011). To fill in the equation, the WHAT and the HOW is the data about the user (WHAT) and the Chinese cultural stereotype (HOW), and these together leads W to formulate hitherto unknown (part '???' of the Dorst equation), the design team's aspired value of what is a healthy user. Dorst (2011) points out that deductive reasoning is a gold standard of reasoning for scientific discovery, and that even in design, rigorous deductive reasoning is necessary to inform justification of the value to be created by the designer.

However, the content of the Dorst equation for deduction in this example is shaped by what cultural knowledge is available to those doing the deduction. Hong, Benet-Martinez, Chiu, and Morris (2003) argue that what makes cultural knowledge available is acculturation, and we know that W has been hired as a cultural expert on China, so he is well acculturated and has this knowledge available.

4.2.2   How deduction is shaped by accessible cultural knowledge

W may very easily come to think about Chinese medicine in the situation, because prolonged exposure to a culture, that is, acculturation, increases the chronic *accessibility* of the shared knowledge in the culture (Hong *et al.*, 2003). In addition, the available knowledge becomes *accessible* to W because he has been primed by the team's ongoing discussion about the design theme 'health' and Chinese users.

W makes this cultural knowledge *accessible* to his Asian and Scandinavian team members by repeating the deduction that the user sleeps early and gets up early (WHAT), which demonstrates the aspect of traditional Chinese medicine in the user's life (HOW), as the user is letting his body detox at night while sleeping, the magic, (the unknown ??? aspired health behaviour).

> *That's actually a little bit of eh: (.) traditional Chinese medicine, that's part of the concept. Your body starts to work itself eh actually: from that time which is eleven at night, your body should start resting before that, so you need to go to bed before that, so that, you know, it can work its magic. [V7, 011]*

W is a bicultural individual who has been exposed to two cultural meaning systems, Asian and Western. Such individuals may provide particularly clear demonstrations of the interaction between availability and accessibility (Hong & Mallorie, 2004). The accessibility of each knowledge system appears to vary as a function of situation. In the above example, W has the cultural knowledge available (Chinese culture) and he makes it accessible because the situation (design discussion about concept of health for

Chinese users) primes him to discuss the user behaviour and its meaning in the Chinese cultural context. Hence, he makes the purely Chinese culture specific knowledge about 'Chinese medicine' accessible to the team members. The deduction process will be meaningless without the cultural knowledge being available and accessible.

4.2.3 How deduction is shaped by appropriate cultural knowledge

To make the cultural knowledge *appropriate* to the design context, W makes shared cultural knowledge *available* and *accessible* by using deduction to explain to the Scandinavian team members that if one was in a Western context one would sleep at twelve or one, but within a Chinese context going to bed before the magic hour of detox is essential (WHAT), and, since the user is traditional (and follows the Chinese medicine concept) (HOW), the concept of being disciplined appears to be an *appropriate* way to think about a health and relaxation in life (the unknown ???).

> So people like us who sleep at twelve. Sleep at one, you have really bypassed that magic hour of where we can actually get that. So ... [V7, 013]

Hong and Mallorie (2004) explain that applicability or *appropriateness* refers to the feasibility of culture-related behaviours in context; the expression of appropriate cultural knowledge in a situation is influenced by the cultural knowledge systems held by partners in the social interaction, the nature of the interpersonal situation, and general behavioural applicability, and more. W is in a situation where most of the others in the design team do not have any Chinese cultural knowledge systems available so he discusses relaxation/health in a Western context to further explain what the concept of 'Chinese medicine' means and signifies in life of a traditional Chinese person. The design team is cross-cultural, so W uses shared cultural knowledge about young people in the West staying up late and AM mentions the trend of partying as an example of relaxation in Western context to make an analogy to 'Chinese medicine': had the user been a young person like the design team members, the user would have been partying after midnight. Furthermore, W is hired as a consultant, so he needs to be polite, and cannot really say more about the partying behaviour, so he ends without finishing the sentence, leaving further interpretation open. In this way, W is repeating his deduction, but from a Western perspective, and by letting cultural knowledge shape the content, W makes the deduction *appropriate* for design thinking in the situation.

4.2.4 How induction is shaped by cultural knowledge

This is followed by an induction process, in which the Asian consultant AM supports W in making the Chinese medicine concept *appropriate* to use. The Dorst equation is WHAT + ??? = RESULT. AM introduces partying late at night as something which is also relaxing and something which Western people do (WHAT), but traditional Chinese people will not do and traditional Chinese person cannot relate to partying (???), when talking about health and relaxation (RESULT).

> But that's interesting here, since how about partying? But I think it kind of (INAUDIBLE), because the other people couldn't relate it with, and they felt that (INAUDIBLE) (.) [V7, 014]

Towards the end of the episode W performs inductive reasoning about the user behaviour to conclude that the user does appear to be aligning to the cultural stereotype of traditional introvert Chinese male [and not aligning to the party going young male in Scandinavia or China]

Yeah. But I also suspect given my – my reading of him, I don't think he's very hard core in partying … [V7, 015].

W suggests that since the user is following traditional Chinese medicine for health and wellbeing (WHAT), he must be an introvert (???), because he appears to be a person who would fit the stereotype, hence he would not enjoy partying as a way to relax (RESULT). A dynamic-situationist cultural theory interpretation could be that of a kind of negotiation situation (Morris & Gelfand, 2004); what W is doing could be trying to keep the Chinese cultural knowledge now *available* to the design team highly *accessible* to the designers by using himself as a role model in the design work, and *appropriate* by using the analogy to partying again.

4.2.5   Summary

In sum, both deduction and induction are shaped by the availability, accessibility and applicability of cultural knowledge in this episode. The content of a Dorst equation for deduction in this example is shaped by what cultural knowledge is available to those doing the deduction, primarily W, who is the Asian culture expert and has this knowledge easily accessible. However, in the situation W needs to repeat and explain his deduction by making cultural knowledge accessible to the Scandinavian design team members, and make it appropriate to use in the design context by using shared cultural knowledge about young people in the West. The content of a Dorst equation for induction is similarly shaped by cultural knowledge about both WHAT they are talking about and the end RESULT of the design thinking about health/relaxation.

4.3   How cultural knowledge shapes abduction and framing

To illustrate how abductive design thinking is shaped by cultural knowledge, we analyse episode 9 from CC2 (V14, 093–112), see appendix 7.2 for the full transcript.

In this episode, the team leader E asks a question to the team with an assumption in his mind and then he *frames* it by reference to an Apple Store example.

093	E	*Mmm, and eh, why, do you think it was important to touch the product?*
094	N	*Eh, because she also said she wouldn't invest in something like she wouldn't believe. So, she wanted to like, try it out, because that's what – was something like with eh, with the price, like if I don't know, if I'm not like sure, like I wouldn't trust it, so I wouldn't invest too much money in it.*
095	E	*Mmm, is it trust of quality or trust in they needed it?*
096	N	*The (.) quality*
097	A	*Yeah*
098	K	*I think it's kind of an idea one of the guys refers to Apple stores, they get kind of this experience that they are (INAUDIBLE) as you get.*

099	E	*But it had- did it have to do with trust, or did it like – I might- this is my crazy assumption, but I assume that people trust Apple, but they still go to the store, it has nothing to with trust, it has to do with I wanna be part of it, I wanna aspire to this culture, hang out.*
100	A	*But that was exactly what they said*
101	E	*Yeah*

This is an example of *abduction with framing*. It is problem solving, since the team members are trying to identify whether the product or the corporate culture is the more important for the design. E asks a question with an assumption in his mind – that physically touching a product has nothing to do with trusting a company – and he explains his assumption by framing the problem as what actions people take in an Apple Store.

In terms of Dorst's equations, the design team is using *abduction 2 with framing* to build a need for a yet unknown product (WHAT is unknown) by using an analogy to what works (HOW is unknown) and framing is used to see whether touching the product or knowing about the company culture in an Apple Store is necessary (VALUE: Trust). However, the analogy itself requires cultural knowledge about the Chinese context to be made available, accessible, and appropriate.

4.3.1   Cultural knowledge comes into play in framing

All the team members, based on the *availability* of their cultural knowledge, reason about if it is important to touch the product or it is the company that bears the trust. Inductive thinking comes into play, while hypothesizing various reasons behind why users would like to touch the product.

One of the Scandinavian team members, A, does not share with E the cultural knowledge – E's "crazy assumption" – that people trust the company, not the products. A reverts to the *available* knowledge about the users that is shared by all in the design team, and tries to use inductive reasoning to argue that the FRAMING suggests that the working principle in the Apple Store is knowledge and experience that builds the VALUE of trust.

102	A	*About they actually wanted to go and see what it was all about*
103	E	*Yeah*
104	A	*Because no one knew them, knew their product*

As it happens, A's inductive reasoning about the actual users is supported by deductive thinking by the Asian consultant AM, who makes her *available* cultural knowledge about 'lack of trust for the products in china' *accessible* to everyone, by telling that people (WHAT) want to touch the product for knowledge, to know whether it is authentic and to not just trust the second hand knowledge (WORKING PRINCIPLE) and this should be the basis for thinking about what VALUE that can be achieved in an Apple Store in China.

| 106 | AM | *I think, knowledge, right? You go there, you see, you experience it. You know and you're authentic of knowing rather than just second hand information.* |

E makes his cultural knowledge about Apple Stores in the West *accessible* to all by stating that there is not much information about the company at an Apple store, it is mainly the products, and still people trust the brand, suggesting that the FRAMING does define the WORKING PRINCIPLE in an Apple Store as building trust in the company which leads to the VALUE of trust in general.

| 107 | E | *And I think that is fantastic thing about the Apple store. There is nothing else there, there's photos of products and products. That's it. There's (INAUDIBLE), just a little text, no nothing, maybe just a little price, or whatever.* |

Then A induces from the data, another idea about the WORKING principle in an Apple Store, and tells that one of the users actually mentioned that company culture of Apple leads to the liking for the brand.

| 108 | A | *And our:- one guy in our group also mentioned about the culture, how the Apple and the culture and what they do, influence them liking the brand.* |

Then E uses deduction based on another Chinese user (WHAT) who commented on the open and approachable culture of the company Panasonic (HOW) to argue that this is part of 'building the product' – and A agrees that the Apple store as a place for building trust in corporate culture is an *appropriate* framing. They kind of agree that corporate culture is important in one way or another.

109	E	*Mmm (.) Yeah, and I think that is also, so: good for us, as THE COMPANY, that it actually has an impact to that. The corporate culture will actually be part of, you know, building the product.*
110	A	*And what parts of the culture?*
111	E	*So the- so the:- In our group they- yeah, in our group they mentioned openness, for example, and they used an example from- from Panasonic. At Panasonic, even the low level assembly dudes can write a message to the CEO and say, I think this is (INAUDIBLE), approachable. So it means that, you know-*
112	A	*They take care of their own*

The slight disagreement between W and A is perhaps an interesting feature of open problem solving such as abduction 2; though the *available* knowledge about the culture may be quite similar among same culture designers, slight changes in priming in the situation may lead to differences in what knowledge becomes *accessible* and is deemed *appropriate*. A is primed by the data about users, but when it comes to the social situation W's deduction is perhaps more convincing to A than AM's deduction, because W and A share the cultural knowledge about Apple Stores in the west, and have little cultural knowledge about what it means to be in an Apple Store in China.

This is a good example of abduction; they know the value to create in the market (improved quality of life), however the product/service and the working principles are not known; it is very open-ended, complex problem solving. By framing the problem in a culturally underspecified context, making it unclear if the Apple Store was in the West or in the East, the designers had to make their available knowledge accessible to the team in order to make the abduction.

4.4   The situational and dynamic aspect of cultural knowledge in the design process

The dynamic constructivist and situation specific theory of culture argues that culture can be understood as knowledge networks that are activated in different ways, depending on the situation. To illustrate how cultural dynamically and situation specifically shapes design thinking, we have illustrated knowledge similarities in episode 5 (V7, 096–119) from the CC1 debriefing workshop, see Figure 9.3.

Figure 9.3 illustrates how the accessible cultural knowledge in two cultural groups in the design team, the Asian and the Scandinavian, varies across a series of events 1–5. Initially the knowledge systems diverge, but towards the end of the discussion topic, they converge, and the team reaches a conclusion. The point is that ethnic culture is not an essential feature of Asians or Scandinavians, but knowledge systems that are available, accessible and applicable, depending on how the situation or event shapes the design thinking.

In Table 9.4 we explain each event depicted in Figure 9.3. Each row in the table discussed the activation of cultural knowledge in the event.

The episode depicted in Figure 9.3 and Table 9.4 illustrates how the design thinking process is shaped by the appearance and disappearance of cultural differences. Towards the end, the two cultural systems are more in agreement, whereas at the beginning they are more divergent. Overall, the episode is a good example of abduction because the value 'well-being/health' is known; the designers are discussing to figure out the

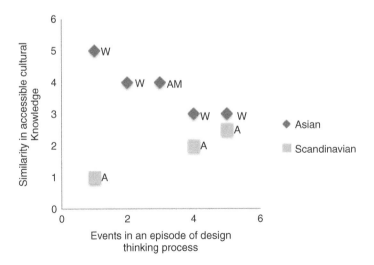

Figure 9.3 Culture X Situation dynamic interaction in an episode of design thinking (V7, 096–119). W: male Asian consultant, AM: female Asian consultant, A: female Scandinavian designer. See Table 9.4 for a detailed explanation of each event.

Table 9.4 Explanation of the culture × situation dynamic interaction illustrated in Figure 9.3.

Events	Design thinking
1	**W** is *accessing* his *available* cultural knowledge about China and describes the commonality of fake products to the team members. He mentions that, in China, people have a hard time trusting the products because cases of forged or dangerous products (e.g., chicken, soy or eggs).
1b	**A**, the Scandinavian designer does not have this particular cultural knowledge *available* and hence she was surprised to see Chinese people voicing their concern about the different products and their sources in the co-creation workshop.
2	**W** makes the shared cultural knowledge more *appropriate* for the Scandinavian team members by suggesting that in the West the issue is to choose the healthy food among the available options, whereas in China the challenge is to find real (and safe) food because chances that something will be fake or unsafe are higher.
3	**AM**, the other Asian consultant, shares the cultural knowledge and agrees with W. AM further makes her *available* cultural knowledge *accessible* and *appropriate* by using an example of the user needs and aligns it to one of the design goals (i.e., environment). She mentions that food is an everyday product that one could choose to buy from a place you like. However, air pollution is not a matter of choice as we cannot choose not to breathe
4a	**A**, the Scandinavian designer brings in her cultural perspective to the table, arguing that knowing a brand for a long time is good and that one would tend to trust something you always buy and have some knowledge about.
4b	However, **W** makes his *available* cultural knowledge accessible when primed by the discussion about the long history of the brand. He mentions that in China it's quite opposite as people don't necessarily trust reputable brands, since people with knowledge of the brand might suggest that a product is not safe and people will easily loose trust. W suggests that generally "there is a lot of mistrust of government, of brands, producers especially when it comes to food". W uses inductive thinking to hypothesize based on his cultural knowledge that health is a complex issue and not strictly limited to food in China. A bigger concern is the environment in general.
5a	**A** brings in her shared knowledge of the Chinese culture to suggest that in China she knows that people do not trust car dealership that they have heard rumours about. She hypothesizes by using inductive thinking and suggests that it is trust is easier to obtain in a situation where you can try out a product since people can see that the product is real and not fake.
5b	**W** adds his cultural perspective to A's example and suggests that it is not about just trust but also about the health and environment because if a person gets cheated it hurts his belief about society and people in general, and that this affects his overall well-being. All the team member converge on this conclusion.

unknown WHAT (products/services) and HOW (working principles of why something would work or not work) in the abduction equation.

5   DISCUSSION

The results indicate how cultural knowledge shapes design thinking within and across the design situations. The core design thinking methods of induction, deduction and abduction were affected by the situation-specific appearances and disappearances of cultural differences among team members across design thinking episodes, as illustrated in Figure 9.2. Our results are in line with previous studies that show how the interaction between accessibility and appropriateness is directed based on who in a concrete situation the person is interacting with: in-group members (same culture) or out-group members (different culture), and that interactions may be primed by cultural stereotypes (Hong, 2000; Wong & Hong, 2003). For an analysis of within multicultural design team conflicts see Paletz *et al.* (2017) Overall, it is worth remembering that design thinking is a sought after problem-solving approach. It is often stated that it is important to "listen to the users" but also that it is perhaps more important to observe the behavior and perceive what users might *not* be able to tell you (Brown, 2009). Our results indicate that the thinking of designers and consultants are biased by their own cultural beliefs and thoughts to varying degrees within and across specific situations.

5.1 Capturing situation specific cultural design thinking

We have argued that culture shapes design thinking, at least as design thinking has been formulated by Dorst (2011). In Figure 9.3 above, we illustrate the dynamic and situation specific nature of how cultural knowledge shapes design thinking within a single episode. The similarity in the cultural knowledge of the team members varies throughout the discussion in the particular episode, that is, either the cultural knowledge accessed by the team member is shared by both the cultural groups, or the knowledge is specific to the cultural group of the member sharing the knowledge.

Following up on our research, we wish to look at a complete design project, by applying a similar framework. Our idea is to obtain average ratings of similarity in group knowledge about the currently discussed topic, and obtain these ratings for individual episodes along the whole design thinking process. The ratings of similarity can be obtained by accessing the available and accessible knowledge structures of the team members in each situation/episode and rate them for similarity (i.e., how similar they are to each other). Note that we suggest that the analysis should mainly be based on what cultural knowledge is made accessible, since this can be observed from design work videos such as the DTRS11 videos. Available knowledge, on the other hand, is in the head and hence hard to see in transcripts or videos or observations, and would require other methods like thinking loud or interviewing. Applicable (appropriate) knowledge is also not of much interest in this context, because when designers decide what cultural knowledge is appropriate to apply in the given social situation, their cultural knowledge systems have in many cases already been closely aligned based on design goals. Overall, we believe that examining the variations in activated cultural knowledge across the complete design process will provide insights about design thinking in general. Further research may consider how situational differences in accessibility might lead to frame-switching, which means understanding the problem with a new perspective, and sometimes also to bring in the whole team on the same page by making the point using an example in a different context. Another interesting topic is to investigate differences in situational applicability of creative design ideas and how these may lead to "culture sampling: the conscious or unconscious selection of cultural normative behaviours that are most appropriate in a given social situation." (Hong & Mallorie, 2004). To understand how cultural knowledge is activated, researchers must go beyond participants' nationality or similar salient cultural features and know much about the situation.

Another future research opportunity is to understand better the material performance of design. Ranging from pen and paper to sophisticated design software and full prototypes in a variety of materials, the work of design can be defined partly as an ongoing manipulation and arrangement of material artefacts (e.g., see Dove *et al.*, 2017). Future research may therefore study in detail how materials, what Eriksen (2009) terms 'micro-materials', shape and choreograph co-design practices in a cross-cultural setting. Based on the cultural contextualization in this paper, we suggest that an in depth examination of the material performance of design thinking can be set in relation to the cultural constraints around communication and cognitive frames. Material performances in design may be considered part of cultural knowledge activation. This can be assessed in relation to the kinds of reasoning in design processes by attending to possible temporal or turn-taking relations between manipulating materials and abductive reasoning.

5.2   Limitations and scope

Our research has some limitations. Dorst mentions the core design reasoning patterns such as abduction to be applicable to the whole design thinking process. However, in our analysis we found that it was possible to identify core patterns even in subsections that are parts of the overall design thinking process. Second, our access to the design thinking process was limited by the pre-selected video recordings provided in the DTRS11 dataset. However, the great advantage of the dataset was its capture of a series of design situations, which allowed us to do a situation specific analysis.

6   CONCLUSION

Our analysis indicates that core design thinking methods of induction, deduction and abduction are affected by the ongoing and situation-specific appearance and disappearance of cultural difference among design team members. The paper presents a novel approach to the understanding of design thinking in the context of cultural knowledge activation. To our knowledge this is the first time Dorst equations have been studied in a cross-cultural context.

REFERENCES

Brown, T. (2009). *Change by Design: How Design Thinking Transforms Organizations and Inspires Innovation.* London: HarperCollins Publishers.

Christensen, B. T. & Abildgaard, S. J. J. (2017). Inside the DTRS11 Dataset: Background, Content, and Methodological Choices. In: Christensen, B. T., Ball, L. J. & Halskov, K. (eds.) *Analysing Design Thinking: Studies of Cross-Cultural Co-Creation.* Leiden: CRC Press/Taylor & Francis.

Clemmensen, T. (2009). Towards a theory of cultural usability: A comparison of ADA and CM-U theory. In: M. Kurosu (ed.) *Human Centered Design: First International Conference,*

HCD 2009, *Held as Part of HCI International 2009, San Diego, CA, USA, July 19–24, 2009 Proceedings.* Vol. 5619. Berlin, Heidelberg: Springer. pp. 416–425.

Dhadphale, T. (2017) Situated Cultural Differences: A Tool for Analysing Cross-Cultural Co-Creation. In: Christensen, B. T., Ball, L. J. & Halskov, K. (eds.) *Analysing Design Thinking: Studies of Cross-Cultural Co-Creation.* Leiden: CRC Press/Taylor & Francis.

Dorst, K. (2011). The core of 'design thinking' and its application. *Design studies, 32*(6), 521–532.

Dove, G., Abildgaard, S. J., Biskjaer, M. M., Hansen, N. B., Christensen, B. T. & Halskov, K. (2017) Grouping Notes Through Nodes: The Functions of Post-It Notes in Design Team Cognition. In: Christensen, B. T., Ball, L. J. & Halskov, K. (eds.) *Analysing Design Thinking: Studies of Cross-Cultural Co-Creation.* Leiden: CRC Press/Taylor & Francis.

Eriksen, M. A. (2009). Engaging design materials, formats and framings in specific, situated co-designing-a micro-material perspective. *Nordes (Nordic Design Research),* (3), 1–10. Available from http://www.nordes.org/opj/index.php/n13/article/view/55

Hong, Y.-Y., Benet-Martinez, V., Chiu, C.-Y. & Morris, M. W. (2003). Boundaries of cultural influence construct activation as a mechanism for cultural differences in social perception. *Journal of Cross-Cultural Psychology, 34*(4), 453–464.

Hong, Y.-Y. & Mallorie, L. M. (2004). A dynamic constructivist approach to culture: Lessons learned from personality psychology. *Journal of Research in Personality, 38*(1), 59–67.

Kimbell, L. (2011). Rethinking design thinking: Part I. *Design and Culture, 3*(3), 285–306.

McDonnell, J. (2017) Design Roulette: A Close Examination of Collaborative Decision-Making in Design From the Perspective of Framing. In: Christensen, B. T., Ball, L. J. & Halskov, K. (eds.) *Analysing Design Thinking: Studies of Cross-Cultural Co-Creation.* Leiden: CRC Press/Taylor & Francis.

Morris, M. W. & Gelfand, M. J. (2004). Cultural differences and cognitive dynamics: Expanding the cognitive perspective on negotiation. In M. J. Gelfand & J. M. Brett (eds.), *The handbook of negotiation and culture* (pp. 45–70). Stanford, California: Stanford University Press.

Paletz, Susannah B. F., Sumer, A. & Miron-Spektor, E. (2017) Psychological Factors Surrounding Disagreement in Multicultural Design Team Meetings. In: Christensen, B. T., Ball, L. J. & Halskov, K. (eds.) *Analysing Design Thinking: Studies of Cross-Cultural Co-Creation.* Leiden: CRC Press/Taylor & Francis.

Pineda, R. G. (2014). *Technology in culture: a theoretical discourse on convergence in human technology interaction.* PhD thesis, University of Jyväskylä. (Jyväskylä studies in computing 191)

Teixera, C., Shafieyoun, Z., de la Rosa, J. A., Cai, J., Li, H., Xu, X. & Chen, X. (2017) Structures of Time in Design Thinking. In: Christensen, B. T., Ball, L. J. & Halskov, K. (eds.) *Analysing Design Thinking: Studies of Cross-Cultural Co-Creation.* Leiden: CRC Press/Taylor & Francis.

APPENDIX A

CC1 episode 1, (Video 7, segment 009–015)

009   W   So, eh, I think there was one guy who, the younger guy, who I think leads a slightly more disciplined life, I mean like, he's not married, he's not, you know, has his own family and whatever. He talks about things like sleeping early:, going to bed by ten:, waking up really early by six:, you know, because your body starts to detox at eleven a clock.

010   A   Yeah.

011	W	*That's actually a little bit of eh: (.) traditional Chinese medicine, that's part of the concept. Your body starts to work itself eh actually: from that time which is eleven at night, your body should start resting before that, so you need to go to bed before that, so that, you know, it can work its magic.*
012	A	*Mhm.*
013	W	*So people like us who sleep at twelve: sleep at one, you have really bypassed that magic hour of where we can actually get that. So-*
014	AM	*But that's interesting here, since how about partying? But I think it kind of (INAUDIBLE), because the other people couldn't relate it with, and they felt that (INAUDIBLE) (.)*
015	W	*Yeah. But I also suspect given my- my reading of him, I don't think he's very hard core in partying.*

CC2 episode 9, China, Co-creation room (Video 14, segment 093–112).

093	E	*Mmm, and eh: why: do you think it was important to touch the product?*
094	N	*Eh: because she also said she wouldn't invest in something like she wouldn't believe. So: she wanted to like, try it out, because that's what- was something like with eh: with the price, like if I don't know:, if I'm not like sure, like I wouldn't trust it, so I wouldn't invest too much money in it.*
095	E	*Mmm, is it trust of quality or trust in they needed it?*
096	N	*The (.) quality*
097	A	*Yeah*
098	K	*I think it's kind of an idea one of the guys refers to Apple stores, they get kind of this experience that they are (INAUDIBLE) as you get.*
099	E	*But it had- did it have to do with trust, or did it like- I might- this is my crazy assumption, but I assume that people trust Apple, but they still go to the store, it has nothing to with trust, it has to do with I wanna be part of it, I wanna aspire to this culture, hang out.*
100	A	*But that was exactly what they said*
101	E	*Yeah*
102	A	*About they actually wanted to go and see what it was all about*
103	E	*Yeah*
104	A	*Because no one knew them, knew their product*
105	E	*So it was about excitement*
106	AM	*I think, knowledge, right? You go there, you see, you experience it. You know and you're authentic of knowing rather than just second hand information.*
107	E	*And I think that is fantastic thing about the Apple store. There is nothing else there, there's photos of products and products. That's it. There's (INAUDIBLE), just a little text, no nothing, maybe just a little price, or whatever.*
108	A	*And our:- one guy in our group also mentioned about the culture, how the Apple and the culture and what they do, influence them liking the brand.*

109	E	*Mmm (.) Yeah, and I think that is also, so: good for us, as THE COM-PANY, that it actually has an impact to that. The corporate culture will actually be part of, you know, building the product.*
110	A	*And what parts of the culture?*
111	E	*So the- so the:- In our group they- yeah, in our group they mentioned openness, for example, and they used an example from- from Panasonic. At Panasonic, even the low level assembly dudes can write a message to the CEO and say, I think this is (INAUDIBLE), approachable. So it means that, you know-*
112	A	*They take care of their own*

Chapter 10

Situated Cultural Differences: A Tool for Analyzing Cross-Cultural Co-Creation

Tejas Dhadphale

ABSTRACT

Designers in the global era are increasingly challenged to design for a diverse cultural context. Both ethnographic and participatory design research methods are integral for conducting cultural inquiries. Co-creation offers an interactive setting for researcher to gain meaningful insights into the tacit and latent aspects of culture. The analysis framework for this study is based on the theoretical construct of situated cultural differences. In the process of co-creation, situated cultural differences become a diagnostic tool used by researchers to categories participants' everyday experiences into meaningful cultural categories. This paper focused on identifying the different situated differences used by participants during the co-creation sessions and creating a typology of differences. The analysis was focused on how the research team interpreted the situated differences to establish the underlying cultural values. Four categories of situated cultural differences emerged from the data: material-observable, material-ideological, behavioral-observable, and behavioral-ideological. The analysis shows that participants preferred the material and behavioral-ideological types of situated differences. This study provides a methodological approach for analyzing cultural differences and integrating diverse cultural aspects into a systematic framework. The theoretical framework of cultural situated differences and the typology of differences provides a framework for designers, design researchers and corporations to identify, categorize and design for cultural differences.

1 INTRODUCTION

With increasing globalization, new emerging markets, technological advancement and worldwide competition, global corporations are trying to expand their market across the world. Products initially designed to serve only the local market are now reaching across international boundaries. As a consequence, there is an emerging interest in the impact of cultural dimensions on the interaction between people and products, both from a professional and an academic point of view (Christensen *et al.*, 2006). Designers and design researchers developing products for new local market need to study user needs, behaviors, practices, rituals, ideologies, and values within a cultural context.

Global corporations expanding business across different local markets have identified cultural insensitivity to be a potent barrier for growth. The degree of acceptance by local consumer cultures has become an integral part of the success and failure of their operations. With the fading influence of cultural imperialism, Americanization or Westernization, local cultures are rejecting homogenized products and services and are demanding culturally relevant and suitable products (Pieterse, 2006; Ritzer, 2006; Tomlinson, 1999). According to Shaw and Clarke (1998), organizations need to understand the degree to which standardized products can be offered unchanged or whether they need to be adapted to local markets. As a consequence, corporations are conducting cultural "deep dive" research to gain meaningful insights into local cultures.

Design research methods and techniques (both traditional and participatory) have been integral for conducting in-depth cultural inquiries. Traditional methods like surveys, interviews and passive observations are largely focused on what people say, think and do (Sanders, 2002), resulting in insights based on explicit (think and say) and observable (actions) aspects of human experience. In contrast, participatory research methods like co-creation can provide insights into the tacit and latent (values) aspects of human experience. In the process of participatory co-creation, participants express their thoughts into visual and tangible materials, and also provide the rationale behind their choices. Co-creation sessions provide an interactive medium for users to express and share their feelings, emotions and values with researchers and designers. The goal of analysis is to explore the process of how the design team uncovered the underlying cultural values of co-creation workshop participants. This study aims to identify the different local, significant and embodied aspects (explicit and observable) of culture and the underlying cultural values (implicit and latent).

This paper describes the different types of situated differences utilized by participants in the co-creation sessions (RQ1), illustrates the different categories that emerged from data and reports instances (excerpts from transcripts) that highlight the core values uncovered by the research team (RQ2). Situated cultural differences are particular practices, rituals, beliefs, ideologies, artifacts, or activities utilized by groups that highlight the core values of groups, and in the process mobilize group identity. Post-it notes were analyzed from co-creation sessions (v08; additional materials), transcripts from debriefing (v07, v08) and insights clustering (v09, v10, v11) sessions (Christensen & Abildgaard, 2017). Participants were engaged in two different co-creation sessions. Each co-creation session was facilitated by two researchers fluent in the participants' native language. After the first co-creation session, researchers from the two teams shared their insights (v08; debriefing sessions) with all team members. This was followed by insights clustering sessions (v09, v10, v11) where the research team focused on uncovering the underlying meanings and values.

Initial coding was guided by typological analysis, reducing and categorizing data based on pre-existing typologies (Hatch, 2002). The analysis framework for this study was guided by cultural layer models developed by eminent scholars like Hall (1976), Trompenaars and Hampden-Turner (1997), Hofstede (2001), Spencer-Oatey (2000), Schwartz (1994) and others. Based on the synthesis of existing models, situated differences were categorized into three layers: material, behavioral, and ideological. The pre-existing layers (or typologies as suggested by Hatch, 2002) guided the early data analysis. Instances (excerpts from transcripts) representing all three layers were identified. After comparing instances (excerpts) from each layer we discovered a strong overlap between the codes. Early data analysis started with pre-existing layers but new categories emerged from the data. The data analysis started with a typology based analysis and ended with data-driven categories (see Figure 10.2). For example, participants discussed using products (like cars, home appliances) both from a purely consumption stand-point and also from an ideological stand-point. Data were then re-coded to reflect the connections between the layers. New hybrid codes were created to reflect the overlap between two layers. Following this, for each instance, the type of situated differences and the underlying cultural values of the group were established.

2   BACKGROUND

2.1   Theoretical stance: culturally situated differences

Culture is one of the two or three most complicated words in English language (Williams, 1976). During the end of the nineteenth century the term culture was associated with a people or nations with *particular distinction*; an aspect I will highlight later in this section. Culture then meant 'a way of life' for particular individuals, groups or nations. This definition laid the groundwork for the contemporary understanding of culture. As defined by Williams (1976), culture is a "description of a particular way of life which expresses certain meanings and values not only in art and learning but also in institutions and ordinary behavior." The two key elements of this definition are the interpretation of culture as 'a way of life' and the 'production and circulation of meanings' through ordinary behavior. It is the clarification of meanings and values implicit and explicit in particular way of life that makes a particular "culture" (Williams, 1976, p. 57).

The work by earlier anthropologists and sociologists focused on the noun form of culture positioning it at the center of every inquiry. For example, describing culture as an evolving entity, the classical cultural evolutionist proposed a grand unilinear theory for evolution of cultures through different stages of savagery, barbarism, and civilization. Following the historical particularist perspective, Kroeber extended the noun usage of the term by defining culture as 'superorganic,' an overarching, autonomous identity that controls individuals and their behaviors. Kroeber defined culture as: "a realm sui generis, or unto itself, separate from psychology and 'above' biology" (Erickson & Murphy, 2006). By defining culture as an autonomous control over individual heredity, psychology and society, Kroeber to an extent reified culture. It wasn't until the rise of psychological and symbolic anthropology that culture was one of the subjective dimensions that mediated human experience and behavior rather than the only factor explaining a phenomenon.

As a guiding principle for conducting cultural studies, Appadurai (1996), McCracken (1986), Miller (2005), Ritzer (2003), Bocock (1993) and others have advocated the adjective usage of culture, that is, *cultural*. The noun form of culture seems to "carry associations with some sort of substance in ways that appear to conceal more than they reveal," but the adjectival usage of culture becomes a force to understand cultural differences, or comparisons (Appadurai, 1996, p. 12). Appadurai (1996, p. 13) advocates "stressing the dimensionality of culture rather than its substantiality permits our thinking of culture less as a property of individuals and groups and more as a heuristic device that we can use to talk about difference." The methodology for this study also emphasizes the adjective usage of culture, that is, *cultural*. The analysis framework for this study is based on the theoretical construct of culturally situated differences.

There are two key aspects to situated cultural differences. First, situated differences are "differences in relation to something local, embodied, and significant" (Appadurai, 1996, p. 12). These are differences a cultural group would utilize as local and distinctive differences that are significant to the group. Second, situated differences are differences that "either express, or set the groundwork for, the mobilization of group identities" (Appadurai, 1996, p. 13). This goes back to the early definition of culture; people, groups or nations with *particular distinction*. Culture then acts as a force that decides the boundaries among different groups. In simple terms, situated cultural differences become a frame of reference for emphasizing local, embodied differences that mobilize group identity. Situated differences are aspects of everyday life that mobilize group identities and represent core values of the group. A group of individuals buying a particular brand of motorcycle (situated difference) creates a homogenous group with distinctive practices, products and ideology. For example, Harley-Davidson owners are a distinctive group (group identity) with shared values that are manifested in different aspects of everyday life. The values of the group are manifested in distinctive practices, products or ideology by different group members.

Analysis of the DTRS11 data set (Christensen & Abildgaard, 2017) shows that the research team invited individuals for co-creation sessions that represented affluent Asian car owners associated with elite and well known corporations. The goal of the co-creation sessions was to uncover the underlying values of this affluent automotive cultural group. For example, this group of individuals identify themselves as "environmentally friendly" – a value that the group projects as a major distinction from other individuals or groups. This particular group of shoppers buy products made out of recyclable materials (situated difference) as a way to imply the core value of "environmental friendliness". In this case, the act of buying from recyclable source (ideology), the act of buying (the practice) and the artifact itself are all situated differences utilized by the group to project the core value of "environmentally friendliness".

2.2   Layers of situated cultural differences

As described above, situated cultural differences could be local, embodied practices, artefacts, ideologies that mobilize group identity and expresses group values. To develop an applicable coding scheme, it was essential to categorize situated cultural differences. This classification of culture into layers (cultural layers) is guided by the work of eminent scholars like Hall (1976), Trompenaars and Hampden-Turner (1997), Hofstede (2001), Spencer-Oatey (2000), Schwartz (1994) and others. The classification of culture into layers reduces the complexity of the term and provides a diagnostic tool for researchers to investigate different aspects of cultures in a systematic manner.

According to Trompenaars and Hampden-Turner (1997) and Hofstede (2001), culture can be classified into layers, like an onion. For Hofstede (2001), culture can be classified into four layers: rituals, heroes, symbols, and values. Hofstede (2001) considers the system of values as the core of any culture. In addition, he considers cultural practices as the fifth layer of culture that connects the layers of rituals, heroes, symbols and values. Similarly, Trompenaars and Hampden-Turner (1997) identified three layers of culture: the outer layer that includes artifacts and products; the middle layer representing norms and values; and the core that represents the fundamental assumption about human existence. Spencer-Oatey (2000) also identified four layers of culture: the outer layer that includes artifacts, products, rituals and behaviors; systems and institutions; beliefs, attitudes and conventions; and the core representing basic assumptions and values. Rose (2004) categorized culture into two categories: cultural mentalities; and cultural environments. According to Rose (2004), the cultural environment (the physical surrounds and structure) constantly shapes the cultural mentalities (the thoughts and behaviors) of a cultural group.

Both Hofstede (2005) and Trompenaars and Hampden-Turner (1997) reference the observable and symbolic attributes of the outer layer of culture. For Hofstede (2005), Trompenaars and Hampden-Turner (1997) and Spencer-Oatey (2000), artifacts, products, art, images and words are not only visible aspects (tools to mediate everyday activities) of culture but act as symbols that carry particular meanings that could be decoded by individuals from a particular culture. The key idea of interpreting the outer layer of culture as both observable and symbolic (interpretive) was incorporated into the four-layered analysis model developed for this paper.

Classifying culture (that complex whole) into manageable layers has encouraged inclusion of cultural aspects into the design discourse. The studies by Hsu, Lin, and Lin (2011), Lin, Sun, Chang, Hsieh, and Huang (2007), Lee (2004), and others have proposed a framework to translate cultural insights into product design. The following table summarizes the different layers of culture suggested by scholars. The levels of culture are listed from the outermost layer to the inner core.

Based on the layered approach (see Table 10.1) suggested by Hall (1976), Trompenaars and Hampden-Turner (1997), Hofstede (2001), Spencer-Oatey (2000), Schwartz (1994), Rose (2004), Lee (2004) and Leong and Clark (2003), culturally situated differences were classified into four layers (Figure 10.1). This four-layered approach was then used as a reference point to identify and classify situated cultural differences from the dataset (see research questions). This four-layered classification is based on Hall's (1976) 'iceberg model' of culture. The iceberg model suggests that the visible aspects of culture only represent the tip of the iceberg. The invisible, under the water, aspect of culture forms the foundation. The intangible aspects of culture are manifested in visible forms. The following paragraphs provide a brief description of each layer.

The outermost 'material' layer: This layer includes tangible aspects of cultural differences including material objects, products, services, materials and processes (Figure 10.1). The studies cited earlier (Hsu, Lin, & Lin, 2011; Lee, 2004; Lin, Sun, Chang, Hsieh, & Huang, 2007) have explored this 'material' layer of culture. Material culture scholars and archeologist have repeatedly highlighted the manifestation of culture into material form. According to McCracken (1986, p. 73), objects are "vital, tangible record of cultural meaning that is otherwise intangible." Material objects, products, materials, manufacturing processes embedded in different cultures carry implicit cultural meanings. Incorporating insights from the models proposed by Hofstede (2005), Trompenaars and Hampden-Turner (1997), the outermost 'material' layer includes both the observable and the symbolic meanings of objects, products, materials and processes within a cultural group.

Table 10.1 Layered classification of culture.

Layers of culture	Hofstede (2001)	Trompenaars and Hampden-Turner (1997)	Spencer-Oatey (2000)	Rose (2004)	Lee (2004)	Leong and Clark (2003)
Level 1 (Outermost)	Symbols	Artifacts and products	Artifacts, products, rituals and behaviors	Cultural environment (the physical surrounds and structure)	Artifact	Tangible (material, artifacts)
Level 2	Heroes	Norms and values	Systems and institutions		Values	Behavioral (practices, rituals, activities)
Level 3	Rituals (Practices connect all layers)	Beliefs, attitudes and conventions				
Level 4 (Inner core)	Values	Assumption about human existence	Basic assumptions and values	Cultural mentalities (thoughts and behaviors)	Basic assumptions	Intangible (values)

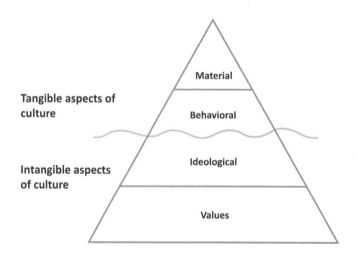

Tangible aspects of culture

Intangible aspects of culture

Material

Behavioral

Ideological

Values

Figure 10.1 Iceberg model of cultural differences (adapted from Hall, 1976).

The mid 'behavioral' level: This layer includes practices, rituals and interactions with material objects and other individuals. Cultural meanings are created through day-to-day interactions between the products and people (Grant & Fox, 1992). Researchers like Margolin (1989), Prown (1993) and Appadurai (1986) have strongly stated that meanings are a result of our social interactions and practices. Meaning is constructed – given, produced – through cultural practices; it is not simply found in objects (du Gay *et al.*, 1997, p. 14). Rituals are collective practices shared by a cultural group that have significant meaning within the group and highlight cultural differences when compared to other cultural groups.

The inner 'intangible level: This layer includes beliefs, attitudes and ideology (a network of ideas) of individuals as part of a group and also the shared group values. This level includes individual and shared beliefs and attitudes. Ideology is a shared system of meanings abstracted from the collectively-held beliefs and attitudes of individuals within a group.

The core 'value' level: The innermost layer refers to the system of values (borrowed from Hofstede, 2001)) that represent a cultural group. Shared cultural values for a group are constantly negotiated by individuals within a group. Individual values are partly a product of shared culture and partly a product of unique individual experiences (Schwartz, 1994). The goal of this study is to apply situated cultural differences as a diagnostic tool to uncover the core values (of the group) manifested in beliefs, attitudes, ideologies, and everyday practices, rituals, interaction and material objects.

3 RESEARCH QUESTIONS, METHODOLOGY AND ANALYSIS FRAMEWORK

From the DTRS11 dataset (Christensen & Abildgaard, 2017), this paper specifically focused on the co-creation sessions (v08), debriefing (v07, v08) and insights clustering sessions (v09, v10, v11). These sessions were particularly selected from a large dataset provided by the DTRS11 organizers. Data analysis was guided by the following research questions:

1 What are the different types of situated cultural differences used by participants during the co-creation sessions and how can they be categorized?
2 How is the interpretation of situated differences used to uncover the core cultural values of the group?

The analysis framework for this study is based on the theoretical construct of situated cultural differences. During the analysis process, situated cultural differences were identified based on two key aspects: (1) the differences that were local, embodied and significant to the group; and (2) the differences that represented group values (i.e., mobilized group identity). To identify and categorize the different types of situated cultural differences, the co-creation (v08) and debriefing sessions (v09, v10, v11) (Christensen & Abildgaard, 2017) were analyzed using the typological analysis framework. The following steps were followed in typological analysis (adapted from Hatch, 2002, p. 153):

1 Identify typologies to be used for analysis.
2 Categorize the data by marking entries related to typologies.
3 Look for patterns, relationships within typologies.
4 Identify instances or part of data to support the typologies.
5 Identify relationship (if any) between different instances and typologies.
6 Select instances (data excerpts) to support each typology.

Typological analysis begins with the process of reducing and categorizing data based on existing typologies (Hatch, 2002). To guide the early analysis, the three-layered classification (material, behavioral and ideology) of situated cultural differences was used. Situated differences classified into layers are different ways to project cultural values of groups (the innermost fourth layer). The process started with two coders separately categorizing parts of data using the following three layers of situated differences: the outer material, the mid behavioral, and the inner ideological. The innermost core values of a group are manifested in everyday material, behavioral and ideological forms. The goal was to study the manifested forms of situated differences and reveal the underlying core values of the group. In this paper, typologies refer to the three layers (used as codes) of situated differences: material, behavioral, and ideological.

The post-it notes and the resulting categories generated during the co-creation sessions were transcribed and analyzed. The post-it notes were categorized into the three layers. In addition, the transcripts from debriefing and insights clustering session were also analyzed using the same three-layered typology. The analysis based on pre-existing typologies was effective in reducing the data, identifying instances to support each typology and to identify inter-relationship between the layers. Instances are parts of the transcript (excerpts) that are representative of each layer. Multiple instances were selected for each layer. With the three-layered typology in mind, each coder assigned codes (material, behavioral and ideological) to the transcripts and then copied the instances (excerpts) into a separate excel sheet. After the analysis, the two coders compared the instances (excerpts) and discussed examples (instances) that represented each layer (typologies). The post-its categorized into three layers were also compared to the excerpts (instances) from the transcripts. This part of analysis answered the first of the two research questions (RQ1): what are the different types of situated cultural differences and how can we categorize them?

The second part of analysis focused on how the research team interpreted situated differences to establish the core values of the group (RQ2). This refers to the second key aspect of situated differences: differences that represent group values (i.e., mobilized group identity). For each layer, we consolidated different instances (excerpts from transcripts). For example, we consolidated data excerpts from different sessions that represented the mid-behavioral level. In this case, all excerpts related to practices, rituals and interactions were consolidated under the mid behavioral level. We then compared the different levels to identify potential overlap between the instances (excerpts). For example, we identified parts of data that support both behavioral and ideological levels. We then created a separate hybrid code (behavioral-ideological) and highlighted supporting instances (excerpt from transcript). At the end, we selected instances for each layer that show how the team identified the core values for the group. These instances are discussed in detail in the finding section.

4   FINDINGS

Data analysis revealed four key categories of culturally situated differences. The four types of situated differences are mapped on the following biaxial map (see Figure 10.2). The horizontal axis represents the continuum from material to behavioral

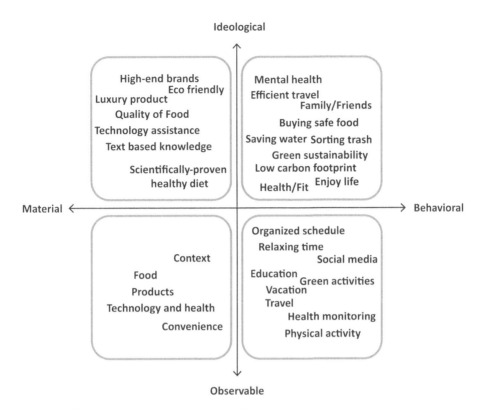

Figure 10.2 Biaxial map of situated cultural differences (includes quotes from the dataset).

manifestations of situated differences. The vertical axis represents the continuum from observable or concrete aspects to the abstract or ideological aspects of participants' everyday experiences. The following paragraphs summarize the four categories of situated cultural differences.

1   Material-Observable: This type of situated differences included reference to material objects, products, services and materials significant to the participants' culture, where the situated differences were manifested in material form. For example, participants discussed (v08, additional materials) products like cars, yachts, purifiers, health supplements, health monitoring bands, security robots, luxury products, and smart phones as the material-observable aspects of their culture.

2   Behavioral-Observable: This type included practices, rituals and activities that are specific to the participants' culture. For example, participants discussed the act of buying safe food or sorting trash, saving electricity, exercising with friends and monitoring health as situated differences that project their core values of environmental sustainability or health and wellness. In this case, the behavioral aspect becomes the situated difference that mobilizes the group identity and acts as a tool that differentiates the participants from other groups or cultures.

3   Behavioral-Ideological: This type included the everyday behavioral aspects (practices, rituals and activities) and the underlying ideology (system of ideas or beliefs). In this case, participants discussed the deep-rooted ideological aspects of their everyday behaviors. This explained the motivation behind participants' everyday behavior. Understanding the system of beliefs or ideas that motivates practices is useful to contextualize behavior specific to particular cultures.

4   Material-Ideological: This category included the symbolic meaning of interactions of participants with material objects, products and services. Participants discussed the abstract ideological meanings assigned to products and services as a situated difference that mobilized group identity. The abstract meaning behind material objects were highlighted in this category. For example, participants discussed the use of cars as a safe haven, cocoon or a place to break free from responsibilities. The following biaxial map illustrates the four types of situated differences identified.

4.1   Types of culturally situated differences

The following section reports instances (excerpts from transcripts) that highlight the core values uncovered by the research team (RQ2). It must be noted that the values representing the group were manifested in different forms. For example, the core value of *environmental sustainability* was reflected in all four categories: material-observable, material-ideological, behavioral-observable, and behavioral-ideological. The following section presents instances from all four categories (Figure 10.2) and highlights the core value discovered by the research team.

4.1.1   Material-observable

In this type, the participants shared examples of material objects, products and services that are used as situated differences that emphasize group identity by reflecting the core values.

For example, in the excerpt below (v09, 21), the research team is discussing (v09; insights clustering session) the role of luxury car brand. The act of buying a particular luxury brand mobilizes group identity and reflects the core value of *safety/security* and *environmental sustainability*. The researcher elaborates how the entire car (from a particular brand) is a reflection of values (like security, sustainability) and the act of buying the car is the only action required to project the values. In simple terms, if you buy the car then you acquire all values associated with the car.

A: [Yeah but I guess it's] almost the opposite, because if we just give them (.) give: them, they need to buy the car of course, (laughter) but if we give them the car with all the ehm: recycled eh: whatever, then: (.) they don't need to do anything, I mean, then they-

E: They did-, they already did it, they bought the car.

AM: I think it's the:- it's like the BMW story, in our group, like they believed that even the:, the metal or the whatever material, or the screw used in the car already reflected that value. Remember?

A: Yeah, yeah, that everything was eh: recycled and so on, but I'm just thinking that (.) that is the only action they need to do. I mean eh:, it's- it's just very very very easy.

In addition, in the co-creation session (v08), participants discussed the role of technology, health supplements, luxury consumer products, health monitoring bands, and smart security alarm systems as observable manifestations for creating group identity. In this case, buying and/or using the aforementioned products was a situated difference that was local, significant and embodied in the culture. By the act of buying or using the products, participants could distinguish themselves into different groups. In both the co-creation session (v08; additional materials) and the insights clustering session (v09) material-observable differences were the least discussed. Both participants and researchers did not emphasize the material-observable differences.

4.1.2 Material-ideological

This category included the symbolic meaning of interactions of participants with material objects, products and services. In this type, the symbolic meaning of product or service was discussed.

Lloyd and Oak (2017) highlight the importance of stories as a way of decoding cultural values held by both the participants and researchers during co-creation. Lloyd and Oak conduct conversational analysis using the 'small story' approach to combine the material and ideological aspects of participants' culture. The quote (v09, 38–58 & 110–122) below demonstrates how the research team creates a story of 'sexy commitment' to combining the tangible aspect of car and how the car facilitates the value of me-time and family-time (v09, 58, 110–119).

For example, the excerpt below (v09, 38) illustrates an example of the material-ideological type of situated differences. The excerpts below also demonstrate the effectiveness of narratives (Lloyd & Oak, 2017) as a way of communicating complex symbolic meanings within a cultural group.

AM (v09, 38): Like, the car becomes this safe place for me to let go, and away from all this burden and >"da-da-da-da"<, but it is still about me, it's just that I'm not performing: what I was expected to do, in a sense.

E (v09, 46): Eh: is it that way you think?

AM (v09, 47): I'm just thinking of another analogy. It's like, you go to a café, because you want that social aspect, you want to be surrounded, that's why you're not at home in your study-room, right?

E (v09, 57): So you want the comfort of the other people around you, but you wanna have your space there.

AM (v09, 58): Yeah, you go there not nece'-, yes, you- the space itself is a good: platform for you to meet other, to see other:, a meeting place, but it's also good enough for my own, my needs. So it's that kind of tension that I think the opportunity lies. *So maybe: going back to the tangible car, it could be "the car itself is catering for me and my family.* When I'm driving, I'm not necessarily the superhero", it's more like The Incredibles where different family have different role in the car, but a different mood from that, is when it is night time, "everyone is all asleep, I need to rest, this is where- this is my cocoon as well".

In this excerpt the research team, discusses the core values of *balancing me-time vs family time*, and *catering to my family*. The car discussed in this paragraph goes beyond its utility and symbolizes the balance between *me-time* and *time for family*. The car provides a safe place for the driver to relax and at the same provides an engaging place for all family members.

4.1.3 Behavioral-observable and behavioral ideological

The observable aspects included practices, rituals and activities of participants that reflected group values and mobilized group identity. The behavioral-ideological aspect focused on understanding the motivation behind participants' everyday behavior. Understanding the underlying system of ideas or beliefs provided cultural context to participants' actions.

For example, the excerpt below (v08, 38) illustrates examples of both behavioral-observable and behavioral-ideological types of situated differences. In the following excerpt the research team is discussing the different manifestations of the core value: *health and wellness*. First, the observable aspects of *health and wellness* are discussed, like exercise, relaxation of the mind and the body, massage, de-stressing and going for drive. The second part discusses the ideological position of participants regarding the role of Chinese medicine in their culture. It is emphasized that in Chinese culture, the role of medicine is to prevent illness. In comparison, Western medicine is to cure a sick person. Also, from a cultural standpoint, health and wellness includes both mind and body.

W: Okay good life ehm:. I think we- we focused on: the top three, ehm: that they have kind of voted on. So: one would be the health b'- min'-, a healthy body and mind, so: capturing things from ah, the most- the most obvious being things like the more things called, like exercise, ah: of course all the way down to relaxation of the mind and the body, and the relaxation there are two parts of course, the body and the mind, so body could be things like massage, ah: mind could be things like de-stressing, you know go for a drive and so on. Ah: traditional Chinese medicine is quite eh: a big thing, there is more for maintenance as opposed to: chronically eh: illness and then what you go for, so it's a little bit different from how: we think- or how they: will think of like western medicine, and we just- when you actually have a: big problem.

E: That's actually a very smart way of looking at it. Western medicine is after: you get sick, you get fixed up, here it's like "drink this tea:, do this thing and you won't: get sick".

W: It's to pull you back in balance or ensure that you are so'- in some form of balance.

In addition, the following activities (v08, transcribed post-its and discussion between researchers) were discussed as ways to achieve the core value of *health and wellness*: positive attitude, health monitoring, travel, deep breathing (mediation), to go out at away time, to go out together (social dining), picnic, jogging, travel, exercise with friends, going on road trips, and swimming.

The analysis of the data shows that core values were manifested in different material, behavioral and ideological forms. For example, *environmental sustainability* (as a value) was manifested in the use of different products and services (material layer) and the rationale for selecting products that project the individual's commitment to environmental sustainability. Participants also discussed practices like recycling, responsible consumption, buying products that are sustainable and choosing brands that reflect a sustainable vision as ways to project the value of *environmental sustainability*. The material and behavioral manifestations were embedded in participants' ideology regarding environmental sustainability. Participants' ideological stance is projected in the following examples: duty of commitment to family and environment, low carbon footprint, non-polluting renewable energy sources, global awareness, and commitment to sustainability. Researchers interpreted the commitment to environmental sustainability as a key distinguishing value for the group, that is, as mobilizing group identity. Using the same example of *environmental sustainability* (as a value), Daly *et al.* (2017) demonstrate the contrasting views of participants' and the research team. From the participants' perspective (the Asian perspective), *environmental sustainability* was viewed holistically that included natural, political and social elements and their interrelationships. The research team (Western perspective) projected *environmental sustainability* as individual acts of consumption of products. This example highlights the challenges of understanding and translating participants' views during a co-creation session. Using situated cultural difference can provide a methodological approach of systematically translating user insights during co-creation.

The analysis also indicated that the material and behavioral-ideological types of differences were heavily discussed by the research team. The research team heavily focused on understanding the system of ideas or beliefs (the rationale) that motivated participants' behavioral and material choices. The analysis of post-its (v08) from co-creation sessions and the discussion during insights clustering sessions shows a greater number of instances (excerpts from transcripts) for the material and behavioral-ideological aspects. Figure 10.2 includes some selected quotes from transcripts as examples of each type. As the figure indicates, a greater number of quotes were categorized in the material and behavioral-ideological quadrants. The preference for ideological type of differences (both material and behavioral) indicates that co-creation was an effective medium for participants to express their inner beliefs and system of values with the research team. Co-creation analysis not only included the observable (both material and behavioral) aspects of cultural differences but also focused on the tacit and latent aspects.

5   DISCUSSION AND IMPLICATION

Comparing the iceberg model with the participatory design model (Figure 10.3) by Sanders (2002) reveals interesting insights. The iceberg model (Hall, 1976) illustrates that the observable aspects of culture only represent the tip of the iceberg. To study the tangible aspects of culture, there are abundant anthropological methods in design research (indicated by the breadth of the inverted triangle in Sanders', 2002, model). Traditional methods like observation, contextual inquiries, focus groups, and interviews are limited to studying the explicit and observable aspects of cultures. The intangible aspects, such as ideologies, values and assumptions form the broad foundation of the cultural iceberg. To gain meaningful cultural insights, it is critical for design researchers to uncover the core values and assumptions associated with cultures. The theoretical construct of situated cultural differences discussed in this paper provides the connection between the two models (Figure 10.3, adapted from Hall, 1976, and Sanders, 2002). Situated differences are particular artifacts, materials, practices,

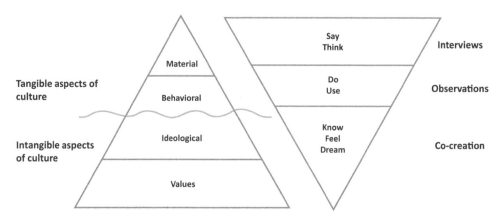

Figure 10.3 Cultural co-creation: Exploring the intangible aspects of culture.

rituals, and activities (the explicit and observable aspects) that highlight the core values of groups and in the process mobilize group identity. Situated differences are a way to theoretically and methodologically connect the explicit (material and behavioral) aspects to the implicit (ideological and values) aspects of culture. The data collected from both traditional and participatory design research methods can be synthesized using the framework of culturally situated differences.

Analysis of the DTRS11 dataset (Christensen & Abildgaard, 2017) also reiterates the importance of designing products and services that reflect core cultural values and assumptions. For example, the following are the broad research questions (v18, picture 02) listed by the design research team. These questions guided the research and ideation phase for the team. To maintain anonymity, the name of the corporation is replaced by the word "brand" in the quotes.

"This brand is for me because its value confirms mine"
"How can our brand help me broadcast my values?"
"Does the brand align with my values?"

These quotes reflect the thinking of global corporations across the world that are interested in developing products and services that resonate with the core cultural values. Corporations are conducting cross cultural studies that are increasingly focusing on the tacit and latent aspects of culture: ideologies, values, and assumptions. These cross-cultural studies conducted on consumer behavior pave the way for corporations around the world to acknowledge the importance of cultural sensitivity towards developing and marketing products for different cultures, but often leaves wanting a consistent framework for implementation.

The DTRS11 dataset and the results from this study indicate that co-creation sessions are valuable for studying the tacit and latent aspects of culture. Cultural co-creation sessions offer an interactive platform for participants to share their needs, wants, aspirations, emotions, dreams, ideologies, values and assumptions with researchers. From a researcher's perspective, co-creation sessions offer insights into both the tangible and intangible aspects of participants' culture. The step-by-step analysis of co-creation, debriefing and insights clustering sessions, presented in this paper, will provide direction for design researchers to methodically analyze cultural co-creation. Culturally situated differences can provide ethnographic practitioners and design researchers with a methodological framework for examining the culture of products, understanding culturally embedded rituals and gauging their impact on consumer behavior. The categorization of situated cultural differences into material, behavioral and ideological layers reduces the seemingly overwhelming task of studying culture into manageable layers. The layered approach should not be misunderstood as an over-simplistic approach for studying cultural aspects. As indicated in the findings section, the cultural layers cannot be studied in isolation. But it is important to recognize the importance of utilizing a layered methodological model that can guide cultural inquiries.

The theoretical framework of cultural situated differences utilizes culture as a dimension (adjective usage; cultural) that mediates our everyday experiences. Situated cultural differences then become a heuristic device for designers, researchers and corporations to discover core cultural values. The methodological framework suggested in this study can provide a yardstick for designers to measure the impacts of their designs on local cultures and to offer culturally appropriate designs. This study provides a methodological approach to analyze cultural differences and integrate diverse cultural aspects into a systematic framework.

This paper identified the different types of situated differences and reports instances that highlight the core values discovered by the research team. This study only analyzed the co-creation and insights clustering sessions. Future work should conduct an in-depth analysis of the ideation and product development phases. Does the ideation and form development reflect cultural values? How do the cultural values influence the design development? Answers to these questions can provide valuable direction for designers to design products and services that resonate will cultural values. Future work should also consider the researchers' cultural background and the potential impact on co-creation. Specifically, how does the researcher's cultural bias impact the co-creation session? Research in this area can provide new ways for investigators to acknowledge their cultural predisposition and how it impacts the co-creation process.

The initial analysis of this study was guided by typological analysis. Applying a pre-existing analysis model (pre-existing typologies, as referred to by Hatch, 2002) had certain positive implications. First, the three-layered approach was helpful to navigate through an overwhelming amount of data and to reduce and identify data most relevant to the study. Second, the situated cultural differences framework provided a systematic way to investigate and categorize data into meaningful chunks. Starting analysis with a pre-existing framework considerably reduced time for analysis. Using a three-layered pre-existing analysis model also posed certain challenges. As researchers, it was challenging to look for new and emerging themes (Hatch, 2002; LeCompete & Schensul, 1999) when the mind was guided by the three-layered classification framework. Researchers conducted line-by-line coding and compared interview excerpts to look for potential overlaps between layers and any additional themes.

6 CONCLUSION

Global corporations expanding business across cultural boundaries are conducting cross-cultural research to gain meaningful insights into local cultures. Corporations are increasingly hiring anthropologist, sociologist and social studies and cross-cultural communication experts for conducting cross-cultural inquiries. As a consequence, rapid ethnography and other anthropological methods are widely utilized for studying social interaction and practices within a culture.

Participatory co-creation provides a unique platform for researchers to gain insights into the tacit and latent aspects of culture: ideologies, values, and assumptions. Cultural values are manifested in explicit and observable aspects like material objects, practices, rituals, and interactions. In this paper, the framework of situated cultural differences provides a methodological direction for researchers and designers to discover cultural values. Situated cultural differences were categorized and mapped into four key categories. The biaxial map of culturally situated differences was based on two axes. The horizontal axis represents material to behavioral manifestation of cultural differences. The vertical axis represents the spectrum of observable to abstract aspects of cultural differences. Four categories of situated cultural difference were discovered: material-observable, behavioral-observable, behavioral-ideological, and material-ideological. The framework illustrated in this paper can help researchers to effectively analyze cross-cultural co-creation. This paper can help researchers utilize cultural differences as a diagnostic tool for studying diverse cultural aspects and categorize insights into a systematic framework.

REFERENCES

Appadurai, A. (1996). *Modernity at Large: Cultural Dimension of Globalization*. Minneapolis: University of Minnesota Press.

Bocock, R. (1993). *Consumption*. London: Routledge.

Christensen, B. T., & Abildgaard, S. J. J. (2017). Inside the DTRS11 Dataset: Background, Content, and Methodological Choices. In: Christensen, B. T., Ball, L. J. & Halskov, K. (eds.) *Analysing Design Thinking: Studies of Cross-Cultural Co-Creation*. Leiden: CRC Press/Taylor & Francis.

Christensen, C. M., Baumann, H., Ruggl, R., & Sadtler, T. (2006). *Disruptive Innovation for Social Change*. [Online] Available from: http://scholar.google.com/scholar?q=disruptive+innovation+for+social+change&hl=en&lr= (Accessed April 2009)

Daly, S., McKilligan, S., Murphy, L., & Ostrowski, A. (2017) Tracing Problem Evolution: Factors That Impact Design Problem Definition. In: Christensen, B. T., Ball, L. J. & Halskov, K. (eds.) *Analysing Design Thinking: Studies of Cross-Cultural Co-Creation*. Leiden: CRC Press/Taylor & Francis.

Du Gay, P., Hall, S., Madsen, A. K., Mackay, H., & Negus, K. (1997). *Doing cultural studies: The story of the Sony Walkman*. Sage.

Erickson, P. A., & Murphy, L. D. (2006). *Readings for a History of Anthropological Theory* (Second edition ed.). Orchard Park, NY: Broadview Press.

Grant, J., & Fox, F. (1992). Understanding the role of designers in society. *Journal of Art and Design Education, 11*(1), 77–88.

Hall, E. T. (1977). *Beyond Culture*. New York: Anchor Books/Doubleday.

Hatch, J. A. (2002). *Doing qualitative research in education settings*. Suny Press.

Hofstede, G. (2001). *Culture's Consequences: Comparing Values, Behaviors, Institutions and Organizations Across Nations*. Thousand Oaks CA: Sage Publication.

Hsu, C. H., Lin, C. L., & Lin, R. (2011) A Study of Framework and Process Development for Cultural Product Design. In: Rau, P. L. P. (ed.) *Internationalization, Design and Global*

Development. IDGD 2011. Lecture Notes in Computer Science, vol. 6775. Springer, Berlin, Heidelberg. pp. 55–64.

Kroeber, A. L. (2006). What Anthropology Is About. In: P. A. Erickson, & L. D. Murphy (eds.) *Readings from a History of Anthropological Theory*. Orchard Park: NY: Broadview Press. pp. 114–124.

LeCompte, M. D., & Schensul, J. J. (1999). *Designing and conducting ethnographic research* (Vol. 1). Rowman Altamira.

Lee, K. P. (2004). Design methods for cross-cultural collaborative design project. *Design Research Society International Conference*. Melbourne: Futureground: Monash University.

Lin, R., Sun, M. X., Chang, Y. P., Chan, Y. C., Hsieh, Y. C., & Huang, Y. C. (2007). Designing "culture" into modern product: A case study of cultural product design. *International Conference on Usability and Internationalization*, 146–153.

Lloyd, P., & Oak, A. (2017) Cracking Open Co-Creation: Categorizations, Stories, Values. In: Christensen, B. T., Ball, L. J. & Halskov, K. (eds.) *Analysing Design Thinking: Studies of Cross-Cultural Co-Creation*. Leiden: CRC Press/Taylor & Francis.

Miller, D. (2005). *Acknowledging Consumption*. London: Routledge.

Pieterse, J. N. (2006). Globalization and Culture: Three Paradigms. In: G. Ritzer (ed.) *McDonaldization*. London: Pine Forge Sage. pp. 278–283.

Prown, J. D. (1982). Mind in matter: An introduction to material culture theory and method. *Winterthur portfolio, 17*(1), 1–19.

Ritzer, G. (2006). *McDonaldization*. London: Pine Forge Sage.

Rose, K. (2004). The development of culture-oriented human machine systems: Specification, analysis, and integration of relevant intercultural variables. *Advances in Human Performance and Cognitive Engineering Research, 4*, 61–104.

Sanders, E. B. N. (2002). From user-centered to participatory design approaches. *Design and the social sciences: Making connections*, 1–8.

Schwartz, S. H. (1994). Are there universal aspects in the structure and contents of human values? *Journal of social issues, 50*(4), 19–45.

Shaw, D. S., & Clarke, I. (1998). Culture, consumption and choice: Towards a conceptual relationship. In: *Consumer Studies & Home Economics, 22*(3), 163–169.

Spencer-Oatey, H. (2000). *Culturally speaking. Managing rapport through talk across cultures*. London: Continuum.

Tomlinson, J. (1999). *Globalization and Culture*. Chicago: University of Chicago Press.

Trompenaars, F., & Hampden-Turner, C. (1997). *Riding the Waves of Culture:Understanding Cultural Diversity in Business*. London: Nicholas Brealey Publishing.

Williams, R. (1976). *Keywords*. London: Fontana.

Chapter 11

Designers' Articulation and Activation of Instrumental Design Judgements in Cross-Cultural User Research

Colin M. Gray & Elizabeth Boling

ABSTRACT

Cross-cultural design practices have begun to rise in prominence, but these practices have infrequently intersected with common user-centred design practices that value the participation and lived experience of users. We identified the ways in which the design team referred to co-creation workshop participants during the design and debrief of the workshop, focusing on how these references invoked or implicated the design team's understanding of Chinese culture. We identified referents to the participants, using occurrence of third-person plural pronouns to locate projection of and reflection on participant interaction. In parallel, we performed a thematic analysis of design and debrief activities to document the team's articulation and activation of instrumental judgments relating to culture. The team's instrumental judgments shifted substantially across the design and debrief session, moving from totalizing cultural references in the design phase to frequent translator-mediated interactions in the debrief phase. Translators "nuanced" the cultural meanings being explored by the design team, while team members attempted to engage with cultural concerns by "making familiar" these concerns within the context of their own culture. Implications for considering culture as a part of standard user research methods and paradigms are considered, along with practical considerations for foregrounding cultural assumptions in design activity.

I INTRODUCTION

Methods that engage potential users in the design process are numerous (e.g., Goodman, Kuniavsky, & Moed, 2012; Martin & Hanington, 2012; Olson & Kellogg, 2014), but it is rare for these methods to articulate the means by which a designer or design team might meaningfully engage with users who draw from different cultural backgrounds. Design processes are inherently power laden (Irani, Vertesi, Dourish, Philip, & Grinter, 2010; Nelson & Stolterman, 2012), driven primarily by the designer with the end goal of bringing about intentional change. Dominant user-centred approaches, such as participatory design or co-design, are intended to equalize power and welcome users into the design process as co-designers. These approaches characterize a push towards egalitarianism on a philosophical level, even while pragmatic concerns about the translational and technical role of the designer – along with the power that these roles assume – can be easily neglected or deemphasized

(Carr-Chellman & Savoy, 2004; Hakken & Maté, 2014; Nielsen, Bødker, & Vatrapu, 2010).

As design for global use becomes increasingly common, there has been a concomitant rise in design principles and practices that are intended to highlight and manage cultural considerations across the project lifecycle (e.g., Barber & Badre, 1998; Bell, Blythe, & Sengers, 2005). Cross-cultural methods exist in parallel with other models of user-centred design (UCD), but rarely directly intersect. Many of the primary examples of poor cross-cultural design can also be seen as examples of failed user-centred design – where Western design imperatives or assumptions often trump cultural realities in a local, non-Western context.

In this chapter, we document how judgments are invoked and supported in design activity, contextualizing these judgments within cultural assumptions of the design team as they planned a co-creation workshop with Chinese consumers. Cross-cultural design practices have risen in prominence, but these practices infrequently intersect with design methods that value user participation. While design judgments necessarily take on a cultural character from multiple stakeholder perspectives, designers' intentional interaction with design methods have not consistently addressed this complexity. While culturally-centric approaches to design activity do exist, these methods are often distanced from UCD approaches that appear on the surface to equally value the user and her lived experience.

2 THEORETICAL FRAMEWORK

2.1 Design judgment

When practitioners engage in design activity, they continuously make judgments about their understanding of the design situation and what steps they should take next (Nelson & Stolterman, 2012). These judgments frequently invoke tacit knowledge (i.e., Polanyi, 1966; Vickers, 1984), and as such are often inaccessible even to the designer herself, yet form the backbone of design expertise (Lawson & Dorst, 2009) and are a recognizable output of "practical knowledge" or *phronesis* (Dunne, 1997, 2005; Stolterman, 2008). We build upon on a theoretical typology of design judgments created by Nelson and Stolterman (2012), which has enriched our understanding of the tacit expertise that practitioners draw upon to direct their design activity. Our prior research using this typology (Gray *et al.*, 2015; Korkmaz & Boling, 2014) has built on the previous scholarship to understand the interplay between judgment and tacit knowledge in design cognition, relating professional expertise both to knowledge of theory and formalized methods as well as a rich repertoire of lived experience and precedents.

While design judgment has often been discussed in theoretical terms, the ways in which these judgments are responsible for directing design activity is unclear, representing a substantial area for further scholarship. We focus our attention on a specific type of judgment proposed by Nelson and Stolterman (2012): *instrumental judgment*. Instrumental judgment refers to a designer's interaction with her tools: "a process of mediation that considers not only technique and which instruments to use, but proportion and gauge, as well" (p. 152). We rely on previous definitions of tool use in design activity that are intentionally broad, encompassing design methods as one class of tools that a designer interacts with to facilitate their design process (Gray, 2016a; Stolterman, 2008). Instrumental judgment is most accessible *in situ*, making this extensive dataset an ideal means of documenting the formation and realization of design judgments over time.

2.2 Design methods and cross-cultural design

Design methods support ways of thinking and acting, used by designers to work through a design process. Methods pertaining to user research and elicitation of needs are one subset of design methods, often existing within larger paradigms of methods (e.g., human-centred design; participatory design). Previous work has shown that practitioners view design methods as flexible "cores" (Gray, Stolterman, & Siegel, 2014) or "ingredients" (Woolrych, Hornbæk, Frøkjær, & Cockton, 2011) which are used to communicate the complexity of practice. However, even in their most prescriptive forms, design methods are generally underspecified (Gray, 2016b), with a wide range of interpretation assumed to be carried out by the designer. The judgments taken in relation to methods are largely *instrumental* in nature, involving questions such as: "What method will give me the information I need next?"; "How will I know when the method has produced the right sort or amount of information to proceed?".

Krippendorf (2005) posits that designers contribute not only explicit and knowable design outputs (e.g., products), but also contribute to design *discourses*. These discourses include designers' tacit (often normative) understandings of the world, which can encompass assumptions of cultural appropriateness, value, or ethical behaviour (e.g., Geertz, 1973; Hofstede, 2001). The term *cross-cultural design* foregrounds these aspects of design behaviour – which are always present to some degree – but are heightened when cultural boundaries are traversed. Cross-cultural concerns can relate to product, designer, and user along cultural dimensions such as geography, ethnicity, nationality, or socioeconomic status (Tai, 2008; see Dhadphale *et al.* (2017) for a comprehensive overview).

While mundane examples of cross-cultural design and use abound, particular instances that foreground issues that arise when cultural concerns are not considered may be especially illustrative of the challenges of designing within a discourse that is not one's own. Chavan, Gorney, Prabhu, and Arora (2009) share an example of the Whirlpool *World Washer*: a washing machine that was designed to serve multiple countries, including Mexico, China, India, and Brazil. In this process of cultural generalization, the impact of the mechanism was not considered for garments that were out of the Western mainstream, such as the Indian *sari*. Due to their length, saris were caught between the agitator and washing drum, resulting in shredded clothing and a lack of trust in the Whirlpool brand.

Design methods are a primary access point into these cross-cultural issues, shaping interactions with users and allowing designers to define and frame appropriate design problems to solve. And indeed, scholars have previously identified ways of using user research methods and paradigms of user engagement such as participatory design to account for cultural concerns (e.g., Hakken & Maté, 2014; Nielsen, Bødker, & Vatrapu, 2010), along with broader interests in designing for interactions and products with culture in mind (e.g., Bell, Blythe, Gaver, Sengers, & Wright, 2003; Horn, 2013).

3 RESEARCH FOCUS

We identify a set of research questions that address how instrumental design judgments are articulated and activated by a design team as they navigate design research methods and the cultural dimension of their specific project. This context included a design team and participants from distinctly different cultures and design goals that specifically engaged cultural concerns.

1 How does the design team reference the participants of the co-creation workshops during the planning of those workshops?
2 How does the design team reference the participants of the co-creation workshops during the debrief sessions following the workshops?
3 In what ways do instrumental judgments function regarding the workshop participants and the translators with respect to co-creation as a method and meaning making as an outcome of this method?

4 METHOD

To investigate these questions, we analysed multiple planning and debrief sessions from the DTRS11 dataset (Christensen & Abildgaard, 2017) using discourse and thematic analysis to build an understanding of how design judgments are invoked, and the cultural implications of these judgments. Through our analysis, we identified key assumptions team members articulate about workshop participants. Building on these assumptions, we then identified how the design team's cultural understanding shaped their professional judgments, and how these judgments related to expected and actual workshop outcomes.

4.1 Data in use

Based on our review of the dataset, we focused our attention on the main design team, spanning the design process (Table 11.1; e.g., design: v2-5; debrief: v7;v14; interviews: v1;v22) to document changing perceptions in cultural assumptions, understandings, and co-creation methods across time. We centred our analysis on the workshop design and debrief, using interviews with Ewan to ground our understanding of the design, execution, and outcomes of the workshops.

4.2 Analysis

We addressed the selected data through three intertwined layers of analysis. First, we built a broad view of the corpus through lightweight discourse analysis, revealing the frequency of referents to workshop participants. Second, we thematically analysed the sessions to reveal how the design team referenced and relied upon notions of culture

Table 11.1 Dataset videos by focus area.

Focus area	Relevant videos
Co-creation workshop design	v02 Designing co-creation workshops
	v03 Iterations on workshop design
	v04 Iterations on co-creation workshops
	v05 Designing co-creation workshop day 2
Co-creation workshop debrief	v07 Debrief co-creation workshop day 1
	v14 Debrief co-creation workshop day 2
Initial and final interviews	v01 Background interview with Ewan
	v22 Final interview

when understanding the actions of the workshop participants. Finally, we synthesized the results of the discourse and thematic analyses to clarify the instrumental judgments of the design team over time.

4.2.1 Lightweight discourse analysis

In this phase, we located points in the discourse where workshop participants were referenced by the team. Because of the size of the corpus (in words: design n = 45557; debrief n = 19249; interview n = 21464), we relied on word frequency for this task. We qualify this instantiation of discourse analysis as "lightweight" to acknowledge our mechanical approach to understanding word usage within the corpus, rather than dwelling deeply on features of the discourse.

First, we identified the frequency of third-person plural pronoun stems (i.e., they, them, their). This allowed us to capture references to participants in the workshops, both projected (design) and retrospective (debrief). All occurrences were identified using a text-parsing PHP script. Many speech acts included multiple instances of one or more plural pronouns, and the referent of the pronoun sometimes shifted within the act.

Second, we identified the referent of each pronoun to determine how instrumental judgments were made. Because the referent of plural pronouns is indistinct without parsing the context, we coded each occurrence, focusing on each occurrence of the target pronouns and coding it discretely using an open coding approach. This resulted in 36 codes, which included a wide range of referents (Table 11.3). The codes were then compared and consolidated, using verification of coded excerpts to ensure a balance between referent precision and patterns of referent use across the dataset. This consolidated set of codes resulted in 21 distinct codes.

4.2.2 Thematic and synthetic analysis

We carefully read the transcripts of all selected videos multiple times, noting themes that related to the workshop participants, cultural assumptions and understandings, and the ways in which these features factored into instrumental judgments. Each researcher identified an initial set of themes based on their individual reading of the transcripts. These themes were then refined, consolidated, and illustrated using exemplars from the data. Reviewing the thematic and discourse analyses together, we established when and how these themes manifested across the design process, representing shifts in instrumental judgments over time.

5   FINDINGS

Based on our analysis, we present three sequential layers of findings. First, we describe the frequency and referent of the third-person plural pronouns used by the design team across the corpus, focusing on workshop participants and translators. Second, we summarize the themes relating to cultural awareness and consideration of cultural issues in design and debrief phases. Third, we synthesize the referents and themes to reveal how the design team shifted in their utilization of instrumental judgments in relation to the cultural content of these judgments.

5.1   Referents to co-design participants

We analysed the frequency of third-person plural pronouns to understand the ways in which the design team referenced workshop participants. Through this analysis, we captured all instances of pronouns that had the potential to reference participants (i.e., they, them, their) to understand more fully how the conversations about and regarding third parties changed across the design and debrief sessions.

5.1.1   Third-personal plural pronoun frequencies

All speech acts were analysed to locate instances of plural pronouns. (Table 11.2). This analysis revealed a substantial increase in use of third-person plural pronouns as the design phase ramped up, with the target words increasing from 1.79% of all words in v02 to 3.47% of all words in v05. In the debrief sessions, the results were mixed, with a jump from 0.98% of all words in v07 to 2.14% of all words in v14. This reveals a tendency to talk more about individual participants in v07, resulting in a higher incidence of first-person singular pronouns (i.e., "he", "she") regarding the actions of specific individuals. This use of first-person referents continued in v14, but there was a substantial increase in less descriptive plural pronouns, similar to the late design videos.

5.1.2   Referents of pronouns

We coded 1844 target pronouns across the selected videos in two stages (Table 11.3). Our final coding scheme included 22 codes, with an average of 87.81 instances per code (SD = 305.19). The highest frequency codes included: workshop participants (n = 1445), company design team (n = 155), company stakeholders (n = 47), design concepts (n = 28), and the product end user (n = 28). Only the workshop participants code appeared in all videos. Most pronouns coded as "workshop participant" occurred in the design of the workshop (i.e., videos 2–5), totaling 1078 instances. This reveals a rapid increase across the four sessions, reflecting both an increase in the design session length (v02 = 4235 words; v05 = 17649 words) and in the frequency of projected references to workshop participants (v02 = 67 [1.58% of all words]; v05 = 571 [3.24% of all words]).

Table 11.2 Word frequency of plural pronouns and other common words.

	Design				Debrief		Interview	
Word	v02	v03	v04	v05	v07	v14	v01	v22
they	45 (1.06%)	129 (1.35%)	206 (1.46%)	444 (2.52%)	27 (0.83%)	264 (1.65%)	27 (0.49%)	172 (1.08%)
them	26 (0.61%)	40 (0.42%)	107 (0.76%)	130 (0.74%)	5 (0.15%)	60 (0.38%)	10 (0.18%)	51 (0.32%)
their	5 (0.12%)	15 (0.16%)	10 (0.07%)	39 (0.22%)	0 (0%)	18 (0.11%)	4 (0.07%)	11 (0.07%)
they+ them+ their	76 (1.79%)	184 (1.92%)	323 (2.29%)	613 (3.47%)	32 (0.98%)	342 (2.14%)	41 (0.74%)	234 (1.47%)
Total words	4235	9574	14099	17649	3271	15978	5505	15959

Table 11.3 Referent of plural pronouns organized by code and frequency.

		Design				Debrief		Interview	
Word	Total	v02	v03	v04	v05	v07	v14	v01	v22
animals	5				5				
assumptions	1			1					
Chinese	10		8		2				
company	8							8	
company stakeholder	47		1				3	25	18
concepts	28			1	25				2
design team	155		19	3				5	128
end user	28						16	3	9
materials	2							2	
other company	22						22		
other people	21						19	2	
outcomes	1								1
process	1								1
product	12				2		4	1	5
project	2							2	
recruitment agency	6		6						
translator	17		17						
translator (previous)	6		6						
workshop logistics	10	2	1		2	1	1	1	2
workshop participant (previous)	18	7	4		7				
workshop participants	1445	67	122	318	571	31	268	2	66
Total	1844	76	184	323	614	32	341	41	234

The referent frequencies reveal the team's expectations about the workshop participants. There is clearly a high occurrence of workshop participant referents, and the specificity and frequency of these referents increased significantly across the four sessions. These referents take on the character of *projection*, where the design team is either articulating the structure or logistics within which the participants will interact, or is anticipating a specific participant response. In the first set of examples, Ewan and

Kenny are suggesting activities to engage the participants, invoking them in a relatively mechanical way (e.g., they will do X, and then Y will happen).

---

v05, 399

E: Yeah 'cause then, that was the question. Should they tell the names while they're in their separate group, which is good because it's faster, we can do it parallel, but it's bad because it will (.) [the other group, the oth'- yeah yeah].

---

v05, 824

K: Yeah, maybe they should just be allowed to have these six hearts or whatever and then put them on, and we: collect the things and then they see the result and they can discuss whether this eh order still makes sense for them. Then they will have a top three from the result, but they can discuss whether it makes sense.

In other examples of projection, the design team moves beyond logistical ordering of the workshop to anticipating the responses of participants. In these examples, E suggests that participants will want to see results from the competition (v05/589) and that one idea will probably win (v05/735).

---

v05, 589

E: Yeah, but still, they wanna see the result, we're building up something and I think it's just- it's a little (.) unfair to- or, should I say it's a little unconclusive to not share that.

---

v05, 735

E: One idea will win, probably, and then they will- we will say okay, the ideas that comes out here are out, but the three top one or two top one or three top one are still in the game.

No Chinese team members were involved in the design phase, which possibly explains the lack of deep cultural consideration when participant interactions were projected. Except for references to "all Chinese people" in v03, participant interactions projected in the design sessions were almost completely free of cultural content or explicitly articulated awareness of cultural appropriateness. In the rare instances where Chinese culture or ethnicity became the centre of conversation, culture was instantiated through stereotypes, as in this example from Kenny in v03:

---

v03, 284

K: And especially Chinese people, [...] I think actually they don't really need to speak Chinese when they go abroad, because they're- I know it's a little bit prejudice, but [they- they are usually in groups, yeah yeah so they basically don't need- need any other language than Chinese].

Analysis of these discourse segments in the data reflect strongly Eckert and Stacey's (2017) observation that the design team was functioning instrumentally in a mode analogous to that of fashion designers "buying an object in a shop" to add to a mood board that serves as a springboard for the next season's designs, not contributing anything specifically inherent to the object or the context from which it came.

5.2 Invocation of culture

In this section, we document some ways in which the design team invoked culture in the design and debrief of the workshop. This description of themes is not meant to be exhaustive, but rather a broad overview of the ways in which culture is discussed, relied upon, or operationalized in design team discussions. A fuller account of how these cultural views are used to build consensus through instrumental judgment is provided in the next section.

5.2.1 *Explicit and implicit efforts to understand elements of culture*

The team made both implicit and explicit efforts to understand the elements of culture confronting them within the statements and behaviours of the workshop participants. They were clearly seeking these elements as a core aspect of the design effort and the team leader mentioned more than once that it was his goal to get "something more than surface level stuff" [v22/104] – presumably the "stuff" that distinguished the Chinese participants as consumers.

5.2.2 *Assuming cultural equivalents*

A number of statements signal that the speaker is tacitly assuming equivalence between what the workshop participants have said and some elements of the speaker's own cultural background. The first example here is straightforward, whereas in the second the speaker *may* be applying an understanding garnered from Phase I but equally may be assuming that a Western trope – the celebrity as a communicator of positive values – will be applicable in this context.

---

v14, 117

E: ... so many things of what they say is like, wow, we- we have that in our company. We just need to tell that story. We need to make sure the product or the service we have eh: the pipelines for telling a story like that.

---

v14, 156–160

E: ... so you have- maybe a foreign brand has some values, but maybe it's difficult to understand that value, because they're foreign values. But how can we find the: local counterpart of those values ... so let's say that THE COMPANY has a value that is (...) safety, but let's say it was difficult for Chinese people to understand what that safety was, but there was a person, a famous person, or an artist or whatever (.) in China, who shares some of that same values (INAUDIBLE), then for them to understand it, we can use that person or that brand or whatever, as this channel ... [160] as a value communicator.

5.2.3   Relating observations to familiar contexts

Attempts to reach cultural understanding are made via two types of relation to contexts familiar to the team members; the first is to contrast what has been expressed with the team member's own experience. Designers highlight the new concept by considering how it differs from the familiar one. In this example, this attempt may be effective in describing the conceptual differences between two views of medicine. It does, however, miss the stated implication of the translator's explanation which is that this is a traditional view of medicine which is nevertheless still a large part of the lives of the young, upwardly mobile and presumably modern, workshop participants.

---

v08, 038–039

W: "Ah: traditional Chinese medicine is quite big eh:" says it is about "maintenance" instead of going to the doctor after a problem occurs
E: "That's actually a very smart way of looking at it. Western medicine is after: you get sick, you get fixed up, here it's like 'drink this tea;, do this thing and you won't: get sick.' "

5.2.4   Asking and explaining from translators

Looking to the translators for answers to specific questions is an appeal directly to the Chinese context for understanding. These appeals are generally for the purpose of clarification: Is this an actual saying?; Are these people "typical" for this city?; What does this expression mean? Arguably, this appeal to the translators is made when there is a clear need for clarification and no ready assumption or analogue is available to address a specific situation.

v09, 253–264

R: He talked about a table of old friends that was like-

...

E: Yeah, yeah, table- is it an expression? [in Chinese]

R: No, but ok I guess the four [characters] is put together quite nicely. (Speaks Chinese) that's exactly what he wrote.

W: (Repeats in Chinese), yeah.

R: But that's exactly what he wrote, it's not an expression, but it literally translates [into old-] table of old friends.

v14, 239

After discussing the personality traits of participants

E: 'cause I was like, is- is she like a typical CHINESE CITY woman, and she was like, oh yeah. And the guy Brian, he was like the typical CHINESE CITY man, like, husband.

5.2.5 Trying out understandings

A common effort at cultural understanding blended with the general design discourse and took the form of paraphrasing, as in the first example, or of a trial statement as in the second (which was followed by further refinement on the part of a translator). While not signposted as efforts to understand culture, as when explanations were requested directly from the translators, these try outs were clearly the primary explicit form of attempts to understand Chinese culture in this design team context.

v22, 114

E: "One of themes ... when we do research in China is kind of the, the whole collectiveness versus – versus individual: and versus the (Guan Shi?) kind of network ... they are individualistic and they are all about money in many ways, but then again they're so traditional and so about the, the society ..."

5.2.6 Explicit and implicit limits on cultural perspectives

During the planning meetings, it was patently clear that the team recognized language as potential barrier to meeting the goals of the workshops and much effort was expended to provide for sufficient, effective translation of what would transpire during the workshops. Beyond the literal translation of what was said by the participants, however, a number of limitations arose regarding the cultural perspectives that the team was able to incorporate into their designing. The limits on cultural perspective noted in our analysis may be understood clearly within Clemmensen *et al.*'s (2017) discussion of the availability of cultural knowledge; the accessibility of this knowledge was limited for the Scandinavian designers by the context provided by the Chinese team members.

5.2.7 A historical discourse positions

While it is not credible to assume that team members are wholly unaware of China's recent history, the Cultural Revolution, significant urban migrations, and even the genesis of Taiwan's claim to status as an independent state – all creating profound disruptions in social order and therefore expectations regarding work – statements like these suggest that what awareness the team members undoubtedly did have was not being applied to the current endeavour.

---

v09, 006

AM says, "In Chinese, the family name comes first … before your own: So I think that carries a lot of weight, right? Like, I am a reflection on my family because it is so intertwined. […]"

E: "And of course this goes back to-this is ultra-European, 'cause Europe had tons of revolutions … to remove themselves from this [the assumption that you would do what your parents do] …"

Expressing the hope that the sentiment expressed in one participant group is actually a trend among their target group in general likewise reveals a position within the team's discourse that takes the immediate situation at face value without connecting it to larger trends. In this instance, signalling social status through patterns of consumption (Martineau, 1958) has long been observed, as has the interaction of brand consumption with such signalling, particularly in developing countries (Batra *et al.*, 2000). The notion that upwardly mobile workshop participants might both be growing out of a knee-jerk fascination with foreign brands and stating that people "despise" those in their society who have not progressed past consumer behaviour that signals "developing country" status can probably be assumed to be pervasive.

---

v08, 222–230

Discussion of respondents shifting from caring about brands ("vulgar;" "people kind of despise these people now") to experiences … E says "I hope that is a trend: that it is bigger than – that it wasn't just planted in our little group."

5.2.8 Awareness of cultural limits on understanding

Alongside the apparently unconscious disregard for historical context in some of the statements made by the team, some of Ewan's explicit statements make it clear that there is a fully conscious awareness that the understandings achieved via these workshops (and presumably the Phase I ethnographic study) are limited.

---

v22, 114

E: "I'm never gonna crack that code . . . it is really difficult for me as a white man to grasp that, and I think, or a white woman . . . a person that is not Chinese or Asian . . ."

5.2.9 Explicit generalizations

Despite a good deal of the discussion that treated workshop participants as individuals during the debrief sessions (in contrast to the planning sessions where the discourse analysis showed referents to the participants in the aggregate as "them" and "they") cultural generalizations persisted into the debriefings.

---

v14, 036

E: ... they are (...) much more individualistic than like, I don't know; maybe "our" view of who the Chinese person are. At least they have a voice, they wanna say something. And I think that's great.

---

v14, 318–320

E: The whole kind of balance metaphor is so powerful and so:-
   (they all speak at once)
E: So Asian!

5.3   Shifts in instrumental judgment over time

One of the shifts that we observed is likely both natural and subliminal on the part of the design team – from using "they" and "them" to refer to the workshop participants, to using names to identify individual participants. While this form of instrumental judgment, impacting what is being foregrounded in terms of methods, does not appear to be an explicit decision on the part of the design team, it is, nevertheless, influential in the design process. What was anticipated to be raw material offered by the participants from which to fashion ideas required by THE COMPANY, turns out to be, after this shift, more piecemeal than expected.

Another shift in instrumental judgment, this one also unreflective on the part of the team, is the role of the translators. The team plans for this role as a somewhat unwonted, or awkward, necessity. Translators are assumed to allow designers to extract the material they need from the workshops even though the designers do not speak Chinese and the participants do not frequently speak English. As the workshops unfold, and are later reviewed, the collective instrumental judgment of the team shifts to positioning the translators as cultural translators, rather than simply English-to-Chinese translators.

In this shift from language translator to cultural translator, we noted a shift in instrumental judgment taking place, moving from a mechanical view of translation as part of the co-creation process to translation as "nuancing" or facilitating the "making familiar" of the workshop participants. Each of these instrumental judgments made by the design team are discussed in more detail in the next two sections.

5.3.1   Nuancing

During the debrief, sharing insights and clustering insights sessions, the translators offered explanations and observations as part of the team discussion. We identified a specific form of explanation we are calling "nuancing." Where this takes place, one of the non-Chinese team members states an understanding of a cultural point and one of the translators follows up with one or more statements that do not contradict the understanding directly but either add layers or gradation, repositioning the understanding.

---

v14, 143–146

R: But- but there's also a lot more trust for the foreign brand, because when the:- I mean for us they were talking about, they were afraid about data, about their privacy being divulged, and then so they want- or some people say that is you have the server in Europe then that's because (INAUDIBLE) they believe that somewhere foreign as we trust versus (INAUDIBLE).

E: Even- yeah and of course, sure, we should focus on that right now, but how many years until that will change. 2 years, 5 years, 10 years, never? Will it change? I feel that we're on the steps of having- I use Chinese, you know:, cloud-services, I buy, like

R: No, but- but- but I don't think they actually think about the actual data on the server itself, but I think they are thinking of that if you have some- if they- they believe in the foreign brand, right?

E: Yeah, the foreign brand, rather than domestic

Rose shifts Ewan from the notion of not trusting Chinese cloud services (which E trusts and uses), to the idea that there is more trust in foreign brands than Chinese brands. Additional information from Rose's lived experience using Chinese cloud services is used to build on Ewan's understanding of these services, adding the complexity not only of the server location (i.e., foreign v. local) but also the security of the data and the "foreign-ness" of the service brand.

These instances of nuancing are sometimes taken up by the team, as illustrated in the previous example, and sometimes not. In the next example, the symbolism of the sun rising and the location that is implied (i.e., city v. ocean or sea) is explored by Will, but not taken up for further conversation by Ewan.

---

v14, 348–349

E: And this one, which was pretty cool, was like, eh: the sun rises, which of course like is very Chinese or Asian ...

W: It must be over – it must be rising from ocean or sea, and not over city, 'cause the city is like- you see a cityscape and you all busy, chaotic and you feel stressed ... The sun rising from ocean is like liberation, freedom, enjoyment, yeah yeah

---

5.3.2 Making familiar

Non-Chinese members of the team, while attempting to understand the perspectives of the workshop participants, often sought to find an equivalence or analogue to those perspectives in their own experiences or understanding of the world.

In the two examples that follow, the design team proposes a cultural analogue to enrich the conversation. In the first example, Ewan compares Will's translation of a Chinese expression to a similar Scandinavian cultural expression. This "making familiar" appears to suggest consensus building on the part of the Scandinavian design team, ensuring that the correct meaning is being made from the interaction with the translator, while also explicitly bridging colloquialisms (and the deep cultural meaning that is implied) across two cultures.

---

v14, 278–285

W: So I guess the whole thing about 先苦后甜 which is the traditional way of saying ... First bitter, and then (...) later sweet

E: Yeah, it's kind of scar- or how do you say in [a Scandinavian language ...] okay I eat this really boring porridge right now, and now I get full, and then I have a little bit of cake afterwards. First you do the crappy part ...

In the second example, Tiffany suggested interactions of Chinese customers with the factory in Scandinavia. She used a Scandinavian cultural reference – a "train that goes inside the factory" – which is then taken up for discussion in a Chinese context after Tiffany explicitly asks "would the Chinese people like that?". In this case, the translators are asked to "make familiar" a Scandinavian cultural scenario and assess its appropriateness for a Chinese audience.

---

v14, 848–857

T: But the Chinese people would then- like, in SCANDINAVIA you can take this train that goes inside the factory.
E: Mhm?
T: Eh: would the Chinese people like that?
W: Yeah, I agree-
. . .
AM: Especially for people like Brian 2, who love this kind of story.
. . .
AM: So for him, having the access to be there (.) is the source of credibility.

---

6 DISCUSSION

Instrumental judgments made by the design team can be traced across the design process, leveraging the articulation of the designers that is occurring in relation to culture and the deeper meaning-making afforded through our analysis. In particular, we call attention to several key moments in the design and debrief process where the articulation of *instrumental judgment* foregrounds moments where the design team tacitly or explicitly guided their use of design methods to increase their understanding of the targeted Chinese consumers.

First, the shift in the defined and implied role of the translators in the workshops can be seen as a manipulation of *translator-as-instrument* in the data collection process, moving from a utilitarian to a nuancing role across the design and debrief sessions. In the design of the workshop, the team treated the selection and coverage of translators as a logistical hurdle to overcome, rather than as a means of entering and understanding a different culture. The translators emerged only in a mediating role during the debrief sessions, insufficiently leveraged in exposing the cultural depth of the workshop participants, but still deepening the design team's understanding of the target user population.

Second, the organization of activities relied heavily on previous workshopping experiences, potentially reducing the conversation about cultural limitations of these activities for the present user context. The instrumental judgments related to the activity selection brought assumptions about the potential user interactions that would be involved, and because of the previous success of some methods, any potential that the cultural fabric of the Chinese participants would be incompatible was largely left unexplored. This set of instrumental judgments in relation to activity selection was due to the initial absence of a cultural sounding board, even though it appears that such resources may have been available from the Phase I ethnographic study. Regardless, as

Daly *et al.* (2017) point out, the team returned to the seven pillars from the first workshop over the course of their discussions and Eckert and Stacey (2017) characterize this return as a changing the participants' views to others or even dismissing them.

Finally, the use of "co-creation" as a paradigm for user engagement took on the quality of an "observed generative activity," implicating a set of instrumental judgments about the depth of engagement that was appropriate or desirable given the cross-cultural context and language barrier. While the design team acknowledged in the design of the workshop that it could barely be considered "co-creation," the team appeared to desire more engagement from the output in the debrief session than was allowed given the highly structured activities that were planned. This set of instrumental judgments prioritized a set of known generative insights over activities that may have been more "messy," yet would have engaged participants more directly.

6.1 Foregrounding the cultural assumptions of instrumental design judgments

Instrumental judgement, tacitly performed through routines of "nuancing" and "making familiar," points to an implicit view of culture which Geertz (1973) identifies, specifically that "men are men under whatever guise and against whatever backdrop" (p. 34). In this view, behaviours or beliefs observed across cultures are understood to be expressions of essentially identical human characteristics that have manifested themselves differently owing to time, place and circumstances. This holds the promise that such an understanding may be possible, and possible in one's own terms – that is, that the *other* may eventually be rendered almost as fully knowable as one's self. Such an implicit view of culture may be termed a "core judgment" (Nelson & Stolterman, 2012), which, although it has general application in a designer's life outside the processes of designing, is particularly relevant to the design method being used in this project and the instrumental judgments being observed. Hess and Fila (2017) make an explicit link between the "figured worlds; the user-centric assumptions the design team bring with them and shape throughout the design act" and both the design of the co-creation workshop and its outcomes.

Pragmatically, "making familiar" can help designers reach an insight which, while not fully accurate, may satisfy the need to move forward in designing when the time and resources to reach a more nuanced understanding are not available or perceived to be necessary. However, if a parallel is fallaciously drawn between an unfamiliar cultural concept and one's own experience, or if the nuancing activated by a culturally-knowledgeable collaborator is misunderstood, the insights thus yielded could miss the mark. Some instances of nuancing on the part of the translators seemed to take place in response to such shortcut understandings when the design team moved too quickly to draw parallels that were flawed or otherwise inappropriate.

The team might also be seen to be using the co-creation method from a cultural studies position in line with the focus of du Gay *et al.* (2013) who address the study of artefacts from multiple perspectives (i.e., representation, identity, production, consumption and regulation) to come to ascribed cultural understandings of meaning, and the relationships that people have with them. However, observing the team engage in meaning making by "making familiar" suggests that cultural relativism manifest in relationships between people and artefacts exists in an unexamined muddle together with a core judgment which regards all people as sharing fundamental needs and desires onto which the layers of culture are overlaid and may be plucked off to reveal an individual much like one's self.

These understandings of the team's culturally-situated behaviours lead us to consider the perhaps unexplored complexity of engaging users in participatory or co-design traditions in cross-cultural contexts, and the ways in which cultural assumptions bound up in instrumental judgments may be more fully articulated.

6.2   User research methods "considered harmful"

Commonly accepted methods for conducting user research have alternately been created by practitioners and adapted from scholarship (Gray, Stolterman, & Siegel, 2014; Stolterman, 2008). Such methods may be at risk of losing potential value because of a strong instrumental focus which keeps engagement along cultural dimensions comparatively shallow. Attending only to commonly accepted user research collection and analysis practices may leave important subjective dimensions, such as culture, inaccessible to the designer. In addition, the adaptation of methods – when carried out without departure from a purely instrumentalist view – may discourage attention to complicating constructs such as culture, thus neglecting assumptions key to the use of user research methods that may in fact yield the most useful and generative insights.

It is possible that user research methods could be "considered harmful"[1] in some sense, in that these methods encourage attendance to the lived experience of users within the paradigm of user-centred design, even while the methods themselves can be too blunt to distinguish areas where intersubjectivity among the designer and users is very difficult or impossible. If key cultural considerations as one source of complexity in design practice are systematically or occasionally neglected, even in a design paradigm where the user is intentionally made more central (i.e., co-creation), perhaps this points to a larger issue of instrumental judgment that so often skirts the ways in which meaning is made and articulated into the design process. Perhaps, echoing Buchanan's (2015) concern regarding the diluting of judgment and complexity in relation to the design thinking movement, current user-centred practices encourage "quick wins" that yield rapid design insights rather than deeper study and analysis that leads to a fuller sense of the complexity of the design space. Thus, an instrumentalist view that does not adequately attend to holistic and societal concerns may weaken design processes in ways that are counter to the grounding principles of user-centred design. The intersubjective dimension of user-centred research methods that allows access to such complexity has been particularly understudied, and the nature of meaning-making among designer and users would be an ideal area for further study.

The ethic of user-centred design is focused on understanding users and engaging them directly in the design process. However, in this series of design and debrief sessions, the conversations regarding participants often relied on unvalidated assumptions about culture, occasionally engaging in explicit cultural stereotypes. While the potential for foregrounding cultural meaning-making was present, the awareness of the importance of such cultural sensitivity did not appear to be a recognized priority. In the early design sessions, the team engaged in detailed discussions that projected participant interactions and possible behaviours, but tropes of culture had largely disappeared by this point in the conversation, and the team did not question these expected responses.

7   CONCLUSION

In this analysis, we traced the cultural dimensions of instrumental judgment across a set of co-creation workshop design and debrief activities, bringing attention to the ways in which the design team activated cultural judgments that shaped their understanding of the participants and target user population. The design team engaged in brief exchanges that were culturally situated, but the invocation of culture was often bracketed within a co-creation approach that did not leave room for deep cultural engagement.

Designers simultaneously work *within* and *shape* discourses, which include a cultural and social dimension (Krippendorf, 2005). However, to understand and potentially alter discourses – which is required if greater cultural awareness is to be present within design activity – the discourse must "[remain] *rearticulable*, [so] that its users can understand, practice, and speak about these changes" (Krippendorf, 2005, p. 12, emphasis in original). This analysis provides a first step towards understanding the ways in which culture may be invoked in design activity, bringing greater awareness of the cultural dimensions that designers must engage *with* and *through* when conducting user research and co-creation projects.

NOTE

1   This term references a critique of programming practices by Dijkstra in 1968. "Considered harmful" now references a critical reading of existing practices in computing, made popular in the HCI literature by Greenberg and Buxton (2008).

REFERENCES

Barber, W. & Badre, A. (1998). Culturability: The merging of culture and usability. *Human Factors and the Web*. Available from: http://www.research.att.com/conf/hfweb/proceedings/barder/index.htm

Batra, R., Ramaswamy, V., Alden, D., Steenkamp, J. & Ramachander, S. (2000). Effects of brand local and nonlocal origin on consumer attitudes in developing countries. *Journal of Consumer Psychology, 9*(2), 83–95.

Bell, G., Blythe, M., Gaver, B., Sengers, P., & Wright, P. (2003). Designing culturally situated technologies for the home. In: *CHI'03: Proceedings of the 2003 CHI Conference on Human Factorsin Computing Systems*. New York, NY: ACM Press. pp. 1062–1063.

Bell, G., Blythe, M., & Sengers, P. (2005). Making by making strange: Defamiliarization and the design of domestic technologies. *ACM Transactions on Computer-Human Interaction (TOCHI), 12*(2), 149–173.

Buchanan, R. (2015). Worlds in the making: Design, management, and the reform of organizational culture. *She Ji: The Journal of Design, Economics, and Innovation, 1*(1), 5–21. doi:10.1016/j.sheji.2015.09.003

Carr-Chellman, A., & Savoy, M. (2004). User-design research. In: D. H. Jonassen (ed.) *Handbook of research on educational communications and technology*. Mahwah, NJ: Lawrence Erlbaum Associates. pp. 701–715.

Chavan, A. L., Gorney, D., Prabhu, B., & Arora, S. (2009). The washing machine that ate my sari – mistakes in cross-cultural design. *Interactions, 16*(1), 26–31.

Christensen, B. T., & Abildgaard, S. J. J. (2017). Inside the DTRS11 Dataset: Background, Content, and Methodological Choices. In: Christensen, B. T., Ball, L. J. & Halskov, K. (eds.) *Analysing Design Thinking: Studies of Cross-Cultural Co-Creation*. Leiden: CRC Press/Taylor & Francis.

Clemmesen, T., Ranjan, A., & Bødker, M. (2017) How Cultural Knowledge Shapes Design Thinking. In: Christensen, B. T., Ball, L. J. & Halskov, K. (eds.) *Analysing Design Thinking: Studies of Cross-Cultural Co-Creation*. Leiden: CRC Press/Taylor & Francis.

Daly, S., McKilligan, S., Murphy, L., & Ostrowski, A. (2017) Tracing Problem Evolution: Factors That Impact Design Problem Definition. In: Christensen, B. T., Ball, L. J. & Halskov, K. (eds.) *Analysing Design Thinking: Studies of Cross-Cultural Co-Creation*. Leiden: CRC Press/Taylor & Francis.

Dhadphale, T. (2017) Situated Cultural Differences: A Tool for Analysing Cross-Cultural Co-Creation. In: Christensen, B. T., Ball, L. J. & Halskov, K. (eds.) *Analysing Design Thinking: Studies of Cross-Cultural Co-Creation*. Leiden: CRC Press/Taylor & Francis.

du Gay, P., Hall, S., Janes, L., Madsen, A. K., Mackay, H., & Negus, K. (2013). *Doing Cultural Studies: The Story of the Sony Walkman* (2nd ed.). Los Angeles, CA: Sage Publications.

Dunne, J. (1997). *Back to the rough ground: Practical judgment and the lure of technique*. Notre Dame, IN: University of Notre Dame Press.

Dunne, J. (2005). An intricate fabric: Understanding the rationality of practice. *Pedagogy, Culture & Society, 13*(3), 367–390.

Eckert, C. & Stacey, M. (2017) Designing the Constraints: Creation Exercises for Framing the Design Context. In: Christensen, B. T., Ball, L. J. & Halskov, K. (eds.) *Analysing Design Thinking: Studies of Cross-Cultural Co-Creation*. Leiden: CRC Press/Taylor & Francis.

Geertz, C. (1973). *The interpretation of cultures*. New York, NY: Basic Books.

Gray, C. M. (2016a). "It's More of a Mindset Than a Method": UX Practitioners' Conception of Design Methods. In: *CHI'16: Proceedings of the 2016 CHI Conference on Human Factors in Computing Systems*. New York, NY. ACM Press. pp. 4044–4055. Doi:10.1145/2858036.2858410

Gray, C. M. (2016b). What is the nature and intended use of design methods? In *Proceedings of the Design Research Society*, Brighton, UK.

Gray, C. M., Dagli, C., Demiral-Uzan, M., Ergulec, F., Tan, V., Altuwaijri, A., Gyabak, K., Hilligoss, M., Kizilboga, R., Tomita, K. & Boling, E. (2015). Judgment and instructional design: How ID practitioners work in practice. *Performance Improvement Quarterly, 28*(3). doi:10.1002/piq.21198

Gray, C. M., Stolterman, E., & Siegel, M. A. (2014). Reprioritizing the relationship between HCI research and practice: Bubble-Up and trickle-down effects. In *DIS'14: Proceedings of the 2014 CHI conference on designing interactive systems* (pp. 725–734). New York, NY: ACM Press.

Greenberg, S., & Buxton, B. (2008). Usability evaluation considered harmful (some of the time). In *CHI'08: Proceedings of the SIGCHI Conference on Human Factors in Computing Systems*. New York, NY: ACM Press. pp. 111–120. DOI=http://dx.doi.org/10.1145/1357054.1357074

Hakken, D., & Maté, P. (2014). The culture question in participatory design. In: *PDC'14: Proceedings of the Participatory Design Conference*, Windhoek, Namibia.

Hess, J. L., & Fila, N. D. (2017) Empathy in Design: A Discourse Analysis of Industrial Co-Creation Practices. In: Christensen, B. T., Ball, L. J. & Halskov, K. (eds.) *Analysing Design Thinking: Studies of Cross-Cultural Co-Creation*. Leiden: CRC Press/Taylor & Francis.

Hofstede, G. H. (2001). *Culture's consequences: Comparing values, behaviors, institutions, and organizations across nations*. Thousand Oaks, CA: Sage.

Horn, M. S. (2013). The role of cultural forms in tangible interaction design. In: *TEI'13: Proceedings of the Tangible Embedded Interaction Conference*, Barcelona, Spain.

Irani, L., Vertesi, J., Dourish, P., Philip, K., & Grinter, R. E. (2010). Postcolonial computing: A lens on design and development. In *Proceedings of the SIGCHI conference on human factors in computing systems*. New York, NY: ACM Press. pp. 1311–1320.

Olson, J. S., & Kellogg, W. A. (2014). *Ways of knowing in HCI*. New York, NY: Springer.

Korkmaz, N., & Boling, E. (2014). Development of design judgment in instructional design: Perspectives from instructors, students, and instructional designers. In: B. Hokanson and A. Gibbons (Eds.) *Design in Educational Technology*. New York, NY: Springer. pp. 161–184.

Krippendorf, K. (2005). *The semantic turn: A new foundation for design*. Boca Raton, FL: CRC Press.

Goodman, E., Kuniavsky, M., & Moed, A. (2012). *Observing the user experience: A practitioner's guide to user research*. Waltham, MA: Morgan Kaufmann.

Lawson, B., & Dorst, K. (2009). *Design expertise*. Oxford, UK: Architectural Press.

Martin, B., & Hanington, B. (2012). *Universal methods of design: 100 ways to research complex problems, develop innovative ideas, and design effective solutions*. Beverly, MA: Rockport Publishers.

Martineau, P. (1958). Social classes and spending behavior. *Journal of Marketing, 23*(2), 121–130.

Nelson, H. G., & Stolterman, E. (2012). *The design way: Intentional change in an unpredictable world* (2nd ed.). Cambridge, MA: MIT Press.

Nielsen, J., Bødker, M., & Vatrapu, R. (2010). Culture and (i)literacy as challenges to Scandinavian cooperative design. In: *ICIC'10*. Copenhagen, Denmark.

Stolterman, E. (2008). The nature of design practice and implications for interaction design research. *International Journal of Design. 2*(1), 55–65.

Polanyi, M. (1966). *The tacit dimension*. Garden City, NY: Anchor Books.

Tai, E. (2008). Cross-cultural design. In: M. Erlhoff & T. Marshall (eds.) *Design dictionary*. Heidelberg, DK: Springer-Verlag. pp. 98–99. Doi:10.1007/978-3-7643-8140-0_67

Vickers, S. G. (1984). Judgment. In: *The Vickers Papers*. London, UK: Harper & Row. pp. 230–245.

Woolrych, A., Hornbæk, K., Frøkjær, E., & Cockton, G. (2011). Ingredients and meals rather than recipes: A proposal for research that does not treat usability evaluation methods as indivisible wholes. *International Journal of Human-Computer Interaction, 27*(10), 940–970.

Part III

Cognitive and Metacognitive Aspects of Design Thinking

Metacognition in Creativity: Process Awareness Used to Facilitate the Creative Process

Dagny Valgeirsdottir & Balder Onarheim

ABSTRACT

The purpose of this study was to investigate the influence of 'process awareness' on the creative process of a design team. Process awareness is a cognitive creativity skill that entails actions derived from instances where individuals in the design team express knowledge of underlying cognitive processes and aspects and utilize it to facilitate their own and their team's creative process. Transcripts from sessions where the design team was working creatively were analyzed both top-down and bottom-up, through quantitative coding, using a coding scheme, and qualitative coding. This was done to ensure capture of all instances of process awareness. Through this iterative process it was revealed that process awareness was predominantly observed in creativity related tasks. Moreover, three distinct facets to process awareness emerged, that is, planning, monitoring and reflecting, which were employed respectively before, during and after initiating a process and/or a workshop. We conclude that process awareness is an important creativity skill, being a crucial mechanism to enhance all stages of the creative process. If a designer is able to plan, monitor and reflect on his or her own cognitive processes, as well as those of other team members, he or she will be able to understand what works and what does not for advancing the creative process. In turn, this enables the designer to become more strategic about which actions are appropriate and at what time they are most usefully deployed, thereby making the use of strategies, methods and tools not just an automatic procedure but a highly conscious and purposeful one.

1 INTRODUCTION

The importance of creativity in any domain has been increasingly acknowledged in recent years; that is, the skill individuals possess and use to produce solutions to any given problem. Here the focus will be on creativity within the domain of design, where creativity has repeatedly been shown to play a crucial role (Christiaans, 1992; Dorst & Cross, 2001). This motivates the research presented here, giving incentive to further investigate how specific creativity skills can be utilized to advance design work. The aim of the presented study is to investigate how *process awareness* (Valgeirsdottir *et al.*, 2016), a cognitive creativity skill, influences the creative process of design teams. More specifically, the focus is on how members of the design team utilize knowledge and awareness of cognitive processes, such as fixation and priming, to advance a creative process and to manage design strategies. The current dataset provided a good venue to study the phenomenon of process awareness, as the core design team consisted of three experienced designers verbalizing their actions and rationale while going through the design process. In a previous study that investigated the skills important for designers when working creatively, we discovered that process awareness was crucial for advancing the design work observed (Valgeirsdottir *et al.*, 2016). Here we want to build on that discovery and further investigate: (1) *how process awareness influences the creative process of design teams*; and (2) *is process awareness only used in creativity related tasks within the creative process of design teams*. We argue that process awareness is a creativity related skill that can advance design work, however we acknowledge the overlap with the well-known concepts of *reflection in action* (Schön, 1983) and *metacognition* (Flavell, 1979). Thus another purpose here will be to elaborate on these overlaps and discuss similarities, as well as clarifying the differences between these concepts.

The main contribution of this paper will be: (1) to clarify and develop the concept of process awareness; (2) to identify more clearly when and where in the design process it materializes; and (3) to relate process awareness to metacognition. Clarifying and positioning the construct of process awareness should enable a more structured way of enhancing individual creativity through the strategic use of process awareness, which will in turn allow for a more efficient process during design work. Finally, we will discuss the importance of future research into the relationship of process awareness and metacognition in creative work.

The following sections will first provide a theoretical overview, framing the key concepts in a broader perspective. Next, a description of the methodological approach is provided, followed by the results of the data analysis. Finally, a discussion of the findings will outline the most interesting conclusions to draw from this study, and the subsequent contribution to the fields of creativity and design.

2   LITERATURE REVIEW

Creativity has widely been accepted as an important ingredient in design processes, which has led to design being a popular domain for creativity research. The 'standard definition' of creativity (Runco & Jaeger, 2012) states that creativity is, as originally stated by Stein (1953), the combination of *novelty* and *usefulness*. However, the standard definition refers to more tangible aspects of creativity, either a creative process or the output from a process, whereas here we will take a more cognitive approach to creativity. How can one manage his or her own creative process on a cognitive level, and what does that actually mean?

2.1   Cognitive aspects of creativity

Two distinct cognitive processes have been proposed to be part of creativity: divergent- and convergent thinking (Guilford, 1950). The two concepts were developed as part of the seminal work of J. P. Guilford – a psychologist often referred to as "the godfather of creativity" as he was the first one to publicly establish creativity as a research field.

We see divergent- and convergent thinking as being related to the two aspects of the standard definition of creativity. Divergent thinking is the process of coming up with *novel* ideas or alternatives, and convergent thinking is the process of combining those into something *useful* (Onarheim & Friis-Olivarius, 2013). Furthermore, these two cognitive processes are related to an operational design process, that is, a process of alternating between 'opening up' (diverging) and 'closing down' (converging), as is operationalized, for example, in the Double Diamond model (Design Council, 2005). Other cognitive aspects can potentially affect the creative process of an individual such as the ability to make remote associations (Mednick, 1962), which is one of the Five Key Concepts of neurocreativity (Onarheim & Friis-Olivarius, 2013b); remote associations, cognitive inhibition, priming, fixation and incubation are all believed to influence the creative process of an individual.

In our previous study (Valgeirsdottir *et al.*, 2016), conducted within the same organization and with the same design team investigated here, we concluded that awareness of the above-mentioned cognitive processes and aspects did influence the design work and the progress of the process. This *process awareness* materialized when individuals were aware of the different stages of their process, as well as the underlying cognitive processes that could influence their own creative abilities. This awareness in turn seemed highly beneficial for advancing and enhancing the team process. Resulting from this previous work was the following definition of process awareness.

Process awareness is a cognitive creativity skill that individuals in a design team use to facilitate a creative process. This creative process can be either their own, their team's or when designing a process for other participants. The individual applies her knowledge of cognitive processes and creativity concepts by being aware of the potential influence of said processes and concepts on the creative process.

Thus process awareness is the skill that allows a person to become conscious of the cognitive processes involved in a creative process so as to be able to adjust this creative process accordingly. This requires both knowledge and self-observation, in order to understand one's own and others' creative processes. The result is recognition of the cognitive aspects, skills and processes that may influence a creative process, how it can successfully flow under certain circumstances, and how it can become hindered in others. The understanding derived of this skill may enable the individual to increase, or develop, their creative potential, for example, by being more deliberate when getting involved in a creative process in order to avoid possible pitfalls (internal or external) and to be able to generate the required conditions to advance the process. An example from Valgeirsdottir *et al.* (2016), which describes an incident where process awareness was employed strategically to use breaks for the purpose of incubation (i.e. utilizing knowledge of a cognitive process to advance a creative process) can be seen here: "*[...] keep thinking about it during lunch while taking a break, keep in mind, and write down if you come up with anything*" (p. 1182). Here an individual recognized that the team was entering a stage of fatigue and fixedness and thus strategically suggested taking a break.

As is apparent from the above, the focus in this paper is on the individual perspective of process awareness; that is, how individuals working within the design team display this skill during a creative process and in turn how it might help with facilitating the process through actions taken as a result. We duly acknowledge that social interactions are important and potentially influential, however, to limit the complexity of the analysis the scope of this paper is restricted to the individual level.

2.2   Reflective practices

From the design research literature the concept of *reflection in action*, first formulized by Schön (1983) and further described in Valkenburg and Dorst (1998), is already widely used. The notion of reflection in action has connotations to process awareness, that is, that reflections are used in design work where designers are being aware of their work and how it progresses in the more tangible design process. It could be proposed that it is design-in-action. Process awareness is, on the other hand, related to cognitive aspects of creativity and how they can simultaneously affect and enhance the creative process of individuals and teams; it is creativity-in-action. We thus see reflection in action relating to the practical task of design, while process awareness is a cognitive concept related to creative thinking. Moreover, design contains creativity, but creativity does not solely relate to design. This paper works with the broader cognitive concept of creativity, thus we will use the concept of process awareness and not reflection in action.

Process awareness is also closely related to a concept elaborated in the psychology literature, *metacognition*, which simply put refers to thinking about one's own thinking (Dunlosky & Metcalfe, 2008; Flavell, 1979). Metacognition provides us with an ongoing overview of own thinking that can help direct the way we do our work; it is cognition about cognition (Dunlosky & Metcalfe, 2008; Nelson & Narens, 1990). There are three different facets to metacognition: *metacognitive knowledge*, *metacognitive monitoring* and *metacognitive control* (Dunlosky & Metcalfe, 2008). Metacognitive knowledge is knowledge about a given type of cognition. Metacognitive monitoring happens when one assesses the current state of cognitive activity. Metacognitive control is when an individual regulates some parts of cognitive activities. As metacognition is indeed a broader concept than process awareness, we see metacognition as being somewhat of an umbrella concept under which process awareness fits. Process awareness is limited to creativity related acts, and not other cognitive processes such as learning and memory.

2.3   The creative process

In creativity research there is normally a distinction made between cognitive creative process and practical creative process, the latter famously formulated as one of the '4P's of creativity' (Rhodes, 1961). In practice the two are entangled, as seen, for instance, in the concept of 'groupthink' where the cognitive processes in a group are negatively aligned and lead to an unproductive creative process (Nijstad & De Dreu, 2002). While the analysis here is focused on individual cognitive creativity, for the sake of the coming analysis it is necessary to outline the key concepts related to the practical creative process in which the designers are engaged. In the dataset analyzed the designers are planning a co-creation process (Sanders & Stappers, 2008), one of numerous prescribed frameworks for a practical creative process. For such processes a key notion in both creativity and design theory is the notion of 'problem and solution spaces', and the co-evolution of these two entities as part of the creative process

Table 12.1 Sessions included in analysis.

Session	Data file name
2	Designing co-creation workshops
3	Iterations on workshop design
4	Iterations on co-creation workshops
5	Designing co-creation workshop day 2
8	Sharing insights from CC1
9	Clustering insights from CC1
11	Linking insights from co-creation to project
15	Sharing insights from co-creation workshops (part 1 and 2)

(Wiltschnig, Christensen, & Ball, 2013). A crucial aspect of the creative process is to focus on both problem and solution spaces, and one typical approach to achieving this is through manipulating the constraints imposed on the task at hand (Onarheim & Biskjaer, 2014).

The interrelatedness of cognitive processes, metacognition and other aspects related to the creative process is of interest to us and we hope to clarify and refine the relationships between the three and elaborate on them while discussing the results in relation to the research questions.

3 METHODS

When initiating data analysis we first reviewed the provided transcripts and then selected those appropriate to the research questions: (1) *how does process awareness influence the creative process of design teams;* and (2) *is process awareness only used in creativity related tasks within the creative process of design teams?* This process excluded sharing and debriefing sessions, as these sessions contained no creative design work and were therefore considered irrelevant for the two research questions. The excluded sessions contained either debriefing by the design team to stakeholders or moderators/translators reporting back to the design team. The included sessions were all sessions where the design team planned co-creation sessions and/or engaged in hands-on design sessions. While it would have been relevant for the research questions also to analyze transcripts from the co-creation sessions, they were unfortunately not part of the provided dataset and thus not available. Eight sessions were included and can be seen in Table 12.1.

The analysis of selected transcripts was twofold and in the following we will refer to these as "quantitative" and "qualitative" coding. First, the quantitative coding was done using an automated word search, applying the coding scheme in Table 12.2. Second, the qualitative coding was performed by reading the transcripts and marking segments falling under the definition of process awareness. The purpose of this twofold analysis was to ensure that all appropriate segments in the selected transcripts were identified. Furthermore, the video recordings from each session were watched as a complimentary part of the analysis to ensure that each identified segment contained an observable occurrence of process awareness. Finally, all identified segments from both rounds of coding were combined for further analysis.

Table 12.2 Coding scheme.

Codes	Example from data	Type of creativity related aspects
We need to	[...] we can't start from everything, **we need to** start from something	Constraints, Problem space
We have to	But there's no doubt that **we have to** utilize the freshness [of new participants]	New perspective, Un-primed
We are	[...] or **we are** biasing them in some way that we didn't really realize	Priming
We want	[...] we didn't elu'- explain what directly and indirectly mean, 'cause **we want** them to ask or discuss what that means	Problem finding, Priming
We should	So I think **we should** do it together, 'cause we are in a different groups [...] we can write it on our own, before we put it out we just share:, to see if anyone- anything spark other thoughts	Brainstorm, Group sharing
I think we	I agree and **I think we**- they should be more or less free but we should make sure that they stay within the realm of [...]	Priming
Open up	[...] once you start to put words on thing, then suddenly you think about new things, I totally agree that it might **open up** some new things that they didn't think about	Associations
Go in	But I'm wondering if we should try to aggregate a couple of more of these before we start **going in**	Diverging, Monitoring
Insight	I think that it's a golden one. That is an **insight** in itself- yeah so we need to decide [...]	Insight, Monitoring
Aware	So they're quite **aware** of what can push them down and what can elevate them [...]	Reflecting
Prim[e]	We want people that can relate, but not people that are so **primed** that they [...] can't see anything else	Priming, Fixation
Break	We can have some features or something lying on the table during lunch **break** [...] just to lie around on the table so they can see them	Priming
Focus	we [...] have very strong **focus,** and then we flesh it out by giving them inspiration or stimulus	Priming, Inspiration
Fixate	[...] for some reason I just got so **fixated** with me-time. I think there's an untapped opportunity there	Fixation
Incubat[e]	I'm just considering to actually utilize the **incubation** inbetween the two co-creation sessions	Incubation

3.1 Quantitative coding

As process awareness is not yet a well-studied concept, there does not exists a prior coding scheme, thus we developed one based on our prior work with the concept. The coding scheme was based on analyzing examples from previous field-notes gathered through a six month observational study of process awareness in two design teams (Valgeirsdottir *et al.*, 2016). From this analysis we generated 18 codes intended to capture formulations used when expressing process awareness. Of the 18 codes, 15 yielded results and in Table 12.2 an overview can be seen as well as examples from the dataset, indicating the type of creativity related aspects the code exemplifies.

Each transcript was coded and analyzed in a two-step process inspired by the steps employed in Ball *et al.* (2010). The steps were: (1) automated word search using the generated codes to identify transcript segments where process awareness was potentially expressed; and (2) the identified segments were read in a linear manner and analyzed for presence of process awareness and coded with "1" if present and "0" if not. This linear analysis was based on the definition of process awareness presented in Section 2.1 in Literature Review.

The automated search using the coding scheme yielded 927 occurrences, where 128 occurrences were manually screened for potentially containing process awareness. After closely examining these instances, 67 of the segments were found to fall under the operational definition applied.

3.2   Qualitative coding

As coding for process awareness is not common and no established coding scheme exists, the quantitative coding was succeeded with a round of qualitative coding through content analysis. The purpose of this was twofold: first, it was done to ensure that all relevant transcript segments were included in the further analysis, and second, it was done to assess whether the proposed new coding scheme for process awareness (Table 12.2) could indeed capture the desired segments.

In the qualitative round all transcripts were read and manually coded based on the expression of process awareness guided by the definition. This resulted in a total of 94 segments found to contain quotes related to the operational definition of process awareness.

3.3   Comparing identified segments

While the quantitative and qualitative coding yielded different number of segments (67 and 94), when comparing the two lists of segments there was a perfect one-way match where all 67 quantitative segments were part of the 94 qualitative segments. Based on this it was concluded that the qualitative analysis had revealed additional instances that were not identified using the coding scheme, thus all the 94 segments was selected for continuation to the analysis.

3.4   Evaluation of coding scheme

The quantitative coding scheme yielded a high number of initial instances, but less than 10% were found to be relevant, and comparing these to the complete qualitative analysis it turned out to still not have been able to capture all relevant segments in the eight transcripts. Based on this it is concluded that the coding scheme proposed in Table 12.2 alone is not sufficient for capturing all relevant instances of process awareness, and that it also gives a large amount of irrelevant segments that have to be screened manually.

3.5   Inter-coder reliability checks

As all initial coding was performed by the first author, who has an in-depth knowledge about process awareness, a second independent round of coding was performed by the second author to ensure inter-coder reliability. The second author redid the above coding process with two out of eight transcripts, and the initial inter-coder match was 87.2% (quantitative) and 67.8% (qualitative), calculated as the shared percentage of the total number of segments selected (shared segments/total segments*100). In reviewing the deviating segments from the two coders it was discovered that the discrepancy was due to the theoretical nature of the definition applied. As pointed out by Amabile (1996) in studies on creativity, it is necessary to have both a theoretical and an operational definition, thus for the purpose of this study we added an operational element to the definition:

Process awareness is a cognitive creativity skill that individuals in a design team use to facilitate a creative process. This creative process can be either their own, their team's or when designing a process for other participants. The individual applies her knowledge of cognitive processes and creativity concepts by being aware of the potential influence of said processes and concepts on the creative process. When verbalized, process awareness appears when the individual *suggests or proposes an action and articulates the value of doing so and/or the danger of not doing so.*

While suited for the verbal analysis applied in this paper the authors acknowledge that the operational addition is delimiting and might not be applicable in other studies of less verbal process awareness.

After redoing the reliability coding with the updated operational definition the inter-coder match was considered satisfactory, only deviating with 1/23 (95.7%) segments for the quantitative and 2/31 (93.5%) segments for the qualitative coding.

4   ANALYSIS AND RESULTS

4.1   Three types of process awareness

In the process of coding and analyzing the data it became apparent that out of the 94 instances it appeared as though there were different facets to process awareness. Through iteratively reading the segments and discussing the different subcategories a proposed tripartite for types of process awareness emerged: *planning, monitoring* and *reflecting. Planning* process awareness occurred when the design team was planning a process, that is, it arose prior to initiating a process. *Monitoring* process awareness occurred during a process, that is, it arose when team members were being observant and verbalizing their thoughts and reasoning at a meta-level. *Reflecting* process awareness occurred after a task had been finalized and the design team was reflecting afterwards on the work they had done. These will be elaborated on further in the discussion section.

4.2   Coding for type of process awareness

After having established the three types of process awareness all 94 segments were analyzed and divided accordingly (Table 12.3). There were no segments that were not considered feasible to be categorized under any of the three types of process awareness, however, 10 of the longer segments were assessed as containing two different types of process awareness.

Table 12.3 Number of instances for type of process
          awareness.

Process awareness	
Planning	67
Monitoring	24
Reflecting	13
	104

Table 12.4 Most frequent creativity related keywords.

Keyword	Occurrences
Priming	17
Constraints	10
Co-creation tool	7
Brainstorming	6
Divergent thinking	6
Incubation	6
Co-creation	5
Convergent thinking	5
Group think	5
Problem focused work	4

4.3   Process awareness and creativity

The other topic of interest was *is process awareness only used in creativity related tasks within the creative process of design teams?* To explore this topic we performed an inverted analysis, investigating the extent to which the selected process awareness segments contained any elements from the central creativity concepts introduced in Sections 2.1 to 2.3 above. By exclusion, all other segments would be 'no creativity' occurrences of process awareness. Based on an in-depth familiarity of the presented creativity concepts all segments were revised and tagged with keywords describing the creativity concept(s) in the segment. If no link to a creativity concept could be applied then the type of process awareness was used. Tagging all 94 segments resulted in a total of 160 keywords applied (e.g., *problem finding, idea pairing, open up*), of which 84 were taken directly out of the segment (e.g., *LEGO, analogy, insight, inspiration, sleep on it*) and 14 were 'no creativity' keywords (*monitoring* and *reflecting*). One example of a direct keyword (*insight*) from the segments is as follows: *"i think that it's a golden one:. That is an insight in itself- yeah so we need to decide "is that something we take action on or-""* (v04, 207).

Removing duplicates left 55 unique keywords related to the creativity concepts, Table 12.4 contains the 10 most frequently used keywords. The additional 45 keywords were applied three times or less (e.g., *cultural probe, role play, trigger, warm up*).

We acknowledge the highly subjective and expertise-dependent nature of this analysis, but in the analysis the authors, both creativity scholars, actively sought to identify any possible relevance to creativity. When the authors could not identify a single link to the creativity concepts in a segment, then the segment is very unlikely to be related to creativity. Thus this inverted analysis is adequate for the overall purpose of identifying instances of process awareness in non-creativity tasks. All three types of process awareness were associated with creativity keywords; however, the 13 segments that did not include creativity keywords were either tagged as monitoring process awareness or reflecting process awareness.

5   DISCUSSION AND CONCLUSION

In this paper we have investigated the concept of process awareness as it materializes during a team design process. The transcripts analyzed contained both examples of the ongoing design process in the observed sessions, and the designer's considerations about how the design process should unfold in the co-creation session planned.

5.1   Inferential discussion in relation to research questions

The first research question of interest was *how does process awareness influence the creative process of design teams?* In the analysis it became apparent that distinctly different types of process awareness were observed in the sessions. A tripartite of these different types was proposed and used in the further analysis: *planning, monitoring* and *reflecting. Planning* process awareness takes place when the design team is preparing and designing a process and/or a workshop. During the planning they express knowledge of underlying cognitive processes and how they might influence the work that will take place and the importance of staying in specific areas of the creative process, for example, the importance of not 'jumping to solutions'. *Monitoring* process awareness takes place during a process and/or workshop when an individual expresses awareness of aspects such as being 'too primed' or needing a 'break to incubate' and the consequences of those aspects. Being 'too primed' might have a negative influence on the ability to generate new ideas (Friis-Olivarius, 2015), and incubation breaks can be used as a method in creative processes (Ellwoo *et al.*, 2009). We see planning and monitoring process awareness having strong references to metacognition as discussed below. *Reflecting* takes place after a process and/or a workshop, where team members discuss the work that has been done in hindsight and reflect on it while discussing aspects that could have influenced it or should have been done differently, or for motivational purposes when expressing satisfaction with the work that has been done. We consider this type of process awareness as relating to reflection in action and we will further discuss this relationship in Section 5.3.

After having re-coded the segments with the three types of process awareness it was revealed, perhaps not surprisingly, that *planning* was the overarching process awareness type (67 instances) in the first four transcripts, before the design team initiated their co-creation process in the field. These first four sessions mainly focused on designing and planning the overall co-creation process and the respective co-creation workshops. Furthermore, it became apparent that the design team utilized process awareness mainly for three types of activities, the first one being how to set the stage and which tools and techniques to deploy in such a process. Here discussions revolved around topics such as which types of co-creation tools to use, 'boundary objects' and how to make sure that the co-creation process was within the intended problem space. In these situations the design team was very diligent in verbalizing different creativity related aspects, for instance, 'priming'. Priming was, in fact, the mostly discussed factor overall with the highest number of appearances (total 17) in the data. An example from the transcript is in the following response (which is part of a larger segment): "*Yeah. And it- yeah its blessing and curses with both of those things of course, we want people that can relate, but not people that are so primed that they can only s'- they can't see it, they can see anything else*" (v03, 117).

The latter is a clear example of an individual utilizing his knowledge of underlying cognitive aspects and *suggesting an action based on the value of doing so and/or the danger of not doing so* as the updated operational definition of process awareness specifies. Priming is a double sided coin, which can be used intentionally for inspiration purposes, however, sometimes it is more appropriate to avoid priming in order to avoid design fixation (Agogue *et al.*, 2011), as expressed in the priming-statement above.

Moreover, it was apparent that process awareness was used for tasks such as how to keep on track and within the desired frame using, for example, constraints. Constraints were the second most re-occurring creativity related aspect that the designers used to plan and monitor their process. An example of a team member being aware of constraints is as follows: "*Yeah, and we want them to gravatize around something. And we want that something to- since we are limited on time we can't start from everything, we need to start from something*" (v04, 240). In fact, the team seemed in general quite aware of constraints and how to use them to their advantage, and the disadvantage of not managing them properly.

In addition, another apparent theme when the team used process awareness was in terms of how to converge and conclude by managing the different stages of diverging and converging. In the latter four transcripts the types of process awareness were more mixed and *monitoring* was the second most expressed one (24) after planning. An example of monitoring when managing the different stages of the process is, for example, verbalized in sentences like: "*Is this helping us forward when it comes to narrowing down?*" (v05, 849) and "*I think we're closing in. We are in the area of solving it, I just think we need to- like, which type- how can we decide what themes come in?*" (v05, 896). An example of *reflecting* can, for example, be found in this exchange where discussions take place about eliminating as part of converging, and the risk of eliminating too much content:

"*...not always to think of eliminating stuff, or eh, also because you need to be aware of what's already there, and not risking of taking something core out. Ehm:, so you don't necessarily think of elimination, but maybe just decrease of some of these variables.*

And I think actually these could also be increased, or expand, or change, or- yeah.

Go deeper, and-

Oh yeah, definitely. It's just that, so you don't think of elimination, like taking something out, it doesn't necessary, but of course you should add something, or change it, but it should arise from the gap which has been changed, right? So, it's not-, the, like the radical outcome doesn't necessarily entitle like a radical elimination or something, so" (v08, 307–310).

However, in relation to the second research question ("*Is process awareness only used in creativity related tasks within the creative process of design teams?*"), it became apparent that some occurrences of monitoring and reflecting process awareness were hard to relate to creativity. One must, though, keep in mind that self-reflective parts might not be as frequently verbalized and team members might simply act while monitoring and/or reflecting upon their own or the team's process. Sessions where the design team was planning a co-creation process for participants outside the core design team proved to be a good place to capture process awareness because then they were being more verbal about different elements that could be used to manipulate the process and explaining/reasoning for it. As previously stated, planning process awareness was an overarching theme in the first sessions where they were designing the co-creation process. The number of creativity related instances was thus higher during that phase, where they needed to be explicit about the aspects they wanted to achieve and/or avoid. It was, however, only in 13 out of 94 segments where an obvious creativity related process or aspect was not identified, and taking in mind the fact that process awareness might not always be externalized in a verbal exchange with another team member when monitoring or reflecting on the process at hand in order to advance it leads us to conclude that process awareness is a creativity relevant skill, applicable throughout the creative process.

5.2 Process awareness and metacognition

The importance of reflective practice of designers has been emphasized in design research. We argued in Section 2.2, that creativity and design are interlinked concepts giving ground to the claim that process awareness and reflection in action are related despite providing support to different constructs. However, we see metacognition being more closely related to process awareness. In fact, we argue that metacognition, the higher level of cognition that provides an overview of all cognitive activities taking place at any given time, is an umbrella concept under which process awareness falls. Process awareness was revealed as being a three layered concept applicable before (planning), during (monitoring) and after (reflecting) a creative process where knowledge about cognitive processes and creativity concepts was strategically used in order to advance a creative process. Through application of process awareness, the uncertainty of design work can be dealt with through a strategic approach to creative processing. As cited in Christensen and Ball (2017) epistemic uncertainty drives a metacognitive switch, meaning that when circumstances are perceived as uncertain a person's metacognition is triggered. Moreover, in some instances where epistemic uncertainty is a stable and overarching backdrop of the whole design context there will be instances with isolated certainty moments that lead to creative processing (Christensen & Ball, 2017) This supports the findings we offer here, the design context in the given dataset had a high level of stable epistemic uncertainty, and thus the metacognition had been triggered and the designers were being self-aware of their work and used process awareness for creative processing.

As introduced in Section 2.2, metacognition has three different facets to it and each can be related to creativity. *Metacognitive knowledge* would be materialized when individuals know about how, for example, priming works and how to improve a process through either using it for inspiration or to avoid it. *Metacognitive monitoring* would take place, for example, when individuals express "I think we are closing into a solution". *Metacognitive control* would be when an individual decides to use a new tactic, such as applying constraints, to help with solving the task at hand. However, as previously stated, metacognition is a broader concept than process awareness and covers other cognitive functions than creativity such as memory and learning. Therefore, we argue that defining a creativity specific facet to it (i.e., process awareness) is a valuable contribution to creativity research. We therefore see the value in investigating the relationship of process awareness and metacognition further, which we will do in our future research. Based on our current work with process awareness, and uncovering the link to metacognition we believe that metacognition should become an important topic for creativity. One possible research direction is metacognition in creativity training, where other metacognitive processes such as learning and memory are normally emphasized. However, the main contribution of this paper is to define process awareness and with that refinement we hope to expand the existing concept of metacognition through this creativity specific facet. Finally, we propose that process awareness could be re-named *creative awareness*, for further clarification of its relationship to creativity, but that will be the object of our future research.

5.3   Concluding remarks

We believe it is a dangerous assumption to consider the cognitive skill of creativity, namely the individual ability to come up with novel and useful solutions, as being important only in a given part of a process, whether it is a creative design process or an innovation process. The findings in this paper point to the importance of using the cognitive skill of creativity at *all* stages of the design process. This is especially the case when considering that if a designer is able to plan, monitor and reflect on her own cognitive processes, as well as those of other team members, she will be able to understand what works and what does not work for advancing the creative process. In turn, this will enable the designer to become more strategic about which actions are appropriate and at what time they are most usefully deployed during creative processing, thereby making the use of cognitive strategies, techniques and tools not just an automatic procedure but a highly conscious and purposeful one.

REFERENCES

Agogue, M., Kazakci, A. O., Weil, B., & Cassotti, M. (2011). The Impact of Examples on Creative Design?: Explaining Fixation and Stimulation. *Iced 2011*, (August), 1–10.

Amabile, T. M. (1996). *Creativity in Context*. Boulder: Westwood Press.

Ball, L. J., Onarheim, B., & Christensen, B. T. (2010). Design requirements, epistemic uncertainty and solution development strategies in software design. *Design Studies, 31(6)*, 567–589. http://doi.org/-10.1016/j.destud.2010.09.003

Christensen, B. T., & Ball, L. J. (2017) Fluctuating Epistemic Uncertainty in a Design Team as a Metacognitive Driver for Creative Cognitive Processes. In: Christensen, B. T., Ball, L. J. & Halskov, K. (eds.) *Analysing Design Thinking: Studies of Cross-Cultural Co-Creation*. Leiden: CRC Press/Taylor & Francis.

Christiaans, H. H. C. M. (1992). *Creativity in design: the role of domain knowledge in designing*. TU Delft, Delft University of Technology.

Design Council (2005). A Study of the Design Process – The Double Diamond. [Online] available from www.designcouncil.org.uk/sites/default/files/asset/document/ElevenLessons_Design_Council%20(2).pdf [Accessed 9.7.2015]

Dorst, K., & Cross, N. (2001). Creativity in the design process: co-evolution of problem-solution. *Design Studies*, 22(5), 425–437.

Dunlosky, J., & Metcalfe, J. (2008). *Metacognition*. Sage Publications.

Ellwood, S., Pallier, G., Snyder, A., & Gallate, J. (2009). The Incubation Effect: Hatching a Solution? *Creativity Research Journal*, 21(1), 6–14. http://doi.org/10.1080/10400410802633368

Flavell, J. H. (1979). Metacognition and cognitive monitoring: A new area of cognitive-developmental inquiry. *American Psychologist*, 34(10), 906–911. http://doi.org/10.1037/0003-066x.34.10.906

Friis-Olivarius, M. (2015). *The Associative Nature of Creativity*. CINC press.

Guilford, J. P. (1950). Creativity. *American Psychologist*, Vol 5(9), 444–454.

Mednick, S. A. (1962). The associative basis of the creative process. *Psychological Review*, 69(3), 220–232. http://doi.org/10.1037/h0048850

Nijstad, B. A., & De Dreu, C. K. W. (2002). Creativity and group innovation. *Applied Psychology*, 51(3), 400–405. http://doi.org/10.1111/1464-0597.00984

Nelson, T. O., & Narens, L. (1990). Metamemory: A theoretical framework and new findings. *The Psychology of Learning and Motivation*, 26, 125–141.

Onarheim, B., & Biskjaer, M. M. (2014). Balancing Constraints and the Sweet Spot as Coming Topics for Creativity Research. *Creativity in Design: Understanding, Capturing, Supporting*, (1), 1–18.

Onarheim, B., & Friis-Olivarius, M. (2013). Applying the neuroscience of creativity to creativity training. *Frontiers in Human Neuroscience*, 7(October), 656. http://doi.org/10.3389/fnhum.2013.00656

Rhodes, M. (1961). An analysis of creativity. *The Phi Delta Kappan*, 42(7), 305–310.

Runco, M. A., & Jaeger, G. J. (2012). The Standard Definition of Creativity. *Creativity Research Journal*, 24(1), 92–96. http://doi.org/10.1080/10400419.2012.650092

Sanders, E. B.-N., & Stappers, P. J. (2008). Co-creation and the new landscapes of design. *CoDesign*, 4(1), 5–18. http://doi.org/10.1080/15710880701875068

Schön, D. A. (1983). *The reflective practitioner: How professionals think in action*. Ashgate Pubishing Limited.

Stein, M. I. (1953). Creativity and culture. *The Journal of Psychology*, 36(2), 311–322.

Valgeirsdottir, D., Onarheim, B., & Li-Ying, J. (2016). Improving Creativity Training: A Study of Designer Skills. In: *14th International Design Conference*. Dubrovnik, Croatia: Design Society. pp. 1175–1184.

Valkenburg, R., & Dorst, K. (1998). The reflective practice of design teams. *Design Studies*, 19(3), 249–271. http://doi.org/10.1016/S0142-694X(98)00011-8

Wiltschnig, S., Christensen, B. T., & Ball, L. J. (2013). Collaborative problem-solution co-evolution in creative design. *Design Studies*, 34(5), 515–542. http://doi.org/10.1016/j.destud.2013.01.002

Grouping Notes Through Nodes: The Functions of Post-it Notes in Design Team Cognition

Graham Dove, Sille Julie J. Abildgaard, Michael Mose Biskjaer, Nicolai Brodersen Hansen, Bo T. Christensen & Kim Halskov

ABSTRACT

The Post-it note is a frequently used, and yet seldom studied, design material. We investigate the functions that Post-it notes serve when providing cognitive support for creative design team practice. Our investigation considers the ways in which Post-it notes function as design externalisations, both individually and when grouped, and their role in categorisation in semantic long-term memory. To do this, we adopt a multimodal analytical approach focusing on interaction between humans, and between humans and artefacts, alongside language. We discuss in detail examples of four different externalisation functions served by Post-it notes, and show how these functions are present in complex overlapping combinations rather than being discrete. We then show how the temporal development of Post-it note interactions supports categorisation qualities of semantic long-term memory.

I   INTRODUCTION

The humble *Post-it note* is one of the most frequently used materials in creative practice. For instance, they play an important role in affinity diagramming as part of contextual design (Beyer & Holtzblatt, 1997), and in the brainstorming activity in Future workshops (Kensing & Madsen, 1991). In this chapter, we use the DTRS11 dataset to investigate the functions these small sticky notes play in design team cognition. In particular, we consider their use in terms of the *cognitive support* they provide for creative design team practice.

Our conception of cognitive activities draws on Hutchins (2000, 2006), who shows us how cognitive processes may be distributed across the members of a social group, through periods of time, and may involve coordination between internal (i.e., mental) and external (i.e., material and environmental) structures. To investigate the role of Post-it notes during these design activities, we draw upon literature discussing the function that design materials and artefacts serve as *externalisations* (Dix & Gongora, 2011), which aid reflection-in-action (Bamberger & Schön, 1983; Schön 1992; Schön & Wiggins, 1992). Because the way in which collections of Post-it notes are clustered and grouped is such an important aspect of their use in creative practice, we also investigate this phenomenon. To help us in this level of analysis, we make additional reference to theories of *categorisation* in *semantic long-term memory* (Collins & Loftus, 1975).

1.1   Post-it note use in design practice

Despite being commonplace, the practices surrounding designers' use of Post-it notes have not, to our knowledge, previously been the subject of close analysis and theorising. Where they have been discussed, it has tended to be in comparison with digital implementations of activities in which Post-it notes are commonly used (e.g., Harboe & Huang, 2015; Hilliges, *et al.*, 2007). This is in contrast to comparable design activities, such as sketching (Buxton 2010; Purcell & Gero, 1998; Stolterman 1999) or prototyping (Lim, Stolterman, & Tenenberg 2008). However, the value of such paper-based informal design tools is commonly acknowledged (Cook & Bailey, 2005). An example of this is *affinity diagramming*, a technique that uses Post-it notes to create meaning out of large amounts of data that are derived from user studies, specifically drawing on the qualities of small, informal sticky paper notes (Beyer & Holtzblatt, 1998). These are qualities that support design thinking and design outcomes (Christensen & Ball, 2017), and mean that Post-it notes can be used in a synthetic or compositional way (Stolterman, *et al.*, 2009). Post-it notes are also commonly used as a tool for brainstorming, being utilised to capture individual ideas and when trying to form categories in response to a design problem.

1.2   Properties of Post-it notes

Post-it notes are characterised in part by their regular shape and small size. This makes them ideal for carrying small, self-contained pieces of information, and results in them typically being used to represent a single idea. A second characteristic property of Post-it notes is the strip of semi-permanent, reusable adhesive along their top edge. This allows them to be positioned (and later re-positioned) relative to other Post-it notes, for example, on larger sheets of card or a whiteboard. This facility to intentionally position and re-position Post-it notes relative to each other also makes it possible for the connections between different Post-it notes to become meaningful. A third characteristic property of Post-it notes is that they are available in a number of different colours, allowing intentional selection of colour to carry semantic value. Finally, individual Post-it notes are of little value in themselves, and can therefore be easily discarded if the idea they contain is no longer considered important. In summary, Post-it notes can be considered capable of flexibly carrying symbolic representations of individual ideas, and which through grouping and associations of position and colour also makes them capable of representing larger, emergent concepts.

2   THEORETICAL BACKGROUND

2.1   Externalisations in design activities

Externalisations are ubiquitous in design work, providing intellectual resources that help make the tacit explicit, facilitate reflection on embodied action and enable designers to shape the world as a resource to support design activities (Dix & Gongora, 2011). As such, externalisations must evolve as complex problems develop and situations are transformed, and therefore be both expressive (mutable) and associative (combinable) (Fischer, Nakakoji, & Ostwald, 1995). In doing so, they serve an important role in contextualising information and creating objects to think with (Arias, Eden, Fischer, Gorman, & Scharff, 2000), and in facilitating a shared understanding of the design work at hand (Warr & O'Neill, 2007).

Dix and Gongora describe externalisation as a process of embodying, representing and exploring thoughts and feelings, with the overlapping purposes of generating, elaborating and communicating ideas (Dix & Gongora, 2011). They then describe four key functions that design externalisations serve, which will provide the core theoretical lens through which we consider the use of Post-it notes in the DTRS11 dataset:

– **Informational:** When serving an *informational* function, an externalisation is communicative, and explicitly expresses one person's idea so that someone else can share a similar understanding. This is one of the most familiar uses we might think of for Post-it notes in collaborative design activities, for example, when one individual writes an idea on a note and shares this with other design team members.
– **Formational:** An externalisation serves a *formational* function through helping to bring an idea into being; that is, the idea is itself being formed during the externalisation process. For example, we might imagine a designer who starts to write one Post-it note, and for whom the act of writing triggers further ideas, which in turn are written down.
– **Transformational:** The *transformational* function of externalisations is to support continuing cognitive activity, that is, the "backtalk" in Schön's (1992) description of design as a conversation with the materials of a situation. An example of this during collaborative design activity would be the way in which an idea donated on a Post-it note by one team member might be modified by a Post-it note that is subsequently added, either by that design team member or by another.
– **Transcendental:** The *transcendental* function of externalisations is meta-cognitive. Here, the act of expressing ideas externally, for example, through symbolic representations, allows us to explore and think about concepts, arguments, criteria and the like as if they were any other kind of 'thing'. A common example of this during design activities would be the way in which collections of Post-it notes are revisited in order to re-conceptualise the ideas contained on individual notes or captured within clusters. Such knowledge and control of cognitive processes has been shown to be important to creativity (Armbruster, 1989).

In our analysis, we highlight examples of each of these different functions served by the Post-it notes used by the design team. However, these four functions should not be considered discrete labels that can be applied atomically to particular use cases; rather they should be considered different aspects of complex overlapping patterns. We also describe how this supports the design team's creative practice, and connect this understanding of the different functions that Post-it notes serve in their role as externalisations of design ideas to those properties described by other theorists.

2.2   Categorisation in semantic long-term memory

Post-it note use during design activities may be similar to sketching in the way it supports a range of cognitive processes, which can include idea access and individual re-interpretation (van der Lugt, 2005), chunking (Suwa & Tversky, 1997), and analogical reasoning (Christensen & Schunn, 2007). In the present paper, we illustrate how the use of Post-it notes during clustering and organising exercises may facilitate category formation through its support for the qualities of semantic long-term memory.

Classically, semantic long-term memory has been conceptualised as being constituted through associations of nodes in a network (Collins & Loftus, 1975). A key feature of this model is that semantic recognition follows patterns of activation that spread amongst nodes (concepts) via pathways stored between associated nodes. No single feature is necessary for judgments of category membership, as concepts have a graded structure, and such judgments are considered contingent on the typicality of the potential member. The semantic relationship between concepts can then be characterised by their conceptual overlap, and the strength and length of associated pathways. This is evidenced by the relationship between the time taken to judge category membership and the centrality of the potential member within that category. It has also been shown that the pathway distance between any two concepts may be estimated dependent on their conceptual overlap. In addition to associative strength within and between nodes, concepts also have hierarchical relations to both superordinate and subordinate level concepts (Rosch, 1978). As a result of these various cognitive qualities related to semantic long-term memory, fuzzy boundaries often exist between categories, and membership may be uncertain. These qualities exist in natural language categories, as well as in *ad hoc* categories constructed for situational purposes (Barsalou, 1983).

3   METHOD

To explore the functions of Post-it notes in design team practice, we analysed video data drawn from the DTRS11 database (Christensen & Abildgaard, 2017). The database consists of *in situ* video recordings of a professional user-experience design team as they work through a specific design task for a worldwide manufacturer within the automotive industry. The data include collaborative design activities at various stages of the design process, and co-creation sessions with lead users in both a Scandinavian and a Chinese context. For this study, we selected video recordings of three different sessions focusing on the design team's use of Post-it notes. In video 3 (V3), we observe the design team forming categories by clustering and grouping Post-it notes containing ideas generated individually in a brainstorm session. In video 15 (V15), we observe the design team generating and sharing insights based on the outputs from the co-creation sessions held with Chinese lead-users. In video 18 (V18) two members of the design team revisit these ideas in preparation for a share-back session with the wider project team. These three sessions enable us to analyse the design team's use of Post-it notes in different activities, and to track their use of Post-it notes across different phases in the design process.

Our analytical approach is multimodal, drawing on ethnomethodology (EM) and conversation analysis (CA) (Goodwin & LeBaron, 2011), and interaction is considered to be more than just speech and text (Kress, 2009). We aim to award Post-it notes a status that renders them analysable without diminishing the significance of human interaction, and through this explore how the design team uses Post-it notes as an integral part of their interaction and communication. Following EM, we highlight the role Post-it notes play in the situational context, as they are actively used by the design team, who shape them, are shaped by them, navigate by them and make sense in relation to them (Garfinkel, 1994). Having first watched the entire collection of videos in the DTRS11 database to gain an impression of the overall activity, we selected three videos to investigate closely: V3, V15 and V18.

To begin with, each Post-it note appearing in a videoed session was given a reference number based on the order in which it was added to the wall or the whiteboard. This resulted in Post-it note 'maps' showing the final position of the Post-it notes in each session, and containing temporal information about their use in that session, such as their placement and movement during the design activities. Selected maps are included, as illustration, in the figures featured in this chapter. This approach is similar to the one Shroyer *et al.* (2017) use in their "Level 3" analysis, where they provide a detailed description of an idea generation session, which includes analysis of Post-it note generation and placements. As part of the map-making process, the content of all the Post-it notes was transcribed and listed according to reference number. In V15, the insight sharing session, a number of Post-it notes had been brought in from the co-creation sessions that were written in Chinese but partially translated into English by the Chinese consultants. In these cases, only the English text was transcribed. Alongside our study of the videos and Post-it note maps, we also investigated transcripts of each session, augmenting these with details of overlapping speech, gesture and gaze. In particular, selected extracts from the transcripts were expanded to include detailed notations of Post-it note use; for example, who wrote the note, when it was written, when it was placed on the wall or whiteboard, where it was placed and when it was referred to (verbally or non-verbally). This enabled us to focus strongly on the role the Post-it notes played in design conversations. This CA-inspired transcription (Jefferson, 1984) allowed the creation of highly accurate and reliable transcripts as a foundation for our subsequent analysis. Adding graphics and gestural observations to these transcripts helped reveal the relationships between the conversation and the recorded interaction (Heath, Hindmarsh & Luff, 2010). When employing a multimodal analytical approach, pointing and gaze direction also become important features to note in each detailed transcript, as they are used to establish when a particular space becomes a shared focus for cognition and action (Goodwin, 2003). Adopting this multimodal analytical approach, in which different kinds of qualitative data are considered, results in a nuanced picture of what is happening during design activities and helps us triangulate particular findings.

Finally, in our analysis, selected examples of Post-it note use were categorised according to our chosen theoretical frameworks, independently by researchers in Copenhagen and Aarhus, Denmark. Agreement was then sought between each group of researchers with regard to the roles and functions served by Post-it notes in each of these examples. Where agreement could not be reached, examples were put to one side, leaving only those that best represent the particular theoretical constructs.

4   FINDINGS

In this section we present examples of Post-it note use for each of four overlapping functions that externalisations serve in design practice: *informational, formational, transformational,* and *transcendental* (Dix & Gongora, 2011). As we show, these are not discreet categories that assign a single function to particular Post-it notes, but are typically present in combination during design activities. In addition, we present examples describing how Post-it notes support the cognitive processes associated with category formation and semantic long-term memory, by paying particular attention to the temporal development of Post-it note placement during design ideation.

4.1   Summary of findings

We find clear evidence of Post-it notes serving each of the four functions ascribed to design externalisations. However, it is rare that these functions are seen in isolation. Rather, the way design activities unfurl temporally seems to require a more complex pattern in which Post-it notes serve multiple functions. This is supported by the physical properties of Post-it notes (e.g., colour, size and the strip of glue that enables them to be temporarily fixed or easily repositioned), which allow the Post-it notes to 'evolve' into more complex forms and groupings, act as objects to think with, and provide grounds for a common understanding of the work at hand. We see least evidence of Post-it notes serving a *formational* function. However, further study is needed before we can say if this is a result of the physical properties of Post-it notes or a reflection of the design team's practice, as the video recordings show few examples of sketching or developing ideas other than through writing, either on Post-it notes or with any other media.

We also find evidence to back up the hypothesis that Post-it note use supports qualities associated with semantic long-term memory, particularly during clustering activities. Here, we see evidence of positioning and moving to indicate membership or dissociation, and of the use of distance to indicate both within and between category associative strength. However, there are limitations, as clustering Post-it notes on a whiteboard results in a two-dimensional layout that makes visualising hierarchical relations within categories difficult. Here, we see that repeated gesturing and dialogue augment the clustering activity, and helps maintain the hierarchical distinctions.

4.2   The functions Post-it notes serve as externalisations of design cognition

To illustrate our findings with regards to the functions that Post-it notes serve in their role as externalisations of design cognition, we take examples from two videos. First, we look at V18, in which Abby and Kenny (two designers) reflect on the Post-it note collections labelled 'Product', 'Sales' and 'Story', in order to distil the insights they contain, and in preparation for a share-back session between the entire project team. Following this, we look at V15 in which Rose (a Chinese consultant), Kenny and Abby, Nina (an intern), and Tiffany (a stakeholder from the company's accessory department) are sharing insights from a co-creation session with Chinese lead-users. In this session, Rose is translating Chinese language Post-it notes.

Figure 13.1   Reflecting on Post-its.

In V18, we first see Kenny and Abby reflecting on the Post-it note collections labelled 'Product', 'Sales' and 'Story' (Figure 13.1). These were generated in an earlier session, when insights gained from the co-creation sessions were shared. Abby and Kenny spend almost six minutes, from 17:35 to 23:20, silently considering the three existing groups of yellow and green Post-it notes, using these as inspiration for a new round of ideation, which is recorded using orange Post-it notes. This activity follows a repeated pattern in which the two designers look carefully at the existing groups of Post-it notes, and then look down to write ideas on new Post-it notes. During this time, their gaze and focus change, as they reflect on different areas of the boards containing the three Post-it note clusters. Here, we see Abby and Kenny creating links between the existing Post-it notes and the new notes they each silently produce. Later, when sharing these new ideas they have written on the orange Post-it notes, they verbalise this process of identifying similarities and linking ideas. During this activity, Post-it notes are objects for reflection-in-action (Schön, 1992), making it possible for Kenny and Abby to identify new connections, and to think and talk about their previous thoughts at a conceptual, meta-cognitive or *transcendental* level.

Later, as the activity unfolds, we see how Abby and Kenny interact with the yellow and green Post-it notes in the 'Product', 'Sales' and 'Story' groupings, simultaneously with the orange Post-it notes they have just been writing (Figure 13.2). Each of the Post-it notes referred to, for example, yellow Post-it note #63 with its circular 'Production -> sales -> consumer -> trash -> recycle ->', and orange Post-it note #4 'Environment & recycling', serve the *informational* function of communicating an idea expressed by a design team member. However, we also see how Post-it notes serve both *formational* and *transformational* functions, often concurrently. For example, we see this in Figure 13.3 as Kenny brings forward the idea of 'environment' in relation to 'product life cycle' from 30:53, and then Abby relates this to their other ideas saying, "It definitely belongs underneath something." In this way, the idea that eventually becomes orange Post-it note #4 'environment & recycling' is expressed.

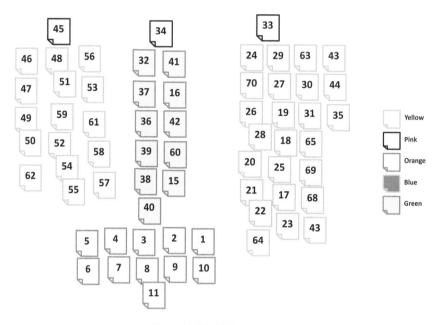

Figure 13.2 V18 Post-it map.

Once this idea is written down, Abby and Kenny continue to reflect on how it fits into the wider collection (Figure 13.4, please note the annotation of colour applies to all the figures), first with reference to orange Post-it note # 2 'Life time companion – personalisation – 3 in 1', and following this with reference to orange Post-it note #3 'Smart living, tech as an improvement of life'. This demonstrates how their current thinking goes beyond the individual idea and Post-it note, and such reflection continues for a period beyond that included in our transcript, ending with Abby re-positioning orange Post-it note #4. It is evident throughout the activity analysed here that both Kenny and Abby are taking a meta-cognitive *transcendental* stance toward not only the particular Post-it notes being discussed, written and moved, but to all the Post-it notes in front of them.

In this example, we also see how Post-it notes are grouped with reference to specific concepts and themes, with properties such as colour and position being given semantic value. This indicates how the physical qualities of Post-it notes are used to create new externalisations, serving different functions that match the development of emerging concepts. We also see that other physical qualities, such as the size of the Post-it notes typically used and the strip of glue that enables them to be temporarily fixed or easily repositioned, are also important in facilitating flexible externalisations of design cognition. In the overlapping of different functions, we see how the Post-it notes 'evolve' into more complex forms and groupings that match and support the current stage of ideation. During this process, Kenny and Abby use Post-it notes (both individually and in groups) as objects to think *with*, and to ground their common understanding of the situation at hand. In doing so, the function served by individual Post-it notes may well shift or gain new layers, as we see with orange Post-it note #4.

30:53	Kenny	Ehh and "recycling" (..) But like all	((Points and makes a circle at English
		of the cycle.	Post-It note #63 'Production -> sales ->
			consumer -> trash -> recycle ->))
30:56	Abby	Yeah. That one-	((Points at the same Post-It note))
	Kenny	Yeah. That one. And maybe it should be	((Looks up at Post-It section 'Story'))
		called something with "environment"	
31:06	Kenny	, but what I MOSTLY saw in it was this	((Points and makes a circle at English
		thing with this "product life cycle".	Post-It note #63 'Production -> sales ->
		(..)	consumer -> trash -> recycle ->))
		(5.5 seconds pause)	
31:13	Kenny	I don't know what we should call- if it	
		deserves its own, but I-	
31:17	Kenny	but we don't really have anything with	((Points at the Orange Post-It notes #1,
		"environment".	#2, #3 on wall))
	Abby	No. Maybe it's good to bring in.	
31:23	Kenny	yeah	((Begins to write on Post-It note #4))
31:24	Abby	Definitely (.) I mean, yes, bring it	
		in, but I think it definitely belongs	
		underneath something, I mean it's the	
		way- there just so many many annoying	
		dust mice everywhere here!	
	Kenny	Yeah. Maybe we can just call it	((Writes on Post-It note #4))
		"environment or recycling"	

Figure 13.3 Transcript extract, Post-it note #4 'environment & recycling' is produced.

31:39	Abby	Yeah. Also, I mean, with it- I	
		think so too, something goes (.)	
	Abby	I mean it might go underneath	((Points at Post-It note #2 'Life time
		that one for example, that's	companion - personalisation - 3 in 1))
		something with,	
31:51	Abby	I mean, and that one it might also	((Points at Post-It note #3 'Smart living,
		belong to.	tech as an improvement of life'))
	Kenny		((Places post-note #4 'Environment &
			recycling' onto the wall))

Figure 13.4 Transcript extract showing how Post-it note #4 is placed.

31:25	Rose	So this was what he was thinking of	((Circles at Post-It note #6 'Online product / service'))
	Tiffany	yes, service is based on a product?	
31:29	Rose	yeah, services based product	
31:31	Rose	then the next was Ying. And then Ying was all about the service actually, he was thinking that it should be an online service that is comprehensive but also modular so that you can add the different types	((Looks up at Post-It note #4 'comprehensive/safety'))

Figure 13.5 Transcript extract showing Post-it notes serve an *Informational* function.

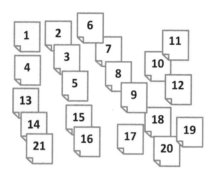

Figure 13.6 V15 Chinese Post-it map.

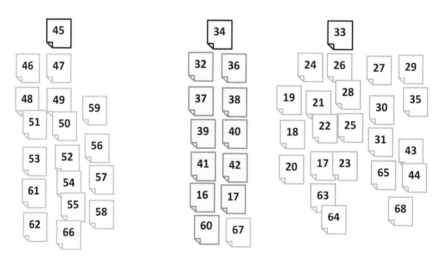

Figure 13.7 V15 English Post-it map.

   In V15, we observe the design team sharing insights from the co-creation work-shops with Chinese lead-users. They work with the Chinese language Post-it notes generated during those co-creation sessions, Figure 13.6, and also the Post-it notes containing English translations of these ideas, Figure 13.7. We use examples from this video to study the *informational*, *formational* and *transformational* functions served by Post-it notes during design activities separately, and in more detail. The transcript in Figure 13.5 shows Rose pointing to, looking at, and then making a circle gesture around the Chinese language Post-it #6 that says 'Online product/service'. It also shows Rose looking at the Chinese language Post-it note #4 that says 'comprehensive/safety' at 31:20. This prompts her to remember that one participant suggested a service that is "comprehensive but also modular". Here, Rose is using the Post-it notes as communicative resources; they provide reference points and affirmations as she communicates the Chinese co-creation session participants' ideas to the current group. They serve an *informational* function.

The transcript in Figure 13.8 shows how the process of writing the Post-it note helps Abby to form and develop the idea that is recorded on Post-it note #26. She starts by writing down the words she has just said out loud, 'Personal Aspect' at 38:36, before writing 'Different Kinds Of Interaction,' an idea voiced by Kenny, at 39:00. Her final addition to this Post-it note is not a direct reference to something that is said out loud, but rather a modifying comment 'Talk' at 39:36. This is likely to refer back to the start of their current discussion, when Rose was relating a story about a Chinese lead user who had introduced the idea of talking with the car as if it were a boyfriend. Here, Abby is finalising her idea. The Post-it note #26, which is then added to the wall, has served a *formational* function.

It is this *formational* function that we see least evidence of when we study Post-it note use within the sessions recorded for the DTRS11 dataset. This may be because the physical properties of Post-it notes, for example, their typically small size, do not facilitate this effectively. However, it may also be a reflection of the design team's practice, as the video recordings show few examples of sketching or developing ideas other than through headline writing, either on Post-it notes or with any other media. Further study is thus needed before we can say anything more authoritative about this.

In the transcript shown in Figure 13.9, we see Kenny write '2-Way Interaction' on Post-it note #28 at 40:08. He then carefully places this Post-it note #28 on the wall, deliberately attaching it to the bottom of Post-it note #26, which had been previously placed there by Abby (see Figure 13.10 and Figure 13.11). We can see that Post-it note #28 has been placed as a response to the content of Post-it note #26, and that it modifies the future meaning of Post-it note #26 for the whole design team. From this point on, we see that the two Post-it notes (#26 and #28) become permanently linked to each other, and are treated as a single idea/note. In this example, Post-it note #28 in combination with Post-it note #26 serves a *transformational* function.

4.3   Post-it note support for qualities of semantic long-term memory

A typical design exercise during brainstorming activities is to attempt to form categories by grouping individual Post-it notes. The Post-it notes may provide cognitive support for reflection and convergence in the categorisation formation process. In V3, the design team is trying to generate ideas for how they will conduct the lead-user co-creation sessions, and after having produced individual ideas silently on distinct Post-it notes, they go through a process of trying to create categories of ideas. The process of clustering takes place on a whiteboard (Figure 13.12 and Figure 13.13),

38:36	Abby	Yeah (..) so it's (.) >I guess it's< more	((Rose looks at the English Post-
		the personal level- personal- yeah- personal	It notes #17-#24))
		(.) aspect or::?	
38:41	Kenny	yeah because I- I think it's eh: it might	
38:41	Abby	be different from person to person how they	((Writes 'Personal Aspect' on
		wanted to realize (.) so, it could be she	Post-It note #26))
		wanted t- wanted to talk to you	((Looks at English Post-It notes
			#17-#24))
38:52	Rose		((Places Post-It #25 on wall))
	Kenny	but other people would be maybe intimidated	
		[by the car talking to you, so they would	
		prefer different] kinds of interaction	
38:54	Abby	(.) exactly	
	Rose	mhh yeah but [then there is k- >some kind	
		of< interaction]I think that was what	
	Kenny	like the person interaction that [makes you]	
		feel like- that this has (.)a relation to	
		me-	
39:00	Abby		((Writes 'Different kinds of
			interaction' on Post-It note
			#26))
	Nina	yeah because that other guy said that he	
		wants to be able to display it, and the	
		other guy- even if you don't wanna display	
		it it sends you a report (..)	
39:12	Rose	yeah (..) that was the: real time -whether	((Points and taps at Chinese
		you have regular or pushed notifications (.)	Post-It note #43 'Alarm' with
		or that you would look at at the same [time.	pen))
		So]	
39:22	Rose	(.) I think (.) that because- that if they	((Points at Chinese Post-It note
		can provide you with push notifications like	#43 'Alarm'))
		your REGULAR REPORTS (.) and then >at the	
		same time< you would be looking at it	
		((touches her wrist)), but it would also:	
		ehm: give me an alarm (.)[or like]you know	
		if something needs to- it's, >I guess, I	
		guess< it is the thing that now most (.) a-	
		mobile product that,	
39:36	Abby		((Writes 'talk' and places Post-
			It note #26 on wall))

Figure 13.8 Transcript extract showing Post-it note #26 serve a *Formational* function.

40:08	Kenny		((Writes '2-way interaction' on Post-It note #28))
	Tiffany	access to eh:: European [eh: (..)] specialists or whatever	((Rose looks at English Post-It notes))
40:14	Rose	[yeah] I think this is good	((Looks at Chinese Post-It notes))
40:32	Kenny		((Sticks Post-It note #28 on the bottom Post-It #26))

Figure 13.9 Transcript extract showing Post-it notes serve a *Transformational* function.

Figure 13.10 Placing Post-it #26.

Figure 13.11 Post-it #26 & #28.

Figure 13.12 The design team brainstorms ideas for recruitment to co-creation sessions.

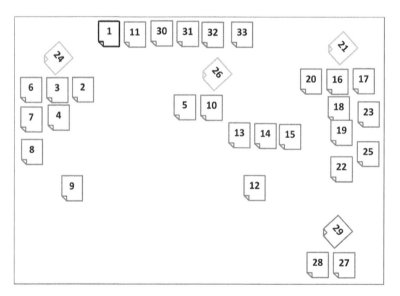

Figure 13.13 Final position of Post-it notes on whiteboard after V3 brainstorm.

where individual team members place their Post-it notes sequentially while simultaneously discussing Post-it note placement and category titles in relation to the emergent categories. Pictures 1–4 on Figure 13.14 illustrate four sequential snapshots from the Post-it note placements, with Post-it numbers representing the order in which they are first placed on the board, and the arrows indicating gesturing or verbal expressions on relations between Post-it notes.

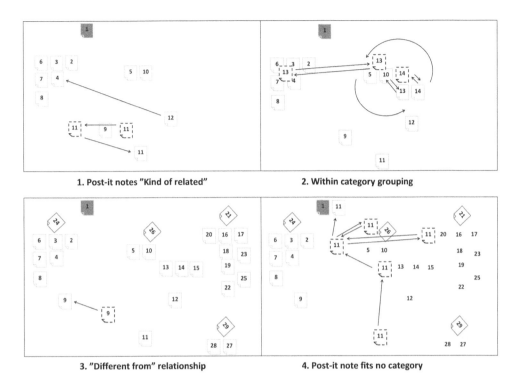

Figure 13.14 Post-it note movement during grouping and category formation.

By analysing the temporal development of the placement of the Post-it notes in relation to the team dialogue and gestures, it is possible to show how Post-it notes support well-known qualities of semantic long-term memory related to categorisation. Several observations can be made:

(1) Categories in semantic long-term memory contain typicality gradients, where individual nodes are more or less typical based on their features or qualities. Although the purpose for the design team is to end up with a number of clearly distinct categories, individual Post-it notes are continually considered 'kind-of' related to the emerging Post-it clusters, illustrating the graded structure behind their placing. Throughout the dialogue, the degree to which Post-it notes fit into a category is expressed with phrases such as (from less to more typical): 'opposite to' (V3, 119) – 'different from' (V3, 119) – 'little bit related' (V3, 122) 'kind of related' (V3, 81), 'basically the same' (V3, 125) – 'pretty much the same' (V3, 129) – 'exactly the same' (V3, 81). Position and distance are then used to indicate how typical category memberships are. For example, Ewan moves his Post-it note #13 stating its connection with #10. He then moves #13 to the group of Post-it notes to the left of the whiteboard, saying 'kind of an opposite to this one'. Ewan then reads the text out loud, explains his point, and places Post-it note #13 below and to the right of #10, since it is 'kind of' related. Abby states that her Post-it

1. Ewan's hierarchical Z gesture   2. Post-It #9 on 'the next level'

Figure 13.15 Placing Post-it #9 on 'the next level'.

note #14 is 'a little bit related to that', so Ewan then holds it up next to Post-it #13. He moves it upwards to the top row of the cluster, but immediately moves it back down and fixes it to the whiteboard saying 'this is kind of the same' and makes a circular gesture around Post-it notes (#5,10,13,14), illustrating the gradients of membership typicality through physical distance.

(2) The chosen medium (Post-it notes on whiteboard) facilitates a two-dimensional representation of clusters, where the addition of a third dimension is more difficult to illustrate. What is known from theory of semantic long-term memory is that categories may contain hierarchical levels (as when the basic level category of 'dog' may have both superordinate levels ['animal'], and subordinate levels ['German Shepard']). When trying to place #9 in relation to a category (see Figure 13.14, Picture 3), Ewan expresses that, "It's kind of an answer element between like the next level of that one in a way", and subsequently goes on to gesture a staircase Z-like motion below the category (Figure 13.15), in order to indicate that the Post-it note is supposed to be understood as a next level in the hierarchy; Kenny places #9 below, and to the category's right. Here, the visual Post-it-note-on-whiteboard format does not support very well the intended categorical relation, and only through dialogue and gesturing do the team recognise the intended two-level meaning. Ewan makes the 'next level' relation clear several times, and further iterates that it is nonetheless related to the category ('just a range (...) a spectrum', v3, 183).

(3) Conceptual distance also pertains to the emerging concepts illustrated through how new Post-it notes are added to the board. In Figure 13.14 Picture 1, we see how the placing of #5, #9 and #11 was done in a way that illustrated their conceptual distinction (away from the reference category to the left). Physical distance is used to show conceptual differences to other reference categories. Over time, the board becomes populated with emerging concepts, and the team tries to re-position some of the lone Post-it notes into the emergent categories. Post-it note #11 is an illustrative example (Figure 13.14, Picture 4), as it is moved from category to category by Ewan to find a fit. David then suggests that Post-it note #11 is placed '*between*' two cluster categories that it seemingly relates to, without quite fitting either; and Ewan acknowledges this

suggestion. This exchange shows that David and Ewan are able to interpret the relative distance of this Post-it note to each of these clusters in a way that suggests its semantic relationship to each of the two categories. Finally, Abby suggests placing it in a new category, seemingly an idea parking lot, where the remaining individual Post-it notes without category headings get placed (#30-#33), thus sticking to the intended output format of fixed categorisations. Nonetheless, the example shows how #11 was initially placed based on considerations of associative strength in relation to two distinct categories, without quite fitting either.

In sum, the properties of Post-it notes support semantic long-term memory in the clustering exercise (moving to indicate membership or dissociation, distance to indicate both within and between category associative strength). However, the two-dimensional Post-it notes-on-whiteboard layout makes visualising hierarchical relations within categories difficult, and had to be done through gesturing and dialogue. The Post-it notes facilitated the process of moving from the typicality gradients apparent in the dialogue and gesturing to the fixed categorisation that the team eventually settled on in order to drive the design process forward.

5   DISCUSSION

The design team practice recorded in the DTRS11 dataset provides many examples of creative cognitive processes that are distributed across different group members, across time, and which are coordinated through both mental and environmental structure (Hutchins 2000, 2006). Our analysis in this paper has focused on a single aspect of this system of cognition, the Post-it notes, and in particular the way they function as design externalisations that support qualities associated with semantic long-term memory. Whilst we consider this to be a particularly interesting and important aspect of the processes under investigation, we also remain aware that it represents just a single factor in the wider cognitive ecology. Our analysis shows the importance of the meta-cognitive functions of the Post-it notes in the design process, which could be studied further in relation to process awareness and reflective practice, which Valgeirsdottir and Onarheim (2017) examine in detail. Alongside Post-it notes, the design team uses a range of other tools and materials, such as spreadsheets in which they list ideas and insights, to support their design practice. The interaction of these different tools and materials, and the way cognitive resources are coordinated and shared across them, is beyond the scope of this paper. However, such a future study would help clarify some of the issues raised here, and improve our understanding of the role Post-it notes play within the system of creative design team cognition.

Our analysis has investigated the functions Post-it notes serve as externalisations of design ideas, and considered why the practice of clustering and grouping Post-it notes might be such an important and effective aspect of design activities. Our examples show that Post-it notes are often referred to in design team discussions, both as representations of single ideas and also as they become clustered into emergent concepts, and that decisions are often manifested in a newly written Post-it note or through placing, moving and organising existing Post-it notes. This strongly indicates the importance of Post-it notes as a design material, and shows how they participate in the on-going conversation with the design situation, supporting the design team's reflection in action (Schön, 1992). Shroyer *et al.* (2017) also note how the design team use Post-it notes to capture and externalise ideas during a later brainstorm activity, but conclude that the discussion seems to go beyond what is actually written on the Post-it notes, supporting the formational and transcendental functions we point out in this chapter. In our analysis we see how Post-it notes are grouped with reference to the emergence of specific concepts and themes, how through this grouping they 'evolve' into more complex forms to match and support the current stage of ideation, how they are used as objects to think with, and how they support a common understanding of the work at hand. In this way, Post-it note use can be seen to scaffold collaborative design cognition, and arguably provide an externalisation that manifests qualities associated with long-term semantic memory.

Many of these factors are supported by the characteristic properties of Post-it notes, notably their regular shape and small size, their strip of reusable adhesive, and their availability in a number of different colours, with individual Post-it notes being very inexpensive. These characteristics mean that Post-it notes are ideal for representing individual ideas, serving an *informational* function. They are also readily available, highly moveable and flexible, and yet can be easily discarded, and therefore serve *formational* and *transformational* functions. They also facilitate grouping and clustering, which enable them to simultaneously represent individual ideas and larger concepts, and therefore serve a *transcendental* function. During these grouping activities, we can see each Post-it note as a node in an emerging semantic network, with relative distance mirroring associative strength within and between clusters. These emergent structures also seem to require an overlapping of the different functions Post-it notes serve as externalisations of design ideas. For example, in order to facilitate the emergence of semantic structure, Post-it notes must simultaneously be *informational* in regard to the individual ideas they represent, and *formational* and *transformational* in regard to the developing semantic network. In order for the design team to recognise this emergent semantic structure, it is clear that Post-it notes must also serve a *transcendental* function. This finding adds another dimension to the growing literature on the ways materials may support design cognition, this time in converging design tasks, rather than the generative, emergent or divergent aspects typically covered in past research.

Finally, the design team would also maintain these externalisations across sessions, that is, by transporting and reusing the actual physical Post-it notes. In doing so, they would sometimes remove all the Post-it notes and reposition them in a new location. Here, they would often maintain the overall cluster structure, but not necessarily the particular relative associations between individual Post-it notes, or within and between clusters. This would indicate that the Post-it note support for categorization was mainly local and situational support, occurring during the process of forming clusters. It appears that the main value to this design team across sessions was in the resulting non-graded clusters. In terms of generalising these observations, it is important to note that the relationships between individual Post-it notes, both within clusters and between clusters, and between clustered groupings of Post-it notes, were not always referenced with regard to levels or degrees of association – although a number were referred to dichotomously as either belonging to a cluster or not, in dialogue.

This study only begins to explore the functions served by Post-it notes in creative design team practice. In particular, we have yet to examine in detail how Post-it notes are used in conjunction with other design materials, including digital resources such as spreadsheets containing ideas and observations. More research is also needed in order to determine more broadly both the purposes and actual uses of Post-it note clustering over time and place, and its relationship to the qualities of semantic long-term memory. However, it seems evident that the kind of activities we witness in the DTRS11 dataset would be very different without the humble Post-it note.

ACKNOWLEDGEMENTS

This research is funded by The Danish Innovation Foundation (grant 1311-00001B, CIBIS).

REFERENCES

Armbruster, B. B. (1989). Metacognition in creativity. In *Handbook of creativity*. Springer US. pp. 177–182.

Arnheim, R. (1969). *Visual thinking*. University of California Press. Berkley CA.

Bamberger, J., & Schön, D. A. (1983). Learning as reflective conversation with materials: Notes from work in progress. *Art Education*, 36(2), 68–73.

Barsalou, L. W. (1983). Ad hoc categories. *Memory & Cognition*, 11(3), 211–227.

Beyer, H., & Holtzblatt, K. (1997). Contextual design: defining customer-centered systems. Elsevier.

Buxton, B. (2010). Sketching user experiences: getting the design right and the right design. Morgan Kaufmann.

Christensen, B. T., & Abildgaard, S. J. J. (2017). Inside the DTRS11 Dataset: Background, Content, and Methodological Choices. In: Christensen, B. T., Ball, L. J. & Halskov, K. (eds.) *Analysing Design Thinking: Studies of Cross-Cultural Co-Creation*. Leiden: CRC Press/Taylor & Francis. Christensen, & Ball.

Christensen, B. T., & Schunn, C. D. (2007). The relationship of analogical distance to analogical function and preinventive structure: The case of engineering design. *Memory & Cognition*, 35(1), 29–38.

Collins, A. M., & Loftus, E. F. (1975). A spreading-activation theory of semantic processing. *Psychological review*, 82(6), 407.

Cook, D. J., & Bailey, B. P. (2005). Designers' use of paper and the implications for informal tools. In: *OZCHI '05: Proceedings of the 17th Australian conference on computer-human interaction: Citizens online: Considerations for today and the future*. pp. 1–10.

Cross, N. (2006). *Designerly ways of knowing*. Springer.

Dix, A., & Gongora, L. (2011). Externalisation and design. In: *Proceedings of the second conference on creativity and innovation in design*. pp. 31–42.

Fischer, G., Nakakoji, K., & Ostwald, J. (1995). Supporting the evolution of design artifacts with representations of context and intent. In: *DIS '95: Proceedings of the 1st conference on designing interactive systems: Processes, practices, methods, & techniques*. ACM. pp. 7–15.

Goodwin, C. (1994). Professional vision. *American anthropologist*, 96(3), 606–633.

Goodwin, C., & LeBaron, C. (2011). *Embodied interaction: Language and body in the material world*. Cambridge University Press.

Harboe, G., & Huang, E. M. (2015). Real-World Affinity Diagramming Practices: Bridging the Paper-Digital Gap. In: *Proceedings of the 33rd Annual ACM Conference on Human Factors in Computing Systems*. ACM. pp. 95–104.

Heath, C., Hindmarsh, J., & Luff, P. (2010). *Video in qualitative research*. Sage Publications.

Hilliges, O., Terrenghi, L., Boring, S., Kim, D., Richter, H., & Butz, A. (2007). Designing for collaborative creative problem solving. In: *Proceedings of the 6th ACM SIGCHI conference on creativity & cognition.*

Hutchins, E. (2000). Distributed cognition. *International Encyclopedia of the Social and Behavioral Sciences.* Elsevier Science.

Hutchins, E. (2006). The distributed cognition perspective on human interaction. *Roots of human sociality: Culture, cognition and interaction*, 375–398.

Jefferson, G. (1984). Transcription Notation. In J. Atkinson and J. Heritage (eds.), *Structures of Social Action*, New York: Cambridge University Press.

Kress, G. (2009). *Multimodality: A social semiotic approach to contemporary communication.* Routledge.

Purcell, A.T., & Gero, J.S. (1998). Drawings and the design process: A review of protocol studies in design and other disciplines and related research in cognitive psychology. *Design Studies*, 19(4), 389–430.

Rosch, E. (1978). Principles of categorization. In: E. Rosch & B. B. Lloyd (eds.) *Cognition and categorization.* Hillsdale, NJ: Erlbaum.

Shroyer, K., Turns, J., Lovins, T., Cardella, M. & Atman, C. J. (2017) Team Idea Generation in the Wild: A View from Four Timescales. In: Christensen, B. T., Ball, L. J. & Halskov, K. (eds.) *Analysing Design Thinking: Studies of Cross-Cultural Co-Creation.* Leiden: CRC Press/ Taylor & Francis.

Schön, D. A. (1992). Designing as reflective conversation with the materials of a design situation. *Knowledge-Based Systems*, 5(1), 3–14.

Schön, D. A., & Wiggins, G. (1992). Kinds of seeing and their functions in designing. *Design Studies*, 13(2), 135–156.

Sonalkar, N., Mabogunje, A., Leifer, L., & Roth, B. (2016). Visualising professional vision interactions in design reviews. *CoDesign*, 12(1–2), 73–92.

Stolterman, E. (1999). The design of information systems: Parti, formats and sketching. *Information Systems Journal*, 9(1), 3–20.

Stolterman, E., McAtee, J., Royer, D., & Thandapani, S. (2009). Designerly Tools. In: *Undisciplined! Design Research Society Conference 2008.*

Suwa, M., & Tversky, B. (1997). What do architects and students perceive in their design sketches? A protocol analysis. *Design Studies*, 18(4), 385–403.

Valgeirsdottir, D., & Onarheim, B. (2017). Metacognition in Creativity: Process Awareness Used to Facilitate the Creative Process. In: Christensen, B. T., Ball, L. J. & Halskov, K. (eds.) *Analysing Design Thinking: Studies of Cross-Cultural Co-Creation.* Leiden: CRC Press/Taylor & Francis.

Van der Lugt, R. (2005). How sketching can affect the idea generation process in design group meetings. *Design Studies*, 26(2), 101–122.

Warr, A., & O'Neill, E. (2007). Tool support for creativity using externalisations. In: *C&C '07: Proceedings of the 6th ACM SIGCHI conference on creativity & cognition* ACM. pp. 127–136.

Youn-Kyung, L., Stolterman, E., & Tenenberg, J. (2008). The anatomy of prototypes: Prototypes as filters, prototypes as manifestations of design ideas. *ACM Transactions on Computer-Human Interaction* (TOCHI) 15.2.

Chapter 14

Fluctuating Epistemic Uncertainty in a Design Team as a Metacognitive Driver for Creative Cognitive Processes

Bo T. Christensen & Linden J. Ball

ABSTRACT

Previous design research has demonstrated how epistemic uncertainty engenders localized, creative reasoning, including analogizing and mental simulation. Our analysis of the DTRS11 dataset examined not just the short-term, localized effects of epistemic uncertainty on creative processing and information selection, but also its long-term impact on downstream creative processes and decisions about what information to take forward. Our hypothesis was that heightened levels of uncertainty associated with a particular cognitive referent (i.e., a post-it note translated from Chinese end-users) would engender: (1) immediate creative elaboration of that referent aimed at resolving uncertainty and determining information selection; and (2) subsequent attentive returns to that cognitive referent at later points in time, aimed at resolving lingering uncertainty and again determining information selection. Our key findings were threefold. First – and contrary to expectations – we observed that increased epistemic *certainty* (rather than increased epistemic uncertainty) in relation to cognitive referents triggered immediate, creative reasoning and information elaboration. Second, epistemic uncertainty was, as predicted, found to engender subsequent attentive returns to cognitive referents at later points in the design process. Third, although epistemic uncertainty did not predict the information that was eventually selected to take forward in the design space, both immediate creative elaboration and subsequent attentive returns did predict information selection, with subsequent attentive returns being the stronger predictor. We suggest that our findings hold promise for identifying more global impacts of epistemic uncertainty on creative design cognition that are possibly mediated through the establishment of lasting associations with cognitive referents.

1 THEORETICAL FRAMEWORK

It is essential that the design process incorporates knowledge of end-users through user-oriented approaches such as anthropological investigations, user-driven design and participatory design. However, understanding users, especially across cultural divides, is a daunting task, as many a failed design artifact illustrates. Although cross-cultural interpretation can be a source of design error and failure, we suggest that it can also act as a catalyst for creative design. That is, because cross-cultural interpretation is frequently uncertain, ambiguous, re-frameable, contextually shiftable and open to exploration, it embodies the essential qualities that provide design objects and pre-inventive structures with creative potential, as captured by dominant theories of design and creativity (e.g., Dorst & Cross, 2001; Finke, Ward & Smith, 1995; Schön & Wiggins, 1992). Indeed, much design-reasoning research has convincingly demonstrated that effective designers are not only at ease with uncertainty but thrive in relation to the opportunities it affords (Alcaide-Marzal, Diego-Más, Asensio-Cuesta, & Piqueras-Fiszman, 2013; McDonnell, 2015; Schön, 1983).

A few previous studies have addressed the importance of uncertainty in design, although not with a focus on cross-cultural interpretation. Beheshti (1993) discussed uncertainty as a key factor influencing design decisions, noting that it is important to minimize its influence so as to increase decision quality. We likewise see uncertainty as a pervasive aspect of design and view it positively since it provides valuable opportunities for creative ideation as part of the process of uncertainty reduction. Designers also view uncertainty as a positive element of their professional self-identity, as shown in Tracey and Hutchinson's (2016) qualitative study of designers who were prompted to reflect on their experiences and beliefs regarding uncertainty. In the present volume D'souza and Dastmalchi (2017) discuss uncertainty in design jargon and slang usage, while Paletz, Sumer, and Miron-Spektor (2017) relate uncertainty to design team micro-conflicts.

Our analysis of the DTRS11 dataset focused on the extent to which uncertainty arising specifically from cross-cultural interpretation elicits creative design reasoning – both in the short-term (e.g., engendering localized analogizing and mental simulation) and in the longer-term (influencing downstream creative processes and decision-making). To address this issue we examined those parts of the dataset that involved the Scandinavian design team comprehending and analyzing lead-user post-it notes written in Chinese. Our overarching assumption was that uncertainties in the interpretation of these post-it notes (pre-inventive structures) would be likely to promote creative processes and subsequent returns to information, eventually predicting what information would be extracted by the team to be taken forward.

1.1 Epistemic uncertainty as a metacognitive trigger for creative analysis

The concept of uncertainty that we draw upon for our analysis is that of 'epistemic uncertainty', which refers to a designer's experienced, subjective and fluctuating feelings of confidence in their knowledge and choices, as measured through phrases in the design dialogue. This epistemic uncertainty is differentiable from 'aleatory uncertainty', which is expressed in natural language via *likelihood* statements (Ülkümen, Fox, & Malle, 2016). Heightened levels of epistemic uncertainty appear to act as a 'metacognitive cue' (Ackerman & Thompson, 2014; Alter & Oppenheimer, 2009; Alter, Oppenheimer, Epley, & Eyre, 2007; Ball & Stupple, 2016; Thompson, Prowse Turner & Pennycook, 2011; Thompson *et al.*, 2013), triggering more elaborate reasoning than might otherwise arise when people feel confident about ongoing processing. Similar ideas are noted by Stempfle and Badke-Schaub (2002) in a study of design-team thinking that coded for expressions of uncertainty. Their findings suggest that simpler design problems are associated with team self-efficacy and rapid and intuitive evaluative reasoning, whereas complex design problems may trigger a shift toward a structured process of effortful idea generation and analysis.

Epistemic uncertainty has previously been associated with creative analogizing (e.g., Dunbar, 1997; Houghton, 1998) and mental simulation (e.g., Nersessian, 2009). Indeed, spikes in expressed uncertainty reliably predict analogizing in engineering design (Ball & Christensen, 2009) and scientific problem solving (Chan, Paletz & Schunn, 2012), with these studies demonstrating that analogizing subsequently reduces uncertainty to baseline levels. Similarly, mental simulations during design have been found to be run in situations of elevated epistemic uncertainty (Ball & Christensen, 2009; Ball, Onarheim, & Christensen, 2010; Christensen & Schunn, 2009) and function strategically to reduce uncertainty through the generation of approximate answers to design issues. In addition, strategic switches between depth-first and breadth-first design moves have been shown to be mediated by epistemic uncertainty (Ball, Onarheim, & Christensen, 2010) and episodes of problem–solution co-evolution also take place under elevated levels of epistemic uncertainty (Wiltschnig, Christensen, & Ball (2013), with solution attempts within these episodes being closely associated with uncertainty above baseline levels.

1.2 Research questions

The underpinning assumption in these aforementioned studies is that a heightened level of epistemic uncertainty immediately sparks off a localized, creative episode (e.g., involving analogizing) aimed at reducing uncertainty. However, what has not been investigated is the degree to which increased epistemic uncertainty might also affect design *across* episodes that extend beyond the localized micro-situation in which it is experienced. No doubt the failure to address this research question reflects the methodological challenge of tracing uncertainty referents over time in naturally-occurring design dialogue. Answering this question requires shifting the unit of analysis from standard, sequential discourse segmentation (e.g., turn-taking in dialogue) to a focus on the *qualities* of the cognitive referents themselves, tracing their occurrence both locally (within micro-episodes) and globally (across episodes).

In our analysis we sought to address head-on the question of whether the epistemic uncertainty initially associated with a cognitive referent predicts repeated referrals back to that referent. This might arise from designers utilizing information that is generated or encountered later in the design process in an attempt to address an earlier, epistemically-uncertain design issue that remains unresolved. A similar phenomenon is found in the classic literature on the function of memory in problem solving and concerns the 'Zeigarnik effect' (Zeigarnik, 1927), whereby people's memory for unsolved problems exceeds that for solved problems, indicating a special 'cognitive alertness' towards unanswered issues. This cognitive alertness might enable later, chance encounters with relevant stimuli to engender productive solution attempts (Christensen & Schunn, 2005; Seifert *et al.*, 1995; Yaniv & Meyer, 1987) according to what has been termed the 'prepared mind hypothesis'. This idea is central to the 'opportunistic assimilation' theory of incubation effects, where incubation is the phenomenon whereby a period of time away from a problem and engaged in unrelated activities leads to enhanced solution likelihood on returning to the problem (Howard *et al.*, 2008; Sio & Ormerod, 2009; Gilhooly, in press).

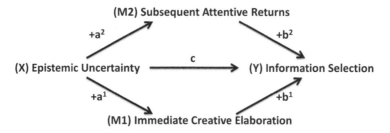

Figure 14.1 The proposed double mediation model.

In light of these findings we propose that it is theoretically plausible that epistemic uncertainty will become associated with its cognitive referent (the object of the uncertainty) so as to influence the design situation at later stages that are temporally remote from the original occurrence of the uncertainty. Epistemic uncertainty was estimated in our analysis based on the initial translation and elaboration of post-its by the design team. We then examined whether the team spent time immediately on local, creative elaboration on a post-it and on whether (and to what degree) the team turned its attention to the post-it at a later time.

1.3 Hypotheses

We hypothesized that elevated levels of epistemic uncertainty would predict local, creative processing (X→M1; Figure 14.1) as well as returns to the cognitive referent over time (X→M2), as per the Zeigarnik effect. We also predicted that both the local, creative micro-episodes and the subsequent returns would predict which information was salient to the team and worth taking forward. In sum, we propose an overarching 'double mediation model' (Figure 14.1) in which epistemic uncertainty (X) on initial encounters with individual post-its predicts: (1) the immediate occurrence of local, creative micro-episodes (M1) that mediate the formation of new post-its (information selection, Y); and (2) the occurrence of subsequent attentive returns across episodes (M2) that also lead to the formation of new post-its (information selection, Y).

Past research investigating epistemic uncertainty in design has focused on the triggering of creative processes. In the present analysis, however, it was also possible to trace the immediate and delayed outcomes of creative processes on the information selected to be taken forward. The focus was, therefore, specifically on whether the epistemic uncertainty in the initial encounter with the cognitive referent (i.e., a post-it or group of post-its) would affect both within-episode and between-episode creative cognition and information selection.

2 METHODS

We applied 'in vivo' analysis (Christensen & Ball, 2014; Dunbar, 1995) to the dataset, which involves studying expertise 'online' as it arises naturally. The in vivo methodology takes a particular stance on data analysis, with verbal data (including data from team discussions) being coded using a similar approach to that deployed when analysing concurrent think-aloud protocols (Ericsson & Simon, 1999).

2.1   Video selection and protocol coding

We selected Videos 15–17 (Christensen & Abildgaard, 2017) since these related to the design team translating into English the post-its that had been written by Chinese lead-users, with the designers sometimes elaborating on these post-its and generating new ideas. The team members subsequently noted down (on new post-its) selected information to take forward. The observed activity clearly involved more than straightforward translation between languages since it also included rich, inferential processes, with the designers mapping across information, making generalizations, deriving cultural meaning and extending information into design ideas. Moreover, for the designers, the process of deciphering what lead-users 'meant' was fraught with uncertainty. The dataset therefore provided a unique testbed for addressing our research questions regarding the impact of epistemic uncertainty on design cognition both within and across episodes.

The selected videos had been recorded back-to-back over a single day, thus varying minimally in temporal factors that might have influenced the design process. In the videos, the Scandinavian design team had finalized co-creation workshops with Chinese lead-users, and then spent approximately 109 minutes going through the Chinese lead-user post-its, moderated by consultants capable of translating the information. The post-its thus supported design team cognition (Dove *et al.*, 2017) Essentially, these sessions constituted iterations of Chinese post-it translations that resulted in some of the translated information being developed by the design team, with the information being selected and documented in English on separate posters to be taken forward.

2.1.1   Coding cognitive referral segments and cognitive referral episodes

The videos were transcribed and segmented by turn-taking of dialogue, resulting in 999 segments. To trace post-it usage and development, all post-its were numbered and categorized according to the poster and poster sub-section they were situated on and when they were added and moved. The data-segments were subsequently sub-divided by coding for post-it referral using gesture and dialogue. When a member of the design team referenced a post-it this was coded as a 'cognitive referral' in the associated dialogue segment, and whenever a segment contained mentions of more than one post-it referent, that segment was subdivided to ensure unique cognitive referents for each segment. This re-segmentation procedure resulted in 1158 segments.

Based on the coding of cognitive referral relating to post-its we then coded for 'cognitive referral episodes', which reflected clusters of segments pertaining to the same Chinese post-it or post-it cluster. This led to 89 episodes. Individual episode-segments contained translations of a post-it together with further elaborative comments aimed at trying to understand its meaning (e.g., by referencing Chinese cultural or contextual information). These cognitive referral episodes constituted our final unit of analysis.

2.1.2   Coding epistemic uncertainty

The coding for epistemic uncertainty followed the coding scheme used extensively in past research (e.g., Ball & Christensen, 2009; Chan & Schunn, 2012; Christensen & Schunn, 2009; Trickett *et al*, 2005). It involved a syntactic approach whereby 'hedge words' are used to locate segments displaying uncertainty (e.g., 'probably', 'sort of', 'guess', 'maybe', 'possibly', 'don't know', '[don't] think', '[not] certain' and 'believe'). Segments containing these words were located and coded as 'uncertainty present' if it was clear from manual screening that the hedge words were not being stated as politeness markers or were otherwise *not* evidence of epistemic uncertainty. All instances of epistemic uncertainty were counted for each cognitive referral episode. Given the cognitive referral episodes contained multiple segments, the measure of epistemic uncertainty was a continuous one calculated by dividing the number of epistemically uncertain statements by the number of episode segments.

2.1.3   Coding immediate creative elaboration

Protocol segments that immediately followed a cognitive referral episode were coded for whether they revealed further generative and creative development of the episode content beyond what was derivable from the cognitive referent (e.g., analogizing, idea generation and old-new information synthesis). Segments were coded in a binary manner as 'immediate creative elaboration present' versus 'immediate creative elaboration absent'.

2.1.4   Coding subsequent attentive returns to the cognitive referent

To measure subsequent attentive returns to a cognitive referent we tabulated the number of segments referring back to each cognitive referent. We then conducted a mean-split to divide the episodes into ones with many subsequent attentive returns versus few subsequent attentive returns.

2.1.5   Coding information selection

Based on the cognitive referral code, all new post-it generation was related to the episodes, allowing for an analysis of which Chinese post-its were linked to the resulting English outcome post-its that would be taken forward by the team. The information selection contained a mixture of notes from the translation, contextual information and further creative elaborations. When counted by episode, this led to a dependent variable that was a cumulative count of the number of outcome post-its deriving from each episode.

2.2   Coding procedure and inter-coder reliability checks

The dataset was coded by two independent student coders who were unaware of the research hypotheses. Each student coded the dataset in four iterations. One coder carried out all post-it categorization, cognitive referral numbering and coding for cognitive referrals, cognitive referral episodes, immediate creative elaboration and information selection. The other student coded for epistemic uncertainty and subsequent attentive returns. The first coder had assisted in the transcription and turn-taking segmentation of the sessions, and was therefore familiar with the content of the cognitive referents and the overall data.

Inter-coder reliability checks were conducted by asking a third coder to independently re-code 10% of the data, with reliability being estimated using Cohen's Kappa. All Kappa coefficients displayed fair-to-good or excellent inter-coder agreement (epistemic uncertainty = 0.79; immediate creative elaboration = 0.83; subsequent attentive returns = 0.75; information selection = 0.58).

3   RESULTS

3.1   Descriptive findings

A total of 173 Chinese post-its formed the basis for the design team's translation, elaboration and generation activities. Eighty-nine unique cognitive referral episodes were identified, constituting 64% of the data segments. These unique episodes were the basic unit of analysis, ranging in length from 1–49 segments (M = 8.4, SD = 7.4). They contained an average of 0.46 epistemic uncertainty phrases per segment (SD = 0.48, Range = 0–2). Overall, 26% of all segments contained uncertainty phrases, which is a high percentage compared to past research, with around 15% of segments containing epistemic uncertainty being more typical (Ball & Christensen, 2009; Wiltschnig *et al.*, 2013). Such elevated levels of epistemic uncertainty perhaps derive from the inherent ambiguities associated with the translational and cross-cultural aspects of the present design situation. Segments arising within cognitive referral episodes contained uncertainty hedge words more frequently than segments arising outside of cognitive referral episodes ($\chi^2(1) = 9.19$, $p = 0.002$). However, uncertainty hedge words did not differ between segments arising within cognitive referral episodes and segments that immediately followed cognitive referral episodes ($\chi^2(1) = 0.63$, ns). Our analysis indicated that 35% of episodes were immediately followed (vs. not followed) by creative elaborative segments, and 55% of the episodes had many (vs. few) subsequent attentive returns.

The design team made 85 notes relating to information selection, with 6 being clearly marked as 'categorical' post-its describing clusters of other post-its. The latter were excluded from the analysis, resulting in 79 post-its, 58 of which were coded as having been generated in reference to prior cognitive referral episodes. The post-it count by cognitive referral episode displayed a Poisson distribution, with 50, 25, 11, 2, 0, 1 counts of 0-1-2-3-4-5 resulting post-its generated on their basis respectively (i.e., the majority of the notes were written with reference to a single episode). In 7 instances information was subsequently added to an existing post-it upon initial production. This adding of information mainly happened as a result of the later classification of the notes and was ignored for the present purposes.

3.2   Mediation analyses

The binary codes for [M1] immediate creative elaboration and [M2] subsequent attentive returns were statistically unrelated ($\chi^2(1) = 0.75$, $p = 0.39$) illustrating independence of the hypothesized mediators, and further indicating that multicollinearity was not a concern in the subsequent regression models. To test the hypothesized relations we followed Baron and Kenny's (1986) step-based procedure for testing mediation effects. It should be noted, however, that because of the different types of dependent variables in our analysis (binary for M1 and M2; Poisson distributed for Y), it was not possible to quantify the level of the direct effect versus the indirect mediation effect since distinct statistical tests were applied for testing individual relations (i.e., logistic regression for the relation between [X] epistemic uncertainty and the mediators [a, b]; GzLM Poisson regression for the relations between mediators [a, b] and [X] epistemic uncertainty on [Y] information selection). As a result, the model test here should be considered as being primarily *conceptual* rather than a precise quantification of the direct and indirect effects. In all models we controlled for the video session the episodes derived from.

Step 1 [X → Y] Epistemic uncertainty and information selection

A GzLM Poisson regression was run to test whether the level of epistemic uncertainty in the initial encounter with a cognitive referent predicted subsequent information selection. Overall, the model displayed acceptable goodness of fit ($\chi^2/df = 1.003$). However, the analysis revealed that epistemic uncertainty did not predict information selection, although the odds ratio of 1.39 (95% CI, 0.87 to 2.23), $p = 0.17$, was in the expected direction. This analysis suggests the absence of a direct effect of epistemic uncertainty on eventual information selection.

Step 2 [X → M1] Epistemic uncertainty and immediate creative elaboration

A logistic regression was conducted to test whether the level of epistemic uncertainty in a cognitive referral episode predicted immediate creative elaboration. Overall the model was significant ($\chi^2(3) = 12.09$, $p = 0.007$, Nagelkerke $R^2 = 0.18$), but epistemic uncertainty did not predict immediate creative elaboration ($p = 0.141$), and with an odds ratio of 0.44 the results go in the opposite direction hypothesized (i.e., *less* uncertainty predicts immediate creative elaboration).

Step 2 [X → M2] Epistemic uncertainty and subsequent attentive returns

A logistic regression analysis was conducted to test whether the level of epistemic uncertainty in an episode predicted subsequent attentive returns. Overall the model was significant ($\chi^2(3) = 12.05$, $p = 0.007$, Nagelkerke $R^2 = 0.17$), with epistemic uncertainty significantly predicting subsequent attentive returns in the expected direction ($p = 0.046$), and with an odds ratio of 2.90.

Step 3 [M1 + X → Y] Immediate creative elaboration and epistemic uncertainty onto information selection

A GzLM Poisson regression was run to predict information selection based on epistemic uncertainty in the initial encounter with the cognitive referent and immediate creative elaboration. Overall the model displayed acceptable goodness of fit ($\chi^2/df = 0.88$). Immediate creative elaboration predicted information selection, odds ratio of 0.52 (95% CI, 0.30 to 0.90, $p = 0.019$), in the expected direction. Epistemic uncertainty approached, but did not reach, significance, odds ratio 1.61 (95% CI, 0.97 to 2.67, $p = 0.067$).

Step 3 [M2 + X → Y] Subsequent attentive returns and epistemic uncertainty onto information selection

A GzLM Poisson regression was run to predict information selection based on epistemic uncertainty in the initial encounter with the cognitive referent and subsequent

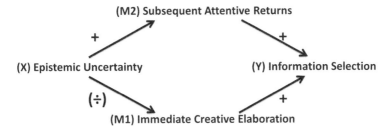

Figure 14.2 The suggested opposing mediation model arising from the analyses.

attentive returns. Overall the model displayed acceptable goodness of fit ($\chi^2/df = 0.93$). Subsequent attentive returns predicted information selection, odds ratio of 0.48 (95% CI, 0.27 to 0.83, p = 0.009), in the expected direction. Epistemic uncertainty did not reach significance, odds ratio 1.19 (95% CI, 0.73 to 1.97, p = 0.485).

Step 3 [M1 + M2 + X→Y] Both mediators and epistemic uncertainty onto information selection
A GzLM model combining both mediators and epistemic uncertainty further illustrated that only subsequent attentive returns significantly predicted information selection (odds ratio: 0.54; 95% CI, 0.30 to 0.97, p = 0.038), while both immediate creative elaboration (odds ratio: 0.62; 95% CI, 0.35 to 1.10, p = 0.100) and epistemic uncertainty (odds ratio: 1.36; 95% CI, 0.80 to 2.31, p = 0.262) were insignificant.

3.3   Interpreting the statistical model

Interpreting these results requires caution, since it was not possible to test for mediation using the same statistical test throughout given the Poisson-distributed outcome variable. As such, we were unable to calculate the direct versus indirect effects, but instead rely on interpreting the overall relations between the variables based on individual test results. Nonetheless, with appropriate caution the illustrated relations can best be described as approximating an 'opposing mediation', whereby the independent variable holds opposing relations to two distinct mediators that subsequently both positively affect the dependent variable (Figure 14.2). In opposing mediation the independent variable does not predict the dependent variable directly since the two mediators operate in opposite directions. In other words, high epistemic uncertainty triggers subsequent returns to the cognitive referent, while immediate creative elaboration is associated with epistemic *certainty* (although not reaching significance, and against the hypothesized direction based on past research). Both mediators positively affected information selection, with subsequent attentive returns proving to be the stronger predictor. Notably, epistemic uncertainty in itself does not significantly predict eventual information selection. In order to understand these patterns of effects in the dataset we present below extended, illustrative examples of the two 'routes' from epistemic uncertainty to information selection.

3.4   Qualitative examples of the two routes from epistemic uncertainty to information selection

3.4.1   Example of high uncertainty leading to subsequent attentive returns and information selection

Table 14.1 exemplifies how an episode with high uncertainty leads to subsequent attentive returns across episodes and ultimately to information selection. We enter the dialogue during a discussion about features for wearable devices connected to online services (Figure 14.3). In this fragment the post-its denoting 'Personal aspect', 'Human', and '2-way interaction' (Figure 14.4) are produced in response to Episode 40, which is related to the cognitive referent 'Interaction'.

The dialogue begins with Rose recalling an observation from the co-creation session. Rose refers, with high uncertainty ("I think" and "kind of") to the analogy 'the car as a boyfriend' to explain product features like interaction and talking back to the user. Nina supplements Rose's comments with her own observation that "it [the car/product] needs to be able to talk to you". Rose confirms what Nina states in a way that links the 'talk' feature to the post-it 'Interaction' by saying "Yeah, that kind of interaction" and using an air-quote gesture while saying the word 'interaction', implying that the word (or idea) belongs to the lead-users (Stivers & Sidnell, 2005). Abby sums up what Rose said and writes 'Personal aspect' on a post-it. Kenny proposes with high uncertainty ("I think", "it might", "it could", "maybe") that preferences might differ from person-to-person, which is confirmed by Tiffany ("mhh") and Abby ("yeah" followed by "exactly"). Kenny repeats the word "interaction", while drawing attention to the post-it and gesturing towards himself "this has a relation to me". Abby

Figure 14.3 Setting (v15 at 39:14).

adds the text 'Different kinds of interaction' to the note. The episode is characterised by a high degree of uncertainty when Rose and Kenny talk, followed by repetitions and confirming utterances, which provides a display of understanding by the others.

In the next episode (Episode 41), Abby completes the post-it note. Nina refers to a statement by one of the lead users and again Rose validates her observations; Rose points at a Chinese post-it with the English text 'Alarm' two times using air-quote gestures while saying "Alarm". She explains the 'interaction feature' of the product using the Chinese post-it note 'Alarm' as a reference point, several times looking and pointing while she elaborates on what the lead-users meant during the co-creation session. Abby condenses the dialogue about product features (the alarm and push notifications) to 'Talk' as a finalizing remark and places the post-it next to another that also contains information on product features and services.

Rose and Tiffany then open a new episode (Episode 42) by referencing a new Chinese post-it 'Home doctor', and continue to talk about this service feature, while Abby writes another post-it with the word 'Human' (Figure 14.4), linking back to Episode 40. At the end of Table 14.1 Kenny writes '2-way interaction' on a post-it, which he sticks onto the bottom of the post-it that Abby previously wrote with the text 'Different kind of interaction'. Kenny's note can be linked to the topic of the product's interaction features from Episode 40, but its content also links to the post-it notes Abby placed on the wall: a 'Human' or 'Personal aspect' of a service, meaning a two-way interaction.

Overall, this example illustrates how uncertain dialogue is followed by subsequent attentive returns across episodes to the cognitive referent 'interaction', resulting finally in three new information selection post-its being produced.

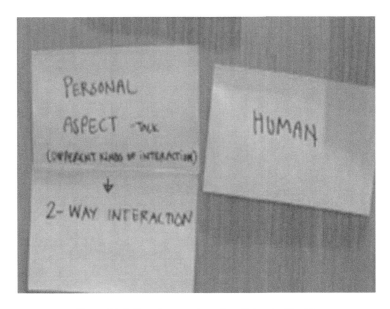

Figure 14.4 Post-it notes based on Episode 40, v15.

Table 14.1 Transcript extract (v15, 227–238, run time 37:58).

		Episode	Post-it reference
Rose	I think this: reminded me of last week in the group, when we talked about (.) yeah I can't really remember what the context was, but some- >about< the car as a boyfriend ehm: (.) I don't really remember, right, but the car as a boyfriend >an interesting thing< but if I had bad day: I would like to be able to talk to my car and then the car is able to emphasize with me like a boyfriend [>that sort of thing<] and then- so I think that was what she was really getting at like that kind of like interaction	Episode 40 Interaction (Chinese Post-it)	Interaction; Personal → family; Personal data
Tiffany Nina	[mhh] ((nods))		
Rose	and also in- >they said that it needs to be able to talk< (.) talk to you.		
Abby	Yeah (.) that kind of 'interaction' ((does an air quote gesture and looks at post-it wall)) so it's (.) >I guess it's< more the personal (.), personal- yeah- personal (.) aspect or::? ((**begins to write 'Personal Aspect' on post-it note**))		Interaction
Kenny	yeah because I- I think it's eh: it might be different from person to person how they wanted to realize (.) so, it could be she wanted to (.) to talk to you ((looks at Abby))		
Kenny	but other people would be maybe intimidated [by the car talking to you ((looks at Tiffany)) so they would prefer different] kinds of interaction		
Tiffany Abby	[mhh] [yeah]		
Rose Nina Kenny	(.) exactly [mhh yeah but] ther- there is k- >some kind of<interaction I think that was what [yeah but, yeah bec-] mhh ((nods)) like the person [interaction that] makes you feel like- that this has (.)a relation to me-		Interaction
Rose Abby Nina	[yes] ((**begins to write 'Different kinds of interaction' on the same note**)) yeah because that other guy said that he wants to be able to display it, and the other guy- if you don't wanna display it it sends you a report (..)	Episode 41 Alarm (Chinese Post-it)	
Rose	yeah (..) that was the: real time ((points and taps at Post-it Alarm' with pen)) -whether you have regular or pushed notifications (.) or that you would look at at the same [time (.) so]		Alarm
Kenny	[mhh mhh]		

Speaker		Code
Rose	(.) I think (.) that they had ((points at Post-it 'Alarm' with pen)) >she was talking about< if they can provide you with push notifications like your REGULAR REPORTS (.) and then >at the same time< you would be looking at it ((touches her wrist)), but it would also: ehm: give me an 'alarm' (.) ((does an air quote gesture)) [or like] you know	Alarm
Kenny	[mhhh] ((nods))	
Abby	((writes 'talk' and places post-it note on wall))	
Rose	if something needs to- it's, >I guess, I guess< it is the thing that now most (.) a- mobile product that, like if you need to:: when is the time to go to >you know< your car to workshop for [a repair for example, right]	
Kenny	[mhh mhh]((nods))	
Rose	if they: they let you know these- this information, but beyond that what other kinds of information can you provide for the-	
Tiffany	yeah, they also mentioned that eh: it needs to be like internet based, and also have access to doctors or receipts if they don't trust the information, they can come to locals	Episode 42
		(Chinese Post-it)
		Home Doctor
Rose	((looks at Post-it wall)) ((nods))	Home Doctor
Abby	((writes 'HUMAN' and places Post-it on wall))	Interaction
Rose	[yeah I think that-]	
Tiffany	[and that] you can adapt also to: eh to the car	
Rose	yeah to the car >exactly< ((nods))	
Tiffany	access to eh:: European [eh: (..)] specialists or whatever	Interaction
Kenny	((writes '2-way interaction'))	
Rose	yeah ((nods)) I think this is good ((looks at Post-it wall))	

Note: Extract showing how low uncertainty triggers immediate cognitive elaboration and information selection. Epistemic uncertainty hedge words are underlined and information extraction is in boldface.

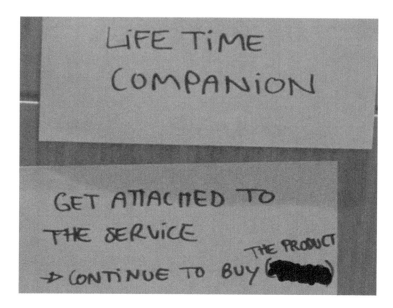

Figure 14.5 Post-it note, v15.

3.4.2   *Example of high certainty leading to immediate creative elaboration and information selection*

This example is taken 3 minutes after the previous example. Episode 44 begins with Abby uttering an observation relating to the earlier co-creation session and the fact that the lead-users mentioned a 'Life time companion'. While Abby is talking and gesturing towards the Chinese post-it notes Rose is confirming her observations by pointing at the Chinese post-it with the translation 'Sustainable, lifelong'. Abby utters that "this is exactly what we need to create", which both Tiffany and Rose agree with. Rose repeats Abby's statement "exactly" in a manner that expresses no uncertainty (unlike the uncertainty hedge words in the previous episodes). Tiffany then goes on to confirm Abby's statement, adding that someone else also talked about this. The approving utterances illustrate that Abby's statement has solid grounding in the group. Rose finalizes the mutual decision; she asks Abby to write the point down.

Tiffany continues to share her observations of what the lead-users mentioned, while Abby is writing 'Life time companion' on the post-it (Figure 14.5). Abby completes the post-it and fixes it to the wall, illustrating a shared representation. The group expresses excitement and certainty about the 'Life time companion' concept. Abby has "ideas popping out" and Kenny finds it to be "an awesome concept". Abby says that "it's so obvious" and "of course", also expressing certainty on the matter. Tiffany adds that it is in fact something that "they" (i.e. THE COMPANY) do already, which, even though the idea is not new, validates the concept further. In the last part of Episode 44 Rose begins to elaborate on the 'Life time companion' concept. Abby adds another post-it at the end of the episode, further elaborating on the idea ('Get attached to the service, -> continue to buy (COMPANY), the product').

The example illustrates how agreement and certainty within the dialogue are followed by immediate creative elaboration in the same episode, resulting finally in two new information selection post-its being produced.

4 DISCUSSION

Previous research on epistemic uncertainty in design has focused on how such uncertainty triggers localized, creative reasoning such as analogizing and mental simulation aimed at uncertainty reduction (Ball & Christensen, 2009; Ball *et al.*, 2010; Christensen & Schunn, 2007, 2009; Wiltschnig *et al.*, 2013). To date, however, no research has examined how epistemic uncertainty may also affect design behaviors beyond the localized micro-episode in which it is experienced, although phenomena such as the Zeigarnik effect (Zeigarnik, 1927), where people's memory for unsolved problems exceeds that for solved problems, suggests that designers may be alert to unresolved issues. The temporally-extended nature of the DTRS11 dataset afforded a unique opportunity to address this gap in existing knowledge through a direct test of the impact of epistemic uncertainty on creative design cognition both within *and* across episodes.

In analysing the dataset for evidence of an association between epistemic uncertainty and creative design cognition we decided not to focus on analogizing and mental simulation, since these strategies were not particularly in evidence. Instead, we examined how the level of epistemic uncertainty associated with a cognitive referent predicts the repeated referral back to that referent in subsequent dialogue. In the present dataset such cognitive referents took the form of the post-it notes deriving from Chinese lead-users. Our analysis was driven by the dual hypotheses that increased uncertainty associated with a cognitive referent would engender: (1) immediate creative elaboration of that referent aimed at resolving uncertainty and determining information selection; and (2) subsequent attentive returns to that cognitive referent aimed at resolving lingering uncertainty and again determining information selection. In sum, we proposed a 'double mediation model' (Figure 14.1) in which epistemic uncertainty on the initial encounter with an individual post-it predicts both the immediate occurrence of local, creative micro-episodes that mediate the formation of new post-its (information selection) as well as the occurrence of subsequent attentive returns across episodes that also mediate the formation of new post-its (further information selection).

Our results challenge some aspects of our a priori assumption that the uncertainty associated with cognitive referents (post-its) would predict localized, creative elaboration *and* subsequent attentive returns, with both types of creative activities mediating eventual information selection. First, it was apparent that increased epistemic uncertainty in relation to cognitive referents did *not* immediately trigger creative cognitive events. Instead, it was epistemic *certainty* in relation to cognitive referents that engendered immediate, creative reasoning and information elaboration. In accounting for this opposite-to-predicted effect we speculate that uncertainty might have been so highly elevated in the present design context as almost to force the design team to opt strategically to make the most of any certain information at hand, immediately latching onto this and utilizing it as a way to establish a stable framework for subsequent work. Achieving a stable framework might be especially critical in design tasks that

Table 14.2 Transcript extract (Video 15, 262–268, run time 42:49).

	Episode	Post-it reference
Abby	I'm thinking that this: whole thing about the lifetime eh: companion ((looks at Post-it wall and does circles with her right hand)) and we want to make sure that (.) that you can get the service ((moves both arms in front of herself and gestures towards Post-it wall on her right))	Sustainable, lifelong	
Rose	((looks at Abby, turns to point and tap at Chinese Post-it 'Sustainable, lifelong'))		
Abby	and so on, I mean this is exactly [what] we want to create		
Tiffany	[yes]		
Abby	so they'll buy a COMPANY PRODUCT again [so it's kind of our company]. and then you have all the accessories (.) (INAUDIBLE)	**Episode 44** Sustainable, lifelong (Chinese Post-it)	
Tiffany	[yeah (.) yeah (.) yeah]		
Rose	yeah! exactly		
Abby	it's- it's really eh: it's like		
Tiffany	and they talk about with the (.) eh the workers-		
Abby	yeah		
Rose	mhhh **I think** you should write that life time companion ((points at Abby))		
Abby	((writes post-it note 'Life time companion')) (1.1 sec pause)		
Tiffany	it was also here they mentioned the:: lifetime companion to manage everybody's health so- they talked about that (.) and connected life and health, and health to life (.)		Life time companion
Kenny	((moves towards Abby and look at her writing on the Post-it))		
	(off task behaviour)		
Abby	((places Post-it on wall))		Life time companion
	(2.1 sec pause)		
Abby	there's so many stories- a lo:: is popping out in my head already and it is (.) it is eh		
Kenny	be an awesome concept for: (.) several products and services		
Rose	mhh ((nods and looks at Post-it))		
Abby	and it's so obvious that the eh: I mean, and that's also what I think we'll hear when we come back, just kind of what we heard a little bit from Phase I (..) kind of (..) yeah okay, kind of (.() we- we kind of knew this, but nothing has happened (..)		Life-time companion

		Episode 44
Tiffany	yeah	Life time companion
Abby	because it is kind of obvious, of course we want them to get into the whole ((does circles with her right hand)) COMPANY story and to buy it again and do the whole thing	
Tiffany	mhh but we are doing a lot of stuff already [it's something already been done]	
Abby	[then it] just doesn't (.) for yeah:	**Episode 44**
Rose	yeah I think that people- I guess that if you could like somehow marked the same things as like the life companion ((points at Post-it)) actually maybe it's the accessories beyond the car. Because if we are ((points at Post-it)) if you want to buy the accessory of the service then you have to choice to kind of buy the car rather that buying the car and then thinking about what you want, but if you are tied to the service ((gestures and looks towards English Post-it's)) (.) so like if I'm so used to using an I-clock for example I have no choice but, but to continually buy an IPhone because I have no choice but I get-	Sustainable, lifelong (Chinese Post-it)
Abby	mhh ((nods)) exactly (.) yeah and I guess, I mean ((turns to Tiffany)) THE COMPANY on call [is one thing] that is going on back the way	
Rose	[yeah that how one-]	
Tiffany	[yeah, yes]	
Abby	yeah exactly	
Kenny	It's the accessory stuff	
Abby	**((Writes post-it note 'Get attached to the service, –> continue to buy (COMPANY), the product))**	Get attached to the service

Note: Extract showing how low uncertainty triggers immediate cognitive elaboration and information selection. Epistemic uncertainty hedge words are underlined and information extraction is in boldface.

involve having to respond to end-user knowledge, where it is important to commence with some certainties as a foundation to build upon. In this way it is possible that epistemic uncertainty triggers a 'metacognitive switch' that is highly bounded by the prevailing context. If epistemic uncertainly is felt to be uniformly high then designers may opt to work creatively in the immediate term with information that feels more certain, whereas if uncertainty shows greater fluctuation then designers may opt to expend more immediate effort on resolving uncertain aspects of the design.

The idea of epistemic uncertainty driving a metacognitive switch mechanism is gaining credibility in the literature on human reasoning (e.g., Ackerman & Thompson, 2014; Ball & Stupple, 2016; Thompson *et al.*, 2011; Thompson *et al.*, 2013). Interestingly, too, it is becoming increasingly clear that the strategic decisions people make on the basis of metacognitive experiences are often determined by *relative* rather than absolute perceptions relating to perceived uncertainty (Wänke & Hansen, 2015). In other words, it appears that dynamic shifts from perceived certainty to perceived uncertainty are critical for determining strategic decisions about the kind of reasoning required at any particular point in time. These important conceptual ideas align well with our view that in some design contexts epistemic uncertainty may fluctuate extensively, with bouts of *uncertainty* leading to immediate, creative processing, whereas in other design contexts epistemic uncertainty may provide a more global and stable backdrop to ongoing activity, potentially leading to isolated moments of *certainty* triggering immediate creative processing, as observed here. It is noteworthy that in a post-study interview the leader of the design team spoke of his perception of post-its as 'emotional triggers', evidencing an understanding that the emotional qualities of post-its could be a driver for subsequent design processes, although he did not specifically focus on epistemic uncertainty.

A second key finding is that we have shown for the first time how epistemic uncertainty can promote subsequent attentive returns to a cognitive referent within the design process, since our analyses demonstrated a predicted correlation between initial, epistemic uncertainty that was linked to cognitive referents and such subsequent attentive returns. This finding corroborates our underpinning proposal that epistemic uncertainty may affect design behavior that extends beyond the localized micro-situation in which the uncertainty is experienced. The idea that epistemic uncertainty can have far-reaching consequences seems novel and important and would be worth exploring further.

A third finding is that while epistemic uncertainty did not directly predict the information that was eventually selected, both immediate creative elaboration and subsequent attentive returns *did* predict information selection, with subsequent attentive returns being the stronger predictor. The fact that both mediator variables were predictive of information selection is interesting, although not surprising. In essence, this finding indicates that once designers have engaged in creative development activity in relation to a cognitive referent this activity then forms the basis of information selection for down-stream design work. Arguably, too, it might be expected that subsequent attentive returns would emerge as the stronger predictor of information selection compared to immediate creative elaboration given that the designers have presumably returned to unresolved issues because of their perceived importance for design success (i.e., the Zeigarnik effect may not only impact on memory for unresolved issues but may also impact information selection for subsequent processing).

Before concluding, we note some caveats in relation to our findings, which primarily center on statistical issues. First, the small sample-size associated with our analysis will have reduced the reliability of our results. Second, we recognize the relatively small effect sizes arising from our analyses. Third, we reiterate the interpretative problems arising from our inability to test our proposed mediation model using the same regression methods because of the particular statistical properties of the variables underpinning the model. Our approach was to undertake a *conceptual* mediation analysis, although this didn't allow us to formalize decisively the evidence for direct and mediated effects between predictor variables in relation to the dependent variable of information selection. Our interpretation of findings therefore needs to be treated with an appropriate degree of caution. That said, we contend that our approach and observations hold promise for identifying more pervasive and enduring impacts of epistemic uncertainty on creative design cognition that are potentially mediated through salient cognitive referents. In addition, our research contributes to a growing appreciation of uncertainty as a salient aspect of design that determines the dynamics of ongoing creative design reasoning and decision making (e.g., Stempfle & Badke-Schaub, 2002; Tracey & Hutchinson, 2016).

REFERENCES

Ackerman, R., & Thompson, V. A. (2014). Meta-reasoning: What can we learn from meta-memory. In: A. Feeney, & V. A. Thompson (eds.) *Reasoning as memory*. Hove, UK: Psychology Press.

Alcaide-Marzal, J., Diego-Más, J. A., Asensio-Cuesta, S., & Piqueras-Fiszman, B. (2013). An exploratory study on the use of digital sculpting in conceptual product design. *Design Studies*, *34*(2), 264–284.

Alter, A. L., & Oppenheimer, D. M. (2009). Uniting the tribes of fluency to form a metacognitive nation. *Personality & Social Psychology Review*, *13*(3), 219–235.

Alter, A. L., Oppenheimer, D. M., Epley, N., & Eyre, R. N. (2007). Overcoming intuition: Metacognitive difficulty activates analytic reasoning. *Journal of Experimental Psychology: General*, *136*(4), 569–576.

Baron, R. M., & Kenny, D. A. (1986). The moderator-mediator variable distinction in social psychological research: Conceptual, strategic and statistical considerations. *Journal of Personality and Social Psychology*, *51*(6), 1173–1182.

Ball, L. J., & Christensen, B. T. (2009). Analogical reasoning and mental simulation in design: Two strategies linked to uncertainty resolution. *Design Studies*, *30*(2), 169–186.

Ball, L. J., Onarheim, B., & Christensen, B. T. (2010). Design requirements, epistemic uncertainty and solution development strategies in software design. *Design Studies*, *31*(6), 567–589.

Ball, L. J., & Stupple, E. J. N. (2016). Dual reasoning processes and the resolution of uncertainty: The case of belief bias. In: L. Macchi, M. Bagassi, & R. Viale (eds.) *Cognitive unconscious and human rationality*. Cambridge, MA: MIT Press. pp. 143–165

Beheshti, R. (1993). Design decisions and uncertainty. *Design Studies*, *14*(1), 85–95.

Chan, J., Paletz, S. B., & Schunn, C. D. (2012). Analogy as a strategy for supporting complex problem solving under uncertainty. *Memory & Cognition*, *40*(8), 1352–1365.

Christensen, B. T. & Abildgaard, S. J. J. (2017). Inside the DTRS11 Dataset: Background, Content, and Methodological Choices. In: Christensen, B. T., Ball, L. J. & Halskov, K. (eds.) *Analysing Design Thinking: Studies of Cross-Cultural Co-Creation*. Leiden: CRC Press/Taylor & Francis.

Christensen, B. T., & Ball, L. J. (2014). Studying design cognition in the real world using the 'In Vivo' methodology. In: P. Rodgers & J. Yee (eds.) *The Routledge Companion to Design Research* Abingdon, UK: Routledge. pp. 317–328.

Christensen, B. T., & Schunn, C. D. (2005). Spontaneous access and analogical incubation effects. *Creativity Research Journal, 17*(2–3), 207–220.

Christensen, B. T., & Schunn, C. D. (2007). The relationship of analogical distance to analogical function and pre-inventive structure: The case of engineering design. *Memory & Cognition 35*(1), 29–38.

Christensen, B. T., & Schunn, C. D. (2009). The role and impact of mental simulation in design. *Applied Cognitive Psychology, 23*(3), 327–344.

Dorst, K., & Cross, N. (2001). Creativity in the design process: Co-evolution of problem-solution. *Design Studies, 22*(5), 425–437.

Dove, G., Abildgaard, S. J., Biskjaer, M. M., Hansen, N. B., Christensen, B. T. & Halskov, K. (2017) Grouping Notes Through Nodes: The Functions of Post-it Notes in Design Team Cognition. In: Christensen, B. T., Ball, L. J. & Halskov, K. (eds.) *Analysing Design Thinking: Studies of Cross-Cultural Co-Creation.* Leiden: CRC Press/Taylor & Francis.

D'souza, N. & Dastmalchi, M. (2017) "Comfy" Cars for the "Awesomely Humble": Exploring Slangs and Jargons in a Cross-Cultural Design Process. In: Christensen, B. T., Ball, L. J. & Halskov, K. (eds.) *Analysing Design Thinking: Studies of Cross-Cultural Co-Creation.* Leiden: CRC Press/ Taylor & Francis.

Dunbar, K. (1995). How scientists really reason: Scientific reasoning in real-world laboratories. In: R. J. Sternberg and J. E. Davidson (eds.) *The nature of insight.* Cambridge, MA: MIT Press. pp. 365–395

Dunbar, K. (1997). How scientists think: On-line creativity and conceptual change in science. In: T. Ward, S. M. Smith & J. Vaid (eds.) *Creative thought: An investigation of conceptual structures and processes*). Washington DC: American Psychological Association. pp. 461–493.

Ericsson, K. A., & Simon, H. A. (1999). *Protocol analysis: Verbal reports as data.* Cambridge, MA: MIT Press.

Finke, R. A., Ward, T. B., & Smith, S. M. (1992). *Creative cognition: Theory, research, and applications.* Cambridge, MA: MIT Press.

Gilhooly, K. J. (in press). Incubation, problem solving and creativity. In: L. J. Ball & V. A. Thompson (eds.) *The Routledge international handbook of thinking and reasoning.* Oxford, UK: Routledge.

Houghton, D. P. (1998). Historical analogies and the cognitive dimension of domestic policymaking. *Political Psychology, 19*(2), 279–303.

Howard, T. J., Culley, S. J., & Dekoninck, E. (2008). Describing the creative design process by the integration of engineering design and cognitive psychology literature. *Design Studies, 29*(2), 160–180.

McDonnell, J. (2015). Gifts to the future: Design reasoning, design research, and critical design practitioners. *She Ji: The Journal of Design, Economics, and Innovation, 1*(2), 107–117.

Nersessian, N. J. (2009). How do engineering scientists think? Model-based simulation in biomedical engineering research laboratories. *Topics in Cognitive Science, 1*(4), 730–757.

Paletz, S. B. F., Sumer, A. & Miron-Spektor, E. (2017) Psychological Factors Surrounding Disagreement in Multicultural Design Team Meetings. In: Christensen, B. T., Ball, L. J. & Halskov, K. (eds.) *Analysing Design Thinking: Studies of Cross-Cultural Co-Creation.* Leiden: CRC Press/Taylor & Francis.

Schön, D. A. (1983). *The reflective practitioner: How professionals think in action.* New York: Basic Books.

Schön, D. A., & Wiggins, G. (1992). Kinds of seeing and their functions in designing. *Design Studies, 13*(2), 135–156.

Seifert, C. M., Meyer, D. E., Davidson, N., Patalano, A. L., & Yaniv, I. (1995). Demystification of cognitive insight: Opportunistic assimilation and the prepared-mind perspective. In: R. J. Sternberg & J. E. Davidson (eds.) *The nature of insight*. Cambridge, MA: MIT Press. pp. 65–124.

Sio, U. N., & Ormerod, T. C. (2009). Does incubation enhance problem solving? A meta-analytic review. *Psychological Bulletin, 135*(1), 94–120.

Stempfle, J., & Badke-Schaub, P. (2002). Thinking in design teams-an analysis of team communication. *Design Studies, 23*(5), 473–496.

Stivers, T., & Sidnell, J. (2005). Introduction: Multimodal interaction. *Semiotica, 2005* (156), 1–20.

Thompson, V. A., Prowse Turner, J., & Pennycook, G. (2011). Intuition, reason, and metacognition. *Cognitive Psychology, 63*(3), 107–140.

Thompson, V. A., Prowse-Turner, J., Pennycook, G. R., Ball, L. J., Brack, H. M., Ophir, Y., & Ackerman, R. (2013). The role of answer fluency and perceptual fluency as metacognitive cues for initiating analytic thinking. *Cognition, 128*(2), 237–251.

Tracey, M. W., & Hutchinson, A. (2016). Uncertainty, reflection, and designer identity development. *Design Studies, 42*, 86–109.

Trickett, S. B., Trafton, J. G., Saner, L. D., & Schunn, C. D. (2005). I don't know what's going on there: The use of spatial transformations to deal with and resolve uncertainty in complex visualizations. In: M. Lovett & P. Shah (eds.) *Thinking with data*. Mahwah, NJ: Lawrence Erlbaum Associates, Inc. pp. 65–86.

Ülkümen, G., Fox, C. R., & Malle, B. F. (2016). Two dimensions of subjective uncertainty: Clues from natural language. *Journal of Experimental Psychology: General, 145*(10), 1280–1297.

Wänke, M., & Hansen, J. (2015). Relative processing fluency. *Current Directions in Psychological Science, 24*(3), 195–199.

Wiltschnig, S., Christensen, B. T., & Ball, L. J. (2013). Collaborative problem-solution co-evolution in creative design. *Design Studies, 34*(5), 515–542.

Yaniv, I. & Meyer, D. E. (1987). Activation and metacognition of inaccessible stored information: Potential bases for incubation effects in problem solving. *Journal of Experimental Psychology: Learning, Memory, and Cognition, 13*, 187–205.

Zeigarnik, B. (1927). Das behalten erledigter und unerledigter handlungen. *Psychologische Forschung, 9*, 1–85.

Chapter 15

Temporal Static Visualisation of Transcripts for Pre-Analysis of Video Material: Identifying Modes of Information Sharing

Andreas Wulvik, Matilde Bisballe Jensen & Martin Steinert

ABSTRACT

When evaluating group performance in the field of group collaboration different forms of information sharing types are utilised as indicators for well performing or under performing groups. A transcript-file of a group conversation provide us with such communication patterns in form of the order of speaker turns and the amount of words assigned to each speaker at a particular instance. In this study we explore how one can utilise transcript meta data and the field of data visualisation to help a researcher to deduce insights for further investigation of the studied conversation faster. This would help qualitative researchers in the pre-analysis of bigger data sets. In Python we wrote a script that – from transcript meta-data – presents Temporal Static Visualisations (TSV) of the group members individual contribution as well as the mean-variance in participation over time. This supports the researcher in identifying sequences of and transitions between monologues, dialogues and group discussions in the observed conversation.

The main strength of TSV lies in the static inclusion of time development that helps the researcher to gain a quick insight into the flow of the observed conversation. However we still face the challenge of providing the design researcher with more contextual information. Thus, proposed future work should focus on how to automatically include additional context and content information such as artefact usage and physical interactions in the visualisations.

I INTRODUCTION

For decades, video has been one of the main data sources for design research experiments, especially on group behaviour (Tang, 1991). Alas, coding transcript data is very time-consuming in terms of educating coders as well as conducting the actual coding itself. Moreover, it can be difficult to keep an overview of a large amount of data. This paper proposes a new method utilising data visualisation of transcript meta-data to help researchers discover different modes of information-sharing in such video material for further content analysis. We present an approach for pre-analysis of transcript data in relation to group dynamic research in the field of engineering design. By extracting meta-data from the transcript file we visualise the conversation contribution over time among the group members as well as plotting mean-variance of group members' participation. We call the approach Temporal Static Visualisation (TSV). TSV allows the researcher to quickly identify different forms of information sharing types, as monologues, dialogues or group conversation, as well as the transition between the types.

Our approach was initiated by the recent trend of dynamic data visualisation, which in recent years has created valuable aids for qualitative researchers. Tools such as Leximancer, Gephi and Cytoscape, for example, allow investigators to gain new insights into bigger datasets (Bastian *et al.*, 2009; Smith, 2003; Smoot, Ono, Ruscheinski, Wang, & Ideker, 2011). However most of the tools only provide the researcher with a holistic overview of the data, like semantic networks, but fail at helping the researcher deduce specific sequences for further investigation. One exception is Dong's (2005) analysis of semantic coherence that investigates the development of a common language within design teams (Dong, 2005).

In the field of audio visualisation several researchers have explored and evaluated the effect of live visualising dialogue flow in a group conversation (Avci & Aran, 2016; Bergstrom & Karahalios, 2007b; DiMicco, Hollenbach, & Bender, 2006). The approach is different compared to a design researcher's focus since the main aim is to actively attempt to affect the discussion and speaking patterns by visualising the live flow of communication. There is a focus on preventing dominant participants to take over the discussion and having less talkative participants to express their opinions (Bergstrom & Karahalios, 2007a). The main assumption in the field of audio visualisation in group discussion is that one gets a more diverse type of information sharing, when all group members contribute to the discussion. Hence, the overall main hypothesis is that there are in fact sufficient and less sufficient dialogue patterns for different contexts. We use this hypothesise to claim that by looking at certain dialogue patterns one can identify interesting video passages of a group conversation in design work as well. Hence, the research question of this paper has been: *How can design researchers apply data visualisation on group conversations for identification of different modes of information sharing?*

First, we present related work. Next, we describe the development of a dynamic network visualisation approach. This we evaluated by interviewing researchers of a research group at the Technical University of Denmark (DTU) also working with the same data set from DTRS11 (Christensen & Abildgaard, 2017). For further reference we will mention this research group as *the DTU group*. With learning that was gained from this evaluation we finally present the suggested approach: Temporal Static Transcript Visualisation (TSV), where time perspective, conversation participation and content are captured in one graph. Finally, we discuss the results and limitations of TSV as well as highlighting opportunities for improving the approach further.

2   RELATED WORK

Our approach is grounded in three different research fields: observational case studies of group dynamics in Design Research, Dynamic Data Visualisation, and finally Investigation of Group Interactions through Visualisations of conversation contribution. Below the three fields will be elaborated.

2.1 Group dynamic analysis and design research

Throughout the history of design research video observation has served as an often-used tool for case based observational studies of group collaboration (Tang, 1991). Sociologists approach group collaboration with the view of seeing it as a system of human interaction. At this degree of abstraction there is no necessary incongruity in comparing cases with each other since the categories of interaction are more general such as "showing empathy" or "ask for clarification". Robert F. Bales defined in 1950 a set of twelve categories for the analysis of small group interaction (Bales, 1950). This set of categories served as a starting point for Bales himself and other socio-psychologists developing more detailed sociological frameworks (Scanzoni, 1983; Sjovold, 2007). Bales' method is called the Interaction Process method and is based on the observation of behaviour act by act. The researcher codes every single act in the categories of the defined interactions such as: 'Shows solidarity', 'Shows tension release', 'Gives suggestion', 'Ask suggestion', 'Shows Tension', 'Shows Antagonism'. The coding mainly focuses on language, but can also take body language (e.g., nodding or gesturing) into account. By giving scores for each act, the researcher will end up with an empirical score-chart of a group interaction which is comparable with other group observations. This provides an insight into what type of group interaction has happened as well as defining roles and team dynamics in the group. Sonnenwald (1996) focuses on the different communication roles in the design process rather than only behaviour. Through evaluating several different case studies of team work all involving the design process they define 13 different types of communication roles appearing in the conversation. Sonalkar, Mabogunje, and Leifer (2013) focused on design interactions in design work and introduced the Interaction Notation Dynamics, which seeks to cover the process of moment to moment concept development and capture the interpersonal interactions. By not only introducing 13 different types of interactions in a concept development dialogue, but also assigning them a specific symbol, they provide the community with a standardised coding scheme for researchers to use. The 13 interactions are 'move', 'question', 'hesitation', 'block', 'support for move', 'support for block', 'overcoming', 'deflection', 'interruption', 'yes and', 'deviation and humour'.

We define the previous three mentioned cases as semi-quantitative since they all require an educated coder to process the empirical data. A more quantitative approach for analysing group interactions was introduced by Dong (2005) and is based on the development of a coherent language throughout a group discussion. The method is called Latent Semantic Analysis (LSA). The baseline theory for LSA is that by looking at the entire range of words chosen in a wide variety of texts, patterns will emerge in terms of word choice as well as word and document meaning. By investigating how participants' choice of words changes throughout design discussions, LSA reveals the development of a coherent vocabulary. This allows for an examination of whether a team manages to create a mutual understanding throughout a discussion, as well as to help identify triggers for development of such patterns. This temporal and quantitative approach has been a great inspiration for the present work.

2.2   Computational approaches and data visualisation for improving data analysis

The use of computational tools is gaining traction in the Design Science community. Several researchers participating in the DTRS11 event applied computational approaches to the DTRS11 dataset (Christensen & Abildgaard, 2017), investigating concepts such as low coherence turns (Menning et al., 2017), conversation topics (Chan & Schunn, 2017) and discourse density (Bedford et al., 2017).

With the domestication of internet and digitalised devices, a new field of data visualisation has emerged in the last decade (Beck et al., 2016). Hans Rosling and his Gapminder project is a perfect example hereof, and seeks to implement a fact-based worldview to anyone interested (Rosling, 2016). With Rosling's dynamic visualisations of demographic historical statistics he argues that the state of the world is not as bad as people might think. Rosling's claim that visualising data can change people's understanding of certain contexts are supported by studies on how the visualisation of data change the comprehensive performance of the user (Ware, 2009). Hence, there is reason to believe that researchers can gain new insights by presenting datasets in a graphical manner. In their review of state-of-the-art data visualisation tools, Beck et al. (2016) conclude that data visualisation has been on the agenda since the early 90's, but the utilisation of the methods has turned into more *evaluating* character rather than only focussing on the *method* or *application*.

2.3   Changing group interactions through data visualisation

Visualising data concerned with language and dialogue patterns has also been an approach in the fields of design research and educational research. This approach aims to investigate how the flow of discussion affects group performance (Avci & Aran, 2016; DiMicco et al., 2006; Ryu & Sandoval, 2015). Moreover, experiments have been conducted to investigate whether one could change group dynamics in a design discussion by providing participants with a visual live feedback of their contribution to a conversation in terms of speaking time and sound level (Bergstrom & Karahalios, 2007a; DiMicco et al., 2004). Bergstrom designed the so-called Conversation Clock that visualises the interaction patterns of up to four individuals, and provides participants with a *communal social mirror*. An interesting finding in their studies is that subjects looking at the visualisation of a group interaction could identify team roles and interactions without observing the actual group discussion.

Similarly, group performance has been observed to correlate with speaking turns, speaking lengths (Dong et al., 2012) and eye-contact rate (Jayagopi et al., 2012). Avci and Aran (2016) developed a method for predicting group performance. Their method takes inputs such as 'group speaking cues', 'group looking cues', 'personality trait statistics', 'individual speaking statistics', 'individual performance statistics', 'group perception statistics', 'influence cues', 'hierarchy cues' and 'frame-based features'. Based on case studies, Avci and Aran's (2016) method succeeded in predicting the performance of groups when making decisions.

In the field of visualisation of group interaction, the goal is to actively affect or predict the discussion and speaking patterns by visualising the live flow of communication.

For us, this means that it is plausible to hypothesise that by identifying modes of information sharing one can gain insights on group behaviour.

3   METHODOLOGY

The strategy of this research has been to leverage as much quantitative information as possible from the DTRS11 transcripts and explore how one could visualise the data in a meaningful way for design observation researchers. First, we explored and identified features from transcript data that are potentially relevant for visualisation. Then, a Python script was developed to extract selected features into a comma-separated value (.csv-file) format. This file served as input to the network graph visualisation tools Gephi and Cytoscape (Bastian et al., 2009; Smoot et al., 2011), enabling visualisation of the transcript. This approach led to the pilot study described in the following section. For evaluation of our first iteration, we conducted a qualitative interview with one of the researchers of the DTU group also working with the DTRS11 dataset (Christensen & Abildgaard, 2017). In this interview we compared results and discussed the potential of mapping transcripts in a new and graphical way. From this evaluation we found some limitations in our pilot study and with these in mind we developed the final approach, Temporal Static Visualisation, based on new features from the transcripts.

4   PILOT STUDY INVOLVING DYNAMIC VISUALISATIONS

With the goal of developing a tool that could visualise conversation patterns, we started with a strategy of classical network analysis. For a pilot study we used the two tools: Gephi and Cytoscape (Bastian et al., 2009; Smith, 2003; Smoot et al., 2011). Gephi and Cytoscape are both open source network analysis tools developed by researchers in the field of computer science (Bastian et al., 2009; Smoot et al., 2011). The software allows researchers to explore and analyse all types of networks. The input is excel-files or .csv-files designed by the researcher including relational information between nodes and edges. Both Gephi and Cytoscape allow animation of the development of a network over time. This was of our specific interest because it allowed us to get temporal insights in the group discussions in our dataset rather than getting one overall view. For the sake of consistency across figures, the same data are used in all visualisations. Video 11 *Linking insights from co-creation to project*, segment 250 to 280, was selected for this purpose (Figure 15.1).

4.1   Creating data visualisations of speaker turns and amount of words spoken

By converting data from the transcripts into a csv.- file (see Appendix A) we were able to both statically and dynamically show the communication flow of a discussion.

4.1.1   Static snapshots

Figure 15.2a shows the total aggregated speaker turns of Video 11 *Linking insights from co-creation to project*. Here, one can see who is contributing to the conversation,

Figure 15.1 Video 11 *Linking insights from co-creation to project* was used as case illustration to secure consistency between the different types of developed visualisations.

and which speakers talk after each other. We see how Ewan and Amanda are contributing the most to the conversation, and show tendencies of talking amongst themselves. The graph also shows that Ewan is central to the discussion, as the most weighted edges connect to this node. Figure 15.2b show a 30 speaker turn sample of the same data, from (v11, 250) to (v11, 280). The previously described pattern of Ewan and Amanda dominating the conversation is even more apparent here. Static visualisations hence provide an overview of the speaker flow, but the actual sequence and quantity is still hidden.

4.1.2 Dynamic visualisation of speaker turns

To visualise the flow of conversation the dynamic possibilities of Cytoscape was utilised. Here one can visualise dynamic graphs of the temporal dimension in forms of videos. Seeing how speaker turns evolve turn by turn gave insights in team dynamics and showed the progression of the conversation. We approached dynamic visualisation by displaying ten speaker turns at the time. Figure 15.3 displays turns v11s250, s265, s280, each depicting the last the speaker turns as edges.

An alternative to displaying the last *n* edges in a dynamic visualisation is to adjust node sizes based on speaker activity. This gives a picture of how much each speaker is contributing to the conversation in the relevant time window. Figure 15.4 shows node sizes based on the subjects' activity in the last 20 speaker turns.

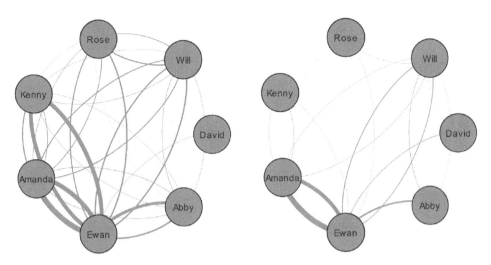

Figure 15.2a and b.   Static visualisations with weighted edges. Left: v11, all sessions. Right: v11, s250-280.

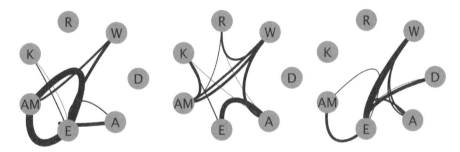

Figure 15.3a, b and c.   Dynamic network visualisation – Edges (v11, s250; v11, s265; v11, s280) https://vimeo.com/172090367 (password dtrs11).

4.2   Evaluation of the automated visualisation approach

To evaluate our suggested approach we initiated a discussion with the DTU group. The evaluation consisted of comparing their manual coded schemes with our visualisations and a qualitative interview with one of the researchers discussing the value and potential of our pilot approach. Our claim regarding the potential of identifying patterns is supported by the work of the DTU group. One of their main findings was that there were statistically significant patterns in design discussions that could be identified through an approach like ours. However, the research group mentioned they had coded for agreement expressions throughout the design group in a combination of speaking cues and body gestures such as nodding. This leads to the criticism that our visualisations lacked detail and did not help the researcher identify relevant sequences for further analysis. However, by adding more contextual information, such as use of artefacts or body language, the approach would have great potential according to the DTU researcher. He saw possibilities for identifying interesting passages in a video, and

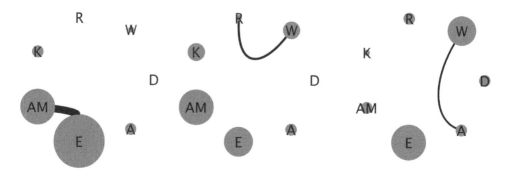

Figure 15.4a, b and c.   Dynamic network visualisation – Nodes (vI I, s250; vI I, s265; vI I, s280)
https://vimeo.com/172090366 (password dtrs I I).

the potential for quantitatively comparing datasets across studies. Another criticism
was that the videos did not provide the intended overview of the dataset, which was
one of the original intentions as well.

5   DEVELOPING THE TEMPORAL STATIC VISUALISATION

The development and evaluation of our pilot test failed to provide information on
interesting sequences for further analysis. Even though the presentation of conversation
flow had become more graphical, it did not provide the intended useful, holistic insights
into the conversations. Hence, we developed a new approach seeking to do exactly this.
By visualising the engagement of group members' participation throughout the group
discussion along a time axis rather than an animate network, we aimed to identify
different modes of information sharing, such as monologues, dialogues and group
discussions.

5.1   Mathematical explanation of TSV

Each individual participant p_i in the conversation is given a score denoting their contribution to the conversation over the last 20 speaker turns t. We define *Score* as a
function of t_j and p_i (Equation 15.1). t_j is a given speaker turn (denoted ### in the
supplied transcripts), for example, $t_j = 25$. and p_i is a given participant, for example,
$p_i = p_k$ meaning Kenny. The calculated score is the product of participant activity and
recent word contribution. *Activity* (Equation 15.2) is a ratio describing how many
times the given participant has contributed to the conversation in the last 20 turns in
relation to the maximum contributions possible, M. Over the time period of 20 turns,
the maximum possible speaker turns of one participant is 10, which would indicate
the person speaking every second time in the sequence of 20 speaker turns. *Words*
(Equation 15.3) is the ratio of words spoken of the given person over the last 20 turns
in relation to the total number of words spoken in the same time period.

$$Score(t_j, p_i) = Activity(t_j, p_i) \times Words(t_j, p_i) \tag{15.1}$$

Figure 15.5 Speaker score calculated over last 20 turns.

$$Activity(t_j, p_i) = \frac{1}{M} \sum_{t=t_j-19}^{t_j} (p_t = p_i \rightarrow 1), \quad M = 10 \tag{15.2}$$

$$Words(t_j, p_i) = \frac{\sum_{t=t_j-19}^{t_j} (p_t = p_i \rightarrow \#words(t))}{\sum_{t=t_j-19}^{t_j} \#words(t)} \tag{15.3}$$

Plotting the *score* of each participant throughout the session of investigation results in a graphical representation as seen in Figure 15.5.

Here, the contributions of each participant to the group conversation are visualised as a coloured curve, ranging between 0 and 1. For example, in Figure 15.5 the turquoise curve (K) indicates one participant speaking a lot in the beginning of the session and later becoming more quiet. The yellow curve (E) shows active participation throughout the conversation.

In order to identify modes of information sharing rather than individual contributions we calculated the variance of the score for each speaker turn to visualise the spread of activity among participants (Equation 15.4). This gives an indication of periods of equal or more one-sided participation, respectively.

$$Variance(t_j) = \frac{\sum_{p_i}^{p} (Score(t_j, p_i) - \overline{Score(t_j)})^2}{\#participants} \tag{15.4}$$

In Figure 15.6 the variance of speaker scores throughout the conversation is shown. The variance calculated corresponds to the difference in activity among the participants for a given speaker turn. This indicates the type of information sharing happening throughout the conversation. When the variance is very low several people are contributing to the discussion at the same time (*Group Discussion* in Figure 15.6). When the variance peaks around 0.1 it indicates one person is dominating the conversation (*Monologue* in Figure 15.6). Hence, the curve in Figure 15.6 provides the researcher with an insight into the occurrences of group conversations, dialogues or monologues and the transition between these modes of information sharing.

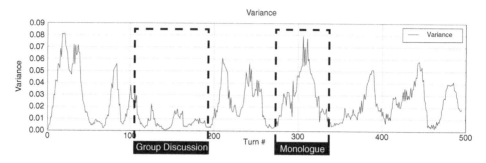

Figure 15.6 Variance of speaker score for each speaker turn. Group Discussion and Monologue is indicated in the dashed boxes.

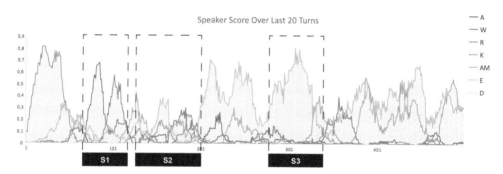

Figure 15.7 Illustration of three sequences (S1, S2 & S3) identified for further investigation.

5.2   Case specific findings

To show the application of TSV we have – with the help of the visualisations – been able to identify three sequences for further investigation. These are illustrated in Figure 15.7.

Sequence 1 (v11, s80-110) was chosen since from the scores we can see how one person is contributing a lot to the overall conversation. Afterwards another person takes over. This could be an indication of a type of information sharing, where participants spend a longer time on sharing a particular opinion on the same topic. Sequence 2 (v11, s110-200) was chosen as a period of equal participation based on scores and variance. This is the situation that most group dynamic researchers seek to create, for example, with the Conversation Clock (Bergstrom & Karahalios, 2007b). Hence, it would be interesting to see the content of the discussion. Sequence 3 (v11, s360) was selected since the variance in this sequence is very high. This indicates that one participant is dominating the conversation. Below we investigate the content of each sequence.

5.2.1   Sequence 1 (v11, s80-110) – Sharing individual concerns

Sequence 1 consists of two peaks of activity from Will and Rose respectively. These are the highest concentrations of activity from these participant throughout v11. Will begins this sequence by voicing concerns about the tone of the framing for the planned workshop. The concern is that the focus is currently on solving pains instead of enjoying good life, and is in his words *miserable*. Some discussion whether or not this is beneficial for the workshop ensues. Rose picks up on Will's concerns and continues the discussion of what the tone of the workshop should be like. She argues for a more positive tone that is more in line with the theme of good life. The conversation pattern in Sequence 1 can be described as consecutive large contributions of the participants.

5.2.2   Sequence 2 (v11, s110-200) – Group discussion and expressions of agreement

Sequence 2 is identified as a period of relatively low variance in the group speaker activity. The start of this sequence is concerned with the notion of *pockets of time to enjoy good life*. The group is discussing whether they should focus on small pockets of time during work life, or the larger pockets of time the workshop participants can enjoy after they are 50 years old. The group seems to agree on placing the focus on smaller pockets of time in work life, what they call the "*pockets of the endurance*". Further, the group is discussing whether or not there will be any translation issues to Chinese and what sort of – and the length of – quotes from the earlier session to give the participants. This sequence is generally characterised by short expressions of agreement.

5.2.3   Sequence 3 (v11, s360) – Facilitating coherence

Sequence 3 is a period of large calculated variance of the speaker score over the previous 20 turns. When looking at the scores, Ewan appears to dominate the conversation. The topics discussed here are: first, how to make the workshop participants think about how to create trust in the company; and second, to include the notion of sustainability in the discussion. This sequence contains elements where Ewan makes long explanations of concepts and is followed up by shorter expressions of agreement from other participants. In this way Ewan can be seen as a so-called *Intergroup Star*, as defined by Sonnenwald (1996). This communication role seeks to facilitate intergroup collaboration and alignment.

5.3   TSV for identifying different forms of information sharing

With the small qualitative investigation of sequence 1, 2 and 3 we show that TSV is able to single out different modes of information sharing. When looking into the actual conversation in each of the identified sequences it appears that group interactions are fundamentally different depending on the activity. We see indication of different team roles (Sonnenwald, 1996) as well as social group interaction, such as agreement and expressing concerns (Bales, 1950). Based on this we find it reasonable to propose that these sequences could be of further importance for design researchers because it is plausible to hypothesise that different types of group interaction result in different types of output in a design discussion. Hence, with TSV we are able to identify sequences with higher reach potential fast, thus allowing researchers to focus their resources better.

6   DISCUSSION AND FURTHER WORK

With TSV, we have showcased a new way to explore video and corresponding transcripts in the pre-analysis phase. The strength lies in the ability to filter large amounts
of information from video and transcript material into one static overview. TSV visualises what is otherwise invisible to the human eye – the actual communication flow and
the historical participation of group members. Researchers can thus see how dialogue
flows, enabling them to identify occurrences of and transitions between monologue,
dialogue and group discussion. This should allow researchers to gain a brief insight of
the group dynamics in the team. We suggest to use these insights as input in identifying critical events the researcher should look further into. The final approach allows
us to see how discussions develop over time at a glance as opposed to tools such
as Leximancer and the network analysis approach described previously. Leximancer
show a summary of the whole discussion, keeping the temporal dimension hidden,
and the network approach still requires the researcher to watch long animations of
the conversation. To summarise, TSV includes the temporal dimension while keeping
the visualisation static. In this respect it resembles the semantic coherence analysis by
Dong (2005).

A limitation of this work is the lack of contextual information in the visualisation.
One of the drawbacks of our approach is exactly what Leximancer offers: content. Yet
we believe that a combination of visualising participant activity with semantic analyses
of different sequences in the graphs in Figure 15.5 would provide the researcher with
an interesting starting point for tackling large data sets.

To illustrate this idea we extracted the text from the speaking cues connected to
sequences S1-S6 in Figure 15.8 and used the software Leximancer to highlight the
topics most frequent in the different sections. This can be seen in Figure 15.8.

When adding keywords to the different passages of the conversation (Figure 15.8)
we show how one can easily provide more semantic content from the transcripts. The
approach does not yet provide any contextual information for the researcher studying
more than group dynamics and conversation flow. This was one of the main criticisms
given by the DTU researcher. As a design researcher studying more than "just" group
dynamics, one would like information on interactions (e.g., artefacts) as well as body
language and directional attention from the participants. This want is supported by

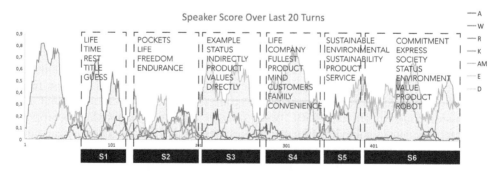

Figure 15.8 Illustration of how adding semantic content analysis to different sequences of the conversation could add more content related information to the temporal visualisation.

the fact that several of the reviewed sources both in the field of design research and group dynamic visualisation include more than only speaking queues (Avci & Aran, 2016; DiMicco *et al.*, 2006; Sonalkar *et al.*, 2013). Hence, we face the challenge of how to include more details in our graphical representations.

Looking at the 13 interactions defined by Sonalkar *et al.* (2013) we see that at least some of them could be captured in a transcript. This concerns questions (by a "?"), interruptions (coded a "-"), yes and (by "yes and") and humour (by "[laughing]"). Also since the LSA method of Dong (2005) already includes semantic development over time it would be an obvious candidate to combine with our temporal visualisations.

7   CONCLUSION

This paper presents the development the Temporal Static Transcript Visualisation – an approach for pre-analysis of video material captured from observational studies of group dynamics. By utilising the meta-data of already written transcripts we map the flow of conversation throughout the video by visualising participation in terms of amount of words said and how often a person talks. Moreover, we visualise how many group members are taking an active part in the conversation over time. In this way we are able to identify different types of dialogue patterns, for example, monologue, dialogue and group discussion. With TSV we help qualitative researchers to gain a faster overview of the overall conversation flow of a studied group collaboration.

As a pilot study we visualised the conversation dynamically with the help of the software Cytoscape. To evaluate our suggested approach, we conducted an interview with a fellow research group also working with the dataset from DTRS11 (Cash & Gonçalves, 2017). First, we compared their manual coding with our visualisations. Further, a qualitative interview discussing the value and potential of the approach was conducted.

Through this evaluation we learned that the pilot test was too "dynamic" and failed at actually helping the researcher to identify different forms of information-sharing types, which was the original idea. Hence we developed the Temporal Static transcript Visualisations (TSV) that over time graphically presents individual participation within the group as well as the overall group activity. With the visualisation we were now able to identify sequences of and transitions between monologues, dialogue and group discussions. We propose to use this as a starting point for analysing bigger sets of group-discussion observations. This approach was illustrated by picking out three different sequences in v11 and analysing them qualitatively. It was confirmed that the sequences were fundamentally different and also included phenomena such as team roles and social group interaction types. Hence we argue that TSV has a relevance in the Design Research community, opening up for quantitative analysis among datasets.

However, we have some concerns on how TSV can become even more valuable for design researchers and not only group dynamic researchers. There are some limitations arising from only providing an overview of communication flow because such a representation might be too simple to enable a design researcher to retrieve valuable insights. Adding details such as the use of artefacts or body language could make our approach into a useful tool not only for identifying moments of interest, but to actually create comparable output and allow for analysis between case studies. Our further work will focus on how we can combine semantic methods with our algorithm and allow the visualisation of more information to provide content-based impressions. In this work we contribute with first attempts to make the process of analysing design observation data less time consuming in the individual cases and comparable across cases.

ACKNOWLEDGEMENTS

We would like to thank the organisers for the interesting initiative of DTRS11 and also Phillip Cash and Milene Goçalves for giving us access to their manual coding schemes as well as participating in a discussion of our suggested approach.

This research is supported by the Research Council of Norway (RCN) through its user-driven research (BIA) funding scheme, project number 236739/O30.

REFERENCES

Avci, U., & Aran, O. (2016). Predicting the Performance in Decision-Making Tasks: From Individual Cues to Group Interaction. *IEEE Transactions on Multimedia*, *18*(4), 643–658. https://doi.org/10.1109/TMM.2016.2521348

Bales, R. F. (1950). A Set of Categories for the Analysis of Small Group Interaction. *American Sociological Review*, *15*(2), 257. https://doi.org/10.2307/2086790

Bastian, M., Heymann, S., Jacomy, M., & others. (2009). Gephi: an open source software for exploring and manipulating networks. *ICWSM*, *8*, 361–362.

Beck, F., Burch, M., Diehl, S., & Weiskopf, D. (2016). A Taxonomy and Survey of Dynamic Graph Visualization: A Taxonomy and Survey of Dynamic Graph Visualization. *Computer Graphics Forum*, n/a–n/a. https://doi.org/10.1111/cgf.12791

Bedford, D. A. D., Arns, J. W., & Miller, K. (2017) Unpacking a Design Thinking Process with Discourse and Social Network Analysis. In: Christensen, B. T., Ball, L. J. & Halskov, K. (eds.) *Analysing Design Thinking: Studies of Cross-Cultural Co-Creation*. Leiden: CRC Press/Taylor & Francis.

Bergstrom, T., & Karahalios, K. (2007a). Seeing more: visualizing audio cues. In *IFIP Conference on Human-Computer Interaction*. Springer. [Online] pp. 29–42. Retrieved from http://link.springer.com/chapter/10.1007/978-3-540-74800-7_3

Bergstrom, T., & Karahalios, K. (2007b). Visualizing co-located conversation feedback. *IEEE TableTop*. Retrieved from http://social.cs.illinois.edu/projects/papers/pdfs/bergstrom-tabletop2007 poster.pdf

Cash, P. & Gonçalves, M. (2017) Information-Triggered Co-Evolution: A Combined Process Perspective. In: Christensen, B. T., Ball, L. J. & Halskov, K. (eds.) *Analysing Design Thinking: Studies of Cross-Cultural Co-Creation*. Leiden: CRC Press/Taylor & Francis.

Chan, J. & Schunn, C. D. (2017) A Computational Linguistic Approach to Modelling the Dynamics of Design Processes. In: Christensen, B. T., Ball, L. J. & Halskov, K. (eds.) *Analysing Design Thinking: Studies of Cross-Cultural Co-Creation*. Leiden: CRC Press/Taylor & Francis.

Christensen, B. T. & Abildgaard, S. J. J. (2017). Inside the DTRS11 Dataset: Background, Content, and Methodological Choices. In: Christensen, B. T., Ball, L. J. & Halskov, K. (eds.). *Analysing Design Thinking: Studies of Cross-Cultural Co-Creation*. Leiden: CRC Press/Taylor & Francis.

DiMicco, J. M., Hollenbach, K. J., & Bender, W. (2006). Using visualizations to review a group's interaction dynamics. In *CHI'06 extended abstracts on Human factors in computing systems*. ACM. [Online] pp. 706–711. Available from http://dl.acm.org/citation.cfm?id=1125594

DiMicco, J. M., Pandolfo, A., & Bender, W. (2004). Influencing group participation with a shared display. In *Proceedings of the 2004 ACM conference on Computer supported cooperative work*. ACM. [Online] pp. 614–623. Available from: http://dl.acm.org/citation.cfm?id=1031713

Dong, A. (2005). The latent semantic approach to studying design team communication. *Design Studies*, 26(5), 445–461. https://doi.org/10.1016/j.destud.2004.10.003

Dong, W., Lepri, B., Kim, T., Pianesi, F., & Pentland, A. S. (2012). Modelling conversational dynamics and performance in a social dilemma task. In: *Communications Control and Signal Processing (ISCCSP), 2012 5th International Symposium on*. IEEE. [Online] pp. 1–4. Available from http://ieeexplore.ieee.org/xpls/abs_all.jsp?arnumber=6217775

Jayagopi, D., Sanchez-Cortes, D., Otsuka, K., Yamato, J., & Gatica-Perez, D. (2012). Linking speaking and looking behavior patterns with group composition, perception, and performance. In: *Proceedings of the 14th ACM international conference on Multimodal interaction* ACM. [Online] pp. 433–440. Available from: http://dl.acm.org/citation.cfm?id=2388772

Low, Y., Gonzalez, J. E., Kyrola, A., Bickson, D., Guestrin, C. E., & Hellerstein, J. (2014). *Graphlab: A new framework for parallel machine learning*.

Menning, A., Grasnick, B. M., Ewald, B., Dobrigkeit, F., Schuessler, M. & Nicolai, C. (2017) Combining Computational and Human Analysis to Study Low Coherence in Design Conversations. In: Christensen, B. T., Ball, L. J. & Halskov, K. (eds.) *Analysing Design Thinking: Studies of Cross-Cultural Co-Creation*. Leiden: CRC Press/Taylor & Francis.

Rosling, H. (2016). *Gapminder – a fact based world view*. Available from: https://www.gapminder.org/ [accessed 20 June 2016]

Ryu, S., & Sandoval, W. A. (2015). The Influence of Group Dynamics on Collaborative Scientific Argumentation. *Eurasia Journal of Mathematics, Science and Technology Education*, 11(3), 335–351. https://doi.org/10.12973/eurasia.2015.1338a

Scanzoni, J. (1983). SYMLOG: A System for the Multiple Level Observation of Groups (book review). Available from: http://www.journals.uchicago.edu/doi/pdfplus/10.1086/227750

Sjovold, E. (2007). Systematizing Person-Group Relations (SPGR): A Field Theory of Social Interaction. *Small Group Research*, 38(5), 615–635. https://doi.org/10.1177/1046496407304334

Smith, A. E. (2003). Automatic extraction of semantic networks from text using Leximancer. In: *Proceedings of the 2003 Conference of the North American Chapter of the Association for Computational Linguistics on Human Language Technology: Demonstrations*-Volume 4. Association for Computational Linguistics. [Online] pp. 23–24. Retrieved from http://dl.acm.org/citation.cfm?id=1073439

Smoot, M. E., Ono, K., Ruscheinski, J., Wang, P.-L., & Ideker, T. (2011). Cytoscape 2.8: new features for data integration and network visualization. *Bioinformatics*, 27(3), 431–432. https://doi.org/10.1093/bioinformatics/btq675

Sonalkar, N., Mabogunje, A., & Leifer, L. (2013). Developing a visual representation to characterize moment-to-moment concept generation in design teams. *International Journal of Design Creativity and Innovation*, 1(2), 93–108. https://doi.org/10.1080/21650349.2013.773117

Sonnenwald, D. H. (1996). Communication roles that support collaboration during the design process. *Design Studies, 17*.

Tang, J. C. (1991). Findings from observational studies of collaborative work. *International Journal of Man-Machine Studies*, 34(2), 143–160.

Van Der Walt, S., Colbert, S. C., & Varoquaux, G. (2011). The NumPy array: a structure for efficient numerical computation. *Computing in Science & Engineering*, 13(2), 22–30.

Ware, C. (2009). *Information visualization: perception for design* (2. ed., [Nachdr.]). Amsterdam: Elsevier [u.a.].

APPENDIX A

A.I　From Transcript to .csv-file (Step 2-3 in Figure 15.1)

In the case of DTRS11 we exploit the fact that the conversations in the source material is already transcribed with numbered speaker turns, initials, and content (Table 15.1). This allowed us to extract chronological development of conversation (last and future speaker) and number of words mentioned in each speaker turn as well.

Based on the sequential format of the data, where each row represents one speaker turn, a Python algorithm has been written to parse transcripts into a directed network graph format (See Appendix 1). Here, nodes represent individual speakers, and edges represent a connection between the current speaker and the next speaker. In Table 15.2, the first data row shows an edge E-W going from node E (Ewan) to node W (Will). The sequence of edges is denoted by the start time column, corresponding to the original speaker turns in the transcript. In addition to translating speaker turns into nodes and edges, the number of times a subject has spoken in the n last turns (Activity), and how many words they have uttered (#words) were extracted. To reduce the impact of #words on the weight calculation of edges, which would affect the visualization dramatically, the natural logarithm of #words were calculated and applied in its place. These are initial suggestions of features used in our first iteration of our suggested approach. All data processing has been done in Python, using the Graphlab Create (Low *et al.*, 2014) and Numpy (Van Der Walt, Colbert, & Varoquaux, 2011) software packages.

Table 15.1 Supplied transcript, stripped of superfluous data.

###	initial	Transcription
255	AM	Optimize them.
256	K	Mhm.
257	E	Yeah, and the meaning is something maybe you find in the end, but you maybe have endured a long life, to find that in the end.
258	A	But I guess, living life to the fullest is, what you're also saying, to doing- enjoying life and doing what matters to them and what they want to do. All of that Is I think what we need with living-
259	W	Yeah, it captures that- probably:- probably Amanda's point is that nobody's saying living life to the fullest has more the connotation of just (.) whole out enjoyment.

Table 15.2 Network graph format, features calculated over the last 20 turns.

Source	Target	Start Time	Stop Time	Activity	Weight	#words	log_words
AM	K	255	275	I	1.10	2	1.10
K	E	256	276	2	1.39	I	0.69
E	A	257	277	3	9.89	26	3.30
A	W	258	278	I	3.74	41	3.74
W	AM	259	279	I	3.37	28	3.37

Table 15.3 Nodes in dynamic graph format.

NodeId	Label	type	fill	size	StartTime	EndTime
1	A	ELLIPSE	#E74C3C	20	255	256
2	W	ELLIPSE	#E74C3C	20	255	256
3	R	ELLIPSE	#E74C3C	20	255	256
4	K	ELLIPSE	#E74C3C	20	255	256
5	AM	ELLIPSE	#E74C3C	20	255	256
6	E	ELLIPSE	#E74C3C	20	255	256
7	D	ELLIPSE	#E74C3C	20	255	256

Table 15.4 Edges in dynamic graph format.

FromId	ToId	weight	fill	StartTime	EndTime
5	4	0.79	#000000	255	265
4	6	1.07	#000000	256	266
6	1	4.59	#000000	257	267
1	2	2.71	#000000	258	268
2	5	2.51	#000000	259	269
5	4	0.79	#000000	255	265
4	6	1.07	#000000	256	266

A.1.1 Requirements for dynamic visualisation

For dynamic visualisation timestamp data needs additional preparation and changes. Cytoscape and the plugin DynNetwork were chosen for visualising dynamic graphs, and the dynamic graph data is processed according to the software's input requirements. Table 15.3 presents node data pre-processed for use in Cytoscape. Each node is represented by a unique numerical node id along with a label. Type and fill are pre-defined values determining the shape and colour of each node. The size variable determines the size of each node, and can be calculated based on e.g. the subject's participation in the previous n speaker turns. With the starting time of each edge from Table 15.2, we can define how long a node is active in the visualisation based on user input.

Table 15.4 shows the corresponding edge data to the nodes described in Table 15.3. For Cytoscape to recognise the data as edges. The two first columns need to be named FromId and ToId. The weight parameter is taken from Table 15.2 and is used for edge width when visualising the data. Fill governs the colour of edges, and Start/End time serves the same function as described above for nodes.

Part IV

Design Talk

Combining Computational and Human Analysis to Study Low Coherence in Design Conversations

Axel Menning, Bastien Marvin Grasnick, Benedikt Ewald,
Franziska Dobrigkeit & Claudia Nicolai

ABSTRACT

This paper presents a mixed computational and manual procedure to systematically probe for distinct low coherent turns in design conversations. Existing studies indicate that focus shifts and their linguistic equivalent, low coherent turns, positively influence ideational productivity. Because coherence is a versatile phenomenon, we contribute a classification of low coherent turns to enable future research to further investigate the influence of low coherence turns on creativity. We analyze the DTRS11 corpus, comprising 16 sessions of design conversation that contain 9830 sentences, with automated Latent Semantic Analysis (LSA) to identify potential low coherent turns. We argue that an additional manual coherence analysis with the Topic Markup Scheme (TMS) further qualifies preselected turns. This mixed method procedure constitutes a promising pragmatic instrument for locating low coherent turns in large corpora. We successfully retained 297 distinct low coherent turns out of a total of 6072 turns. The selected data contain twice as many turns that shift the focus of attention within an existing design issue as turns that interrupt and introduce a new design issue. Based on an interpretative analysis of low coherent turns, we suggest distinguishing between turns that interrupt the focus of attention and turns that shift the focus of attention through either diversifying, reframing, or selective tendencies.

I INTRODUCTION

1.1 Why study low coherence in design conversations?

Advancing in the process of designing is ultimately connected to alternating between divergent and convergent thinking and acting (Guilford, 1967; Cross 2006). Divergence and convergence are reflected in the number of topics discussed. Either a design team creates a variety of new topics (diverging) or it synthesizes many topics into less topics (converging). The ability for divergent thinking – going diverse and creating many possible alternative – is necessary to create options to choose from. It is both an important indicator of overall creative ability (Christensen & Guilford, 1958) and, in various framings, a condicio sine qua non for creativity and innovation processes (e.g., abductive thinking – cf. Dorst, 2015; Endrejat & Kauffeld, 2016; lateral thinking De Bono, 1967; associative thinking – Mednick, 1962).

Such lateral movement is especially susceptible to the phenomenon of getting stuck. This "writer's block" of designers has become conceptualized under the notion of design fixation. It has been described as "blind adherence to a set of ideas or concepts limiting the output of conceptual design" (Jansson & Smith, 1991) or as a "mental state" making it impossible "to move beyond an idea or set of ideas to produce new ideas" (Howard-Jones & Murray, 2003). This state has been reproduced and confirmed in these and several other studies (see Crilly, 2015, for a review). Design fixation can result in a literal halt or premature departure from the divergent stage. Generally speaking, it is desirable to defocus at certain stages of the process through shifts in perspective in order to prevent or overcome design fixation. The notion of focus shifts is strong in both design theory and cognitive creativity research (Schön, 1993; Dorst 2015; Guilford, 1950; Onarheim, B. & Friis-Olivarius, M., 2013).

Literature on the so-called "incubation effect" explores moments of insight that happen after having shifted away from an unsolved problem (for a review see Sio & Ormerod, 2009). Finke, Ward, and Smith (1992) suggest that broadening the focus of attention supports overcoming fixation and Suwa and Tversky (1997) confirm that focus shifts "allow for a lateral variety of design topics/ideas". How global shifts in perspectives (e.g., diverging/converging, reframing etc.) work has received extensive attention, but it is still unclear what these shifts look like and how they function at a micro-level (cf. "Future Work" in Suwa & Tversky, 1997). The exploration of disruptive stimuli (i.e., things that cause focus shifts) at a micro-level would provide insights into the kinds of contexts and dynamics of idea generation processes. Low coherent turns represent such disruptive stimuli because they increase the inference load among discourse participants (cf. Grosz, Joshi, & Weinstein, 1995). Either they are the linguistic equivalent of focus shifts (as a state of attention) or they may cause focus shifts.

In its most general meaning, coherence specifies how the elements of a perceived whole are connected. In language, coherence specifies how text and context transport meaning. In this study, we explore the opposite – the moments of 'disconnectedness' in design conversations, that is, where the conversation shifts from one prominent topic to a different one. "Micro-level" in our case means turn-to-turn level. A speaker turn begins when a speaker begins to speak and ends when the speaker ends their articulation deliberately or is interrupted. By definition, coherence cannot be assigned to a single turn but it always refers to the topical relation of two turns. Thus, when we write about low coherent turns, we actually refer to turns which lower the coherence in relation to their directly precedent turn. While *cohesion* (Halliday & Hasan 1976) describes the lexical and grammatical relatedness in text and talk, *coherence* is always a perceived relatedness and depends on the knowledge and perception of the receiver. Discourse participants infer relations in texts not only based on cohesive devices but also based on their individual knowledge and pragmatic context. Without further information or explanation by the speaker, discourse participants and observers can only rely on their personal perception of the relatedness of two statements, which is shaped by the inexactness of language, the personal background knowledge and the history of the discourse itself (cf. van Dijk, 1977a, 1977b).

Achieving coherence is a collaborative endeavor (cf. Tanskanen, 2006). It requires mutual assessment of the communicative situation. The producer needs to infer the receiver's state of knowledge in order to place cohesive devices accordingly to deliver the message in a coherent manner (Lambrecht, 1994; van Dijk, 2014). The receiver, on the other hand, may signal their state of knowledge by showing understanding or not. This means that discourse participants always have to take an active role in presuming the intention of the sender through interpretation. This is especially the case for statements with a perceived low coherence. Due to their non-embeddedness in the local topic, they offer a high degree of interpretative freedom inviting associative behavior and can therefore help in getting unstuck.

We hypothesize that there is a similar mechanism at work when designers with a high degree of shared domain knowledge and contextual information actively bridge meaning gaps by making inferences. This aligns with the reasoning by Dong and MacDonald (2017) about the nature and genesis of hypothetical inferences in design. Especially rich for inference making are movements "up- and down a semantic scale ranging from concrete details to decontextualized features" (Dong & MacDonald, 2017).

1.2 Low coherent turns as creative stimuli

There are several reasons to believe that some types of low coherent turns can be seen as creative stimuli. If coherence gaps exist, the inference load amongst discourse participants will increase (cf. Grosz, Joshi, & Weinstein, 1995). This act of inferring meaning to "bridge coherence gaps" may trigger creative and associative thinking. Similar positive cognitive effects of perceiving low cohesive texts have been shown by McNamara (2001). McNamara manipulated texts by inserting or deleting cohesive devices and found a positive correlation between less cohesive text and learning success for high knowledge readers:

> "*Cohesion gaps require the reader to make inferences using either world knowledge or previous textual information. When inferences are generated, the reader makes more connections between ideas in the text and knowledge. This process results in a more coherent mental representation. Hence, cohesion gaps can be beneficial for high-knowledge readers because their knowledge affords successful inference making.*" (Graesser, McNamara, Louwerse, & Cai, 2004)

We assume that in a similar way, new ideas may occur if discourse participants with a high degree of domain knowledge and contextual information actively bridge meaning gaps with inferred meaning. Thus, we suppose that coherence gaps lead to creative stimuli.

The relation of low coherence and creative thinking has not been studied so far and cannot be studied right away because the kinds of low coherence that occur are too versatile. Our intention is to identify low coherent turns in design conversations, to explore the different characteristics and to offer a classification of low coherent turns. This classification also reflects how low coherent turns contribute and interact with design issues, which are topical entities on a higher level of discourse. A design issue is the most salient topic or question of a discourse segment, with a discourse segment being a sequence of turns. A design issue is a "controversial question, about which people may have differing points of view" (Noble & Rittel, 1988). Once a new design issue gets introduced, members of the design team ideally contribute their positions in order to satisfy and conclude it. The perceived topical relation between two design issues (or discourse segments) is called global coherence and the perceived topical relation between two lower topical entities (utterances, sentences, turns) is called local coherence.

1.3   Related work on coherence in design conversations

The connectedness of speaker turns was firstly analyzed by Goldschmidt using Linkography (Goldschmidt, 1990, 2014). A Linkograph is a matrix-based representation of topical links of moves in design conversation. Such a graph allows for the identification of critical moves (turns with number of links above a certain threshold) and for statements to be made regarding the local and global coherence of analyzed data.

In line with this, Topic Markup Scheme (TMS; Menning, 2015, 2016) is a structured coding procedure based on Centering Theory. It is used to determine the perceived topical relation between two consecutive turns. Low coherent turns that shift the focus of attention range in terms of impact. TMS is able to describe three different instances of focus shifts (TMS transitions states: Drift, Integration, Jump) and also acknowledges the continuation of a topic. The resulting perceived relatedness of conversation segments can be visualized in the form of topic threads. Linkography and TMS depend on human analysis and are thus very limited in the amount of data that can be processed.

Dong (2004, 2005) took coherence measures of design conversations using Latent Semantic Analysis (LSA; Landauer & Dumais, 1997). He compared the cognitive coherence of individual team members' utterances with the team's overall coherence and conceptual focus. This analysis led him to the proposition that coherent communication indicates a shared understanding among team members and that successful knowledge convergence appears for high performing teams. Chan and Schunn (2017) also explore the feasibility of computational topic models for design conversations. They propose Latent Dirichlet Allocation (LDA; Blei, Ng, Jordan, & Lafferty, 2003) as a "cost effective supplement to traditional methods in providing a 'quick and dirty' bird's-eye view of what designers are talking about and how these topics change over time" (Chan & Schunn, 2017). Furthermore, Wulvik et al. (2017) introduce Temporal Static Visualisations (TSV) to identify transitions between monologues, dialogues and group discussion. Although this is not related to coherence measurements the approach is similar to ours. They support qualitative researchers with a computational pre-analysis of bigger design conversation datasets using a sliding window approach in order to reduce noise (see Section 2.2 of this paper).

2   METHOD: IDENTIFYING AND CLASSIFYING LOW COHERENT TURNS IN LARGE TRANSCRIPT CORPORA

We are using a mixed method procedure for the meaningful probing of speaker turns that shift the focus of attention in design conversations. The initial automated analysis of transcripts is based on LSA and extracts potential low coherent turns. These turns are then analyzed and classified with the help of TMS. The flowchart

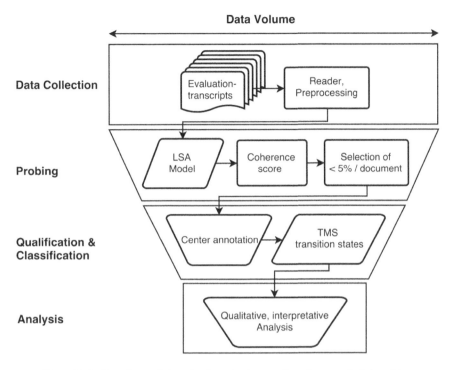

Figure 16.1 Flowchart of the mixed method procedure for meaningful probing.

(Figure 16.1) illustrates the single steps of the complete probing procedure. Each step will be explained below.

2.1 LSA for probing low coherent turns

LSA is a statistical procedure which applies singular value decomposition (SVD) to a word-document matrix, counting word co-occurrences. SVD is used to reduce this large-dimensional matrix to a smaller number of "topic" vectors. These vectors constitute the LSA model and contain the probability of occurrence of single words for each topic. We used the resulting LSA model to compare text entities based on calculating the cosine similarity of their topic probability vectors. LSA is usually applied to structured, written text corpora. Analyzing the transcripts of spontaneous, spoken conversations poses some challenges that affect accuracy. In our approach, we performed additional data cleansing by filtering out stop words, symbols (e.g., "(..)" for pause) and meta-comments (e.g., "the design team brainstorms on post-its"). Words were stemmed in order to reduce the sparsity of the word vectors and thus improve model accuracy. Because the turn length varies heavily from only a few words to multiple sentences, an analysis on the sentence level provides more accurate results. For this reason, documents were split into sentences, using the python library Natural Language Toolkit (Bird *et al.*, 2009). The first 100 LSA dimensions were retained when training our model on all sentences, as this amount of topics was shown to result in satisfactory performance (Hill *et al.*, 2002). In line with other research analyzing design conversations with LSA (Dong, 2004, 2005), the data that constitutes the LSA model were limited to the analysis data itself. The reasoning behind this is the assumption that the common ground that was built up by the design team is the only "true" data that represent the mental model of the team. Thus, any other training data would "delude" the LSA vectors.

2.2   Sliding window and extraction of low coherent turns

The resultant LSA "coherence" (sentence-to-sentence similarity) is usually determined by the cosine similarity of the LSA topic vectors of two subsequent discourse entities (Landauer & Dumais, 1997). Even after the preprocessing of the data, the coherence score was fluctuating heavily. This happened because of little interruptions (which we consider as "noise") between coherent segments of a conversation. In order to overcome this issue, we developed a weighted sliding window (WSW) approach to soften the strict sentence-to-sentence cut off, as shown in Equation 16.1. With this approach, the current sentence is not only compared to its single predecessor but to all preceding sentences, within a sentence distance of N, where N is the window size. To determine the WSW score for a sentence, the cosine similarity to each predecessor is calculated. Next, each obtained similarity is weighted by its sentence distance using a negative exponential function. Lastly, all similarities are summed up and normalized in order to obtain results between -1 and 1. We used a negative exponential function to significantly downregulate the influence of more distant sentences in such a way, that the nearest neighbor still has the biggest weight for the coherence evaluation.

$$Sim(i) = \frac{\sum_{n=1}^{N} 1/n * \mathrm{cosine}(\bar{s}_i, \bar{s}_{i-n})}{\sum_{n=1}^{N} 1/n} \tag{16.1}$$

The low coherent turns were selected by using an adaptive threshold of the lowest 5% of sentences. These sentences were mapped with their corresponding turns. The script for extracting low coherent turns as described in this section can be retrieved here: https://github.com/DTRPVisualDiagnostics/DTRS11-LSA.

2.3   Coherence results improve when using a mixed method approach

An additional qualification of preselected low coherent turns is necessary in order to obtain low coherent samples for two reasons. First, topic determination largely depends on contextual information. Humans mostly infer meaning of terms, sentences or sequences based on semantic and pragmatic contextual information (van Dijk, 2014). This contextual information is "hidden" and thus the meaning of a text may not exist in the text itself, or as van Dijk puts it, "contexts are called 'contexts' precisely because they are not 'texts' [...] Context models and their properties remain largely implicit and presupposed" (van Dijk, 2008). Compared to lexical approaches, such as WordNet or language modeling, LSA has the advantage that it is able to cope with context knowledge by calculating semantic topic dimensions. The accuracy of these topic dimensions result from a sufficient corpus size (Landauer *et al.*, 1998). Since we do not want to "delude" the analysis data with additional training data as argued before, we assume that a reanalysis of the LSA coherence results is necessary. Many transcription sets from design meetings fall into this "midsize corpus dilemma", where they are too big for considering all data for manual analysis and at the same time too small for obtaining accurate results using machine learning approaches like LSA.

The second factor that limits the stand-alone application of LSA for design conversations is that a large portion of textual meaning is influenced through bodily experience, deictic references and interaction with material, sticky notes and sketches (Schön, 1983). LSA results can get as meaningful as the meaning of a communicative situation is close to the sum of meaning expressed directly or partly implicit by words during this situation. Considering these ways of making meaning, it is useful to reanalyze LSA results. Of course, any kind of reanalysis cannot cover all contextual aspects, but we assume that it is more accurate to some extent and even more so when including video analysis at points where analysts show disagreement. We acknowledge the existence of false negatives (low coherent turns which are not identified by LSA). Therefore, we cannot claim to find all turns with the lowest coherence relative to all documents, but we can call this mixed method procedure a meaningful probing for low coherent turns because false positives (non low coherent turns among the preselected <5% LSA results) can be filtered out by using an additional analysis approach. Thus, a human evaluation instance is useful to approximate the 'real' coherence within the design conversation led by humans. For this reason, we suggest the mixed method approach of LSA in combination with the Topic Markup Scheme (TMS) as especially apt for design conversations. The result of this procedure is a structured probing for some distinct low coherent segments. "Some" because possibly not all low coherent turns have been retained with LSA. "Distinct" because the retained turns have shown low coherent characteristics according to two different analysis procedures. Furthermore, the application of TMS allows for classification of these turns.

2.4   TMS – reanalysis and classification of low coherent turns

TMS enables human analysis of perceived semantic relatedness of turns. As mentioned in the introduction, the assessment of coherence is subjective because it is a psychological concept, which depends on the knowledge and contextual interpretation of the observer. Also, coherence between two subsequent statements is not a binary state which either exists or not. Coherence rather happens on a quasi-continuous scale from low to high. For example, Botta and Woodbury (2013) state that every utterance is shifting the focus of attention "somewhat". Coherence analysis poses the following question to the human observer. Which reference of a statement has which distance to the preceding reference? At least two kinds of assessment have to be performed. First, analysts have to infer the most salient topic for each statement; second, they need to find ways to describe the distance between both statements. Clearly, the mutual assessment of coherence needs a well-defined procedure and benchmarks, otherwise it would lead to uncertainty and a high disagreement between analysts. In TMS, the assignment of a topic center to each turn is based on Centering Theory (Grosz *et al.*, 1983; 1995). The description of the distance between two centers is formalized by rules that make benchmarks unnecessary.

2.4.1 Analysis procedure

Three analysts were asked to reanalyze the preselected segments with TMS to have an additional qualification of the results that allows meaningful probing of some distinct low coherent turns. TMS, like CT, follows the assumption that each turn at a certain time t_n carries exactly one entity, a center C, which is topical more central than others. This center can either be explicit, $C_e(t_n)$, or implicit $C_i(t_n)$.

1 $C(t_n)$ is explicit when grammatical, lexical or close semantic overlap (e.g., references, substitutions, repetition, synonymy) from t_n to t_{n-1} exists.
2 If t_n has no C_e, we define that t_n has a C_i. The $C_i(t_n)$ is either a content word or phrase that shows syntactic or semantic overlap to a C prior to t_{n-1} or we define the $C_i(t)$ similar to the preferred center of CT. In this case, it is the highest ranked entity according to salience ranking.
3 If t_n has no content word, which is the case for one word turns (e.g., "yeah"), we define that the last $C_{e,i}(t_k)$ gets assigned, with t_k being a turn prior to t_{n-1}.

All centers were assigned through consensus coding of at least two analysts. With consensus coding we mean that reasoning was necessary in case of disagreement between analysts. When all analysts agreed after reasoning, the center got assigned. The determination of transition states between two turns is based on formal rules and is, therefore, not a matter of interpretation. Figure 16.2 presents how the coherence of subsequent turns can be described in the form of transition states (Continuation, Drift, Integration and Jump).

A turn continues a preceding turn if the center is explicit and shares the same semantic value with $C_{e,i}(t_{n-1})$. A turn t_n is drifting if its center is explicit but does not share the same semantic value with $C_{e,i}(t_{n-1})$ but with any other content word of t_{n-1}. A turn t_n is defined as Integration if its center is implicit and relates to a center of a turn prior to t_{n-1}. If the center of t_n is implicit and cannot be matched with the center of a turn prior to t_{n-1}, t_n is discontinuing and jumping. For the purpose of this study,

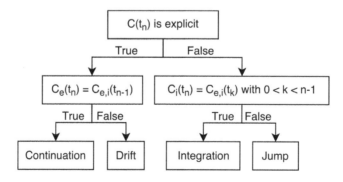

Figure 16.2 Overview of the determination of transition states.

we treat Drift and Integration as similar because they shift the topical focus within an existing design issue. We treat Jump separately because a Jump creates a new design issue.

In conclusion, the TMS reanalysis of the preselected LSA data adds two qualities. It adds the quality of human assessment of coherence and it further classifies low coherent turns.

3 CASE STUDY

With the described mixed method procedure to probe low coherent turns, we conducted a case study on the DTRS11 corpus (Christensen & Abildgaard, 2017).

3.1 Empirical data and data selection

The DTRS11 dataset consists of 21 *in situ* recordings, transcripts and background material of design sessions. The material shows a professional design team solving a specific design task for an international manufacturer within the automotive industry with a regional department in Scandinavia. This study is based on the transcript data of 16 design sessions in which the core design team and two to four additional persons who work closely together with the design team, are recorded. All sessions circle around typical design activities, including concept creation, iteration, sharing insights, clustering insights, etc. Sessions with external persons, who play no active role in the actual design process, were not included, because we assume that the core team would speak differently, which would blur the LSA topic model. The 16 transcripts consist of 6072 turns, 9830 sentences and the vocabulary of the preprocessed corpus contained 3228 unique words. The core design team has a working experience of approximately eight to ten years and has been working together for approximately two and a half to three years. Some text passages contained Scandinavian spoken language with English translations. In this case, the latter were chosen for the analysis to obtain a language consistent model. Moreover, the anonymized term "THE COMPANY" was replaced with a fictional company name to avoid wrong co-occurrence with the word "company".

3.2 Samples that validate and limit the LSA topic model at the same time

By reviewing some samples of the LSA results we found that many turn pairs with low cosine similarity validated the LSA topic model, while at the same time introducing its limits. Our model found several segments which show lexical breaks but are perceived as coherent. This is the case for many turns that introduce metaphors or examples.

(example 1)

1 E: [The first exercise is about to get it- yeah]. Open up, warm up exercises, literally like going [to the gym together].
2 N: [You need to connect].
3 E: It's a muscle. You need to: flex it and it's warm, and of course that is our first step into it....(...)

In this case, the LSA model identified <muscl> with the third lowest cosine similarity. Our LSA model did well in detecting it as a low cohesive turn, because it is technically an anatomic term and thus creates topical distance to the preceding turns. Turning to the perceived connectedness of these sentence pairs, it helps speaker E to solidify his proposition. We know from Schön's generative metaphor study that designers and other knowledge workers use metaphors to "construct meaning" (Schön, 1993). With manual TMS analysis, turn 3 (in example 1) was assessed as continuation of the preceding turn. In a similar sense, LSA marked up several sequences in which a team member "acts" a situation. Indeed, there is change in the conversation mode happening, yet it was most often perceived as objectifying a preceding statement. Take the following example from the transcript, in which turn 3 <know make decis kind thing> was detected with the fifth lowest cosine similarity in the document, but was not low coherent according to TMS analysis.

(example 2)

1 **AM:** Yeah. He's very independently minded. He always drives like "no one tell me what to do"
2 **A:**   Mhm.
3 **AM:** You know, "it's me making my own decision" kind of thing.

Another type of foreign particles in the text have been onomatopoeia like <chk chk> (10th lowest cosine similarity in document). Moreover, the LSA model often missed relations that stem from grammatical cohesion, such as co-references and pronouns. Take the following example in which <max> was detected with the 42th lowest cosine similarity in the document. In this case <that> refers to <one minute> and thus connects both utterances.

(example 3)

1 **A:**   Yeah we can say one minute per person.
2 **AM:** Yeah. That's max!

These examples prove our assumption that it is necessary to validate LSA coherence data with manual analysis. Nonetheless, LSA has been proven to limit the corpus to sequences that have low co-occurrence of words with a similar meaning.

3.3 Results

Through the application of LSA, we pre-selected 455 potential low coherent turns within a total range of 6072 turns. TMS further refined the LSA pre-selection down to 297 (65%) distinct low coherent turns (see Table 16.1). Within the total number of the LSA pre-selection, 104 turns interrupt the current topic in such a way that a new design issue is created (Jumps, 23%). The majority, 193 turns of all preselected low coherent turns, expand an existing design issue (Drift or Integration 42%). While the creation of a completely new topic is a binary event, the diversification of a topic as the "typical" occurrence of low coherence exhibits different characteristics that will be addressed in the discussion section.

Table 16.1 Mean percentage of TMS results.

TMS-Code	Coherent Continuation	Low coherent Drift & Integration	Jump	Total
Count (Percent ± SD*)	158	193	104	
	35% ± 12%	42% ± 7%	23% ± 8%	455
	158	297		100%
	35% ± 12%	65% ± 12%		

*Standard deviation among documents.

4 DISCUSSION AND INTERPRETATIVE ANALYSIS

TMS has proven to add human perception to computational results about whether a statement is low coherent or not. It is a well-functioning guideline for making statements about the coherence of subsequent discourse segments and articulate reasoning. It was interesting to realize that TMS provides a common "vocabulary" (centering terminology) to support the reasoning during center determination. It helped analysts to justify their perception and led to well-structured discussions even in difficult cases when transcripts consisted of complex multilayered semantics. The reasoning was often based on grammatical instances of the turn. For example: "I think [pronoun] is the center, because it refers to the subject of the preceding sentence." Or reiteration "I think because it is used multiple times in this turn". Or the semantic relations were the determining factor, for example: "I think [x] is the center because it is semantically close to the preceding center". Hence, TMS can be used as a manual qualification of coherence measurements. Together, TMS and LSA succeed in identifying low coherent turns in mid-size transcript corpora and provide a basic classification of these turns. Coming back to our initial question of what language representation of divergent thinking on a granular level looks like, we were especially interested in how low coherent turns can be further classified. For the evaluation of these diversifying statements, we went back to the text and applied interpretative analysis.

4.1 Characteristics of low coherent turns

The application of the method mix presented in the first part of this paper led to 297 low coherent statements labelled as Drift, Integration or Jump. Our aim was to characterize these turns further with regard to their impact on the conversation. We therefore probed into creating categories based on conversation analysis (CA). CA is a qualitative method, which identifies and describes the practices that interactants use in talk-in-social-interaction and it uses these results to understand and describe underlying structural organization of social interaction (Stivers, 2015). All sequences preselected with LSA and reanalyzed with TMS were reviewed several times by four analysts individually as well as together. In our analysis, we focused on finding recurring types of low coherent turns, which we could then summarize into categories of higher abstraction. Figure 16.3 situates these categories in a hierarchical scheme.

Low coherent turns in Design Conversation

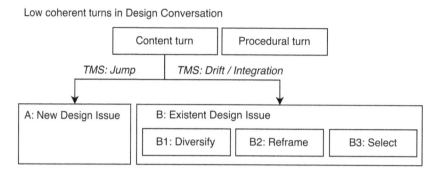

Figure 16.3 Classification of low coherent turns in design conversation.

In general, statements in conversations can be of different levels and qualities, for example, problem-focused, procedural, socio-emotional or action-oriented (Meinecke & Lehmann-Willenbrock, 2015; Klonek *et al.*, in press). We simplified this scheme to distinguish content-related and all other statements, which we named "procedural" (for the sake of simplicity). This category then comprises action-oriented or organizational statements like "Yeah? Alright. (..) Then:, alright. Let's start with you." or "…if we have the time…". We have thereby crystallized design-related statements from all other statements, as the problem under investigation is exactly the design. The next differentiation is based on our observation that low coherent turns range from a slight shift within the same discourse segment to completely cutting off the topical thread. This quasi-continuous spectrum is confined by its boundaries of complete continuation of the topic (the repetition of the same sentence) and complete unrelatedness.

This leads to our first distinction of low coherent content turns. These turns can either interrupt the current design issue and introduce a new one (A) or these turns expand the current design issue (B).

4.1.1   *Low coherent content turns that open up a new design issue (A)*

The design process consists of a set of interwoven design issues. Ideally, a singular design issue has a commonly agreed and well defined scope, it gets opened, treated and closed within a limited amount of time. The interdependencies between issues are threatening the structuredness of the course of conversations. Goldschmidt states that the analysis of sequences of issues (their amount, their size) reflects the structuredness and efficiency of the design process. A distorted conversation which has a high frequency of new and recurring issues might present a poorly structured process and indicate "inefficient design thinking and reasoning" (Goldschmidt, 2014).

Low coherent turns that open up design issues seem to represent the cornerstones of the design process because they shape the global topical structure. These turns are signaled by the TMS transition state 'Jump'. A high frequency of Jumps would indicate an ill-structured design process. 23% of all preselected low coherent turns were coded as Jumps with a standard deviation of 8% across all design sessions. The following example illustrates a Jump in the DTRS11 dataset.

(example 6)

Context of this segment:

This session is about linking insights from previous workshops. After discussing a "How might we" question that includes <comfort>, Ewan brings up a new "How might we..." question.

1 **A:** "Equals comfort". (laughter)
2 **K:** Not an uncomfortable sofa.
3 **E:** Yeah. (..) Okay:. Then "environmental sustainability. How might we ensure that our product service benefits the environment and is created in a sustainable way?"...

Coding justification:

$C_i(t_3)$ is <environmental sustainability> and is not equal, nor intersects with $C_e(t_2)$, which is <uncomfortable sofa>.

4.1.2 *Low coherent content turns that contribute to an existent issue (B)*

It was interesting to realize that the majority of retained low coherent turns in the DTRS11 dataset are happening within the same general issue. This shows the non-binary nature of topic affiliation and coherence and opens up a whole spectrum of intermediate states that can be evaluated. What all low coherent turns that contribute to an existing issue have in common is that they shift the focus of attention, while turns that open up new issues (Jumps) interrupt the focus. We are especially interested in classifying shifts because they theoretically make discourse participants infer meaning. They have associative qualities and help to get unstuck.

Shifting turns are a central aspect of creative team interaction and idea generation activities, such as brainstorming. For example, one often cited principle during ideation is to build on each other's ideas during brainstorming sessions (Paulus & Brown, 2007). More specifically, existent externalized information is taken and gets further evolved. This makes some responses topically distant to the initial statement. We refer to the following definition of topic shifts (Botta & Woodbury, 2013):

> "First, considering how every sentence shifts the topic somewhat, with a chain of sentences [A, B, C], B is a relevant response to A, and C is a relevant response to B, but C is not necessarily a relevant response to A—likeliness of relevance of the last sentence to the first sentence drops with the length of the chain."

Transposed to turns in design conversations, a topic shift from one turn to another would be: Speaker 1 makes a statement A and Speaker 2 makes a statement consisting of B and C. The statement of Speaker 2 is shifting, if B correlates with A, but C is the topical center of the statement and shows no relation to A. As mentioned in Section 2.2, shifting turns correlate with the TMS transition states Drift and Integration. Throughout all preselected turns, we have identified 42% of shifting turns with a standard deviation of 7%. In particular, Drifts have been assigned most frequently to low coherent turns (179 turns, 93% of all retained shifting turns). Thus, we wanted to further specify the techniques of diversifying an existing topic. After re-reviewing all shifting turns again, we created subclasses of shifting turns, which are (B1) diversifying shifts, (B2) reframing shifts and (B3) selective shifts.

4.1.3 B1: Diversifying turns

Diversifying turns take the topical center of the preceding turn and enrich it. This can happen via adding new components or layers to the topic, for example, by building on each other's ideas or consciously enlarging the solution space.

(example 7)

Context of this segment:

In a couple of turns the issue of "what happens at a co-creation session during lunch?" comes up. Someone suggests that the workshop participants should have a break and "not think too much". Ewan responds with turn 1:

1   E: loose the connection, 'cause we invest pretty heavily in getting them into a place, we want them to stay in that zone
2   K: yeah, we were planning to show them some examples here (point at the white board) to kinda show the range of what health is

Coding justification:

T_2 drifts from t_1 because $C_e(t_2)$ is <show them examples>, which is not exactly picking up $C_e(t_1)$ <stay in that zone>, but intersects with it.

4.1.4 B2: Reframing turns

Reframing turns shift the focus of attention by offering a new frame on the same design issue, that is, by giving analogies, choosing (potentially richer) synonyms or rephrasing statements.

(example 8)

Context of this segment:

In this session, the team discusses workshop insights. They review statements from different workshop participants.

1   W: Did virtual reality come from Brian? (laughter)
2   A: That came from Yen?
3   AM: Maybe they are soul buddies! (laughter)
4   E: Yeah, so there's a-, ask for a DNA-test next time. (laughter)

Coding justification:

$C_e(t_3)$ is <they> (Brian and Yen) while the $C_e(t_4)$ is <DNA-test>. <DNA-test> relates to the phrase <soul buddies>, therefore the focus from t_3 to t_4 drifts. We define this as a reframing because t_4 situates the topic in the frame of workshop design (although humorously) while t_3 only makes a statement about Brian and Yen.

4.1.5   B3: Selective turns

This category might sound counterintuitive to be related to low coherence, but we found out that many turns shift the focus from one turn to another by taking a certain aspect of the preceding turn and intensifying it or singling it out.

(example 9)

Context of this segment:

In a debrief session, the design team discusses means to inspire other people in a humble way. Then Abby gives the following example:

1   A: I think there's also the thing about ehm:. I think it was Yen saying something about ehm:. I think it- it was when he was buying some things ehm from a company in Europe who was taking care of the environment. He felt better, he felt superior, but it was unspoken of.
2   E: Yeah how- how did-
3   R: (happens many times?)-
4   E: How did he solve it? How did he: get- how did he get the credit he: felt he wanted?

Coding justification:

In this example, t_4 reacts to t_1 because t_2 and t_3 have no center. $C_e(t_1)$ is <taking care of the environment>. $C_e(t_4)$ is <get the credit> which relates to <unspoken of> in t_1. This means t_4 drifts from t_1. As t_4 picks only the credit aspect of t_1, it picks up one specific part of t_1. Such a drift is a selective drift.

5   LIMITATIONS AND FUTURE WORK

With regard to the insufficient size of the training data for the LSA topic model, it is likely that some low coherent turns (false negatives) were not retained. Hence, the procedure presented in this paper can only be applied for structured probing.

In general, it is very difficult to create a sufficiently big training corpus that leads to an accurate topic model for conversations. This is even more difficult for design conversations. They consist of terms from the design domain, contain very specific vocabulary from the design target domain and exhibit a big portion of vague language (cf. Glock, 2009). To build up a robust multi-dimensional topic model that "gets" design language and is combinable with training data from the design-target domain would be intriguing future work for the design research community. In the present study, the retained LSA turns have been reanalyzed for finding false positives and to make a first distinction between turns that shift the focus of attention within a design issue, and turns that discontinue an existing design issue and open up a new one. This reanalysis with TMS was not blind, meaning that analysts knew that they were dealing with LSA-preselected segments. Therefore, we cannot claim that TMS validates (confirms or rejects) LSA results. But TMS is so far a useful additional qualification of the preselected data and together with LSA comprises a structured probing procedure. Compelling future work would be a blind reanalysis of LSA using TMS by testing random segments against LSA pre-selected segments. Depending on the results, this could lead to a unified human perception procedure of coherence, which also validates computational coherence measurements.

The sub-categorization of low coherent turns that shift the focus of attention within an existing design issue was the outcome of an interpretative approach. This categorization still needs to prove its usefulness and could be enhanced by assigning the categories to other low coherent conversations samples. For future research we suggest to investigate the influence of low coherent turns on ideational productivity using the classification presented in this paper.

6   CONCLUSION

Research on design fixation and related experiments lead us to the assumption that some forms of low coherence can be seen as creative stimuli. It seems that experienced designers are comfortable with sending and receiving low coherent turns in order to move forward. They creatively cope with some forms of low coherence and bridge coherence gaps with innovative meaning. To understand and confirm this process, a classification of low coherence turns is necessary.

The mixed method procedure presented in this paper works well for efficient and structured probing of low coherent turns in transcribed conversation data. For the DTRS11 corpus, consisting of 6072 turns, one would need approximately 6000 minutes to evaluate every turn manually (with TMS). With the LSA-TMS mixed procedure, only the 455 most promising turns need to be analyzed manually. TMS has proven to enable analysts to externalize, discuss and justify their perception concerning topical relatedness of speaker turns.

TMS distinguishes between continuing turns and low coherent turns. Low coherent turns can either expand the current design issue (Drift, Integration), or interrupt the current design (Jump). Mapped to the intentional state of discourse participants, we can say that Continuation does not change the focus of attention, Drift and Integration shift the focus of attention within the same design issue, and Jumps interrupt and turn the focus from one design issue to another. The results indicate that most of the low coherent turns in design conversations are shifting turns (B, 42% ± 7%). Jumps (A) take up 23% (±8%) of the preselected turns. Based on interpretative analysis we have created subclasses for shifting turns: B1: Diversify, B2: Reframe and B3: Selective (given theses turns were content turns, not procedural turns).

This is the first classification of low coherent turns in design conversations. Considering that some low coherent turns cause focus shifts and lead to the inference of innovative meaning, this classification is a promising source to continue exploring contexts and dynamics of idea generation processes.

ACKNOWLEDGEMENTS

We gratefully acknowledge the various ways in which the HPI Stanford Design Thinking Research Program has enabled this research. We would like to thank the DTRS11 organizers for providing a rich data set and hosting an inspiring conference.

REFERENCES

Baillie, C. & Douglas, E. P. (2014). Confusions and conventions: Qualitative research in engineering education. *Journal of Engineering Education, 103*(1), 1–7.

Bird, S., Klein, E., & Loper, E. (2009). *Natural language processing with Python*. O'Reilly Media, Inc.

Blei, D. M., Ng, A. Y., Jordan, M. I., & Lafferty, J. (2003). Latent Dirichlet Allocation. *Journal of Machine Learning Research*, 993–1022.

Botta, D., & Woodbury, R. (2013). Predicting topic shift locations in design histories. *Research in Engineering Design*, 24(3), 245–258.

Chan, J. & Schunn, C. D. (2017). A Computational Linguistic Approach to Modelling the Dynamics of Design Processes. In: Christensen, B. T., Ball, L. J. & Halskov, K. (eds.) *Analysing Design Thinking: Studies of Cross-Cultural Co-Creation*. Leiden: CRC Press/Taylor & Francis.

Christensen, B. T. & Abildgaard, S. J. J. (2017). Inside the DTRS11 Dataset: Background, Content, and Methodological Choices. In: Christensen, B. T., Ball, L. J. & Halskov, K. (eds.). *Analysing Design Thinking: Studies of Cross-Cultural Co-Creation*. Leiden: CRC Press/Taylor & Francis.

Christensen, P. R. & Guilford, J. P. (1958). *Creativity/fluency scales*. Beverly Hills, CA: Sheridan Psychological Services.

Crilly, N. (2015). Fixation and creativity in concept development: the attitudes and practices of expert designers. *Design Studies*, 38, 54–91.

Cross, N. (2006). *Designerly Ways of Knowing*. London: Springer.

De Bono, E. (1967). *The use of lateral thinking*. London: Cape.

Dong, A. (2004). Quantifying coherent thinking in design: a computational linguistics approach. In: J. S. Gero (ed.) *Design computing and cognition '04*. Dordrecht, Netherlands: Springer Netherlands. pp. 521–540.

Dong, A. (2005). The latent semantic approach to studying design team communication. *Design Studies*, 26(5), 445–461.

Dong, A. & MacDonald, E. (2017). From Observations to Insights: The Hilly Road to Value Creation. In: Christensen, B. T., Ball, L. J. & Halskov, K. (eds.) *Analysing Design Thinking: Studies of Cross-Cultural Co-Creation*. Leiden: CRC Press/Taylor & Francis.

Dorst, K. & Cross, N. (2001). Creativity in the design process: Co-evolution of problem-solution. *Design Studies*, 22(5), 425–437.

Dorst, K. (2015). *Frame innovation: Create new thinking by design*. MIT Press.

Endrejat, P. C. & Kauffeld, S. (2016). *Über innovationsverhindernde und innovationsfördernde Denkweisen* [About Innovation Impeding and Innovation Facilitating Mindsets]. Gruppe. Interaktion. Organisation. Zeitschrift für Angewandte Organisationspsychologie (GIO). Advanced online publication.

Finke, R. A., Ward, T. B., & Smith, S. M. (1992). *Creative cognition: Theory, research, and Applications*. MIT Press.

Glock, F. (2009). Aspects of language use in design conversation. *CoDesign*, 5(1), 5–19.

Goldschmidt, G. (1990). Linkography: assessing design productivity. In: *Cybernetics and System '90, Proceedings of the Tenth European Meeting on Cybernetics and Systems Research*. World Scientific. pp. 291–298.

Goldschmidt, G. (2014). *Linkography: unfolding the design process*. Cambridge: MIT Press.

Graesser, A. C., McNamara, D. S., Louwerse, M. M., & Cai, Z. (2004). Coh-Metrix: Analysis of text on cohesion and language. *Behavior research methods, Instruments, & Computers*, 36(2), 193–202.

Grosz, B. J., Joshi, A. K., & Weinstein, S. (1983). Providing a unified account of definite noun phrases in discourse. In: *Proceedings of the 21st annual meeting on Association for Computational Linguistics*. Association for Computational Linguistics. pp. 44–50.

Grosz, B. J., Weinstein, S., & Joshi, A. K. (1995). Centering: A framework for modeling the local coherence of discourse. *Computational linguistics*, 21(2), 203–225.

Guilford, J. P. (1950). Creativity. *American Psychologist*, 5: 444–454.

Guilford, J. P. (1967). *The nature of human intelligence*. New York, NY: McGraw-Hill.

Halliday, M. K. & Hasan, R. (1976). *Cohesion in English*. London: Longman.

Hatchuel, A. & Weil, B. (2002). *La théorie C-K: Fondements et usages d'une théorie unifiée de la conception. Colloque Sciences de la conception*. [Online] Available from: http://www.spatial-computing.org/~michel/lib/exe/fetch.php?media=documents-ro:hatchuelweil2002latheorieck.pdf [Accessed February 21, 2017]

Hatchuel, A., Masson, P. Le, & Weil, B. (2004). C-K Theory in Practice: Lessons from Industrial Applications. In: Marjanovic, D. (ed.) *8th International Design Conference*. Dubrovnik. pp. 245–257.

Hill, A. W., Dong, A., & Agogino, A. M. (2002). Towards computational tools for supporting the reflective team. In: Gero, J. S. (ed.) *Artificial intelligence in design '02*. Dordrecht, Netherlands: Springer Netherlands. pp. 305–325.

Howard-Jones, P. A. & Murray, S. (2003). Ideational productivity, focus of attention, and context. *Creativity research journal*, 15(2–3), 153–166.

Jansson, D. G. & Smith, S. M. (1991). Design fixation. *Design Studies*, 12, 3–11.

Lambrecht, K. (1994). *Information structure and sentence form: A theory of topic, focus, and the mental representations of discourse referents*. Cambridge University Press.

Landauer, T. K. & Dumais, S. T. (1997). A solution to Plato's problem: The latent semantic analysis theory of acquisition, induction, and representation of knowledge. *Psychological review*, 104(2), 211.

Landauer, T. K., Foltz, P. W., & Laham, D. (1998). An introduction to latent semantic analysis. *Discourse processes*, 25(2–3), 259–284.

Maher, M. L. O. U., Poon, J., & Boulanger, S. (1996). Formalising Design Exploration as Co-Evolution: A Combined Gene Approach. In: Gero, J. S. & Sudweeks, F. (eds.) *Advances in formal design methods for CAD*. London: Chapman and Hall.

McNamara, D. S. (2001). Reading both high-coherence and low-coherence texts: Effects of text sequence and prior knowledge. *Canadian Journal of Experimental Psychology*, 55, 51–62.

Mednick, S. (1962). The associative basis of the creative process. *Psychological Review*, 69(3), 220–232.

Meinecke, A. L. & Lehmann-Willenbrock, N. (2015). Social dynamics at work: Meetings as a gateway. In: Allen, J. A., Lehmann-Willenbrock, N. & Rogelberg, S. G. (eds.) *The Cambridge handbook of meeting science* New York, NY: Cambridge University Press. pp. 325–356.

Menning, A., Scheer, A., Meier, B., & Nicolai, C. (2015). Designing as Weaving Topics: Coding Topic Threads in Conversations. In: Popovic, V., Blackler, A. L., Luh, D., Nimkulrat, N.,

Kraal, B. & Yukari, N. (eds) *Proceedings of the IASDR Conference 2015*. Brisbane, Australia. pp. 1460–1468.

Menning, A., Scheer, A., Nicolai, C., & Weinberg, U. (2016). The Topic Markup Scheme and the Knowledge Handling Notation: Complementary Instruments to measure Knowledge Creation in Design Conversations. In: Plattner, H., Meinel, C., & Leifer, L. (eds.) *Design Thinking Research*, Berlin: Springer. pp. 291–307. (Reprint of figure 16.2 with permission by Springer Nature)

Noble, D., & Rittel, H. W. (1988). Issue-Based Information Systems for Design. *Computing in Design Education* [ACADIA Conference Proceedings].

Onarheim, B. & Friis-Olivarius, M. (2013). *Applying the neuroscience of creativity to creativity training. Frontiers in Human Neuroscience*, 7 (October), 656.

Paulus, P. B. & Brown, V. R. (2007). Toward more creative and innovative group idea generation: a cognitive social motivational perspective of brainstorming. *Social and Personality Psychology Compass*, 1(1), 248–265.

Schön, D. A. (1983). *The reflective practitioner: How professionals think in action*. Basic books

Schön, D. A. (1993). Generative metaphor: A perspective on problem-setting in social policy. In: Ortony, A. (ed.) *Metaphor and Thought*. London: Cambridge University Press. pp. 137–163.

Sio, U.N. & Ormerod, T.C., 2009. Does incubation enhance problem solving? A meta-analytic review. *Psychological Bulletin*, 135(1), 94–120.

Stivers, T. (2015). Coding social interaction: A heretical approach in conversation analysis? *Journal Research on Language and Social Interaction*, 48(1), 1–19.

Suwa, M., & Tversky, B. (1997). What do architects and students perceive in their design sketches? A protocol analysis. *Design studies*, 18(4), 385–403.

Tanskanen, S. K. (2006). *Collaborating towards coherence: Lexical cohesion in English discourse* (Vol. 146). Amsterdam, Netherlands: John Benjamins Publishing.

Van Dijk, T. A. (1977a). Context and cognition: Knowledge frames and speech act comprehension. *Journal of Pragmatics*, 1(3), 211–231.

Van Dijk, T. A. (1977b). Semantic macro-structures and knowledge frames in discourse comprehension. *Cognitive Processes in Comprehension*, 3–32.

Van Dijk, T. A. (2008). *Discourse and context. A Sociocognitive Approach*. Cambridge University Press.

Van Dijk, T. A. (2014). *Discourse and knowledge. A Sociocognitive approach*. Cambridge University Press.

Wulvik, A., Jensen, M. B., & Steinert, M. (2017). Temporal Static Visualisation of Transcripts for Pre-Analysis of Video Material: Identifying Modes of Information Sharing. In: Christensen, B. T., Ball, L. J. & Halskov, K. (eds.) *Analysing Design Thinking: Studies of Cross-Cultural Co-Creation*. Leiden: CRC Press/Taylor & Francis.

Chapter 17

"Comfy" Cars for the "Awesomely Humble": Exploring Slang and Jargons in a Cross-Cultural Design Process

Newton D'souza & Mohammad Dastmalchi

ABSTRACT

Every profession breeds its own set of vernacular language that seems foreign to the outside. Designers are no exception as they use expressive words all too often in the form of slangs (e.g., awesome lighting, pop of color) or jargons (e.g., feasibility study, materiality). Slangs are playful descriptors while jargons are institutionalized and technically savvy. However, not much is known about what role slangs/jargons play in the design process. Understanding this unconventional form of language, especially in its cross-cultural mutation, is important to how design is communicated in today's global context. Because the DTRS11 dataset consisted of Scandinavian and Asian teams, it lent itself well to a cross-cultural study of design jargons. Using a content analysis approach, this paper presents instances of slang/jargon and their role in aggregation and accumulation of concepts in the design process. The findings also reveal unique usage of slangs/jargons by the Scandinavian and Asian teams and their mutations in the cross-cultural collaboration.

1 THEORETICAL FRAMEWORK

1.1 Slangs and jargons

Designers are notorious for using expressive words specific to their trade. For example, phrases institutionalized in the design profession such as 'feasibility study' or 'materiality' are all too common. These are called design jargons. Other phrases, called slangs, might be playful in their construction and include words such as 'pop of color,' or 'awesome lighting'. While the use of these phrases are noticeable in design conversations, not much has been written about how they are used or affect the design process. If the design profession perpetuates its own set of jargons and slangs what role do they play in design? Are they mere fillers or facilitators of design? Do they have any performative functions? These questions formed the initial motivation for the paper.

While jargons and slangs can be considered as unconventional forms of language, it is useful to clarify their similarities and differences. According to Nash (1993), slang is akin to jargon's neighbor while metaphor is its distant cousin. Jargons are technically savvy in their usage, whereas slangs can appear as rousing clichés, eloquent shop-talk and arresting gobbledygook.

In linguistics, slangs and jargons seem to have both negative and positive connotations. In terms of negative connotations, jargons are considered to be pretentious, excluding, evasive, or otherwise unethical and offensive uses of specialized vocabulary (Hirst, 2003). Jargons are used to exclude (or even hide from) outsiders by the virtue of being unintelligible (Chaika, 1980). Similar negative connotations are ascribed to use of slangs as well. According to Oxford English dictionary, slangs are a special vocabulary used by any set of persons of a low, disreputable character or a language of a low, vulgar type (e.g., prison slang or school slang). Because of their incomprehensibility to people outside their circle, slangs are considered as a species of crime against language and clean thinking because they are pretentious and dreary.

However, slangs and jargons are also attributed positive functions. For example, in science, jargons are scientific terminologies that are essential for designating new entities for which the language has no name. It makes for economy and for the accuracy and precision required in scientific research. According to Chaika (1980), jargons are used only in specific settings and for particular purposes, and are practical because they help busy people to do their jobs efficiently. Their use is also social because they reinforce in-group camaraderie and alleviate the tedium of shared labor. In jargon, there is something beyond identifying oneself in social circumstances. It is more about facility and efficiency of speech than about linguistic defiance or class-identification. Similar positive roles are ascribed to slangs. Despite their unconventional form, slangs are formed through complex processes of compounding, affixation, shortening, and functional shift (Eble, 1996). Hence, social groups and subgroups are fertile breeding grounds for an idiosyncratic type of lexis such as slang and jargon in order to reinforce the group solidarity and identity (Amari, 2010).

According to Nash (1993), the difference between slang and jargon can sometimes be difficult to reveal. Jargoning often resorts to metaphorical expressions and certain words that are used in some jargoning styles are close to slang. Whereas metaphor clarifies, jargon obscures, and whereas slang displays an intention to amuse with a jest, jargon is brandished in earnest. While they can have a different purposes, Adams (2009) suggests that whether it is slang or jargon has little to do with the nature or quality of words, but with whatever social purposes the words serve. Hence, instead of analyzing them separately as slangs and jargons, it is the intention of this paper to deliberately study these two terminologies as a type of unconventional language. Our thesis is that, while we might not be able to create a 'theory' of their usage based on the scope of this paper, the study might reveal at least a preliminary insight into their usage. Hence, there are two purposes for this paper. First, to examine their specific characteristics of use in design; and second, to explore their role in the design process, especially in a cross-cultural context. Understanding these unconventional forms of language, especially in their cross-cultural mutation, is important to how design is communicated in today's global context and to investigate their role in building communities of knowledge.

1.2   Design as language

Language conception and production has a long tradition in linguistics, first reflected in the works of structuralist theorists, including: Saussure's (1959) *dialectal structuralism*, which evokes a strong system of relationships between the signifier and signified; Lévi-Strauss' (1976) *cognitive structuralism*, which focuses on structures and social rules brought about by human classifications); Chomsky's (1965) *structural linguistics*, which examines the idea of deep structures in language formation; and Eco's (1979) *semiotics*, which focuses on meanings of signs that are structured by cultural codes. Common to these theories is the distinction between three elements in structuralism: syntactics, semantics and pragmatics. Syntactics deals with the formal grammar or the structure as a self-relational system. Semantics deals with meaning (specific meaning assigned to the structure) and pragmatics deals with how these meanings are interpreted in practice (in other words, collective meanings that evolve over time and use).

The structuralist approaches gave way to post-structuralist interpretations at the end of the 21st century, propagated by the insufficiency of stable and precise structures to study language, which was considered fragmented and ambiguous. Instead of an identifiable structure, deconstructive theorists such as Derrida (1978) describe language as a 'event' and a kind of 'play' one which is fundamentally unstable (cf. the notion of différance), multiple, contradictory, and subject to change across settings. Barthes (1978) points that any literary text has multiple meanings, and that the author was not the prime source of the work's semantic content.

In the design literature the application of language has been studied through different theoretical frameworks: structural linguistics and latent semantic analysis (Dong, 2005, 2006, 2007); discourse analysis (McNair *et al.*, 2014); and conversational analysis (Oak & Lloyd, 2014; McDonnell, 2010; Glock, 2009). While no studies have dwelt on unconventional forms of language such as slangs and jargons, recent works have focused on other conventional language forms such as use of metaphors and hedge words (Christensen & Schunn, 2007, 2009) and polysemy (Georgeiv & Taura, 2014). The lack of studies specifically on slangs and jargons might indicate that either they are not considered legitimate forms or language, or that their role is not considered to be significant. Only one specific study, which indirectly addresses jargons, concludes that they can be a barrier for collaborative communication, especially in interdisciplinary design processes (Kleinsman & Valkerburg, 2008).

Nonetheless, the studies conducted on language in design provide a useful starting point. Dong (2006) uses the linguistic process of appraisal to explore the affective aspects of design. Critiquing the prevalent notion of information process models in design, Dong measures the notion of 'feeling' through positive and negative values of appraisal, specifically focused on judgment and appreciation, and applying it to a taxonomy of products, people and processes. Further attempts to understand the 'feeling' aspects of design are illustrated in computational linguistics. These studies are useful to move the focus of the use of language in design from a rational role to an affective one. In some ways, we could think of the role of slangs, because of their playful nature, to be more of an affective one, while the role of jargons, because of their institutional character, to be a rational one.

In another study, Dong (2007) attributes the role of language as an instrumental one. Rather than language 'facilitating' design, Dong advocates the role of language as one of 'becoming' design, a role which he considers 'performatory'; that designing is a language on its own, partly performing what cannot be conversed and enacted through designing; and in doing design, linguistic descriptions are not mere bystanders, but become an active, functional instrument of design. Given the instrumental role of language in design, the recent shift from studying precise formal language and parts of speech to more informal language based on specific contextual use, points out to the attempts of designers to curtail the effect of veridicality, where formal language is inadequate to capture or make sense of all the cognitive content of designing.

Adams, Mann, Jordan, and Daly (2007) observe that "imprecise" language distinguishes a type of everyday language that lacks technical precision or specificity (e.g., "wobbly bits" when referring to the part of the pen design that holds the print-head). This mixing of languages refers to linking together different kinds of language in uninterrupted talk and may signify efforts to translate, bridge, or integrate multiple perspectives. As evident, recent analysis of language in the design process has made use of Jeffersonian transcription, which not only coded the formal language but also micropauses, overlapping language, pace, raise in volume, intonation, interest level, stretch and continuation (Oak & Lloyd, 2014).

Subsequent initiatives to understand unconventional forms of language have led to studies on hedge words that contain expressions of uncertainty, which include terms like 'probably', 'sort of', 'guess', 'maybe', 'possibly', 'don't know', and 'believe' (Christensen & Schunn, 2007, 2009). Based on these word, text segments containing these words or phrases are coded as 'uncertainty present' or as 'uncertainty absent'. This 'epistemic uncertainty' refers to a metacognitive state that arises during a design process on occasions when a designer is unsure about some aspect of their on-going design work such as their understanding of elements of the problem or their confidence in the effectiveness of solution ideas (Ball & Christensen, 2009). The manifest expression of epistemic uncertainty by designers is often associated with analogizing (e.g., Ahmed & Christensen, 2009; Ball & Christensen, 2009; Christensen & Schunn, 2007, 2009).

Similar to use of analogies in design, other efforts have been made to study unconventional language in terms of polysemy in design conversations. Georgiev and Taura (2014) elaborate on the idea of using metaphors that allow words with specific meanings to have additional (related) meanings and when the meaning of text is extended beyond its original meaning. Polysemy also has been seen as "an essential manifestation of the flexibility, adaptability, and richness in meaning potential" (Fauconnier & Turner, 2003). Some examples include concept abstraction (identification of similarity, e.g., a violin as a dress), concept blending (identification of similarity and dissimilarity, e.g., an object that is a combination of a violin and a dress, i.e., a wearable violin) and concept integration (identification of thematic relations not physically related, e.g., a dressy lady who plays a violin).

There is also the challenge of local and global understanding of linguistic structures. Trickett and Trafton (2009), for example, illustrate how linguistic codes might differ from global codes for the same conversation based on the context. Giving an example of how meteorologists handle uncertainty in a forecasting task, they illustrate a range of linguistic pointers to uncertainty, such as markers of disfluency (e.g., um, er, and so forth), hedging words (e.g., "sort of," "maybe," and "somewhere around"), and explicit statements of uncertainty (e.g., "I have no idea" and "what's that?"). In the example, a forecaster was highly uncertain overall, but the utterances contained fewer linguistic markers of uncertainty. Conversely, another forecaster was highly certain overall, but used a number of uncertain linguistic markers. This mismatch between linguistic and global coding schemes might sometime distort the way language is analyzed.

The flexibility of understanding unconventional forms of conversation is also addressed in studies which contend that vague, natural language terms can replace precise numbers and introduce interpretative flexibility (Glock, 2009: Pinch & Bijker, 1987) and the idea that ambiguity is essential to the design process (Bucciarelli, 1994).

Just as with any indexical term, the meaning of terms such as 'kind of small', or 'big enough' depends on context, the use of indexical, natural language requires contexts for understanding and may serve several functions. Vague expressions are not just poor but good-enough substitutes for precise expressions, but are preferable to precise expressions because of their greater efficiency. Vagueness is not only an inherent feature of natural language, it is crucially an interactional strategy (Jucker *et al.*, 2003). In a study of design negotiation, McDonnell (2010) describes how vagueness, hesitation and delay are part of a designer's conversation that allows for design negotiations and contends that vague scenarios are able to contribute valuable information.

Similarly, in reference to the current DTRS11 dataset, Menning *et al.* (2017) have explored the relationship between low-coherence terms and creativity. They argue that experienced designers are comfortable with sending and receiving low-coherent turns in order to move forward and bridge coherence gaps with innovative meaning. In another paper from this DTRS11 dataset Bedford *et al.* (2017) consider that team designing is essentially an abduction process and identify "how" adverbs as most prevalent. Their linguistic register for within and across sessions identified and characterized adverbs, with more than 50% being adverbs of manner ("how") – although they observe that evidence for collaboration in the form of actual collaborative language, consensus building and reaffirmation was relatively weak throughout the design process.

The above studies point out that unconventional forms of language are just as useful and can complement formal analysis of language. Along with recent studies on analogies, metaphor, hedge words and uncertain terminologies, the use of slangs and jargons might reveal the aspects of design that formal analysis might be unable to. To examine slang and jargon usage, the following research questions are posed in this study:

–   What are the variety of instances of usage in slangs and jargons in a design process and for what purposes are they used?
–   What can slangs and jargons reveal about the essential qualities of cross-cultural interaction?

2   METHODOLOGY

The DTRS11 dataset provided us with an opportunity to examine the cross-cultural slangs/jargons for two main reasons. First, the dataset was large enough for us to analyze the instances of slang/jargon usage and the dataset consisted of two distinct cultural groups, Asian and Scandinavian teams. The Scandinavian team included the *Design Team (SD); Accessory Department (SA) and Design Specialists (DS)* whereas the Asian team included *External Consultants (AE)* and *Lead Users (AL)*. Since we wanted to focus mainly on design, rather than product development or user analysis, ideally we would have liked to see a robust interaction between the *Scandinavian Design Team* (SD) and the *Asian Lead Users* (AL). However, the involvement of the *Asian Lead Users* (AL) was much internalized and the transcription was unavailable due to proprietary reasons. Hence, we decided to focus on the *Scandinavian Design Team* (SD) and *Asian External Consultants* (AE) as the subjects of our study, with the assumption that the AE played the proxy role of translators between SD and AL.

To trace how design jargons were communicated within the SD and AE groups, we identified specific days in which the SD-AE interaction occurred the most. As we were also interested in the mutation of jargons and slangs through the process, we conducted a phase-wise analysis of how jargons and slangs unfolded. To capture both the breadth and depth of the slang/jargon usage we focused on samples that include the front, middle and backend of the dataset: Co-creation Workshop 1 (video 07, 08, 09, 10, 11, and 12), Co-Creation Workshop 2 (video 14, 15, 16, 17, and 18).

We used a content analysis technique (Hsieh & Shannon, 2005), which involves coding the transcript based on instances from prior understanding of slangs and jargons between the SD and AE. The coding scheme we used is as per the descriptions used by Nash (1993) and Adams (2009) as our guiding framework (Table 17.1). The procedure involved systematically working through the transcript looking for instances of slang and jargon usage which were most prevalent. As described earlier, we did not make a distinction between slangs and jargons. Two coders reviewed the transcripts, and each coder created a list of slang and jargon phrases that the SD and AE had used. Then the two coders cross checked the lists by comparing the results that were independently derived. The final list was counted and sorted into a table, based on the teams and sessions where the phrases were uttered. Although traditional quantitative content analysis requires one-to-counts of the number of instances (number of responses that fit a specific code), we were more interested in the variety of instances, that is, the different types of slang/jargon usage. Hence, duplicate instances were not counted. Further investigation extended the analysis to include latent themes and the similarities/differences in how slangs and jargons were used by SD and AE teams. The following table presents the coding categories used to identify slangs and jargons.

Coding Scheme for Slangs and Jargons:

Slang and Jargon Features	Examples
Buzz words and catch phrases	State-of-the-art
Showy modifications	*Thrust* (instead of 'tendency')
Noun pairs	Innovative measures
Prefix	*Problematize* (instead of problem)
Suffix	So done
Shortened words	Comfy
Hyphenated words	Consumer-led
Elongated Hyphens	Once-in-a-lifetime-opportunity
Premodified words	*Feasibility study* (instead of study of feasibility)
Overtly descriptive phrases	Pedal extremity protection device (instead of shoe)
Epithets	Wholesome
Compound verbs	Thrillingly
Idioms	Keep an eye out
Clichés	Time will tell
Superlative words	Coolest
Blend words	Micropause
Word mutations	*Chillax* (from chill, to hang, to hang-out, to meet)
Playful puns and metaphors	*Shopaholic* (derived from alcoholic)
Adjuncts	Having said that
Expletives	f+ing great
Words of technical savvy/shop talk	Materiality

3 FINDINGS

3.1 Slang/jargon usage by Scandinavian and Asian teams

Our analysis of the DTRS11 dataset revealed a total of 196 different uses of slangs/jargons in which SD accounted for 100 and AE accounted for 96. This showed a near identical variety in slang/jargon usage between the two teams. An example of slang and jargon usage by the Asian team is illustrated in Table 17.1.

In the example one can notice hyphenated words (clan-guy), buzz words (care-free), clichés (go with the flow), and words of technical savvy/shop talk (brand, constraints, barriers). Other examples of slangs/jargons include *Me-time, Asian-thing, retarded seats, gold-plated humbleness environmentally conscious, classic cue, tipping the balance, car is a womb, boring porridge, stereotypical mother* among others.

We mapped the variety of slangs and jargons in the design process using the Interact software (Figure 17.1). The analysis of the process reveals that AE used slangs and jargons with higher intensity at the beginning of the process (video 7 and 8) while the SD used them most at the latter half (video 11, 12, and 14). What these results mean is difficult to ascertain, but perhaps they show the different ebb and flow between the SD-AE interaction. For example, AE might be attempting to express complexity of cultural issues in the beginning of the process, which the SD is able to absorb and react to it at the end. Nonetheless, it is an interesting finding that the most intense SD-AE interaction is confined to video 14 of the design process.

Table 17.1 An example of slang and jargon usage by the Asian team.

Legend	File Reference	Speaker (Team)	Transcription
▨ Slang ▨ Jargon	(V8, 66)	W (AE)	"I guess, even for the single guy I think he's, he may not be totally effective of singles in general, ah he's slightly more **clan-guy** I think, but even then there's some differences between him and ah: those who've got children for example, so, overall mindset more short term:, **care-free, go with the flow**, ah: they don't have to worry as many things because it doesn't really hit them in pocket, in the mind as much, whereas for the family, if nothing else, just the simple fact of having to buy or consider buying something for the child, will already trigger many things. "I should pay attention to where it's made, where it's from, what is used in it:, what's the brand."
		AM (AE)	They become the constraints.
		W (AE)	Yeah, so it becomes a- a barrier. It's like "I can't be:- It's almost like mentally I can't be free to just pick whatever I want."

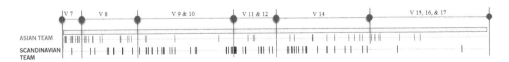

Figure 17.1 Usage of jargons/slangs by Asian and Scandinavian team across the design process.

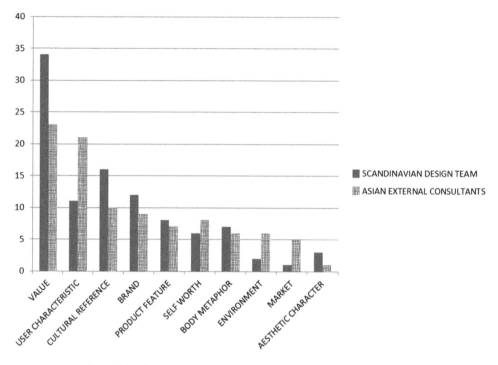

Figure 17.2 Jargon and slang themes by Scandinavian and Asian team.

Further analysis of slang/jargon usage revealed ten themes under which they could be categorized (Figure 17.2). Themes include references to value (positive value such as *gold-plated humbleness* or negative values such as *crappy reality*), user characteristic (*care-free*), cultural descriptors (*Scandinavian-ness*), self-worth (*me-time*), body metaphor (*soul buddies*), brand (*evolving status symbol*), product feature (*feasible display*), environment (*sustainability*), market (*priority banking*) and aesthetic character (*understated*).

3.2   The unfolding of three major themes of slangs/jargons in the design process

As observed in the Interact chart on Figure 17.3, we wanted to explore the three main themes, namely *values, user characteristics,* and *cultural references,* in the context of the design process. Interestingly, SD used more *value* and *cultural reference* slangs than AE, while AE used more *user characteristic* slangs than SD. This finding could be a result of how SD and AE saw their roles in the design process – with SD trying to extract values and cultural codes from a group foreign to them (Asian clients), while AE tried to feed specific user characteristics of clients from a group which was culturally familiar to them. Figure 17.3 also reveals that while the value theme pervaded throughout the process, user characteristics dominated in the first phase of the process (video 7) and cultural references preceded the later phase (video 11, 12, and 14). In other words, the

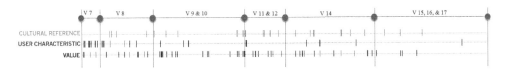

Figure 17.3 Major themes of jargons/slangs by Scandinavian and Asian team across the design process.

discussions were more specific in the beginning and later led to the revealing of more global and cultural characteristics. These three themes are described below in further detail.

3.2.1 Values

Value seems to be the most vital of these three themes and it consistently showed up during the entirety of the process. Value can be defined as one's judgment of what is important in life. The prevalence of values in the current dataset has been extensively recorded in other DTRS11 papers. For example, using their linguistic register, Bedford *et al.* (2017) identify a heavy representation on values (119) while Eckert and Stacey (2017) refer to explicit/implicit in the form of phrases such as *"taking care of our family," "Eco friendliness," "Chinese dream"* among others. Our analysis shows that when slangs and jargons were considered in the creation of values, interestingly, they were expressed differently by SD and AE. For example, SD used exaggerated noun pairs such as *awesomely humble* (positive value) or *crappy reality* (negative value) whereas AE slangs/jargons referring to values were usually muted and in the form of idioms such as *a bowl of good life* and *it can work its magic.* Table 17.2 shows some examples of this usage.

3.2.2 User characteristics

While values are one's judgment of what is important in life, user characteristics refer to the personalities and lifestyles of users. The references to user characteristics, however, are not simplistic. In one of the papers of DTRS11, Daly *et al.* (2017), for example, observes that when discussing *"status"* as a user characteristic, it seems to be a more prominent driver in the beginning, but as the team's work continues the data from the co-creation session do not emphasize status. They observe that ultimately when recording the translation from participant data to the design decisions made by the team, it did not align with the evolved understanding of the solution requirements in the problem space.

In our study, as shown in Table 17.3, slang and jargon references to user characteristics also reveal interesting distinctions between SD and AE. While references made by AE focused on group user characteristics (*traditional-minded, community-minded, masses*), SD references were focused more on self and individual users (*autonomy, in control*).

3.2.3 Cultural references

Cultural references are references that relate to shared cultural traits, belief systems, rituals and ideas that are unique to specific groups. Interestingly, cultural references

Table 17.2 Values.

File Reference	Speaker (Team)	Transcription
(v14, 516)	E (SD)	And I think that is one of the reasons behind the success of Tesla, yeah (..) And it's like yeah, it's so: the right thing to do, eh:: because it is environmentally friendly *and I look awesome*, it is **awesomely humble** in a way, like (..) yeah, I'm doing this for the environment you can say, and people have to believe you, even though you can think you own thoughts in your head, but you can't argue against that.
(v14, 714)	R (AE)	It's just- what will (INAUDIBLE)- something else that they were talking about the advertising, the marketing message and somebody talked about the ad and said like, you can (be official?) the comparison between if you don't have the products or what (your **crappy reality** currently is) and then when you have the product it's like, you're enjoying life while other people are stuck in their **crappy reality**, so that they gives you a sense of comparison and (INAUDIBLE) other benefits versus the cost.
(v08, 294)	R (AE)	yeah that was when "**good life**" went together with society, and all four of them surprisingly talked about that, like, once there is-, they really saw it as like the whole society moving up together and having a more harmonious: kind of-
(v11, 003)	K (SD)	And eh:: we try to- to take it from top down. So, going from **good life**, and somewhere in enjoying life, the theme. So, I think we- we: found it actually pretty hard to crack open. We kind of started from scratch, but to base in the old files that you wrote, and eh tried to take something from one file, take something from another file, but I think everything turned out to be a little bit messy and we couldn't really navigate through all the things. And after a while I think we realized that we had to: base it in the last file that you wrote about enjoying life, and then: I think we wrote some things that we used. For example the beginning about where we came from to simplify the **bowl of good life** things, because it's not really the essence of eh: of the theme. So we wanted to quickly go through the- the introduction of eh of the overall- overarching theme, **good life**, and how it leads to enjoying life. So I- I don't know, should I just read it?

were made both to express one's own cultural characteristics (*Asian nationalistic-kind-of-thing, Asian-thing, Oriental mentality*) but also to probe the others (*so Asian, ultra European, super Scandinavian* Western; see Table 17.4).

In other instances, the cultural references pointed to a hierarchy of scales, that is, *national, group* and *product user* references. For example, the *national* references include *Asian-thing, Scandinavian-ness, Scandinavian Salmon, so-western, and Typical Asian*, the *group* references include *clan-guy* and *tree-hugger* and the *product user* references include *Ikea-thing* and *X-box people*.

The usage of slangs/jargons for the three identified themes (i.e., values, user characteristics and cultural references) points to the contrasting styles of applying expressiveness and restraint among Western and Asian cultures, respectively. The results show that while in today's global context the dichotomy of Western and Eastern cultural traits is blurring, the cultural differences seem to align with interesting studies

Table 17.3 User characteristics.

File Reference	Speaker (Team)	Transcription
(v08, 045)	W (AE)	some of the other things that were captured also include things like tradition. This came a little bit more from Susan and: especially Yen, so some of the more: **traditional minded** eh feelings, deeper emotions and connections and all that matter. New experiences, doing stuff that they have always wanted to do:, ah but they can't, ah: ehm: not repeating routines every day, you know, ah: for the younger guy who: recognizes that he's still quite young and new to society he's making sure that he will learn something new as quickly as possible, and as often as possible, alright? Ah: the knowledge accumulation and all that, because, that's also a very practical thing, it's- your: stepping stone to accumulation of, you know, networks, connections and you know, rising up in the corporate ladder for example.
(v07, 150)	W (AE)	Demonstrated that friend:. Okay so, they are expressing some- no, the one thing that stood out eh was about tradition. Eh and when this actually came mostly from Yen, eh I think (.) a couple of points. I think he came from a more- I think he came from a background where he stayed in like eh whole houses, where it's more **community minded**.
(v14, 472)	W (AE)	So a guy in our group were saying like, you know, if you are able to obtain a certain level of fitness or health, that is (me?) of habit, or the habit of the: **masses**. Why wouldn't you want other people know about it.
(v11, 021)	K (SD)	"Becomes a manifestation of **great autonomy**. The freedom-" maybe it should say "that: freedom of having **pockets of enjoyment** reflects a **sense of achieve'- achievement** towards a good life".
(v08, 004)	W (AE)	A little bit more yeah, so: he's got more knowledge in terms of-, he likes to feel that he's **in control**:, to think his own decision to know that, so he say "oh: don't tell me what to do, don't try to sell me stuff. I will decide what I want" so, there's a lot more: of, his own decision making from: his own network, from online and so on, and not wanting to feel like he has been pressured into making some kind of commercial decision or: purchase decision, or: don't want other people to influence his choice, ah: he likes to: find time, you know at home, well either at home or out, you know, at night, because that's the only time he can find some escape from his child:, from his wife:, and everything else, so he spoke a little about, you know, going out on a cruise in a car for an hour, on highway:.

of differences in Asian and Western cultures in terms of individualism and collectivism (Gudykunts & Ting-Toomey, 1988; Haugtvedt *et al.*, 2008; Hofstede, 2011). In Western cultures there is more prevalence of self-expression and individuality whereas in Asian cultures one can see a prevalence of formality and guarded emotional expressions. A rigid stereotyping of Western and Eastern traits might be too broad-brushed in today's heterogeneous context, but studies point out that in collectivistic cultures people tend to be group-oriented and more interdependent, while in individualistic cultures, people tend to be self-confident and more independent (Haugtvedt *et al.*, 2008).

Table 17.4 Cultural references.

File Reference	Speaker (Team)	Transcription
(v08, 296)	R (AE)	Yeah, this is like everybody going- really progressing together and then, society will be more trusting, ah more civilized more, ah: less dece'- less war safer, that was how they really saw it as.
	E (SD)	Yeah. It was like now they were part of a movement, in a way.
	R (AE)	Yeah. Which I thought was really interesting and, I don't think others-, at least I feel like this is a pretty sophisticated way of thinking, in a way.
	E (SD)	Yeah.
	W (AE)	Well it's a bit- a bit of a **Asian nationalistic kind of thing**, I thought. I feel-
	R (AE)	But even now it is a little, I mean the-
	W (AE)	But there's this big thing about **Asian nationalism** now, and the [**Asian dream:**]
	AM (AE)	[Yeah, "I drink the tea" and eh:-]
	W (AE)	Asian dream now, and they- they are starting to feel proud again because technologically brands or whatever, culturally they are getting there, so they can afford to think this way now.
(v09, 006)	E (SD)	And of course this goes back to- this is **ultra-European**, 'cause Europe had tons of revolutions and stuff like that to remove: themselves from this. Your father was a blacksmith, you are a blacksmith, that kind of thing, like no, it is not about that, it is about you as an individual, you are a blank canvas, you could do whatever you want, your past should not follow you.
(v14, 811)	W (AE)	Yeah yeah, there were two interesting part, okay one is not that special, one was really about, you know, ehm: specialty storehouse that is able to carry across an experience. So this is not just a product but it's also experience, that you need some (INAUDIBLE) on it, some extra experience before you can- you can say I buy it because of the specifications.
(v11, 243)	E (SD)	And it's important that of course we totally respect that it is: about family and it is about, you know, enduring and stuff like that, but it's also about enjoying and that is kind of living their own life to the fullest.
	AM (AE)	This concept is- I don't know what you guys feel. It's a **very Western-**
	E (SD)	Yeah, it's super: Western. "Be all you can be".
	AM (AE)	"Live to the fullest" kind of sense. (..) What'll be way too-
	W (AE)	But may'- maybe not to that extreme. I mean, now they mentioned it, maybe just to the actual of, you know, that puts you in a better: physical and mental shape to (.) do: what really matters to you, or what you ma'- what matters to you, or what you enjoy.

The use of cultural references, however, should be seen in the context of temporal aspects of the design process. As one of the DTRS11 papers suggest (Gray & Boling, 2017) cultural understandings and "making familiar" help designers reach an insight which, while not fully accurate, may satisfy the need to move forward within the constraints of time and resources. However, they observe that there is also a danger of misunderstanding in that it could encourage "quick wins" that yield rapid design insights rather than deeper study and analysis that leads to a more nuanced and fuller understanding.

Table 17.5 The role of slangs/jargons in aggregation.

File Reference	Speaker (Team)	Transcription
(v14, 472)	W (AE)	So a guy in our group were saying like, you know, if you are able to obtain a certain level of fitness or health, that is (me?) of habit, or the habit of the: masses. Why wouldn't you want other people know about it.
	AM (AE)	Yeah it's to inspire others like
	E (SD)	Inspire others, that's what-
	AM (AE)	You have to prove it first yourself, and then others hopefully will find inspiration in you.
(v14, 480)	E (SD)	Yeah (..) I think it's really cool the thing about ehm: you know you is- inspire others. That's also like a **humble thing to do**, but it means so **much about how f*ing awesome you are** (everybody's laughing)
	E (SD)	Yeah, I talked to these guys to do that:, and I inspired them to that, you know, the reason why they're doing that is because, I inspired them to do that, like, I was- I wa- I started this.
	K (SD)	But also the humble
	W (AE)	**Awesomely humble**
	AM (AE)	**Awesomely humble** (everybody's laughing)
	E (SD)	**Gold-plated humbleness**
	W (AE)	**With a medal**
(v14, 480)	E (SD)	And I think that is one of the reasons behind the success of Tesla, yeah (..) And it's like yeah, it's so: the right thing to do, eh:: because it is environmentally friendly **and I look awesome**, it is **awesomely humble** in a way, like (..) yeah, I'm doing this for the environment you can say, and people have to believe you, even though you can think you own thoughts in your head, but you can't argue against that.

3.3   Slang/jargon mutations in cross-cultural design process

While the prior section provides us insights into the usage of slangs/jargons in terms of content, how the slangs/jargons mutate during the design process is also important to understand. We found that the mutations occurred in the design process through what Dong (2007) characterizes as *aggregation* and *accumulation*. In his performative descriptions of language, Dong observes a restless nature to design concepts and for them to mature they need both aggregation (gathering and building-up) and accumulation (framing and scaffolding). Aggregation requires synthesizing diverse elements into a whole, whereas accumulation involves coming up with new ways of representation. Aggregation and accumulation have a similar role to some observations made by Menning *et al.* (2017) on how low-coherent phrases can either interrupt the current design issue and introduce a new one or these turns expand the current design issue.

3.3.1   The role of slangs/jargons in aggregation

Aggregation requires the synthesis of multiple ideas into a cohesive one. Our analysis of slangs/jargons point to two examples presented in Table 17.5 and Table 17.6. In the first example, instances of exaggerated words are used as a vehicle for aggregation – the idea of 'humble' aggregates into 'gold-plated humbleness with a medal.' This slang/jargon

Table 17.6 More examples that show the role of slangs/jargons in aggregation.

File Reference	Speaker (Team)	Transcription
(v08, 293)	K (SD)	Mhm. I like this **human stability**. I'm not sure that's the right way to say it, but I think it's more about that, because that includes both the environmental issue and how we solve that, but also for example wars:, political eh: situations, refugees, things like that. And that's all the things that we heard, and I think human stability or [something like that].
	E (SD)	[And even telling] the story that the accessory you buy or the car you buy is sourced from material from conflict-free areas for example.
	K (SD)	That's kinda sexy.
	E (SD)	Yeah! (laughter) And I think that's- those are all day is ***sprinkling the king***! 'Cause those are all the small things that are your ***sprinkling- your sprinkling the king.***
	AM (AE)	What do you **sprinkle the king** with?
	K (SD)	It's not being said here.
	E (SD)	You're ***sprinkling the king*** with one of the things is buying eh: material from conflict-free areas, that's ***just a little sprinkle***, it's one of the ***fifty sprinkles***. (laughter)
	K (SD)	One of the fifty sprinkles (..) yeah, how many **sprinkles** do we want?
	E (SD)	Only fifty.
	K (SD)	Only fifty, alright.
	E (SD)	It's the ***max-fifty sprinkle king.***

mutation can be seen as a building-up from simple commentary to recognition, where the recognition is amplified by the use of expressive words. Lloyd *et al.* (1995) refer to commentary characterized by a stable voice tone when a designer methodically works on an aspect of the problem, and recognition which is characterized by an amplified voice tone when a part of a solution 'drops' into place, alerting the designer to a 'good fitting' solution. As Lloyd *et al.* point out, these indicators happen quickly and unexpectedly – there is a build-up of information before a period that is often termed *incubation*, but there seems to be no general rules as to when these insights occur. Interestingly, the aggregation process is not dominated by one team. As seen in the example, AE starts with a rhetorical question on 'why wouldn't you want other people to know about it?' and that 'others hopefully will find inspiration in you.' This forms the starting point of the build-up. SD follows it up with 'humble thing to do', AE further reinforcing this idea and capping it with a stamp of approval with the slang 'with a medal.'

In another example from video 08, the design team uses '*sprinkle the king*' slang as an aggregation tool to describe the greater purpose of their product. Here the slang of 'sprinkle the king' is used to move from unclear concepts (human stability) through aggregation to global concepts (environmental, political, material). Once again, the slang words build up into expressive words as an aggregation tool.

3.3.2   The role of slangs/jargons in accumulation

In other examples we found the pervasiveness of slang/jargon usage as an accumulating tools. Accumulation goes beyond aggregation in that they lead to new ways of thinking. In the following example, the initial generator of '*commitment*' was explored by the usage of slangs over four days (video 7 to 12 and 14 to 17). In the preliminary descriptions of video 8 (Table 17.7), one can observe that attribution of commitment to

Table 17.7 The role of slangs/jargons in accumulation.

File Reference	Speaker (Team)	Transcription
(v08, 282)	R (AE)	Brian he called it "**commitment**". He said the English word first and then he put it more like, taking up the role or promise, so then it's a little bit softer: than- than how she is, but I think there was a general sense that they saw the importance of it and, I think, just maybe not so much like her.
(v9, 116)	K (SD)	And maybe that's alright, but then we need to maybe **spice it with the sexy commitment**
	E (SD)	(laughs)
		Yeah.
	K (SD)	No not **sexy commitment**, but- but with the **global awareness** responsibility.
(v9, 184)	E (SD)	Yeah, mhm. And also the level like "how can we:" (..) so maybe this is what I'm drawing here, it's my- the- so I have this base layer which is kind of called "I'm aware, I have a surplus, I **have commitment, I take commitment**"...
(v9, 274)	E (SD)	I almost had this golden thing up here, like advertisement wise. 'Cause like everyone can buy a diamond, **no one can buy commitment.** (laughter)..
	A (SD)	It's a **beautiful commitment**!
	E (SD)	...Yeah, it's embodied, it is yeah **tangible commitment** (laughter)
	E (SD)	Yeah yeah, yeah. And the thing is- was that we elevate the eh, the kind of the exclusivity and the **premiumness** to the- it's not tangible anymore, it is the, it's a single value,
	A (SD)	Mhm.
	E (SD)	Eh: which is far beyond the golden diamond, which is like in the excessive part.
	A (SD)	(..) Yeah.
	E (SD)	So I write here on the- yeah?
	A (SD)	But I: kind of agree with Will or whoever it was, asking if commitment and responsibility are- or **how to make responsibility and commitment sexy**.
(v9, 381)	E (SD)	Ehm:. I would- it's almost like I would like to discuss "sexy commitment" with them.
(v9, 007)	AM (AE)	So then when it comes to commitment, how:- how like- okay the question is do we want to enable that, or do we actually wanna give them also (.) the space of not having to feel that burden?
(v9, 023)	E (SD)	She was the one who said it, "**we need this commitment no matter what** ".
(v10, 106)	A (SD)	...Exactly. "Please eh:- we need it to be **environmental friendly**, it's a new:- it's a **hot spring**" (laughs)
	E (SD)	Yeah, exactly yeah, "turns out- shit! Stop the (INAUDIBLE)! (laughter) **Sexy commitment** is all- what it's all about in Europe!"
	A (SD)	So, that was how I was actually thinking it. Because, I think it's important to: get the new (.) people in already here, to- when they are forming the- or creating the company.
(v11, 350)	W (AE)	(..) **Sustainability**.
	E (SD)	(Writes on computer) (..) Okay:. Okay. It's okay? And then, **sexy commitment**. How might our product service's story enable our customers to feel enlightened, aware, responsible, and elevated in regards to the environmental sustainability?" And this is the one we are most unsecure about, so how can we:-
(v14, 346)	E (SD)	...that like why is this thumb there, it like because when your friends see that you use the product or service, they give you the thumb
		(everybody's laughing)
	E (SD)	I mean that is **sexy commitment** right there. Like wow. And then he feels like, so. (..) Eh:: and this one, which was pretty cool, was like, eh: the sun rises, which of course like is very Oriental or Asian (INAUDIBLE) the rising sun, and it was like new hope. New beginning. New possibilities.
(v11, 351)	W (AE)	Actually we were: talking about that actually during our discussion, but because (.) what is- that was my impression that when we talked about sexy commitment, right? This was (.) this all arose from the whole discussion about the environment and sustainability.

a serious quality that had the 'role of promise'. Later in video 9 and 10, team members discuss 'how to make commitment sexy?'. Here, commitment, mutates from a rational word to an affective and embodied one (sexy commitment, tangible commitment, beautiful commitment). Similar observations are made in other DTRS11 papers in the current dataset (Lloyd & Oak, 2017; Ylirisku *et al.*, 2017). For example, Lloyd and Oak observe how this concept encapsulates for the group an ethical idea about what the project initially defined as 'the good life' and how this concept is associated with complex social ideas – politics, war, refugees – and the development of its meaning especially the idea of 'being good' adds an explicitly moral tone.

The differences in expressiveness and restraint are clearly demonstrated by SD (sexy, beautiful, tangible) and AE (commitment, environment, sustainability), respectively. By framing commitment as more playful and expressive, the design team provides alternative ways of exploring the concept of commitment and connects it to related concepts such as 'global awareness' and 'surplus' value. Later on, commitment is connoted to 'environmental friendliness' and 'environmental sustainability.' Here the role of slangs/jargons is a generative one. Schön (1983) and Darke (1978) have similarly pointed out how concepts are dimly apparent at the beginning of the project, and explored as generators which mutate and change based on the merits of the concept.

4 DISCUSSION

The DTRS11 dataset provided us an opportunity to analyze the design vernacular of slangs/jargons. It complements other studies of language in design that echo a recent shift from studying precise formal language and parts of speech to more informal language (Glock, 2009; Oak & Lloyd, 2014; McDonnell, 2010). It also sheds light onto unconventional forms of language that have been gaining more importance in the recent past that include hedge words, metaphors, uncertainty and polysemy (Christensen & Schunn, 2007, 2009; Trickett & Trafton, 2009; Georgeiv & Taura, 2014). While analogies and metaphors can be considered as slangs/jargons, the study shows that slangs/jargons are much more than mere use of analogies and metaphors. They are a product of improvisation and spontaneity that other formal parts of language do not afford. Hence, studying slangs and jargons, as if they were a structured form of language can be fraught with difficulties, particularly because of the immediacy of use. As Mcnair *et al.* (2010) point out, literal speech alone does not inherently imply the level of quality within design. Tone, discourse, and timing are all contextual factors that impact how the discourse is interpreted. In this study, the use of specific transcripts in relation to the phases of the design process helps alleviate some of the contextual issues, but not completely.

While conventionally slangs/jargons are seen as detrimental to language, the study points out to some evidence of their performatory function, specifically in terms of aggregation and accumulation. Hence, one might argue that slangs and jargons are useful to the design process because they provide a heuristic to the design process. However, it will be naïve to consider that aggregation and accumulation occur in a linear order in the design process. As presented in several design process models over the last few years, the iterative nature of the design process and its subsequent 'wickedness' is well known. As has been pointed out in the past, the language of design consists of its own unique way of production, what Cross (2006) maintains as 'designerly ways of knowing, doing and being'.

Moreover, questions of aggregation and accumulation assume a focus on convergence, which might be the prerogative of relatively rational disciplines of design (e.g., engineering). Will it apply to more creative fields such as architecture and product design, where divergence might be as important as convergence? Additionally, one might argue that the dataset is more suited to the study of branding and programming rather than design proper. Clearly further studies that are focused on conceptual design phases are needed to clarify slang/jargon usage. It was also a deliberate attempt in our study to use slang and jargon as a single type without making a rigid distinction between them. Future studies can be conducted where slangs and jargons are studied separately since their specific nature might reveal more insights into the design process.

The cross-cultural analysis of slangs/jargons in this study reveals the traditional dichotomy of Eastern and Western characteristics of culture that include individual versus collective and expressive versus restrained world-views and are reflected in the slang/jargon usage. The design process also demonstrates the predominant focus on values and cultural characteristics by the Scandinavian team, and the focus on user characteristics by the Asian team, which might have been a consequence of their respective professional roles: the former as designers and the latter as cultural brokers. However, the usage of slangs and jargons was not dominated by one group alone. As demonstrated in the examples presented, there were several instances in which reciprocal communication occurred in advancing ideas. Further distinction between Scandinavian and Asian users might need to address finer nuances in sub-cultures and regional/linguistic differences that were not dealt in the study.

Finally, while considering a cross-cultural study in design, the analysis of slangs and jargons has to be considered in the light of 'language' itself as a variable. According to Smith and King (2013), persons fluent in a second language can usually think aloud in that language even while thinking internally in the oral code of their native language. Whereas certain contexts do allow designers to verbalize processes in their native language, the challenges of interpretation do influence the composition of protocols and their semantic content might be easily missed if we don't pay attention to how a specific language is constructed. This observation is made in one of the DTRS11 papers by Hess and Fila (2017) in their interesting reference to a telephone game, where user understanding was passed from one company representative to another, and the team's inability to speak the user's language forced them to rely upon their translators and technology to facilitate their user-centric and empathic understanding.

The usage of slangs and jargons can also be seen in the light of a sociogenetic perspective proposed by Jornet and Roth (2017) referring to the DTRS11 dataset, which considers thinking and communicating as different. Jornet and Roth argue that what participants in social relations say and do are not revelations of what they individually and internally think but integral aspects of a jointly produced thinking practice. They further argue that the generative nature of communication in design could be attributed to design thinking not so much as a form of natural intelligence but as a form of natural intelligibility and one which is emergent in the design process. Slangs and jargons can then be seen as a mechanism of improvisation that gives form to such emergent meanings.

One of the limitations of the present study is the lack of opportunity to directly examine the Asian lead users (AL) who might have provided us a more unfiltered window into their process. Rather, we had to contend with Asian external consultants (AE), who although performing well as effective cultural brokers, might also have come with a disadvantage of having to work with their hybrid cultural experiences. While the usage of slangs/jargons in this collaborative design process is highlighted as largely positive, the question of slangs/jargons becoming a barrier is a real one, especially when a design process does not consist of cultural brokers. It would be interesting to see how these barriers are overcome in the usage of slangs/jargons and what effect they might have on the creative process.

REFERENCES

Adams, R., Mann, L., Jordan, S., & Daly, S. (2007). Exploring the boundaries: language, roles, and structures in cross-disciplinary design teams. *Proceedings of the 7th Annual Design Thinking Research Symposium (DTRS7)*, London.

Adams, M. (2009). *Slang: The People's Poetry*. Oxford: Oxford University Press.

Ahmed, S., & Christensen, B. T. (2009). An in situ study of analogical reasoning in novice and experienced design engineers. *Journal of Mechanical Design*, 131(11), 111004.

Amari, J. (2010). Slang Lexicography and the Problem of Defining Slang. In: *The Fifth International Conference on Historical Lexicography and Lexicology*. Oxford.

Ball, L. J., & Christensen, B. T. (2009). Analogical reasoning and mental simulation in design: Two strategies linked to uncertainty resolution. *Design Studies*, 30(2), 169–186.

Barthes, R. (1968). *Elements of Semiology*. New York: Noonday Press.

Bedford, D. A. D., Arns, J. W., & Miller, K. (2017). Unpacking a Design Thinking Process with Discourse and Social Network Analysis. In: Christensen, B. T., Ball, L. J. & Halskov, K. (eds.) *Analysing Design Thinking: Studies of Cross-Cultural Co-Creation*. Leiden: CRC Press/Taylor & Francis.

Bucciarelli, Louis L. (1994). *Designing Engineers*. Cambridge, MA: MIT Press.

Chaika, E. (1980). *Jargons and language change*. Anthropological Linguistics, 22(2), 77–96.

Christensen, B. T., & Schunn, C. D. (2007). The relationship between analogical distance to analogical function and pre-inventive structure: The case of engineering design. *Memory and Cognition*, 35, 29–38.

Christensen, B. T., & Schunn, C. D. (2009). Putting blinkers on a blind man. Providing cognitive support for creative processes with environmental cues. *Tools for innovation*, 48–74.

Chomsky, N. (1965). *Aspects of the Theory of Syntax*. MIT Press.

Cross, N. (2006). *Designerly ways of knowing*. Springer, London. 1–13.

Daly, S., McKilligan, S., Murphy, L., & Ostrowski, A. (2017). Tracing Problem Evolution: Factors That Impact Design Problem Definition. In: Christensen, B. T., Ball, L. J. & Halskov, K. (eds.) *Analysing Design Thinking: Studies of Cross-Cultural Co-Creation*. Leiden: CRC Press/Taylor & Francis.

Darke, J. (1979). The Primary Generator And The Design Process. *Design Studies* (1:1) 1979.

Derrida, J. (1978). *Writing and Difference*. Alan Bass (trans.), Chicago: University of Chicago Press.

Dong, A. (2005). The latent semantic approach to studying design team communication. *Design Studies*, 26(5), 445–461.

Dong, A. (2006). How am I doing? The language of appraisal in design. In: Gero, J. S. (ed.) *Design Computing and Cognition '06 (DCC06)* Kluwer, Dordrecht, pp. 385–404.

Dong, A. (2007). The enactment of design through language. *Design Studies*, 28(1), 5–21.

Eble, C. C. (1996). Slang and Sociability. In: *Group Language among College Students*. Chapel Hill/London.

Eckert, C., & Stacey, M. (2017). Designing the Constraints: Creation Exercises for Framing the Design Context. In: Christensen, B. T., Ball, L. J. & Halskov, K. (eds.) *Analysing Design Thinking: Studies of Cross-Cultural Co-Creation*. Leiden: CRC Press/Taylor & Francis.

Eco, U. (1979). *A Theory of Semiotics*. Bloomington: Indiana University Press, 1979.

Fauconnier, G., & Turner, M. B. (2003). Polysemy and conceptual blending. In: Brigitte Nerlich, Vimala Herman, Zazie Todd, & David Clarke (eds.) *Polysemy: Flexible Patterns of Meaning in Mind and Language*. pp. 79–94.

Gentner, D., & Bowdle, B. F. (2001). Convention, form, and figurative language processing. *Metaphor and Symbol*, 16, 223–247.

Georgiev, G. V. & Taura, T. (2014). Polysemy in Design Review Conversations. *Design Thinking Research Symposium 10*. Purdue, IN.

Gray, C. M. & Boling, E. (2017). Designers' Articulation and Activation of Instrumental Design Judgments in Cross-Cultural User Research. In: Christensen, B. T., Ball, L. J. & Halskov, K. (eds.) *Analysing Design Thinking: Studies of Cross-Cultural Co-Creation*. Leiden: CRC Press/Taylor & Francis.

Gudykunts, W. B., & Ting-Toomey, S. (1988). *Culture and Interpersonal Communication*. Newbury Park, CA: Sage.

Glock, F. (2009). Aspects of language use in design conversation. *CoDesign*, 5(1), 5–19.

Haugtvedt, C. P., Herr, P. M., & Kardes, F. R. (Eds.). (2008). *Handbook of Consumer Psychology*. New York, NY. Lawrence Erlbaum.

Hess, J. L., & Fila, N. D. (2017). Empathy in Design: A Discourse Analysis of Industrial Co-Creation Practices. In: Christensen, B. T., Ball, L. J. & Halskov, K. (eds.) *Analysing Design Thinking: Studies of Cross-Cultural Co-Creation*. Leiden: CRC Press/Taylor & Francis.

Hirst, R. (2003). Scientific jargon, good and bad. *Journal of technical writing and communication*, 33(3), 201–229.

Hofstede, G. (2011). Dimensionalizing cultures: The Hofstede model in context. *Online Readings in Psychology and Culture*, Unit 2. Available from: http://scholarworks.gvsu.edu/orpc/vol2/iss1/8. [Accessed June 17, 2012].

Hsieh, H. F., & Shannon, S. E. (2005). Three approaches to qualitative content analysis. *Qualitative health research*, 15(9), 1277–1288.

Jornet, A. & Roth, W. (2017). Design {Thinking | Communicating}: A Sociogenetic Approach to Reflective Practice in Collaborative Design. In: Christensen, B. T., Ball, L. J. & Halskov, K. (eds.) *Analysing Design Thinking: Studies of Cross-Cultural Co-Creation*. Leiden: CRC Press/Taylor & Francis.

Jucker, A. H., Smith, S. W., & Lüdge, T. (2003). Interactive aspects of vagueness in conversation. *Journal of Pragmatics*, 35(12), 1737–1769.

Kleinsmann, M., & Valkenburg, R. (2008). Barriers and enablers for creating shared understanding in co-design projects. *Design Studies*, 29(4), 369–386.

Lévi-Strauss, C. (1976[1973]). Anthropologie structural deux. In Monique Layton (trans.) *Structural Anthropology, Vol. II*. University Chicago Press.

Lloyd, P., Lawson, B., & Scott, P. (1995). Can concurrent verbalization reveal design cognition? *Design Studies*, 16(2), 237–259.

Lloyd, P. & Oak, A. (2017) Cracking Open Co-Creation: Categorizations, Stories, Values. In: Christensen, B. T., Ball, L. J., & Halskov, K. (eds.) *Analysing Design Thinking: Studies of Cross-Cultural Co-Creation*. Leiden: CRC Press/Taylor & Francis.

McDonnell, J. (2010). Slow Collaboration: Some uses of vagueness, hesitation and delay in design collaborations. In *International Report on Socio-Informatics: Workshop Proceedings of the 9th International Conference on the Design of Cooperative Systems*. Vol. 7, No. 1, pp. 49–56.

McNair, L. D., Paretti, M. C., & Davitt, M. (2010). Towards a pedagogy of relational space and trust: Analyzing distributed collaboration using discourse and speech act analysis. *IEEE Transactions on Professional Communication*, 53(3), 233–248.

Menning, A., Grasnick, B. M., Ewald, B., Dobrigkeit, F., Schuessler, M., & Nicolai, C. (2017) Combining Computational and Human Analysis to Study Low Coherence in Design Conversations. In: Christensen, B. T., Ball, L. J. & Halskov, K. (eds.) *Analysing Design Thinking: Studies of Cross-Cultural Co-Creation*. Leiden: CRC Press/Taylor & Francis.

Nash, W. (1993). *Jargon: Its Uses and Abuses*. Wiley-Blackwell.

Oak, A., & Lloyd, P. (2014). 'Wait, wait, Dan, your turn': Authority and Assessment in the Design Critique. In: Adams, R. & Siddiqui, J. A. (eds.) (2016) *Analyzing Design Review Conversations*. West Lafayette, IN: Purdue University Press.

Pinch, T. J., & Bijker, W. E. (1987). The social construction of facts and artifacts: Or How the Sociology of Science and the Sociology of Technology Might Benefit Each Other. In: Bijker, W. E., Hughes, T. P. & Pinch, T. (eds.) *The Social Constructions of Technological Systems: New Directions in the Sociology and History of Technology*. p. 17.

Saussure, F. de. (1959 [1916]). Cours de linguistique generale. In: *Course in General Linguistics*. New York: Philosophical Library.

Schön, D. A. (1983). *The reflective practitioner: How professionals think in action*. New York: Basic Books.

Smith, P. & King, J. R. (2013). An examination of veridicality in verbal protocols of language learners. *Theory and Practice in Language Studies*, 3(5), 709–720.

Trickett, S. B. & Trafton, J.G. (2009) A primer on verbal protocol analysis. In: Schmorrow, D., Cohn, J. & Nicholson, D. (eds.) *The PSI Handbook of Virtual Environments for Training and Education* 1, 332–346.

Ylirisku, S., Revsbæk, L., & Buur, J. (2017). Resourcing of Experience in Co-Design. In: Christensen, B. T., Ball, L. J. & Halskov, K. (eds.) Analysing Design Thinking: Studies of Cross-Cultural Co-Creation. Leiden, CRC Press/Taylor & Francis.

Wiltschnig, S., Christensen, B. T., & Ball, L. J. (2013). Collaborative problem–solution co-evolution in creative design. *Design Studies*, 34(5), 515–542.

Chapter 18

Design {Thinking | Communicating}: A Sociogenetic Approach to Reflective Practice in Collaborative Design

Alfredo Jornet & Wolff-Michael Roth

ABSTRACT

Design thinking has been characterized as a social phenomenon in the pioneering studies of Bucciarelli and Schön. A challenge remains, however, to theorize design thinking without having to begin from the individual to arrive at the social. In this paper, we describe a sociogenetic approach, which theorizes design thinking as an inherently social communicative process that is generative of its individual aspects. Accordingly, there is but one design thinking practice that generates both persons and contexts. We use this approach to examine the reflective work of a design team aiming to develop a design and marketing concept-package for a company targeting the Chinese market. The analyses focus on the joint work the participants produce to generate, frame, and re-frame design concepts as they reflect upon prior co-creation workshops throughout the design trajectory. Our findings reveal constitutive features of design thinking as primarily social, including: (a) the dual nature of descriptions as instructional devices; (b) the irreducible receptive aspect inherent to social relations; and (c) the unity of thinking and affect. The analyses allow us to discuss design thinking not so much as a form of "natural intelligence" but as a form of natural intelligibility, a view that has implications for how researchers and practitioners may approach their own reflective practice.

1 INTRODUCTION

Existing research on reflective design practice tends to be divided between those studies that focus on individual (mental) aspects and those that focus on social (discursive) aspects (Paton & Dorst, 2011). Reconciling the individual and social approaches remains an unresolved and fascinating challenge in the current literature (Alexiou, 2010). A challenge is that when the social is indeed theorized, then this tends to be in a trivial sense – people designing together versus people designing on their own. When the social is theorized in a deep sense, it is constitutive of thinking and reasoning practices (Livingston, 2008); in other words, even when individuals do something on their own, their actions are social through and through (Vygotsky, 1989). A sociogenetic approach to design thinking shows how any individual action first is a social relation.

One difficulty concerns overcoming the division between thinking and doing generally, and between thinking and communicating in particular that still dominates much of the existing theorizing about (design) thinking. In this study, we challenge such a division by describing a *sociogenetic* perspective, which considers thinking and communicating as different manifestations of one and the same design practice (Roth & Jornet, 2017). What participants in social relations say and do are not revelations of what they individually and internally think – their "meanings" of the situation subsequently shared with others – but integral aspects of a jointly produced thinking practice that transforms individual participants and materials. Accordingly, there is a *design thinking practice* that is inherently social.

The present chapter takes a perspective diametrically different from that exposed in other chapters in this volume that approach creativity and (meta-) cognition as ultimately residing within the individual designers (Valgeirsdottir & Onarheim, 2017). The purpose of the analyses below is to show how design thinking manifests itself in the work of publicly and jointly establishing con-text relations, of framing and reframing prior accounts throughout a design trajectory. In the case analyzed in this study, the *thinking* nature of design praxis is exhibited in the fact that changes in the designers' framing and reframing of descriptions involve changes in the logical type to which the statements belong: During the first phases connections are established between identified particulars from the workshops and general narratives on consumer values; later in the trajectory of the participants' framing work, these acquire an increasingly general character, concerning first relations between concepts, and then between these and narratives about the company's knowledge and expectations. This increasing generalizing character – a process of concept formation that is characteristic of the mental–is here described as a feature of the social setting. Furthermore, rather than a purely rational process, our analyses describe the development of reflective accounts as the development of the intellectual and dispositional orientations that participants exhibit towards emerging collective *needs* or goals, thus manifesting the social nature of bodily and affective aspects of thinking that rarely make it into research publications. This is based on the late Vygotsky's Spinozist Marxian foundations, according to which the development of the means of production, the needs, and the persons are seen as three irreducible aspects of the development of consciousness (Marx & Engels, 1978). In producing a social context, the participants simultaneously produce the need or motive, and the material conditions for that production. In so doing, the participants produce themselves as the members of the unfolding design *thinking practice*; as (design) thinkers but also and at the same time as sensible designers, living participants of living thinking practice. Our findings thus add to and expand those by other authors who emphasize the role of *empathy* in design (Hess & Fila, 2017).

We begin by describing this approach that builds on the foundational work of Vygotsky. We then use the approach to examine the reflective practices of a design team intending to develop a design and marketing concept-package for a company aiming to expand its Chinese market.

2 THE {THINKING | COMMUNICATING} UNIT

Any particular view on (design) thinking has a theory of communication associated with it, for it is in *communicating* that something like thinking can be made intelligible as a real object suitable for empirical examination by practitioners and researchers alike. Semiotics, the study of signs in communication, has had increasing presence in the cognitive and the learning sciences, but has received relative little attention in the design thinking literature. When semiotic aspects are considered, design researchers tend to follow a constructivist or interpretative approach, where "social life is analyzed in terms of how individuals construct meanings and identities and so make sense of their everyday lives and interactions" (Glock, 2009, p. 9). In taking such an interpretive stance, however, we risk beginning with the individuals' thinking, who "construct meanings," and arrive at the social as a result of the individuals' co-constructing, sharing, or negotiating those meanings. But, as some critics point out, such processes can occur only when the participants already have something to be shared, while everything utterly individual could never be shared (Radford & Roth, 2011). There is therefore an *additive logic* in such approaches to the social (see also Ingold, 2015). In this logic, the thinking mind of the individual remains "an externally bounded entity ... divided off from other such minds and from the wider world in which they are situated" (p.10). People communicate between them, but what they communicate does not change them in communicating. If this were so, however, the possibility to be surprised, of sudden creative insight – a most characteristic feature of design thinking (Cross, 1997) – would be problematic. Because individual participants are described as *intentionally constructing* what they say and what they hear, sudden insights also should be the result of their intentional construction. But if designing were intentional construction, surprise and astonishment over unanticipated design results would not exist (Roth, 2016).

Our approach to the relation between thinking and communicating builds on the work of Russian psychologist Lev S. Vygotsky, who, during the last 18 months of his life, rejected his earlier work on signs as mediators (Vygotsky, 2010) and began developing his new framework, thinking with the Spinozist (rather than Hegelian) Marx, who showed that the supersensible ideal is the result of relations between sensible material things that reflect soci(et)al relations between people (Marx & Engels, 1962). Although Marx exhibited the origin of the ideal in his study of commodity exchange, the exact equivalent was shown to exist for communicative exchanges (Roth, 2006). As a result, any ideal form (e.g., higher psychological function, personality) first was a social or societal relation. This point was not lost on Vygotsky (1987). In this approach, any sensible material form in a communicative exchange – word, drawing, or hand gesture – has a supersensible dimension, which reflects a living relation between real people. Vygotsky takes up this point when, paraphrasing the philosopher L. Feuerbach, he writes: "in consciousness, the word is what ... is impossible for one person but possible for two" (p. 285), where the sound-word simultaneously resonating in the mouth of the speaker and the ear of the listener inherently is social. Even when persons write in a personal journal, the relation is social, a relation to the self as to another. Because the word is a relation with others, and the other within the self (Mikhailov, 2001), "thought is never the direct equivalent of word meanings" (Vygotsky, 1987, p. 282); thinking is "not expressed but *completed* in word" (p. 251, emphasis added). Thinking (meaning) and communicating (word, sketch), then, are not two sides of the same sign or idea, as semiotic studies in the design literature have suggested (e.g., Medway & Clark, 2003). Instead of a static entity that has two sides, one material and one immaterial, in the moving of thinking to speaking there is *change*. Thinking and communicating form a "true unity/identity" (Vygotsky, 1987, p. 251), where thinking and communicating cannot be understood independently because each is only a manifestation of a unitary and dynamic phenomenon that transcends the different manifestations. We use the form {thinking | communicating} to emphasize the provenance of thinking and communicating from this higher unit: *thinking praxis*. Rather than being mediators, signs (words) are shared forms of thought and this "language *is* consciousness that exists in practice for other people and therefore for myself" (Marx & Engels, 1978, p. 30, emphasis added). In the final paragraphs of his last book, Vygotsky included this statement in the articulation of a research agenda to come.

The beginning and endpoints of thinking and speaking lie outside the person (Il'enkov, 1977); in fact, beginning and endpoints exist in and as relations. Participants in design are not just agential subjects of practice but also and simultaneously are subject and subjected to (thus patients). As patients, they undergo events as much as producing them. Here lies the reason for our use of the term *sociogenetic*: anything that may be ascribed to individual thinking has its genetic origin in the social relation with others. A crucial implication for the study and practice of design is that, from this view, the possibility opens for the coming of an *in-sight* that is not a construction in the individual mind but which rather is *given* to the individual's experience without her having intended it, a view that aligns with current arguments of creativity as an "opening to the unknown" (Ingold, 2014, p. 124), where "imagination leads not by mastery but by submission" (p. 124).

3   EMPIRICAL BACKGROUND

In this study, a sociogenetic approach is employed for understanding the relation between (design) thinking and (design) communicating. Signs and sign complexes (communication) inherently are social, especially when employed in communicating understanding, acting upon others, and eliciting certain actions from others (Schütz, 1932). In the remainder of this paper, we use this approach to examine the joint work that a team of designers performs to develop a concept package for a European company targeting the Chinese market for accessories. The purpose of such analyses is to exhibit the nature of design thinking as an essentially social phenomenon that cannot be reduced to the thinking of individuals.

The data are drawn from the DTRS11 dataset (Christensen & Abildgaard, 2017), which documents three months of a co-design project lying "at the border between marketing and design" (p. 5). The design team's task involves conducting an exploratory study for the purpose of developing design and marketing concepts to improve a company's presence in the Chinese market of premium product accessories. In documents generated as part of their preparatory work, the designers characterize their task as involving a process of moving *from insights to concepts*. The team designs and implements co-creation workshops (CC1 and CC2) with informants, where the designers hope to find answers to questions such as: "How do [the informants] make decisions? What is important to them? How do they think? ... What stories do they pull up?" (v12, 005). To that aim, the designers organize *reflective sessions* in which they generate insights based on their experiences in the co-creation workshops. The intended end product was defined as: "[a] concept package ... that will outline and exemplify how a regional relevant holistic approach will increase take rates and brand penetration in China" (v1, 42).

The present paper is based on the method of interaction analysis (Jordan & Henderson, 1995). This involves conducting individual and joint data sessions where we screen the materials, examining the turn-taking and tool-facilitated conversational design practices. We do not interpret what the participants utterances mean or what they individually think, but, interested in the internal dynamic that drives the work from beginning to end, we follow the means and consequences of actions as these are made relevant by, and become relevant to, the participants themselves. We thereby generate ethnographically adequate descriptions, where "the ethnographer's adequate account of what natives do together must follow from the way in which the natives structure a situation to allow their participation" (McDermott, Gospodinoff, & Aron, 1978, p. 246). This requires ethnographers to hear participants as these participants hear each other.

4 THE UNITY OF INDIVIDUAL AND SOCIAL IN REFLECTIVE DESIGN: A NATURAL HISTORY OF {THINKING | COMMUNICATING}

In the following subsections, we provide a sociogenetic account of reflective design practice. We focus on history-constituting features that stem from and make visible the unity of thinking and communicating in collaborative design. We emphasize three such features, which themselves embody the unity of the individual and the social: the unity of describing and instructing (Subsection 4.1), the unity of doing and undergoing (Subsection 4.2), and the unity of thinking and feeling (affect) (Subsection 4.3). In presenting these three features, we select episodes from three different periods of the team's conceptual development, which correspond to three levels of generalization: (a) the initial emergence of statements and design concepts; (b) the framing of those concepts by means of clustering and diagrams; and (c) the re-framing of previously established concepts into new ones. To mark this dual character of our presentation, each section is entitled with regard to: (i) the aspect of the unity of the individual and the social that it makes apparent; and (ii) the aspect of the history of design development it illustrates.

4.1 {Describing | instructing} and the emergence of concepts

Design thinking has been conceptualized as an abductive process in which defining the problem space and finding its solution are not two different phases but occur concurrently (Dorst & Cross, 2001). Common ways to investigate how this process takes place have included individual protocol analyses, coding (e.g., Valkenburg & Dorst, 1998), and interviews (Paton & Dorst, 2011). Few studies have examined the situated practice of framing situations in their micro-genetic detail (but see Wiltschnig, Christensen, & Ball, 2013). In this subsection, we expand on the latter type of studies by offering a micro-genetic account of the work of generating initial descriptions (i.e. signs) of the co-creation workshops. Our analyses exhibit how this work is not first individual and then social, but inherently social.

The first example takes place during Session 8, in which the team debriefs CC1, which had taken place the day before. The participants formulate the work that is to be done during the first part of the session as a "quick sum up of some sentiments and comments that ... represent each individual" (v8, 002). The members then provide descriptions in which we find generic categories along the telling of particular anecdotes that took place during CC1.

Fragment 1

```
002 W:  ... so we've got Yen, I think he's eh:, in terms of
        his affluence and so on he's a bit more like a
        self-made man,
003 E:  Self-made man. yeah, mmm.
004 W:  A little bit more yeah, so: he's got more knowledge
        in terms of-, he likes to feel that he's in control:,
        to make his own decision to know that, so he say
        "oh: don't tell me what to do, don't try to sell
        me stuff" ...
```

The excerpt illustrates a form of *membership categorization* (Sacks, 1972), where an informant (Yen) is presented as a member of the "self-made man" category, which has implications for everything associated with him. But this categorization is not self-evident; instead, it involves *work*. The term self-made man is preceded by a clarification concerning the domain for the class category, namely "in terms of his affluence." The term is further taken up and re-marked in the next turn, where it is selected and re-produced as an invitation for Will to continue. Will further specifies that Yen has more "knowledge," that he "likes to feel more in control." These, in turn, are not given as final descriptors but are followed by yet another descriptive item: a form of ventriloquized speech in which Will speaks as if he were the informant. That is, the produced description is elaborated by means of a number of items that qualify each other. How, we may ask, does an intelligible conceptual organization among these diverse items emerge?

The fact that a given term is accented in and through being taken up in the relation (v8, 002-003) already gives clues to the social nature of the process. There is not a specification of the relation between the items, but their relation exists in and as the relation between the words and the turns at talk, which are produced with differential prosody and order for the purpose of being *intelligible* for the conversation. To competent listeners and participants alike, there is in the description a formal connection between a concrete particular from the original event being debriefed (Yen) who is described in terms of a general term, self-man made, which in turn is modified by a number of descriptors for the class. In events such as this, where an order seems to be transparent to the participants, the social and bodily (not just intellectual) nature of the work involved tends to remain invisible (Suchman, 1995). The situation changes in events where there is trouble and conversational repair work is initiated and produced. It is in those instances that the conditions for intelligibly producing and hearing descriptions become visible.

Consider the following fragment, where a stakeholder (Hans) who has participated in CC2 but only partly contributed in the rest of the trajectory remarks, as an account, how participants expressed "values" related to environmental concerns (v14-I, 38). He is invited to expand, to list some of those "particular values," an invitation that is accepted in an the articulation of a list of items: the "environmental things," the "impact ones," "that for marketing." One such item is taken up in the next turn (v14-I, 041), but the fact is marked that there were several mentions, inviting further recollecting. The request is thereby specified: not just a list of items in general, but some items "in particular." There are two more items that are not subsequently taken up. Instead, there is a long silence after which Ewan elaborates what "was interesting" (v14-I, 047–049). He describes how the informants, who had been creating an imaginary company, "talked about the location but also the building," concerned that all their activities should be sustainable and environmentally friendly.

Fragment 2
```
041 E:  So the environmental things? They mentioned several
         things in regards to the environment.
042 H:  Yeah
043 E:  [Anything in particular?]
044 H:  [But tho:se] (.) yeah (4.0) quite a lot of examples on
         the health topic.
045 E:  Yeah
046 H:  The project as such.
         (4.5)
047 E:  Any: eh:- any- cause you know, they- I think- what I
         thought was really interesting, in my group,
         was that- ...
```

In the fragment, a description provided by a stakeholder, who happens to also be an executive connected to communication and who was present and therefore may have been expected to "know" what happened during the workshop, fails to live up to the conversation. Why would Hans hesitate if he knew what happened? Apparently, having been there, and knowing and being able to describe what happened, are not the same. A difference is created in the movement from Being (there, in the workshop) to the consciousness of such Being, a movement from consciousness to consciousness of consciousness. Thus, describing involves work, which is inherently tied to the fact of conversation, of having to address a situational context that is not just incidental to but is constitutive of the knowledge being described.

In what follows after the sequence, Ewan provides an alternative account that indeed identifies a particular event *and* offers a reading thereof, which concerns "the type of [informant] thinking" that such an event exemplifies. There is here again categorization work, where a particular item and a general class, a "type of thinking" are connected. The two can now be heard as a unit-pair. Important to our discussion, this then becomes a form of modeling, a tutoring, an occasion for learning, in which *having to communicate* not just externalizes some already formed meaning, but itself is generative of whatever analysts may identify as such in the participants' discourse. In the future, the participants may develop and adjust expectations concerning how accounts of particular are to be performed. But here, not only Hans learns; Ewan, as project leader who often assumes a role of moderator, also learns to adapt his questioning practices so as to elicit discussions that generate general concepts suitable for further developing an understanding of the Chinese market. The evolution of practice here is primary, living in and through change in the participants' mutual orientations and actions.

In summarizing, we can say that explicit and implicit conceptions of communication in design tend to treat individual utterances in terms of statements that have a dual character: sensible material thing (e.g., a word) and supersensible ideal ("meaning"). The excerpts above show how designers produce descriptions by connecting yet more descriptions rather than expressing externally internal "meanings," ideas, or memories. The thinking here is not individual but social through and through. In producing initial descriptions of the CCs, the participants' produce work to create the frame that makes the description intelligible. But we do not observe two types of work here, one first establishing a frame for providing descriptions later. Rather, the frames are established in the exhibitable and exhibited *joint* work of connecting words and statements. In the act of *describing* (communicating), participants also and at the same time instruct how to hear and see the description, which in turn shapes their way of further producing descriptions. Because what participants end up saying is not up to them alone, there always is the possibility of learning in and through communicating. Describing here is not just expressing, but inheres the possibility for learning and developing.

4.2 {Doing | undergoing} and framing concepts

In the preceding subsection, we exemplify the work of producing accounts of prior being (events). We observe that, rather than descriptions being formed before speaking, there is but *one framing practice* where descriptions and their context of description emerge in and through one single social unit or event: conversation. Jointly producing descriptions is also and at the same time a process of producing attitudes and expectations towards each other and the unfolding subject matter in the meetings and as the team's work progresses. Concept-related statements emerge that the participants unproblematically refer to subsequently without requiring explicit connections between descriptors and the particular events of the workshops. The need to make such relations explicit recedes and changes into a higher logical type: It then becomes possible to *reframe*.

Framing is crucial for creative work and is associated to individuals' sudden insights, ways of *seeing-as* (Schön, 1983) that open new inquiry paths to design teams. From an interpretative or social constructivist approach, the individual thinker constructs or interprets meanings of statements, making it difficult to account for the *emergence* of sudden, unforeseen insights, an unforeseen character that otherwise is well recognized in the literature. A way to resolve this contradiction involves acknowledging that *doing* (saying) always also is *undergoing* (what one is saying). Communicating therefore never only is about what already is known or thought, but also involves a radical uncertainty. In every experience, there is "an element of undergoing, of suffering in its large sense" (Dewey, 1934/1980, p. 41). As designers communicate, they become not just subjects but also subjected to their thinking, which is the setting's thinking.

To illustrate the significance of this observation, we examine a session where concepts are connected to each other and reframed as the designers cluster insights with the help of charts and diagrams. Our analyses focus on the prospective character of the participants' descriptions.

Fragment 3

032 AM: So: the way I was thinking, well actually, you have the story, which is the overarching (.) storyline

033 E: Mhm?

034 AM: And then you have the product service experience which is the opportunity areas. So for example one plus ten thousand is part of that and there should be: three or four or five more ((*taps on center magic chart*)).

035 E: Mmm. And I'm pretty sure that these ((*pointing over first column first, then over second column*)) opportunity areas will need different types of- of way of selling it.

Fragment 4 presents a sequence in which two participants jointly produce a description. Initially, the description concerns what different aspects of the chart *stand for*. Amy formulates what she is "thinking" (v18-I, 032) by offering a tutorial on how to read the table of rows and columns that she has just drawn (Figure 18.1). But, although Amy initiates drawing and talking about the diagram, the thinking involved cannot be reduced to one single participant. Amy's initial description is responded to with eventual continuers ("mmm,") that both allow and invite her to continue exhibiting what she "was thinking." Amy's description identifies namable things in the chart. But the description does not only does that, but has a prospective character as well. Thus, for the opportunity areas, "one plus ten thousand" is one of the concepts that the team has developed throughout previous reflective sessions. The statement anticipates that "there should be three or four or five more," as Amy taps over the places where those other yet-to-be examples are to be inscribed. The same prospective character is observed in the next turn, where Ewan refers to the "opportunity areas" of the first column in the future tense (v18-I, 035). Opening with the conjunction "and," Ewan points over the second column, thereby connecting the two, and adding to Amy's formulation. In adding, Ewan accepts and acts upon the categories that Amy has offered: we can say that he accepts because we can see in Ewan an exhibition of Amy's exhibited method.

The sequence evidences the joint and material nature of the thinking process. Amy is articulating what she says she "was thinking," but Ewan in fact immediately

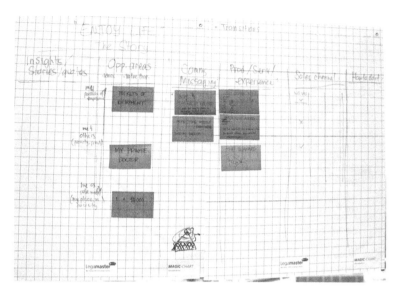

Figure 18.1 Magic chart pictured at the end of session 18.

partakes in the form of reasoning that Amy's exhibits. The immediacy in which participants jointly relate to each other and to the chart in concerted ways suggests that whatever they are doing, is not the external expression of an internal creation. Rather than each of the participants having to interpret and create for themselves representations of what the others are saying, the *situation* makes it possible for them to jointly participate without time-out for interpreting. Indeed, what we observe is a form of reasoning that leads to prospects that are beyond the current knowledge of the individual participants. Moreover, this reasoning is not just verbalized, but becomes an organizing field that orients joint action. Alternatively, we may say that both Amy and Ewan partake in an unfolding thinking practice that is being materially produced in and through communication, and which leads to new insights not from within but rather from without the individual.

The designers are producing a frame that puts the concepts in place, but they do not know whether the frame is adequate before they put the pieces in place. Precisely because the thinking is not of any single person, the individual participants cannot know what their thinking will be before they come to partake of the thinking they are producing. They do not just think but also *undergo* thinking.

Once a form of reasoning for connecting concepts becomes intelligible in and through the designers' actions, there emerges the possibility not of giving definite solutions but of asking new questions. In fact, a few turns later Amy asks whether the concept "mobility" should be located at the level of storyline or at the level of opportunity areas (v18-I, 38). Abby brings up yet another concept, "accessibility" and the sequence becomes an occasion to contrast the latter concept with "transition of activities," and to clarify whether accessibility should be categorized below the [communication] category (column 3) or as an "opportunity area" (column 1) (v18-I, 48).

It is in this circumstance that, while the designers are staring silently at the chart, a sudden insight brings about a new way of reading the chart.

Fragment 4

```
050 AM: Or it could be the other way too, this could be a
        message. (3.5)((wipes off magic chart))
051 K:  I see this (.) like a matrix; this should be here,
        so this one transcends this this this, [or maybe all]
        three of them, so maybe there are only two, so I see
        it as a matrix.
```

```
052 AM: [Ah::. Mhm]. (..) Interesting. OK, let's try again.
        ((wipes off part of magic chart))
```

Amy wipes off the second item in column 1, [transition of activity], and verbally offers an alternative: "this could be message" (v18-I, 049). Kenny then offers a new way of *seeing-as*: he sees the chart *as* "a matrix" (turn 051). The participants then display surprise, and describe this form of seeing as "interesting," and as an occasion to "try again," which here means wiping off large parts of the diagram. The members make it visible that Kenny's insight is intelligible; something all recognize immediately as Kenny articulates, and which becomes not Kenny's, but the team's new way of seeing. In fact, for Kenny to see the chart differently, he did not have to go inside but had only to look at the chart. The analyses show that participants develop ways of orienting to and expectations in and through participating in the team's conversations. We can thus expect such an attunement to underlie Kenny's seeing the chart differently. What before were listed as opportunity areas appear now under the label [insights/stories/quotes]. Members of this class will be modified in terms of all other classes (columns), which include [opportunity areas] along with others such as [communication] and [sales channels]. Precisely a form of reasoning that had emerged as a means to prospectively orient to relations between concepts as relations rows and columns in the prior turn is what affords an alternate way of organizing the team's thinking that qualitatively changes how concepts will relate to each other.

In this section, we observe changes in the material setting and the bodily orientations within it. These changes are associated with a shift in the level of generality in which insights gained from the co-creation workshops are connected. The shift exists as the material organization of the setting, which can be seen as manifesting a form of thinking that becomes think-able by virtue of its being exhibited and made recognizable (intelligible) in and through communicating. Even when we observe individual statements, these can be seen as a manifestation of the same thinking practice that the participants are jointly producing and which involve an attentive, prospective orientation to the setting, a setting the participants are as much producing as they are undergoing, sudden insights here being one more manifestation of the emerging thinking practice.

4.3   {Thinking | feeling} and re-framing

In the preceding subsection, we describe the work participants perform to frame and relate concepts to each other. As the participants produce a recognizable and instructable thinking practice, the material and semiotic conditions also change and therefore the basis on which further experiences can be had. The designers not only are active agents, but also patients of their doings, sayings, and perceiving. The evolving design thinking practice manifests not only intellectually and perceptually, but also *affectively*. To illustrate this latter point, we focus on the occurrence of *laughter* as a constitutive aspect in the team's concept development trajectory. A paradigmatic form of affective expression, laughter has been shown to have social and not just individual functions (e.g., Roth *et al.*, 2011). In the analyses presented here, we examine its relation to the collective production of conceptual development in design.

The excerpt below takes place during a session previous to CC2, in which the team members are discussing a text they have assembled about the concept "good life." As it will turn out, this concept will develop throughout the entire trajectory into another concept more inclusive or general, initially labeled as "sexy commitment," which later on will become "conscious commitment," described by one of the designers in a late session as a "collectivis[t] value that you are reinvesting in society and doing the right thing, but ... as you do this you get social recognition that would elevate your status" (v21-I, 071). However, in the excerpt "good life" still is treated as a less inclusive category that is being defined. A participant (Kenny) has just finished reading what their current definition working definition of the "good life" concept based on observations and descriptions from CC1, and Ewan now asks in plenum whether the others recognize this definition. Will takes the word.

Fragment 5

```
068  W   Eh, I just have two thoughts. one is that; I think,
         (0.5) well; my first thought is that it feels:: (.)
         a little bit=eh:: °how should I put it° (0.6)
         <mi:se:rable>.
         ((strong generalized laughter))
069  R   <<laughing>this is WHAT I FELT as well:.>
070  W   yeah BECAUSE WE=re trying to- a little to- he:y:,
         we are searching for good life, ((several utter
         acknowledgement: "mmm")) what pocket-dimension
         it is:, and you know, this is what we want:, and
         >good life< <<pretending another's voiceZ<oh let me
```

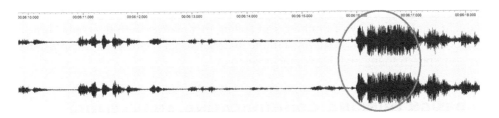

Figure 18.2 Sound wave displays abrupt shift in intensity following turn v10-068.

```
        ask you this and that>, (0.4) eh:: (0.5) but, (0.3)
        >I mean< we HAVE A LIttle bit of that, but it's a
        little bit more towards the end, it's almost like
        °oh, the pockets°, which is true,
071  E  yeah?=
072  W  =you know, but it sounded like, oh, we have so-
        no much choice that we only- can only rely on them.
073  E  Okay okay.
074  K  Mhm.
075  AM Yeah, I can see that.
```

Will formulates that he is going to describe "two thoughts," the first being that the description given of the good life concept "feels" "miserable," an assertion that leads to a generalized burst in laughter that is remarkably louder than the surrounding talk (Figure 18.2). There is then an explanation, announced with the particle "because" (v11, 070). But, by the time Will provides this explanation, turn 068 has already been treated as laughable by all and explicitly remarked as recognizable by some, one participant asserting that she "felt" the same (v10, 069), and others marking having got news, so that now they "can see that" (e.g., v10, 75).

From a formal perspective, the sequence can be described as one in which a concept, good life, is presented as a class or category, along with a definition (offered by Kenny) as a set of items that modify or belong to the class. Will then describes the same category using an item, "miserable," that results in laughter, a laughter that in turn is justified through the following turns as involving a contrast between positive expectations associated to the concept "good life," and the negative feeling being described. We observe here the genesis of the challenge to overcome their current conceptualization and its evolution through the trajectory towards the development of a new class (sexy commitment first, conscious commitment later) that is more inclusive: it allows for the contrast between enjoying life and do sacrifices for the common. The tension between these two aspects of the concept, however, does not exist as mere formalism in the social relation. In the team's consciousness, it emerges first as a lived feeling that is recognizable (and therefore shared) immediately in the conversation, without mediating interpretative work. It is recognizable and in fact undergone before it comes to be formulated explicitly in talk, as the emotive burst and subsequent appreciations show. A cognitive, conceptual leap (an act of *thinking*) exists here in and as the social

relation, which manifests as an affect at the same time as it manifests as a logical contradiction. Rather than a thought that leads to a feeling, or a feeling that leads to a
thought, there is a unity of thinking and feeling that evolves with the teams' history.
It is because the two are intertwined in the material social relation that the possibility
for concept reframing and development emerges.

5   DESIGN THINKING, COMMUNICATING, AND FEELING: DESIGN AS NATURAL INTELLIGIBILITY

In this study, we describe a sociogenetic approach to the collective achievements and
development of sign relations according to which design thinking and design commu-
nicating form an irreducible {thinking | communicating} unit. The study is part of a
research program aimed at understanding how design thinking practice can be thought
of as inherently social (Roth *et al.*, 2016), where individuals and materials shape each
other in mutually constitutive relations (Jornet & Steier, 2015). Consistent with but
also expanding prior research describing design practice as a recursive and iterative
process of framing and reframing (e.g., Wiltschnig *et al.*, 2013), we provide a *natural history of the production of descriptions* in a conceptual design trajectory. This
involves examining the embodied, irreducibly *joint* work that participants to reflective
design practice perform as they move from doing design to becoming conscious of
design, generating descriptions and sign relations that once established can further be
framed and reframed. The analyses show how a conceptual hierarchy of descriptions,
concepts, and connected concepts emerges not as the result of individual constructions
but as the lived joint work of producing them. Without this work, there would be no
material basis for the accomplishment of design practice, for if design thinking were
to be in the mind of the participants alone, it could not affect practice. Rather than
thinking individuals, we see a *design thinking praxis*, a praxis driven by a dynamic
that is internal to the group and that cannot be reduced to the autonomous individuals somehow *inter*-acting. In producing relations, participants produce the conditions
for their own existence as designers; in producing their social relations, they produce
themselves, intellectually as well as affectively.

In this study, we thus observe a movement from Being (workshop events) to
consciousness. Such a movement had been anticipated in the realization that "consciousness [Bewußtsein] never can be anything other than conscious Being [bewußtes
Sein], and the Being of men is their real life-process" (Marx & Engels, 1978, p. 28).
Because prior Being (workshop event) no longer is present, the participants need to
make it present again. But this does not involve simply retrieving and externalizing
internally stored memories. Descriptions of prior Being are *produced* accountably in
and through situated, embodied, communication work. Throughout the trajectory, initial descriptions are *reframed*, mainly by means of storytelling and the production of
clustering/diagrams. The whole trajectory may then be seen as a history of development
of propositions, where relations among statements emerge, stabilize and further transform. Instead of a *technical rational* account of such process, however, a sociogenetic
approach proposes a *natural history of the emergence and development of these relations,* a "natural history of descriptive propositions, information, injunctions, abstract
premises and the aggregate networks of such ideas" (Bateson & Bateson, 1987, p. 156).

In this natural history, any simple phrase taken up by another person in the reply not only constitutes a question, invitation, instruction, order, etc. but also constitutes the materiality of the relation between speaker and recipient (Roth & Jornet, 2017). A sociogenetic approach thus allows examining the *thinking* of design thinking not as something that primarily happens inside participants' minds as the result of being with others as a group, but as something that exists first *as* the social relation, which cannot be reduced to any of the individuals.

Taking a sociogenetic perspective implies positing and respecting the primacy of social relations, which are generative of both these relations and the individuals that form part of them. This has implications for the way designers may want to conceive their own reflective practice. Design thinking has been defined as *a natural form of intelligence* (Cross, 1998), an "ability" that can be "possessed by everyone" (Cross, 2006, p. 20). But the present study suggest that the inherently creative aspect of collaborative action may be best described not so much as a form of intelligence, but as a form of achieving *intelligibility*, a particular form of *relating* to others and to the self. In our analyses we show how features of design thinking that often are attributed to individuals (where the social is taken to be a sort of context), exist first *as* social relations. These relations, in turn, are not the product of but rather produce individual intellectual and affective dispositions and orientations. Having to communicate whatever we may say we have "thought" itself transforms thinking, and it does so because communicating never can be reduced to one, but involves a material relation with another, with the world. It is because we come to be part of changing context that is broader than our own intellect that sudden insights and affects come upon us. Individuals' creativity thus results not from their internal intentions but from "their attending upon a world in formation" (Ingold, 2014, p. 124). The findings in this paper thus suggest that design thinking practice may benefit from the practitioners' attention to *receptive* qualities that stem from being subjected to relations with others and with the environment. Competence for *responsibility, faith, anticipation* (Wong, 2007), or *astonishment* (Roth, 2016), all of which are qualities that derive from our being inherently embedded in social relations in excess of our own individual intentions, may find a more central place in future research concerned on describing what design ability is and how it can be fostered.

ACKNOWLEDGEMENTS

This study is supported by a grant co-funded by the Norwegian Research Council and the European Commission's Marie Curie Actions, grant number 240246. We also wish to thank David Socha for his valuable comments on prior versions of this chapter.

REFERENCES

Alexiou, K. (2010). Coordination and Emergence in Design. *CoDesign, 6*, 75–97.
Bateson, G., & Bateson, M. C. (1987). *Angels of fear: Towards an epistemology of the sacred.* New York: Macmillan.

Christensen, B. T. & Abildgaard, S. J. J. (2017). Inside the DTRS11 Dataset: Background, Content, and Methodological Choices. In: Christensen, B. T., Ball, L. J. & Halskov, K. (eds.) *Analysing Design Thinking: Studies of Cross-Cultural Co-Creation*. Leiden: CRC Press/Taylor & Francis.

Cross, N. (1997). Analyzing and modeling the creative leap. *Leonardo, 30*, 311–317.

Cross, N. (1998). Natural intelligence in design. *Design Studies, 20*, 25–39.

Cross, N. (2006). *Designerly ways of knowing*. London: Springer.

Dewey, J. (1980[1934]). *Art as experience*. New York: Perigee Books.

Dorst, K., & Cross, N. (2001). Creativity in the design process: Co-evolution of problem-solution. *Design Studies, 22*, 425–437.

Glock, F. (2009). Aspects of language use in design conversation. *CoDesign, 5*, 5–19.

Hess, J. L., & Fila, N. D. (2017) Empathy in Design: A Discourse Analysis of Industrial Co-Creation Practices. In: Christensen, B. T., Ball, L. J. & Halskov, K. (eds.) *Analysing Design Thinking: Studies of Cross-Cultural Co-Creation*. Leiden: CRC Press/Taylor & Francis.

Il'enkov, E. V. (1977). *Dialectical logic: Essays on its history and theory*. Moscow: Progress Publishers.

Ingold, T. (2014). The creativity of undergoing. *Pragmatics & Cognition, 22*, 124–139.

Ingold, T. (2015). *The life of lines*. Abingdon, UK: Routledge.

Jordan, B., & Henderson, A. (1995). Interaction analysis: Foundations and practice. *Journal of the Learning Sciences 4*, 39–103.

Jornet, A., & Steier, R. (2015). The matter of space: Bodily Performances and the emergence of boundary objects during multidisciplinary design meetings. *Mind, Culture, and Activity, 22*, 129–151.

Livingston, E. (2008). *Ethnographies of reason*. Aldershot, UK: Ashgate.

Marx, K., & Engels, F. (1962). *Werke Band 23: Das Kapital* [Works vol. 23: Capital]. Berlin: Dietz.

Marx, K., & Engels, F. (1978). *Werke Band 3: Die deutsche Ideologie* [Works vol. 3: The German ideology]. Berlin: Dietz.

McDermott, R.P., Gospodinoff, K., & Aron, J. (1978). Criteria for an ethnographically adequate description of concerted activities and their contexts. *Semiotica, 24*, 245–275.

Medway, P., & Clark, B. (2003). Imagining the building: Architectural design as semiotic construction. *Design Studies, 24, 255*–273.

Mikhailov, F. T. (2001). The "Other Within" for the psychologist. *Journal of Russian and East European Psychology, 39*(1), 6–31.

Paton, B., & Dorst, K. (2011). Briefing and reframing: A situated practice. *Design Studies, 32*, 573–587.

Radford, L., & Roth, W.-M. (2011). Intercorporeality and ethical commitment: An activity perspective: An activity perspective on classroom interaction. *Educational Studies in Mathematics, 77*, 227–245.

Roth, W.-M. (2006). A dialectical materialist reading of the sign. *Semiotica, 160*, 141–171.

Roth, W.-M. (2016). Astonishment: A post-constructivist investigation into mathematics as passion. *Educational Studies in Mathematics*. Doi: 10.1007/s10649-016-9733-4

Roth, W.-M., & Jornet, A. (2017). *Understanding educational psychology: A late Vygotskian, Spinozist approach*. Dordrecht, The Netherlands: Springer.

Roth, W.-M., Ritchie, S. M., Hudson, P., & Mergard, V. (2011). A study of laughter in science lessons. *Journal of Research in Science Teaching, 48*, 437–458.

Roth, W.-M., Socha, D., & Tenenberg, J. (2016). Becoming-design in corresponding: Re/theorizing the co- in codesigning. *CoDesign*. DOI: 10.1080/15710882.2015.1127387

Sacks, H. (1972). An initial investigation of the usability of conversational data for doing Sociology. In: Sudnow, D.N. (Ed.), *Studies in Social Interaction*. New York: Free Press. pp. 31–74.

Schön, D. A. (1983). *The reflective practitioner: How professionals think in action.* New York: Basic Books.

Schütz, A. (1932). *Der sinnhafte Aufbau der sozialen Welt: Eine Einführung in die verstehende Soziologie* [The phenomenology of the social world]. Vienna: Julius Springer.

Suchman, L. (1995). Making work visible. *Communication of the ACM, 38*(9), 56–64.

Valgeirsdottir, D., & Onarheim, B. (2017) Metacognition in Creativity: Process Awareness Used to Facilitate the Creative Process. In: Christensen, B. T., Ball, L. J., & Halskov, K. (eds.) *Analysing Design Thinking: Studies of Cross-Cultural Co-Creation.* Leiden: CRC Press/ Taylor & Francis.

Valkenburg, R., & Dorst, K. (1998). The reflective practice of design teams. *Design Studies, 19,* 249–271.

Vygotsky, L. S. (1987). *The collected works of L. S. Vygotsky, vol. 1: Problems of general psychology.* New York: Springer.

Vygotsky, L. S. (1989). Concrete human psychology. *Soviet Psychology, 27*(2), 53–77.

Vygotsky, L. S. (2010). Two fragments of personal notes by L. S. Vygotsky from the Vygotsky family archive. *Journal of Russian and East European Psychology, 48*(1), 91–96.

Wiltschnig, S., Christensen, B. T., & Ball, L. J. (2013). Collaborative problem-solution co-evolution in creative design. *Design Studies, 34,* 515–542.

Wong, D. (2007). Beyond control and rationality: Dewey, aesthetics, motivation, and educative experiences. *Teachers College Record, 109,* 192–220.

Chapter 19

Unpacking a Design Thinking Process With Discourse and Social Network Analysis

Denise A. D. Bedford, Jennifer Weil Arns & Karen Miller

ABSTRACT

Design thinking has become a common topic of conversation since Tim Brown published *Change by Design* (2009). The literature is rich in discussions of various approaches to design (Kimbell, 2011), case studies demonstrating the value of co-creation and the rewards of exploiting ideas in new ways, and the power of knowledge, empathy, and trust among design team members (Brown & Wyatt, 2015; Cross, 2011; Dorst, 2011; Kolko, 2015). Less attention has been given to the inner workings of design teams and the roles that factors such as intellectual capital, personality, culture, and intrinsic knowledge sharing may play during design interactions (Badke-Schaub, Roozenburg, & Cardoso, 2010). This paper seeks to shed new light on these questions from two perspectives that focus on the activities and learning experiences of the design team members. The first draws from the literature and theories of philosophy and communication. The second draws from methods and theories including discourse-linguistic analysis and network analysis. Linguistic registers, knowledge engineered semantic profiles, and network structures were constructed and applied to all DTRS11 transcripts. The results suggest that communication and language are productive perspectives for understanding the inner workings of design teams, including the nature of the team's discourse, the nature of the design process, the roles and contributions of individual team members to the process, and the evolution of the team as a community of practice. Furthermore, the results suggest that discourse analysis, semantic analysis, and network analysis may be effective tools for developing a broader understanding of design thinking.

1 INTRODUCTION

Design thinking has become a common topic of conversation in the design and business literature since Tim Brown published *Change by Design* (2009). Both provide rich descriptions of various design approaches (Kimbell, 2011), case studies demonstrating the value of co-creation and the rewards of exploiting ideas in new ways, and the power of knowledge, empathy, and trust among design team members (Brown & Wyatt, 2015; Cross, 2011; Dorst, 2011; Kolko, 2015; Hess & Fila, 2017). Less attention has been given to the inner workings of design teams and the roles that factors such as intellectual capital, personality, culture, and intrinsic knowledge sharing may play during design interactions (Badke-Schaub, Roozenburg, & Cardoso, 2010).

This paper seeks to shed new light on these questions from two perspectives that focus on the activities and learning experiences of the design team members. The first draws from the literature and theories of philosophy and communication, beginning with Peirce's concept of abduction (Peirce, 1960). Within this context, we conceptualize the design process as a complicated and somewhat messy decision (Cross, 2001) or problem space (Nelson & Stolterman, 2003; Schön, 1993) where instinct and reason function as complementary powers in co-inquiry and the attainment of knowledge. As described by Peirce, this type of reasoning shades perceptual judgment without any sharp line of demarcation. The abductive suggestion comes as a fallible flash. Although different elements of the hypothesis may have already been in one's mind, "it is the idea of putting together what we had never before dreamed of putting together which flashes the new suggestion before our contemplation" (Peirce, 1960, p. 113).

Our second perspective speaks to the methods and theories that we use to examine this process, beginning with the concepts that support and encourage discourse and linguistic analysis. Although this approach varies considerably among disciplines, it shares the premise that "language is a 'machine' that generates, and as a result constitutes, the social world [and] extends to the constitution of social identities and social relations" (Jorgensen & Phillips, 2002, p. 9). A similar premise is provided by Dong (2006) and Schön (1988) with the propositions that language and speech may be seen as formative agents and articulation as the final component of ideation (Vygotsky, 1987). The network perspective "help[s] small group researchers gain traction in investigating phenomena that have proved difficult to pin down" (Katz, Lazer, Arrow, & Contractor, 2004, p. 324) while providing a meaningful spatial element (Decuyper, Dochy, & Van den Bossche, 2010) that complements discourse and linguistic analysis of the design thinking process.

1.1   Research goals and questions

The research reported in this paper explores the feasibility of using discourse and social network analysis to explore the design thinking process and the interactions that characterize a design thinking team's ideation process. The work is also intended to provide an example of the rigorous use of these methods and a discussion of how these methods might be used in the future.

– Research Question 1: What are the characteristics of the language and discourse of this design thinking project?
– Research Question 2: Working from this perspective, what type of a design process does the DTRS11 case study represent?
– Research Question 3: Is it possible to characterize the design team's interactions using Brown's design qualities?
– Research Question 4: Is it possible to characterize the DTRS11 design team's knowledge exchange practices?
– Research Question 5: How might we describe the DTRS11 design team's interactions from a community of practice perspective?

The questions were selected in order to create an opportunity to build a deep characterization of the DTRS11 project. Given that the design process is necessarily situational and dependent upon the goals, objectives, and challenges of a given design project, we do not suggest that this characterization is generalizable. We do, however, suggest that our approach to exploring the questions may be profitably applied in design thinking research, provide a complementary look at similar research questions, and contribute to a better understanding of the design process.

1.2 Literature review

This research draws inspiration and guidance from four disciplines, including discourse analysis, semantic analysis, social network analysis, and knowledge management. Our linguistic and semantic analysis research is grounded in the early work of Fairclough (1995), Fairclough and Wodak (1997), Hart *et al.* (2005), Halliday (1978; 2004), Jorgensen and Phillips (2002), Marcu (2000), and Wagner, Greene-Havas, and Gillespie (2010). Much of the peer-reviewed literature categorized as semantic analysis falls into two categories, including: stochastic analysis using statistical algorithms to dynamically analyze bodies of text, and sentiment analysis that uses natural language processing to characterize tone or attitude expressed in text. This research employs advanced semantic approaches, including knowledge structures and knowledge engineering, part-of-speech tagging, concept extraction, and rule-based categorization (Thomas & Bedford, 2014; Bedford, 2011; Bedford & Gracy, 2012; Martin & Jurafsky, 2000; Schank & Abelson, 2013).

We have also drawn guidance from the field of knowledge management, in particular, research on knowledge sharing (Al-Alawi *et al.*, 2007; Bock & Kim, 2001; Cheng *et al.*, 2009; Chin *et al.*, 2006; Connely *et al.*, 2012; Dyer & Nobeoka, 2000; Fullwood *et al.*, 2013; Gundel, 1985; Law & Ngai, 2008; McAdam *et al.*, 2012; Powers, 2004; Riege, 2005; Ryu *et al.*, 2003; Seba *et al.*, 2012; Sandu *et al.*, 2011; Wang & Noe, 2010; Webster *et al.*, 2008), communities of practice and situational learning (Sonnenwald, 1995; Sonnenwald, 1996; Wenger, 1999; Wenger & Snyder, 2000; Wenger *et al.*, 2002), and intellectual capital profiling (Amidon *et al.*, 2005; Andriessen, 2004). Finally, our research draws on the field of social network analysis (see, e.g., Baer *et al.*, 2015; Fruchtermann & Reingold, 1991; Katz *et al.*, 2004; Obstfeld, 2005).

2 METHODOLOGY

The complexity of the design process suggests the need for mixed methods perspectives and multiple lines of inquiry. The research reported in this paper employs three: linguistic analysis; semantic analysis; and network analysis. The first of these, linguistic analysis, is the scientific study of language. In this case, we focused on the syntactical branch of linguistic analysis. This approach recognizes that language is the tool we use to communicate and to share tacit and explicit knowledge. Although there is no serious characterization of the language of discourse in the literature at this time, each linguistic subject domain has a distinct linguistic register and this analysis reveals how the DTRS11 team members use and process language. The linguistic register of language use across the 20 sessions is described in the research results for Question 1.

Semantic analysis leverages knowledge elicitation and engineering methods to construct rule-based machine profiles, including construction of grammar-based profiles (e.g., the use of adverbs to describe the "how" of design) and of categorization profiles (e.g., deep description of collaboration, experimentalism, mentoring, risk-taking behaviors). We distinguish this approach to semantic analysis from stochastic approaches. The primary distinction is that stochastic methods commonly referred to as text analytics leverage statistical approaches to discover patterns in word usage (i.e., discovering terms which co-occur with are adjacent to the word, experimentation). Semantic analyses underlying the DTRS11 research reported here involves research into the language representation of particular factors (i.e., what do the language experts tell us about what experimentalism looks like when people talk about it, and what are the language markers that tell us that one person is listening to and has understood what another person has said?), the construction of matching and discovery rules in semantic technologies, and the application of those profiles to each discourse record. Profiles are constructed distinct from the research corpus – a confounding effect that can result when using stochastic methods. Those profiles generate semantic markers that are indicators of the existence of that factor. The collective generation of markers is the source of analysis for each research question.

Network analysis describes the interactions (links) among individuals (nodes) in a network and is used to create a quantitative, objective representation of the design team's communication patterns across and within all stages of the design process. The detailed session transcripts provided the raw data of verbal exchanges that were analyzed and coded to reflect the flow of incoming and outgoing exchanges between session participants. Socio-matrices were constructed for each session from the coded communication flows. Further analysis of the matrices using the open source NodeXL Basic software from The Social Media Research Foundation (http://nodexl.codeplex.com/) allowed for construction of directed network graphs for each session. Each node or design session participant in the network graph was weighted by in-degree (the number of incoming communications) and out-degree (the number of outgoing communications). The visualization algorithm selected for constructing the network diagrams was the Fruchterman-Reingold force-placed algorithm, which produces uniform edge lengths and conceptually intuitive models representing network structures and the relationships between network nodes (Fruchterman & Reingold, 1991). Eigenvalues, which quantify the relative centrality or strength of each design session participant within a session's social network, were calculated using the open source R Project SNA: Tools for Social Network Analysis package (https://cran.r-project.org/web/packages/sna/index.html).

3   DATA COLLECTION AND ANALYSIS

3.1   Research question 1: Language and discourse characteristics

Atlas.ti (Scientific Software Development, 2016) software was used to generate a single linguistic register across all of the design sessions. The register provides a listing of

Table 19.1 Types of adverbs.

Type of adverb	Interpretation
How – manner	How the action occurs
Where or place	Where the action occurs
Time	At which time the action occurs
How much	Intensity of action
How many	How many times action occur

words used throughout the sessions, the total number of occurrences of each word across all sessions, and the number of occurrences of words in individual sessions. Each word in the register was grammatically tagged to provide a syntactical characterization of the discourse.

3.2   Research question 2: Process characteristics

Dorst's (2010) characterization of three types of design processes – deduction, induction, abduction – serves as the model for this research question. Kimbell (2011) defines abduction to include the "what" (i.e., development of values to guide the design and the establishment of a framework for design) and the "how" (i.e., the process the team follows to arrive at the result). "What" is linguistically represented by nouns. The research team reviewed the register for nouns representing values and framework elements. "How" is represented by adverbs, so the full list of adverbs was extracted from the register and categorized according to the five commonly accepted types of adverbs (Table 19.1). Similarly, the subset of adverbs of "manner" was extracted for analysis.

In addition, the full register of verbs was analyzed and classified as: (1) predication-identification – states or simple association; (2) design actions; (3) resolutionary actions; (4) marketing actions; (5) knowledge sharing actions; and (6) actions that suggested diverse opinions.

3.3   Research question 3: Optimal design thinking behaviors

This analysis draws from Brown's (2008) suggestion that design thinking is optimized in the presence of five types of skills and behaviors: empathy, integrative thinking, optimism, experimentalism, and collaboration. In this research, each of these characteristics is represented as a semantic profile. *Empathy* was interpreted to include references to others (use of pronouns), listening, sympathy, and understanding. *Optimism* was interpreted to include the use of language that had a positive tone, higher degrees of expression, and abstraction. *Experimentalism* was interpreted to include use of conditional language, projection language – supposition or suggestion, and reference to alternatives. *Integrative thinking* was interpreted to include use of comparative language, and interrogatives and questions. *Collaboration* was interpreted to include indications of collaboration, use of consensus building language, and reaffirmation language.

Each semantic profile was generated based on linguistic research into each concept. For example, the semantic profile for "Listening" was constructed around verbally expressed semantic markers that indicate the person listened to and heard what was being said, discovered through a concept extraction exercise of texts that were "about" listening. Each semantic marker was morphologically expanded to include all forms that might be found in the text. For example, the semantic profile for "Consensus" was constructed around a base set of 157 concepts, each then expanded morphologically. Each profile was embedded in the SAS Content Categorization Suite TK240 client, threshold rules for matching against texts were set and the profiles were applied to each discourse text.

A total of 14 semantic profiles were constructed and applied to explore this research question. Each interpretation was used to construct a semantic profile. Additionally, 14 semantic profiles were constructed to represent Brown's five qualities. Semantic profiles generated semantic markers or matches when applied to the transcripts of individual sessions. Both the instance of individual markers and their frequencies were recorded for data analysis.

3.4   Research question 4: Knowledge exchange practices

In addition to Brown's personality profile, the research team was interested in understanding the DTRS11 team's knowledge exchange practices. The characterization of knowledge exchange was interpreted to include mentoring, idea generation, and knowledge sharing. Three semantic profiles were constructed and applied to explore this research question.

3.5   Research question 5: Communities of practice

Research Question 5 focuses on individual team members and their interactions in a community networking context. Network analysis methods were used to graphically represent and analyze the interactions. Those sessions with the most dense knowledge exchange were assessed as communities of practice.

4   RESEARCH RESULTS

Taken together, these five research questions were expected to provide a robust test of the suitability of linguistic, semantic, and network analyses for design thinking research. Quantitative results and interpretation are presented for each question.

4.1   Research question 1: Language and discourse characteristics

The primary goal of Research Question 1 was the development of a linguistic profile that would support analysis of the discourse that occurred during the design thinking project. The first section of the analysis focuses on determining the general density of the discourse across sessions (Table 19.2). The second section takes the form of a grammatical characterization of the discourse (Table 19.3). The third section provides a semantic interpretation of the utterances found throughout the discourse (Table 19.4).

Table 19.2 Density of the discourse by session type.

Session	Session type	Discourse density	Ranking	Length (mins.)
1	Background Interview with Ewan	5,374	13	32
2	Designing co-creation workshops	4,150	16	21
3	Iterations on workshop design	9,413	5	55
4	Iterations on co-creation workshops	14,356	2	77
5	Designing co-creation workshop day 2 (CC2)	105	20	100
6	Co-creation workshop day 1	3,249	18	10
7	Debrief co-creation workshop day 1	9,245	7	18
8	Sharing insights from CC1	9,368	6	56
9	Clustering insights from CC1	8,116	8	52
10	Iterations from CC2 design	6,597	11	34
11	Linking insights from co-creation to project	2,810	19	39
12	Briefing stakeholders on CC1	6,638	10	16
13	Co-creation workshop day 2	4,955	14	10
14	De-brief Co-creation workshop day 2	15,714	1	78
15	Sharing insights from co-creation workshops part 1	9,479	4	54
16	Sharing insights from co-creation workshops 2	4,955	15	34
17	Sharing insights from co-creation workshops part 3	3,665	17	21
18	Clustering insights from CC1 and CC2	11,912	3	86
19	Recap with consultants	6,274	12	38
20	Recap with stakeholders	6,660	9	37

Table 19.3 Language and discourse of design.

Part of speech	Incidence	% Total discourse
Pronouns	24,714	30.09%
Verbs	16,399	19.97%
Adjectives	13,258	16.14%
Adverbs	13,176	16.04%
Conjunctions	7,709	9.39%
Nouns	2,704	3.29%
Interrogatives	2,704	3.29%
Interrogative Punctuation	1,459	1.78%

4.1.1   Density of the discourse

The sessions represent a range of concentration or density of language – the number of distinct words used in the text segment or fragment. The actual co-creation workshops are not at the high end of the density scale. We assumed the sessions that focused on brainstorming and creation would have demonstrated the most intense discourse, but this is not the case. The sessions which were most dense were those which focused on de-briefing, iterations, insight sharing and clustering of ideas. Sessions with stakeholders were of midrange density. It is also interesting to note that there was not a direct relationship between the length of a session and the density of the discourse.

Table 19.4 Semantic interpretation of utterances.

Expression interpretation	Frequency	% Total discourse
Agreement	137	42.95%
Disagreement	40	12.54%
Fun	35	10.97%
Uninterpretable	21	6.58%
Thinking	18	5.64%
Learning	17	5.33%
Regret	11	3.45%
And so on....	10	3.13%
Pleased	8	2.51%
Surprise	7	2.19%
Frustration	5	1.57%
Greeting	4	1.25%
Concern or Regret	3	0.94%
Not Understanding	3	0.94%
Total	319	

4.1.2   Grammatical characterization

This analysis goes to the heart of the question often asked – is Design Thinking special? Our research suggests the answer is yes (Table 19.3). In contrast to other domains and contexts where nouns are the most common part of speech, pronouns – representations of people – are the most prominent part of speech in a design process. This reinforces the critical role of and focus on people in design. Additionally, verbs – reflecting the process of design itself – rank second in incidence in the discourse. Adjectives, particularly descriptive adjectives, and adverbs, also play a prominent role. What is particularly interesting in terms of characterizing the discourse of design is the low rank that nouns assume in the discourse. The focus in this particular design project is clearly about the process and not entities or products.

4.1.3   Semantic interpretation of utterances

Linguistic analysis often extends beyond recognizable words. Although the number of utterances was not significant in this case in comparison to the total discourse (Table 19.4), we decided to conduct a high level interpretation to determine whether this approach might pertain to other research questions. The two most frequently occurring types of utterances were those related to agreement and disagreement. Together, they align with aspects of Brown's integrative thinking, as well as to the overall tone of the team's discourse. We also observed that they were suggestive of team fun, thinking, learning, expressing surprise or concerns and regrets. All of these kinds of expressions suggest some kind of knowledge exchange among the team members.

4.1.4   Initial observations

The linguistic register provided a strong foundation from which to explore the basic nature of the DTRS11 design discourse. We understand the analysis is limited to this case study; however, it confirms our understanding of the nature of design as focused on

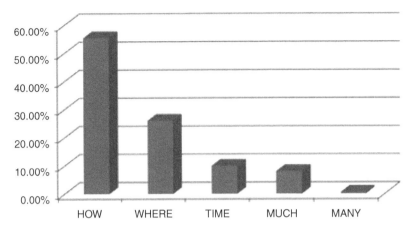

Figure 19.1 Evidence of "How" adverbs in linguistic register.

people and process. The analysis also suggested that there is no correlation between the length of a working session and the density of discourse. Discourse was denser in sessions that focused on recaps, insights, or follow-up. Although results are not generalizable to other design projects, they suggest that the discourse of design thinking may be linguistically special.

4.2   Research question 2: The design process characteristics

4.2.1   Evidence of "how" in the process

Analysis of the adverbs in the linguistic register indicates that more than 50% were adverbs of manner, referring to process and actions (Figure 19.1) – a central abduction-oriented design processes.

Analysis of the verbs on the linguistic register also indicates that 32.84% of verbs describe aspects of the design process. Figure 19.2 lists in rank order the top 100 design verbs extracted from the register. The deep focus on problem-framing verbs reflects the mental processes of the design team. The frequency of the top 20 framing verbs is an additional indication of the intensity of their use throughout the process.

4.2.2   Evidence of "what" in the discourse

The research team manually reviewed and categorized the nouns on the linguistic register, paying particular attention to values- and framing-oriented nouns. We provide sample lists of the value nouns (Figure 19.3) and the framework nouns (Figure 19.4). The research team discovered 246 framing-oriented nouns and 119 values-focused nouns. The large number of framework- and values-focused nouns is a strong indication that the DTRS11 is an abduction design effort.

need	move	moving	pick	design
make	express	probe	need	move
say	looking	thinking,	make	express
see	end	hear	say	looking
mean	run	control	see	end
thinking	add	close	mean	run
guess	connected	explain	thinking	add
said	mentioned	finding	guess	connected
mean	wrote	focusing	said	mentioned
start	help	follow	mean	wrote
needs	push	introduce	start	help
use	rest	capture	needs	push
find	said	meant	use	rest
sort	saw	pull	find	said
think	using	pushing	sort	saw
show	break	reflect	think	using
look	include	says	show	break
create	aware	writes	look	include
made	end,	aligned	create	aware
focus	fit	gain	made	end,
making	split	lead	focus	fit
understand	guess	needed	making	split
write	heard	represent	understand	guess
saying	sitting	running	write	heard
think	starting	say	saying	sitting
bring	build	showing	think	starting
keep	imagine	starts	bring	build
say	read	turn	keep	imagine
writes	sounds	written	say	read
used	speak	compared	writes	sounds
makes	building	explore	used	speak
started	continue	focused	makes	building
design	means	mention	started	continue
means	explain	pull	lead	turn
moving	finding	pushing	needed	written
probe	focusing	reflect	represent	compared
thinking,	follow	says	running	explore
hear	introduce	writes	say	focused
control	capture	aligned	showing	mention
close	meant	gain	starts	pick

Figure 19.2 Top 100 design verbs listed in rank order.

4.2.3   *Initial observations*

The linguistic analysis produced considerable evidence to suggest that the DTRS11 case study represents an abduction design process. The team's discourse is devoted to defining the design process they will follow. While there are references to previous

accessibility	commitment	endurance	humbleness	safety
accomplish-	company.	engagement	humour	satisfaction.
ment	company-brand	enjoyment.	image	security
accumulation	company-car	enlightened	incentives	service
activities	company-guy	enthusiasts	infusion.	sociability
adventure	company-	environment	integrity	society
aesthetics	product	exercise	intelligence	sophistication.
affluence	compensation	expectation	kids	stability
age	confidence	experience	liberation	stress
animals	consumption	expertise	lifestyle	superhero
aspirations	convenience	extrovert	luxury	sustainability
assurance	courage	family.	mindset	symbol
authenticity	creativity	fashion	mobility	technology
balance	credibility	food	mood-things	tree-huggers
bike	culture	freedom	mothers	unification
brand	dedication.	friends	peace	uniqueness
budget	devices	fun	pedagogs.	urbanites
business	diamonds	function	people	wastefulness
cellphones	diet	gadgets	persona	well-being
children	drivers	gratification	personalities	wisdom
choices	dudes	gut	personalization	yen-state
city	durability	guy	pollution	
club	eco-ego	gym	quality	
coffee	eco-system	health	responsibility	
comfort	emotions	home	rewards	

Figure 19.3 Value-oriented nouns extracted from the design discourse.

projects, this is a new team that needs to define its own approach. There is also evidence that the team is actively engaged in defining the values and developing a framework for the design product. This result has implications for the composition of the design team and the design team's qualities.

4.3   Research question 3: Team member characteristics

Fourteen detailed semantic profiles were constructed to represent five personality characteristics associated with design thinking: *Empathy, Optimism, Experimentalism, Integrative Thinking,* and *Collaboration.* The profiles were then used construct a picture of the play of these qualities as they could be seen in individual session discourse. Each semantic analysis produced two data sets: (1) the number of unique matching semantic markers; and (2) the average incidence of those markers. These data were analyzed to determine the presence and strength of each quality.

4.3.1   Empathy

Empathy was defined as the reference to others, evidence of listening, sympathy, and understanding. While acknowledging that additional time and effort could produce a more robust representation of empathy, of the three aspects, listening appears to be the

analogy	dialogue	hook	mindset	questions
analysis	differences	ideation	mission	refinement
approaches	dimensions	identification	models	reflection
artifact	disconnections	image	mood	relatedness
aspects	edge	imitations	motivation	relevance
aspirations	effects	immersion	nuances	representations
assessments	elements	impact	objects	rewards
assumptions	embodiment	implications	opportunities	risks
assurance	ethnography	impressions	options	sacrifice
authenticity	exemplification	infusion	outcome	satisfaction
background	s	ingredients	patterns	scenarios
balance	expectations	innovation	people	scope
benchmarks	experiences	insights	perception	stakeholders
boards	expertise	inspiration	performance	storyline
categories	explanation	instance	persona	strands
cause	expressions	integration	perspectives	strategy
challenges	facilitation	intention	phase	streams
concept	features	interaction	pillars	symbol
connections	feedback	iterations	placement	themes
containers	form	knowledge	portfolio	tone
context	foundation	labels	power	tools
customization	frame	layers	priority	traits
decisions	functionality	location	profiles	transitions
definitions	goals	manifestation	proposition	types
deliveries	habits	measurements	props	
demonstrations	hierarchy	methods	prototype	
development	history	milestones	purpose	

Figure 19.4 Nouns on linguistic register representing framework development.

weakest. Listening is more prevalent in sessions where the team is debriefing, iterating or clustering insights (Sessions 3, 4, 14 and 18). Sympathy and understanding are stronger across the sessions, but they are also highly variable. Sessions with the greatest instances of sympathy and understanding are also those where debriefing and insights are shared (Sessions 18, 14, 10) and the sessions with stakeholders (Session 10). The strongest aspect appears to be references to others.

4.3.2 Optimism

Optimism was interpreted to include the use of language that had a positive tone, a higher degree of expressiveness, and abstraction. Here, we also acknowledge that more time and effort could produce a more robust representation of optimism. However, more semantic markers related to optimism were found in the discourse than for empathy (Table 19.5). Markers for Abstraction and Expressiveness were stronger than those representing a Positive Outlook. The sessions producing greater levels of Abstraction involved iterations, sharing and clustering insights, and debriefings

Table 19.5 Semantics of optimism (use of semantic markers).

Session	Abstraction (201)	Average incidence	Expressiveness (2,063)	Average incidence	Positive outlook (153)	Average incidence
1	18	2.72	24	3.67	7	4.71
2	14	3.21	16	5.50	7	4.14
3	19	6.21	28	6.75	5	5.20
4	23	5.30	27	6.48	8	3.63
5	28	0.04	35	0.00	14	0.00
7	16	2.94	11	0.18	6	2.33
8	23	5.17	22	2.23	11	2.00
9	18	4.67	23	7.91	9	3.33
10	19	3.68	26	1.73	6	2.83
11	12	3.00	19	2.63	15	1.67
12	14	3.29	21	2.38	8	5.50
14	31	6.87	34	2.82	20	1.60
15	20	3.75	24	1.41	12	1.08
16	15	3.00	21	13.10	6	1.83
17	9	8.00	17	1.88	3	6.33
18	23	6.04	25	6.96	13	3.00
19	24	3.08	17	6.12	10	3.80
20	20	3.00	25	4.32	11	3.64

(Sessions 4, 8, 14, 18). Expressiveness was most intense when the team was sharing insights (Session 16). Positive outlook was generally weaker across sessions, but was also stronger when the team was sharing insights or working through iterations (Sessions 3, 12).

4.3.3 Experimentalism

Experimentalism was interpreted to include use of conditional language, projection language – supposition and suggestions – and language that referred to alternatives. Of the three aspects, the use of conditional language was the strongest across the team and sessions (Table 19.6). We see the same higher rates of occurrence in sessions involving sharing of insights and debriefing. Suggestions and suppositions were comparatively weaker in the discourse. Again, they were most evident in those sessions that represented sharing of insights and debriefing (Sessions 14, 18). By definition the number of alternative semantic markers (if-then, if-else, if-when, and otherwise, among others) is limited. The number of markers found appears to be low in the table but is actually proportionally higher than for the other two aspects. The incidence rate for these markers is greatest in Sessions 14 and 4.

4.3.4 Integrative thinking

Integrative Thinking was interpreted to include the use of comparative language and interrogatives. The use of comparative language was strong across sessions, but particularly evident in Sessions 4, 14, 12 and 9 (Table 19.7).

Table 19.6 Experimentalism: Conditionals and consciousness of projection (use of markers).

Session	Conditional (120)	Average incidence	Consciousness of projection (50)	Average incidence	Alternatives (36)	Average incidence
1	18	8.56	3	5.33	9	15.56
2	23	6.78	7	4.00	11	14.64
3	31	12.77	12	4.58	12	36.08
4	28	19.07	15	4.87	12	55.75
5	32	0.19	14	0.14	1	7.00
7	19	4.26	8	4.88	11	9.82
8	25	9.16	15	4.20	12	37.83
9	24	10.67	11	5.18	12	27.17
10	21	10.71	8	6.13	11	31.18
11	22	3.55	10	1.70	11	11.00
12	15	13.13	5	8.80	11	20.00
14	29	15.10	10	10.40	12	70.08
15	21	2.04	8	4.73	0	0
16	21	7.14	7	5.00	10	22.50
17	20	3.78	5	3.98	0	0
18	25	14.08	12	6.58	11	46.91
19	22	7.64	8	3.63	11	0.58
20	18	10.00	7	3.86	11	0.55

Table 19.7 Breakdown of integrative thinking (use of markers).

Session	Comparatives (242)	Average incidence	Interrogatives	Average incidence
1	23	5.35	7	176.43
2	26	7.58	7	4.29
3	25	16.20	7	8.57
4	44	9.77	7	12.00
5	31	0.00	7	25.57
7	27	6.78	7	24.86
8	39	12.15	7	8.29
9	25	13.72	7	17.43
10	23	10.39	7	5.86
11	24	3.33	7	11.14
12	17	14.94	7	22.43
14	42	17.52	7	4.71
15	29	9.23	7	8.00
16	21	9.86	7	14.71
17	22	7.45	7	10.14
18	34	11.91	7	12.14
19	25	10.24	7	15.57
20	27	9.15	7	4.14

4.3.5 Collaboration

Collaboration was interpreted to include indications of collaborative actions and activities, use of consensus building language, and reaffirmation language. The evidence for consensus building and reaffirmation was relatively weak in the discourse (Table 19.8).

Table 19.8 Breakdown of collaboration semantics (use of markers).

Session	Collaboration (127)	Average incidence	Consensus (153)	Average incidence	Reaffirmation (95)	Average incidence
1	8	4.63	4	5.00	11	2.09
2	6	4.50	4	1.50	7	1.57
3	10	5.20	6	8.00	11	3.00
4	8	8.75	8	6.13	12	5.00
5	9	0.11	11	0.00	15	0.07
7	0	7.00	2	1.00	9	2.00
8	8	6.50	10	2.30	12	2.58
9	6	6.67	5	2.40	4	2.75
10	6	13.17	6	1.83	8	3.00
11	5	3.20	6	1.17	14	0.36
12	7	3.43	2	6.50	3	14.00
14	12	5.25	12	2.25	18	3.00
15	8	2.78	8	1.67	11	2.31
16	5	3.80	6	1.17	10	3.10
17	2	1.02	5	1.24	5	2.87
18	12	5.75	11	3.36	11	4.64
19	3	12.33	7	2.14	12	2.17
20	9	3.44	5	1.40	11	2.09

Evidence of collaboration was stronger, but in general this quality was weaker for the DTRS11 team than others.

4.3.6 Initial observations

Semantic analyses suggest the presence of some strong *Empathy* qualities, but a weakness in the area of listening. *Optimism* qualities appear to be mixed – neither notably strong nor weak. The analysis also suggests the presence of strong *Experimentalism* qualities, although these are also mixed. In terms of *Integrative Thinking*, the design team appears to have strong qualities, but these results also suggest the possibility of weakness in *Collaboration* area.

4.4 Research question 4: Knowledge exchange

Semantic analysis was also used to characterize patterns of knowledge exchange, including *mentoring*, *idea generation*, and *knowledge sharing*. Three semantic profiles were constructed and applied to explore this question. The results for mentoring are summarized in Table 19.9.

4.5 Research question 5: Characteristics of engagement and interactions

Social network analysis was used to depict interactions among team members and actor engagement in knowledge flows. The social network graphs, such as the graph of

Table 19.9 Breakdown of mentoring, ideation and knowledge sharing attributes (use of semantic markers).

Session	Mentoring (29)	Average incidence	Idea generation (167)	Average incidence	Knowledge sharing (129)	Average incidence
1	1	1.00	23	3.00	16	3.63
2	1	7.00	20	3.95	17	3.65
3	1	7.00	33	4.03	19	5.89
4	1	1.00	32	11.13	20	6.45
5	1	0.00	27	0.22	2	3.00
7	1	1.00	14	5.93	12	5.17
8	1	0.00	32	6.94	17	9.24
9	1	4.00	21	10.10	17	4.06
10	1	18.00	22	8.91	13	11.69
11	1	2.00	26	2.00	19	4.95
12	1	1.00	10	16.00	13	8.62
14	3	0.00	30	13.10	21	2.62
15	1	1.00	23	5.48	24	3.79
16	1	19.00	13	26.85	14	12.71
17	1	4.00	11	16.18	12	3.08
18	1	8.00	38	4.00	25	2.16
19	1	5.00	21	1.14	21	3.10
20	1	0.00	23	15.17	17	8.00

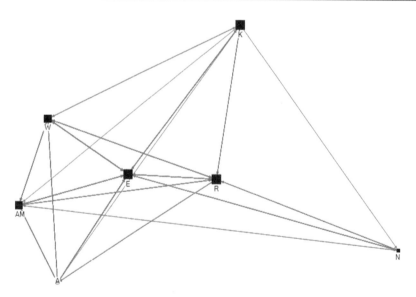

Figure 19.5 Session 8 social network graph.

Session 8 (Figure 19.5), suggest that actor interactions and engagement varied by session (vertex shapes are weighted by in-degree and vertex sizes are weighted by out-degree). Despite that variation, socio-matrices such as the Session 2 matrix in Table 19.10 generally show that, when the corporate design team leader, Ewan, participated in a session, Ewan was likely to assume a facilitator role.

Table 19.10 Session 2 sociomatrix.

Actors	A	D	E	K	N
A	0	8	24	8	0
D	6	0	11	1	0
E	19	10	0	13	4
K	6	2	19	0	0
N	1	0	7	0	0

Table 19.11 Eigenvector centrality, selected design sessions.

Role	Actors	Session 2	Session 3	Session 4	Session 7	Session 8	Session 9	Session 14	Session 18
Corp.	E	0.59	0.63	0.57	–	0.65	0.64	0.65	0.52
Design	A	0.58	0.51	0.57	0.69	0.27	0.53	0.35	0.53
Team	K	0.44	0.48	0.65	–	0.10	0.39	0.14	0.52
Outside	R	–	–	–	–	0.42	0.03	0.29	–
Consultants	AM	–	–	–	0.26	0.37	0.39	0.31	0.41
	W	–	–	–	0.68	0.43	–	0.42	–
Interns	N	0.14	0.19	–	–	0.07	–	0.09	–
	D	0.30	0.29	0.50	0.00	0.00	0.06	0.04	0.00
Stake-	T	–	–	–	–	–	–	0.25	–
Holders	H	–	–	–	–	–	–	0.07	–

Eigenvector centrality is a method of computing the importance or "centrality" of each node – in this case, a person – in a network graph. An individual's eigenvector centrality score increases as his or her direct network connections increase and as those connections increase their own connections within the network. Visual inspection of the network graph depicted in Figure 19.5, indicates that Ewan is the most central actor during Session 8, which is confirmed by Ewan's high eigenvalue centrality score as shown in Table 19.11. However, the eigenvalue centrality scores of some sessions reveal more nuanced connections between actors in the network – even revealing that the eigenvector centrality of some team members were similar to E's. For example, while visual inspection of Session 2's network graph indicated that E facilitated the session, Abby's eigenvector score is nearly equal to Ewan's. By Session 18 (the clustering of insights session), the three corporate design team members are equally central to the network and an independent consultant's eigenvector score is only 0.11 lower than Ewan's score. Based on eigenvector centrality scores, it does appear that the team achieved a community of practice as the project progressed and individual actors achieved more balanced social network centrality.

Social network analysis results reinforce the roles and interactions of team members described in the semantic analysis. Although Ewan played a core facilitator role across sessions, eigenvector centrality analysis reveals that other actors achieved alignment with Ewan's contributions as the project progressed, indicating the development of a community of practice among the individual actors.

5   RESULTS AND CONCLUSIONS

The research team developed five research questions, four of which were explored through knowledge-engineered semantic analysis and one of which was explored with social network analysis. An implicit goal was to test the viability of using semantic analysis methods to quantitatively characterize what are typically subjectively observed aspects of design work. Our results suggest that semantic analysis methods are effective in surfacing and quantifying elements that might not be amenable to human observation.

Our linguistic and semantic analyses suggest that the DTRS11 project can be considered an abduction design process. The team's discourse is devoted to defining the design process they will follow. While there are references to previous projects, this new team appears to be defining its approach during the sessions. *Process* and *how* adverbs are central to this abduction-oriented design processes. The linguistic register identifying adverbs characterized more than 50% as adverbs of manner ("how"). There is also evidence that the team is actively engaged in defining the values and developing a framework for their product. The linguistic register demonstrated a heavy representation on values- and framing-oriented nouns (119 and 249 nouns, respectively). These results support categorizing the DTRS11 project as an abductive design process.

Based only on our subjective human interpretation of the team's verbal and physical interactions, we initially thought that the team would be dominated by one or two individuals, perhaps to the detriment of the whole team's experience and the end game design work. However, review of the quantitative characteristics of the team – paying particular attention to Brown's desirable qualities of design team members – caused us to shift our thinking. We ultimately came to understand that this was a complex abduction design process that required strong experience and expertise from the design team.

The research team was also interested in the knowledge exchange practices among the team members. Given the flow of the DTRS11 design project, we initially assumed the sessions that focused on brainstorming and idea creation would have demonstrated the most intense discourse and knowledge exchange. However, this was not the case. The sessions that were most dense in terms of verbal exchange were those which focused on de-briefing, iterations, insight sharing, and clustering of ideas – demonstrating a rich exchange of knowledge. However, it was also observed that a small set of the design team members actively contributed to the exchanges. The semantic analyses further suggest that mentoring was weak across the sessions – in some cases no instances of mentoring language were detected. In contrast, idea generation is very strong although it varies across sessions. Knowledge sharing is moderately strong but also varies across sessions.

Regarding the analysis of Brown's optimal design thinking skills and behaviors, we offer several observations. *Sympathy* and *Understanding* were found across the sessions, though they were variable. The greatest instances of *Sympathy* and *Understanding* were observed in debriefing sessions and settings where insights were being shared. Semantic markers for *Abstraction* were also found in sessions which involved iterations, sharing, clustering of insights, and debriefings. Similarly, *Expressiveness* was most intense when the team was sharing insights. Of the three types of semantic markers used to represent *Experimentalism*, conditional language was the strongest across the team and all sessions. *Integrative Thinking* was most evident in the team's use of comparative language. This evidence was found across all sessions, but was stronger in sessions focused on debriefing and sharing of insights. Evidence for *Collaboration* in the form of actual collaborative language, consensus building, and reaffirmation was relatively weak throughout the design process. This was initially surprising to the research team. However, it was a reasonable observation considering the abductive nature of the project and the levels of experience and expertise among the team members.

Regarding the research methodology, the linguistic analysis was straight-forward and revealing. Assuming that DTRS11 is representative of design efforts, the discourse of design is clearly different than the discourse of other domains. If this holds for other design examples, design discourse is a domain worthy of further exploration.

Brown's personality profiles presented a challenge for knowledge engineering and semantic analysis. The semantic profiles were sufficiently representative to highlight individual variations, strengths, and weaknesses. While the semantic profiles were mentally intense to create, they provided a quantitative and objective tool for consistent assessment of the team's and individual's discourse. The availability of detailed transcripts of the design discourse enabled social network analyses. This aspect of the research allowed us to see how the team interactions evolved over the course of the project and provided an orthogonal view of team interactions and relationships – demonstrating the evolution of a community over time.

5.1 Future work

The DTRS11 shared materials provided a rich context for discourse and network analyses. However, there are several additional aspects of the Design Team interactions that could have been explored had time and page limitations not argued for a more parsimonious approach. These include, but are not limited to, factors related to the adoption of concepts and ideas across and by team members over the period of the project (e.g., initial and subsequent use of concepts across sessions and by team members), the use of framework and design concepts across sessions by team members, and the use of unique concepts and terms by individual Design Team members. The likely influence of interpersonal skills and behaviors also suggests opportunities for deeper definition. Lastly and quite importantly, little light could be shed using the materials available on the relationship between Design Team practices and the *effectiveness of the design process*. Additional and detailed descriptions of the final products and records of user feedback might address this problem and lead to valuable conclusions in the future concerning the relationship of values, framework, personal qualities, and team processes to the quality and acceptance of design thinking products.

REFERENCES

Al-Alawi, A, Al-Marzooqi, N. Y., & Mohammed, Y. F. (2007). Organizational culture and knowledge sharing: Critical success factors. *Journal of Knowledge Management, 11*(2), 22–42.

Amidon, D. M., Formica, P., & Mercier-Laurent, E. (Eds.). (2005). *Knowledge economics: Emerging principles, practices and policies*. Tartu, Estonia: University of Tartu, Faculty of Economics and Business Administration.

Andriessen, D. (2004). *Making sense of intellectual capital: designing a method for the valuation of intangibles*. London: Routledge.

Badke-Schaub, P., Roozenburg, N., & Cardoso, C. (2010). Design thinking: a paradigm on its way from dilution to meaninglessness. *Proceedings of the 8th Design Thinking Research Symposium (DTRS8)*. pp. 39–49.

Baer, M., Evans, K, Oldham, G. R., & Boasso, A. (2015). The social network side of individual innovation: A meta-analysis and path-analytic integration. *Organizational Psychology Review, 5*(3), 191–223.

Bedford, D. A. D. (2011). Using semantic technologies to analyze the semantic orientation of religious sermons – A validation of the early work of McLaughlin. *Advances in the Study of Information and Religion (ASIR), 2011*, 68–91.

Bedford, D. A.D, & Gracy, K.F. (2012). Leveraging semantic analysis technologies to increase effectiveness and efficiency of access to information. *QQML, 1*, 23–36.

Bock, G. W., & Kim, Y. G. (2001). Breaking the myths of rewards: An exploratory study of attitudes about knowledge sharing. *Pacis 2001 Proceedings*, 78.

Brown, T. (2008). Design thinking. *Harvard Business Review, 86*(6), 84.

Brown, T. (2009). *Change by design: How design thinking transforms organizations and inspires innovation*. New York: Harper.

Brown, T., & Wyatt, J. (2015). Design thinking for social innovation. *Annual Review of Policy Design, 3*(1), 1–10.

Cheng, M. Y., Ho, J. S. Y., & Lau, P. M. (2009). Knowledge sharing in academic institutions: A study of Multimedia University Malaysia. *Electronic Journal of Knowledge Management, 7*(3), 313–324.

Chiu, C. M., Hsu, M. H., & Wang, E. T. (2006). Understanding knowledge sharing in virtual communities: An integration of social capital and social cognitive theories. *Decision support systems, 42*(3), 1872–1888.

Connelly, C. E., Zweig, D., Webster, J., & Trougakos, J. P. (2012). Knowledge hiding in organizations. *Journal of Organizational Behavior, 33*(1), 64–88.

Cross, N. (2001). Designerly ways of knowing: Design discipline versus design science. *Design Issues, 17*, 49–55.

Cross, N. (2011). Understanding how designers think and work. *Design thinking*. New York, NY: Bloomsbury Academic.

Decuyper, S., Dochy, F., & Van den Bossche, P. (2010). Grasping the dynamic complexity of team learning: An integrative model for effective team learning in organizations. *Educational Research Review, 5*, 111–133.

Dong, A. (2007). The enactment of design through language. *Design studies, 28*, 5–21.

Dorst, K. (2010). The nature of design thinking. *Proceedings of the 8th Design Thinking Research*. [Online] Available from: from http://www3.nd.edu/~amurniek/assets/DTRS8-Dorst.pdf

Dorst, K. (2011). The core of 'design thinking' and its application. *Design studies, 32*(6), 521–532.

Dyer, J. H., & Nobeoka, K. (2000). Creating and managing a high-performance knowledge-sharing network: The Toyota case. *Strategic Management Journal, 21*(3), 345–367.

Fairclough, N. (1995). *Critical discourse analysis: The critical study of language*. London: Longman.

Fairclough, N., & R. Wodak, (1997). Critical discourse analysis. In T. van Dijk (ed.), *Discourse as Social Interaction. Discourse Studies: A Multidisciplinary Introduction, vol. 2*. London: Sage. pp. 258–284.

Fruchtermann, T. M. J., & Reingold, E. M. (1991, November). Graph drawing by force-directed placement. *Software – Practice and Experience, 21*, 1129–1164.

Fullwood, R., Rowley, J., & Delbridge, R. (2013). Knowledge sharing amongst academics in UK universities. *Journal of Knowledge Management, 17*(1), 123–136.

Gundel, J. K. (1985). Shared knowledge and topicality. *Journal of pragmatics, 9*(1), 83–107.

Halliday, M. A. K. (1978). *Language as social semiotic.* Arnold: London.

Halliday, M. A. (2004). The spoken language corpus: a foundation for grammatical theory. *Language and Computers, 49,* 11–38.

Hart, C., Rymes, B., Souto-Manning, M., Brown, C., & Luke, A. (2005). Analysing political discourse: Toward a cognitive approach. *Critical discourse studies, 2*(2), 189–201.

Hess, J. L. & Fila, N. D. (2017) Empathy in Design: A Discourse Analysis of Industrial Co-Creation Practices. In: Christensen, B. T., Ball, L. J. & Halskov, K. (eds.) *Analysing Design Thinking: Studies of Cross-Cultural Co-Creation.* Leiden: CRC Press/ Taylor & Francis.

Jorgensen, M., & Phillips, L. J. (2002). *Discourse analysis as theory and method.* Thousand Oaks, CA: Sage Publications.

Katz, N., Lazer, D., Arrow, H., & Contractor, N. (2004). Network theory and small groups. *Small group research, 35*(3), 307–332.

Kimbell, L. (2011). Rethinking design thinking: Part I. *Design and Culture, 3*(3), 285–306.

Kolko. J. (2015, September). Design thinking comes of age. *Harvard Business Review.* Retrieved from https://hbr.org/2015/09/design-thinking-comes-of-age.

Law, C. C., & Ngai, E. W. (2008). An empirical study of the effects of knowledge sharing and learning behaviors on firm performance. *Expert Systems with Applications, 34*(4), 2342–2349.

Marcu, D. (2000*). The theory and practice of discourse parsing and summarization.* Cambridge, MA: MIT Press, 2000.

Martin, J. H., & Jurafsky, D. (2000). Speech and language processing. *International Edition, 710.*

McAdam, R., Moffett, S., & Peng, J. (2012). Knowledge sharing in Chinese service organizations: A multi case cultural perspective. *Journal of Knowledge Management, 16*(1), 129–147.

Nelson, H. & Stolterman, E. (2012). *The design way: Intentional change in an unpredictable world* (2nd ed.). Cambridge, MA: MIT Press.

Obstfeld, D. (2005, March). Social networks, the tertius iungens orientation, and involvement in innovation. *Administrative Science Quarterly, 50*(1), 100–130.

Peirce, C. S. (1960). *Collected papers volume V: Pragmatism and pragmaticism.* Cambridge, MA: Harvard University Press.

Powers, V. (2004). Virtual communities at Caterpillar foster knowledge sharing. *T AND D, 58,* 40.

Riege, A. (2005). Three-dozen knowledge-sharing barriers managers must consider. *Journal of Knowledge Management, 9*(3), 18–35.

Ryu, S., Ho, S. H., & Han, I. (2003). Knowledge sharing behavior of physicians in hospitals. *Expert Systems with applications, 25*(1), 113–122.

Schank, R. C., & Abelson, R. P. (2013). *Scripts, plans, goals, and understanding: An inquiry into human knowledge structures.* Hillsdale, NJ: Psychology Press.

Schon, D. A (1988) Designing: Rules, types and words. *Design Studies, 9,* 181–190.

Schon, D. A. (1993). *The reflective practitioner: How professionals think in action.* New York, NY: Basic Books.

Seba, I., Rowley, J., & Delbridge, R. (2012). Knowledge sharing in the Dubai police force. *Journal of Knowledge Management, 16*(1), 114–128.

Sonnenwald, D. (1995). Contested collaboration: A descriptive model of intergroup communication in information system design. *Information Processing and Management, 31*(6), 859–877.

Sonnenwald, D. (1996). Communication roles that support collaboration during the design process. *Design Studies, 17*(3), 277–301.

Thomas, M., & Bedford, D.A.D. (2014). A case study in the effective use and quantitative validation of performance advisers. *Problems of Management in the 21st Century,* Vol. 9.

Vygotsky, L. S. (1987). *The collected works of L. S. Vygotsky, vol. 1: Problems of general psychology.* New York, NY: Springer.

Wagner, L., Greene-Havas, M., & Gillespie, R. (2010). Development in children's comprehension of linguistic register. *Child Development*, *81*(6), 1678–1686.

Wang, S., & Noe, R. A. (2010). Knowledge sharing: A review and directions for future research. *Human Resource Management Review*, *20*(2), 115–131.

Webster, J., Brown, G., Zweig, D., Connelly, C. E., Brodt, S., & Sitkin, S. (2008). Beyond knowledge sharing: Withholding knowledge at work. *Research in personnel and human resources management*, *27*, 1.

Wenger, E. (1999). *Communities of practice: Learning, meaning, and identity*. Cambridge University Press.

Wenger, E. C., & Snyder, W. M. (2000). Communities of practice: The organizational frontier. *Harvard Business Review*, *78*(1), 139–146.

Wenger, E., McDermott, R. A., & Snyder, W. (2002). *Cultivating communities of practice: A guide to managing knowledge*. Harvard Business Press.

Part V

Framing in Design

Design Roulette: A Close Examination of Collaborative Decision Making in Design From the Perspective of Framing

Janet McDonnell

ABSTRACT

Examining what designers say as they work together, this chapter uses data from a single design meeting to present an account of the design process as a failure to benefit from the potential of design framing to support design moves and reflection. It draws attention to the role of framing in shaping design, exposing some aspects of the skills that operationalizing framing effectively entail to contribute to better understanding the role of disciplined framing in characterizing expert design practice.

I INTRODUCTION

Design research has an established history of using the concept of framing as a way of understanding some of what goes on at individual and team levels during a design process. Framing is one way of conceptualizing the necessity in design of imposing order on complex, uncertain, unstable situations, in order to develop an intervention to serve some purpose(s). Framing is a concept that acknowledges that designers select what they "will treat as the 'things' of the situation, [we] set the boundaries of [our] attention and [we] impose upon it a coherence which allows us to say what is wrong and in what directions the situation needs changing" (Schön, 1983, p. 40). Within a reflective practice paradigm for designing, situations are 'framed' and the things that will be attended to within that framing are 'named'. Decisions, usually termed design moves so as to embrace experiment and excursions that can be revoked, ensue within a particular framing. Reflection on moves within the discipline a frame imposes supports assessment of moves' values, leads to further moves within frame and can also lead to appraisal of the utility of a frame and thus to reframing.

Framing is a concept that serves at many different levels of granularity. However, giving an account of a design process by calling upon the concept of framing does not imply frame awareness on the part of those engaged in it. Frame awareness is advocated as a skill for serving society in dealing with intractable, controversial problems through enabling collaboration between parties whose world views are incompatible (Schön & Rein, 1994). Furthermore, procedures in which framing is explicitly instigated have been advocated as a means to innovate (Dorst, 2015). The better understanding of design processes has also been served by looking at framing at meso- and micro-levels. Studies that use the notion of framing as an instrument for understanding what is happening in team-work show that successful establishment of shared frames can effect positively design progress and the quality of outcomes (e.g., Valkenburg, 2000). This is hardly surprising at one level because shared perspectives (framing) on what the task is, and how to proceed effectively with it, are likely to serve collaborators better than not sharing a common conception of the situation to be addressed. However, we also know that agreement can lead everyone happily down the wrong track, so a shared perspective itself is insufficient for effective outcomes. Positive appraisal can be associated with lack of critical inspection of assumptions, whereas negative appraisal invites justifications which are subject to scrutiny (Dong *et al.*, 2009).

Whist using the concept of frames to inspect a design process may have explanatory value, the identification of framing phenomena are not, of themselves, indicators of the 'health' of a design process. This is rarely acknowledged. Not all frames are good frames. The value of a certain framing is established post-hoc. 'Good' framing supports effective progression with designing; it includes frames that are discarded having invoked moves and reflection that increase understanding of what is required and what is possible. In an era when the adoption of certain characterizations of 'design thinking' is being advocated as a readily acquirable life skill (Brown, 2009; Stanford University, 2016), it is important for design research to contribute to better clarification of what goes on during designing, and what particular skills experienced designers draw upon. And to what extent these skills rest on the mobilization of domain knowledge and varieties of expertise. This chapter presents an interpretation of one of the meetings from the DTRS11 dataset based on imposing the notion of framing on the design team's interactions to draw attention to some aspects of framing practice.

2   DATA AND NOTATION

The DTRS11 dataset tracks a project to deliver a design concept within a large international manufacturing company. The project's aim is to develop concepts for a product range specifically targeting emerging markets in China. The brief is to deliver a package comprising products, sales channels, and marketing, that will increase brand penetration in China. Video recordings and transcripts from meetings to design two co-creation workshops form part of the dataset. The co-creation workshops are to engage 'lead users' in activities to generate material for use by a team from the company's user involvement department and *their* internal clients, the company's special products unit, in delivering the concept package. It is the user involvement department team that are designing the two co-creation workshops.

The description provided here assumes the reader is familiar with the dataset description (Christensen & Abildgaard, 2017) This chapter analyses meeting v5 (duration 110 minutes), which is concerned with the design of a co-creation workshop (CC2). A preceding workshop (CC1) has been largely designed by the time of meeting v5 and some details of this second workshop, CC2, have already been discussed in outline. The interpretation presented in the following makes use of the numbered turns-at-talk in the dataset transcript preceded by a speaker identifier (e.g., [A:990]); direct quotations from the transcript are italicized. The term 'participant' refers to someone present at the meeting; the term 'respondent' refers to someone taking part in the co-creation workshop being designed.

3   METHOD

The approach to the research comprised two distinct phases. First, survey and selection of candidate material of potential interest from the DTRS11 dataset. Second, detailed examination of the selected data, meeting v5, by moving attention between the transcribed data and successive constructed interpretations of these data. In the first phase all of the DTRS11 data were viewed, summarized, and assessed for their potential for a study focusing on how designers shape, or fail to shape, their task by various means which are apparent in their verbal interaction.

3.1   Survey and selection

The surveying of the dataset was conducted from a standpoint which included the following interests: prior research on a number of phenomena including accounts of *frame conflict* (Schön & Rein, 1994) and its consequences for design collaboration (Stumpf & McDonnell, 2002; Valkenburg & Dorst, 1998); the creative potential of (self-) *imposed design constraints* (McDonnell, 2011; Biskjaer & Halskov, 2014); *constraint balancing* (generators vs. obstacles) (Onarheim & Biskjaer, in press)); and expert designers' capacity to use incommensurate requirements to generate novel designs through *re-framing* (Dorst, 2015). This broad range of sensitivities served to focus attention initially but did not act as a simple filter on the dataset. There was interplay between successive readings of the data, the set of interests, and the lenses these afforded. Several potential lines of enquiry were hypothesized and discarded as familiarity with the data intensified. The process of survey and selection resulted in the identification of a sub-set of material for close inspection and a refinement of the constructs with which to pursue interpretation; the outcome was the decision to focus on constructing an account of meeting v5 in terms of framing.

3.2   Detailed interpretation

Two of the resulting interpretations are included here. Section 4 is a descriptive, chronological account that sets up the interpretation in Section 5, which makes use of prior notions about designers' uses of framing. Instruments for close examination of the data were not determined a priori. The emerging concern to give an account of progress in terms of a failure to establish effective framing drove the selection of instruments. The interpretation in Section 5 uses some content analysis: counting turns where certain terms are mentioned and comparing the number of occurrences across participants. As the numbers of turns involved are small ($N < 100$) it has been possible to inspect the conversational context in which each instance appears. Interpretation in Section 5 relies on this. A comparison of the transcript with the video recording of the meeting shows that a finer grained transcript would include further turns at talk if turns containing para-linguistics (e.g., 'uhuh') were transcribed. Therefore, turn counts are used directly so that measures are not compromised by the omission of paralinguistic turns. Multiple references to a term within a turn at talk are counted as a single turn reference. Turns at talk are relatively short as the meeting does not have any monological passages. In building a plausible account, descriptions appeal to evidence acceptable from the perspective of the tradition of conversation analysis (ten Have, 2007) in which the meanings of turns at talk are constituted by their practical effects among interlocutors; this relies on inspection of the sequences of turns at talk.

4 INTERPRETATION: DESCRIPTIVE CHRONOLOGICAL SUMMARY

The participants, a team from the 'User Involvement Department' of the company, are assembled in a room at their work location to design the second of two co-creation workshops, that will be held in China with a small group (N = 9) of 'respondents' who are Chinese citizens representing a specific demographic. A first workshop, CC1, has already been designed by the team (leader Ewan, Abby and Kenny) and shared with consultants engaged for their expertise in Asian markets. These consultants will facilitate CC1 and CC2. Ewan, Abby and Kenny know each other well and have worked together frequently in the past. A researcher, David, is also present and contributes occasionally.

4.1 First half hour

Post-it notes from previous sessions are arrayed in the room. The meeting starts with team members looking at these, Kenny refers to there being some '*start things*' [K:007]. The start things are that each group (respondents who have worked in two groups during CC1) will have its own '*little company*' [K:013]; and the conceit for CC2 is that respondents will be challenged [E:023] to express ideas for a product (where to sell, price, its story) [K:015]. The task for the team is to design how this will happen. Within the first five minutes it becomes apparent that an additional requirement is to assimilate two new respondents joining CC2 but not present at CC1 [A:047]. Talk over how to make use of the newcomers and integrate them surfaces problems with what actually will be the starting point for CC2 – what constitutes the '*little company*' on the post-it notes.

Just under 10 minutes in Abby [A:085] points this out, '... *we're ending here with the 'now imagine you're a CEO in a company wanting to invest in these themes' so now I'm thinking we haven't made the companies yet, so they can't present a company ...*'. The next 20 minutes are spent trying to address several incommensurate requirements including: how to integrate the newcomers; maximize the benefits of the newcomers' inputs; establish focused proposals variously described as a company, theme, product; and involve respondents (old and new) in a process to achieve focus. We are 30 minutes in, there is no breakthrough, part of what was previously decided about CC1 is now in doubt.

There is evidence that the team are aware that there is little progress. Here are two examples: Abby [A:164], referring to CC1 says, '*but we don't have any more time* (in it) *that's the problem*', followed by Ewan [E:165] saying, '*no no so I – no I'm just saying something needs to go ...*'. There ensues consideration of redesigning some of CC1, that is, back-tracking on the '*start things*' with which the meeting opened. Ewan [E:175] says, '... *maybe I'm repeating myself I'm coming back to is it possible that they come here and we spend I don't know half an hour forty minutes on them becoming a company ...*'. However, this proposal creates adverse conditions for dealing with some other requirements.

4.2   Second half hour

After the first half hour there is an eight minute break. Recording resumes with a change of focus: Ewan starts to document on the white board practicalities for the start of CC2, [E:283] *'welcome and split out'*. 'Split out' refers to the separation of the newcomers from the other respondents so that they can complete an ice-breaking exercise that the original respondents completed during CC1. Discussion is entailed by practicalities such as: how introductions will be effected; how much time to spend on things including re-cap on CC1; seating arrangements; who will be in each of two groups; and so on. This is relatively easy ground and runs smoothly until Kenny [K:533] raises something they have apparently discussed at an earlier occasion, that is, whether to change respondents' group membership for CC2 from that established for CC1. This is dismissed by Ewan at this point but raising the matter reminds the team of the *'major issue'* [E:557] unresolved in the first half hour, namely, at what stage of 'company establishment' the workshop respondents will be at the start of CC2. The first hour concludes with documentation of the plan of activities for CC2 up to noon with *'pitch and investment'* [K:562] noted as the place-holder for this problematic. This encapsulates (McDonnell, 2012) the still undecided workshop activities, which must deliver the prioritised ideas that respondents will develop through co-creation activities.

4.3   Start of second hour

It rapidly becomes apparent that there is a conflict between how CC1 concludes – designed to leave respondents able to spend the week between CC1 and CC2 thinking about the collectively most popular ideas *'think about your investment, what is good … what makes sense'* [E:564] – and the current plan to get respondents individually to pitch and then vote on their favourite ideas part way through CC2 [A:374]. Kenny articulates what they are all aware of *'we have a result from last time (CC1), but maybe we'll get a different result now'* [K:575]. The team grapple at length with how to accommodate their desire to have all respondents engaged with working on a focus they are personally enthused by, whilst at the same time having the respondent groups collaborating around a shared theme. In designing activities for establishing focus, they want to build in opportunities to elicit participants' individual accounts of personal preferences. This is information the team are to deliver to their company internal clients [E:754, 756, 758]. Trying to design this outcome occupies the team for more than half an hour in discussing the challenges of 'pitching' an idea one is not invested in, and in elaborate tinkering with voting mechanisms.

Towards the end of this period, the team reconsider the boundary (of progress with focusing the respondents' attention) between CC1 and CC2. About 25 minutes into the second hour, Ewan [E:777] says, *'I think we should continue as we planned* (for CC1)' but five minutes later Abby [A:820] is suggesting that more focus needs to be agreed during CC1 to ease the pressure on CC2. Ewan raises a series of concerns about this ranging from re-iterating that it is demotivating to have to 'pitch' something one is not convinced about [E:838, 840, 875, 877], through to practical questions about whether there is time in CC1 to accommodate more activities [E:870]. He doesn't seem convinced one way or the other himself, saying, *'still don't feel hundred percent we have it in the bag ehm because I'm – in principle I'm very much for them deciding already there* (at end of CC1)' [E:864]. There is a lot of side-stepping issues by re-naming, we get it with the voting – it's putting notional money on things, with investing – its putting three hearts on one or several favourites, or three stars, or six hearts. Maybe respondents don't need to 'sell' a theme, they can 'explain' it [K:878]. Establishing foci for development with the respondents in CC2 is variously characterized as conceiving a 'company', developing 'theme/s', or identifying a 'product'.

4.4   Final twenty minutes

About 100 minutes in, the major issues that the team started the meeting with remain unresolved. Ewan, to test his belief that individuals will respond differently to a given prompt, runs a quick '*simulation*' [E:904] by asking his colleagues what '*eco*' means to them. This is an attempt to overcome an impasse. He concludes that, '*we learned that we have four fairly different perspectives at least on it*' [E:929]. But Abby immediately questions the validity of his experiment, '*but we didn't yeah and we didn't discuss it the whole* (topic for) *a full day in advance*' (unlike the respondents during CC1) [A:930].

The team are tiring, Kenny's comment, '*I think it's getting late now*' [K:914], is repeated minutes later [K:949] after yet another round of inconclusive discussion about the questionable value of pitching something to which one is not committed [E:938]. Ewan agrees with Kenny's sentiment [E:950]. There is a sense that the team need new ideas, Kenny suggests they may need to keep things open [K:947], that there may be a '*third option*' [K:959], with Abby saying twice that maybe a different exercise is what is needed [A:954, 958]. There is a little more conversation about the boundary between CC1 and CC2 and finally, about what the period between CC1 and CC2 will deliver and whether the scope for respondents' thinking in this period will be sufficiently broad. But it seems, as the talk peters out, that the team are not clear, together or individually, about what respondents are to do between CC1 and CC2. David asks whether establishing a focus at the end of CC1 will make the scope of respondents thinking (too) narrow [D:1001, 1005]. Abby suggests that in the interim respondents might look up facts or solicit opinions from other sources [A:1006]. Ewan doesn't want this to happen and links ensuring this to keeping the 'pitching' of ideas as a CC2 activity. He wants the incubation period between CC1 and CC2 to be characterised as '*divergent*' [E:1014], but in almost the last contribution of the meeting, he reiterates what they started with almost two hours earlier, musing back on the dilemma of where to place focusing, '*but they (the respondents) know about their theme, and I think this was kind of our original (.)*' [E:1011].

4.5   Hindsight

After this meeting decisions are made which result in a change to the boundaries between CC1 and CC2. We know this from instructions for the workshop facilitators which are included in the DTRS11 dataset along with recordings of the workshops themselves. These show that a voting activity takes places during CC1 to assist in narrowing down the scope of the topics/ideas that will be developed in CC2. No voting occurs in CC2. CC2 works with the most popular foci from CC1 and entails respondents working in two groups to generate ideas for a company (products, markets and branding) developed collectively then shared (pitched and justified) through presentations. Respondents' contributions throughout CC2 comprise the resource material for the user involvement team and their internal clients. Apart from agreement about the timing of some relatively unproblematic initial tasks which survive into the facilitators' instructions for CC2 it is difficult to see what contribution this particular meeting made directly to the final design of CC2.

5 INTERPRETATION: THE MEETING AS FRAMING FAILURES

Prior to the meeting the team have already decided that in CC2 respondents will work in two groups on tasks that will allow observers and facilitators to gather material which will form the basis of the 'concept package' they have been tasked to produce. To elicit what they need they are planning to use the conceit of having each respondent group form a 'company' which will embody, in product lines and marketing materials, a set of consumer values and ideas previously established (via activities involving discussion, clustering and voting in CC1). The clustered sets of ideas from which companies will be generated are also referred to as selected 'themes'. Here a close inspection of references to the two terms company/companies and theme/themes (singular and plural forms conflated below) is used to expose the team's failure to make progress during the meeting, viewed as failure to operationalize design framing.

5.1 Frame discipline: moves

The term 'company' appears in 42 turns-at-talk in the transcript of meeting v5. Occurrences are distributed among the team members as follows: Kenny 4 (including once in reported speech), Abby 17, Ewan 21. The term appears predominantly in the first 25 minutes in 32 turns. It then crops up in three further episodes: first [A:547 and E:552] as a synonym for a group of respondents; second about an hour and 20 minutes into the meeting [E:724, E:737, E:750, A:751 and A:753]; and finally in three turns towards the end of the meeting [A:975, E:995 and A:996]. Looking more closely at these episodes we see a number of interesting things.

Right at the outset there are issues with the notion of 'company'. On the positive side, the idea of 'company' as a conceit for the respondents to work with – Kenny, reading from a post-it note, '*each group is its own little company*' [K:13] – is valued as presenting a clear focus for respondents' group-work. A company will be pitched to investors, have associated products, target markets and so on. However, there are two major drawbacks to using the idea of 'company' to frame the design of some workshop activities. First, this sense of 'company' relies on its prior establishment by the respondent groups in CC1. This is incompatible with the (prior) design of CC1. Second, the collective working on forming a 'company' undermines the teams' objectives to devise activities for respondents that will elicit their personal preferences and the rationales for each of these.

We now look at how each team member makes reference to the 'company' notion. Kenny's references to 'company' all relate directly back to his first use in reading from the post-it note [K:038, 040, 143], so while he does use the term he does not build on it, or from it, during this meeting. Abby's references are more numerous, however looking at them individually we can see that only in the first one [A:070] is she concurring with the 'company' concept – assenting with its appearance on the post-it note. Thereafter all her references to the 'company' concept are to question it on grounds of one of the two drawbacks described above, or to try to conflate it with the term 'theme' that carries a different set of connotations that she prefers.

If we think of 'company', then, as a 'naming of things to be attended to' in the sense of establishing a context for designing (making design moves) it does not seem to work well for two of the team. Is this simply a case of a failure of the team to agree on, and to share a frame, a case of frame conflict as characterized by Schön and Rein? Or is it a frame dispute (within frame differences of interpretations)? Does the 'company' notion work for Ewan then, who uses the term most frequently? Refer to Extract 1.

Whilst Ewan likes the 'company' idea for the positive quality identified above [E:027] he is aware that it presents the problem of loss of opportunities for eliciting respondents' personal preferences – the second drawback we have already noted. At [E:033] he says this directly. The dissociation (Stumpf & McDonnell 2001) between 'company' and the personal is problematic. Kenny proposes a repair by re-associating 'company' and the personal via the suggestion that they tell respondents that customers of the company should be just like themselves, [K:034]. Ewan's response of 'yeah' [E:035] is not taken as unqualified assent as Kenny continues to reinforce his proposal [K:036].

5.1.1   Extract 1: Trouble with the concept of 'company' v5:027-048

027	E	And I- and the whole idea of the the – kind of the mini company, I think is a cool approach because-
028	K	Yeah it's a cool approach and they will like it
029	E	Yeah because it makes it very easy for them to- it's very logical for them to think in story and sales and
030	K	Yeah
031	E	'Cos basically they are:- they're doing their- the job for us in a very natural way
032	K	Yeah, so they can focus on the: fun things rather than trying to understand the task
033	E	It does of course create another challenge, which is the opposite challenge of if it was very personal for them, because then they will be coming from a personal side, now they come from a company side so now maybe the focus is not around what's important for them, but what's important for the company which is typically money. So that is our challenge. How can we keep them within their own personal realm
034	K	Yeah I guess they need to know that target, customers needs to be themself
035	E	Yeah
036	K	Yeah, so like in the first session they should imagine that there are millions of you, and you need to sell this product to them
037	E	And I think that is the cool part, that we don't introduce this before the very end of the first session, or maybe in the first eh, at the start of this session, so we already have that captured, so that stick is in the ground
038	K	Yeah. And they already have this company thinking from the first session probably, because they needed to invest as a company
039	E	Yeah. So we- the moderator needs to always:, and we need to always, if they go out on a tangent and think only money or whatever, we need to draw them back into "this is what we know is important, this is what we agreed on"

040	K	Yeah. And it gives a pretty good flow that we come from this company investment thing, and now we need to sell the product, we need to market the product
041	E	Maybe one thing that they should make is some sort of company slogan, which is the company values
042	K	Yeah, hmm
043	E	And every- and then, when they talk about storytelling or product or sales, it needs to make sense for their company slogan
044	A	Yeah. Maybe we can just write the time:, just like we did last time, so we'll start- [it's- it will be the same]
045	E	[Eight forty-five]
046	K	Yeah
047	A	Yeah. Ehm: and we also need this eh: kind of warm up exercise to include the two new guys
048	K	Oh the two new guys, yeah

There is still trouble, though, because at [E:037] Ewan enumerates alternative design paths in a reference to the first drawback we identified. This may be a deferral mechanism to accommodate potential disagreement (McDonnell, 2012). He leaves options open, and Kenny in the next turn acknowledges the contingent nature of where their decisions stand with 'probably'.

What happens at [E:037] is interesting from a framing perspective because, whilst deferral of design decisions can serve designing constructively (McDonnell, 2012), here, lack of either resolution, or systematic pursuit of some conjecture(s), leaves the team unable to make progress. To express it differently, we can see here that 'company' as a named thing for attention **does not support design moves** at this moment, in this situation, even for Ewan. Instead, we see a few more exchanges between Ewan and Kenny in which Ewan tinkers by suggesting a design fix, namely [E:039] strong moderator intervention during the workshop to keep respondents on track. Kenny's intervention [K:040] does not address the negative aspect of company but reinforces the positive one they have identified already. Abby has not participated in this episode, at [A:044] she introduces a topic change which is effectively established by the assent of the others [E:045, K:048].

Neither the tinkering and the reinforcement of a positive assessment drawn to attention in Extract 1, nor all the later tinkering during the meeting (e.g., extensively with voting mechanisms as described in Section 4.3) compensates for **failure to establish a way of seeing** that is free from the major drawbacks (incommensurate requirements) identified by the team themselves early on in their meeting. Nor does it serve in setting these aside temporarily to enable progress with designing. Effective framing imposes a discipline. It invites and supports certain moves and rules others out. Here it is not some frame that is at fault, then, but rather the team's **inability to work generatively and selectively within the constraints a particular framing imposes**. The notion of 'company' neither serves as a frame to impose a 'what if' discipline for a series of conjectural moves, nor does it serve as instigator of re-framing (which we consider in Section 4.3). It is not strongly defined, and relies significantly on negative assessments in relation to some important concerns of the team.

5.2   Frame discipline: reflection

The term 'theme' is more pervasive than 'company', appearing in 84 turns distributed throughout the meeting. It is only absent for the period when attention is focused on how to deal with new respondents joining for CC2. Looking more closely at the distribution of references to 'theme' we see that the majority come from Abby (59). It is Abby who first uses the term, and like Kenny's first use of 'company', she does this by referring to the team's documentation from their design of CC1. Refer to Extract 2.

5.2.1   Extract 2: Trouble with the concept of 'theme' v5:085-096

085	A	But I think, I mean, we're ending here, with the: "now imagine that you're a CEO in a company wanting to invest in these themes". So I'm thinking, we haven't made the companies yet, so they can't present a company. So-
086	E	So, wait, okay so, maybe this- could that happen that the first thirty minutes or something, eh: the two new people, they are making name tags over here, and the groups, so group one and group two, (*draws on whiteboard*) are making the company over here and here?
087	A	I guess- I- yeah if-
088	E	And then they- we are putting them together here. So then they have time to, for thirty minutes or something like that, they have time to make the company: and make the kind of get together
089	A	So I'm thinking that the company can't really be:- I mean, here it- the company can be based on values only then. Not any products or anything because that's kind of what we are doing later. So I'm thinking that maybe it's kind of the first half of the day we'll use- we'll spend on making it more concrete, and then we'll make the companies based on something a little bit more concrete than just the values and themes
090	E	Yeah, that could be- this would be the company values that they would create here
091	A	But isn't it hard to- if you have no idea what product you are going to sell, but you just have an idea of "okay, I want to create a company with these values"
092	E	Yeah?
093	K	And I think- I think it's- maybe it doesn't feel so nice for this person, or [these two]
094	E	[To be separated?]
095	K	To be separated so obviously, from the beginning, when they need to part of the rest of the process later
096	E	Oh no, no I agree. Eh: let's scrap this one. Eh:. But there's no doubt that we have to utilize the freshness, or the virginity of these two people. Eh and how do we do that in the best what? 'Cos these- the people who come here, even though they've been with us one time, they're still a little raw and a little virgin themself, so how can we make sure that they can communicate something fairly straightforward, together, err as a unit, where the other person is kind of watching. Or commenting

Abby reminds her colleagues [A:085] how things stand at the end of CC1, quoting from the team's records. The team have been discussing how to handle the new respondents; they want to use their fresh perspectives but also to integrate them with the respondents from CC1. Extract 2 shows an attempt to associate 'theme' and 'company' by conceptualizing 'company' as a collection of values rather than an enterprise with products. Ewan attempts this but at [E:096] concedes both to Abby's objection to (a) 'company' that only comprises values [A:089, 091] and to Kenny's concern that new respondents may feel uncomfortable if kept apart for a long time from the others [K:093]. As a candidate notion for framing, we see that 'theme' serves no better than 'company' to support the generation of design moves because rather than imposing order on the design task, 'theme' generates further problems for the team to solve: first, how to integrate the newcomers, and second, something that subsequently occupies a great deal of time, how to get respondents to select from the 'themes' through pitching and voting. At the end of Extract 2 [E:096] Ewan appears to set aside his recent proposal but about 10 minutes later the team are still pushing the same issues round. Refer to Extract 3.

5.2.2   *Extract 3: Trouble with the 'company-theme' combo v5:170-175*

170	K	So they can say "alright, we saw these three, four themes, and we invested in this one, because of >this and this and this and this". Because that's what they- that's the point where they are. They know why they invested in it, but they don't know exactly how it can be turned into a product and how it can be turned into a business yet
171	A	But it will be:- so it's not a group thing yet, so this is eh
172	E	Individual
173	K	[Yeah, individual. For each person represent the- yeah]
174	A	[Individual. Eh it could easily be three or four] or five eh themes that-
175	E	Different themes, easily. They could- it could be- hopefully they will align a little bit, but it could definitely be that. So, and this is why, and of course maybe I'm repeating myself, I'm coming back to, is it possible that(.) they come here and we spend, >I don't know<, half an hour, forty minutes on them becoming a company, like agreeing on the values of the company, while we're doing the other exercise, the name tag exercise, with the others in for example the sc'- so if they, they come a little asynchronously. So these go in to this room and then the other people come into this room, but there is like a twenty minute gap between them or whatever

More trouble is brewing because conflict between the requirement to elicit person preferences from respondents and the setting up of activities that encourage group endeavor is back as a problem to be addressed. The lack of development of groups' notional companies is back on the agenda too. This leads to more tinkering, with Ewan suggesting the respondents should arrive at different times [E:175] so that those from CC1 can start before the new respondents join. Not only can we see a similar pattern here, namely that working with a design task framed in terms of 'theme' fails to prompt moves, but we also see that the **failure to confront problematic issues head-on by evaluating the consequences of a line of reasoning (reflecting on moves within frame)** leads to a lot of time (more than 30 minutes) discussing how to solve problems that the team have *generated for themselves* by failing to evaluate – here failing to confront the requirement to elicit personal preferences which they identify as conflicting with designing activities than work to establish group consensus.

A reluctance to undo decisions from the design of CC1 (which is after all a *self-imposed* constraint) generates the need to address a number of issues such as mechanisms for getting people to pitch ideas they are not interested in, and how to maintain motivation among respondents whose preferred ideas are not carried forward. By setting themselves the task of narrowing down from a broad set of themes to a narrower focus based on popularity among the respondents, the design team spend a lot of time designing activities including very detailed voting mechanisms which are subsequently entirely discarded. There are some symptoms of fixation at play here if we define fixation as a reluctance to accept that another line of development is possible. But, staying with the construct of framing, we have a **failure to evaluate moves in a timely manner during the design process.** This failure to reflect incurs cost – the effort of designing details which will have to be discarded (cf. novice designers' tendencies to commit too soon, e.g., see Christiaans & Dorst, 1992). There are occasions during the meeting when the possibility of revising the design of CC1 is mentioned, but it is never actively pursued. There may be many reasons for this: face-saving between the team members, or between the team and the consultants with whom they have already shared the plans for CC1; or a reluctance to confront the shortcomings of something in which they have already heavily invested. Iteration in designing or the suspension of (some) constraints to pursue conjectures only serves the design task usefully when there is assessment of what those moves amount to – without such reflection iteration is simply going round in circles, and ignoring constraints leads to effort expended on pursuing dead ends.

Examination of the nine references by Ewan to 'theme' indicates another possible source of trouble with 'theme' as a concept to support framing. Four of Ewan's references to 'theme' come immediately after turns in which themes are referenced by a colleague. His use of the term may simply indicate a micro-level conversational strategy – of using the same term to signal topic continuity (relevance preference). Scrutiny of his five other uses of the term show him referring to 'themes' in the sense they are established during CC1. This draws attention to the fact that what the term 'theme' refers to drifts during the meeting because the participants are discussing how 'themes' evolve. During CC2 themes output from CC1 will be reduced in number and fleshed out by the respondents through activities the team are designing – as they are voted on, selected, elaborated and pitched. Themes will become fewer and pick up detail through the transformation into 'companies' (a term Ewan uses most positively as we have seen above) with storied products and their markets. 'Theme' seems to be **a poor device for framing because it is too slippery to impose order.** Many courses of action are discussed, but with no guide rails in place, many workshop activities (design alternatives) are possible and there is **nothing** clear enough **to serve evaluation purposes.**

5.3   Reframing

One way to make progress in designing is to deliberately set aside certain constraints, to 'start somewhere', perhaps to start in several places; and through design moves, experimentation and iteration, to become more familiar with the situation – to understand it better by setting to and tackling something (Glegg, 1971). As Dorst points out, viewed from outside, this particular characteristic of designer behavior can seem aimless (2015, p. 68). The team know what desired outcome they want from CC2, they know neither what apparatus to use nor how to use it (Dorst, 2012, 2015) and they do not exhibit the skills for either compromising over the conflicting requirements they have been given or created for themselves, or for creatively reframing to make the conflicting requirements generative of a new perspective. Instead of forcing the contradictions to help them think differently (reframe by dissociation) they tend to plaster over these by association to create vaguer, broader concepts. The example of Kevin doing this at [K:034] has already been mentioned above. In the long period in the meeting when the team talk about voting they side-step the issue of dealing with respondents whose preferred ideas are voted out by moving from voting with notional money (investments) to voting with 'hearts' and then later 'stars', never addressing the substantial issue. Refer to Extract 4.

5.3.1   Extract 4: Trouble with voting v5:655-663

655	A	So let's assume that they are choosing (.) one, eh: one theme
656	E	I think they should- yeah?
657	A	And: they write it down on a post-it, they share back and say "okay I chose this one because of this and this and this", and it can't just- I mean, Rose and Will needs to really ask why and get deeper into it so that they know even better themselves why they actually chose it. To make them reflect over this
658	E	So you still think we should only do it for one, not- they don't- they shouldn't have three hearts. They can do all on three or two on one (.) and. 'Cos I kind of wanted that discussion, "why is this still here? Why- what is the value in this one? You didn't want to put it on top, but it still- you're still dragging it with you. What are the features and the: the values of this?"
659	K	I think it can be done in the moderation. That- that Rose or Will can just ask "why did you not choose that instead?" or "why did you not choose that?"
660	A	Yeah. And, I mean, we have like (.) two minutes, per person, for this, to share back, in this eh session here
661	E	Yes, and it's a fairly important part, it's the conclusion of this in many ways
662	A	Yeah yeah, I-
663	E	So we can't down prioritize it. It's kind of this is the fruit of our labour that day. It will be what sustains us to three days here. Not a hundred percent, but in many ways it-

Extract 4 gives a flavor of the discussion about voting; here we see that Ewan's concern to elicit from respondents what underlies their preferences **does not prompt the team to re-conceptualize** what the activities might be. Kevin [K:659] folds into the voting arrangements as they currently stand [A:657], an accommodation to Ewan's concern expressed at [E:658] by associating the information he wants to capture with the planned activity of respondents' speaking about their top choice. Kevin tinkers with the design to do this. **Here would have been an opportunity to reframe** through the dissociation of the need to focus the respondents on a smaller set of themes with the requirement to gather rich detail about their individual preferences. This is one of several occasions when a **conflict is not resolved creatively by reframing, nor is it solved by negotiating one of the requirements to be set aside to be reintroduced later, nor is it solved by deliberately deciding to deem one or other of the requirements unobtainable** (see Dorst, 2015, p. 51, for a description of precisely these three options when getting stuck in this way). Any of these three outcomes would be evidence of framing supporting evaluation of (reflection on) design moves. Instead we have avoidance of dealing with the issue effectively and this in turn gives rise, inevitably, to a reinforcement of a sense of circularity in this design process for an observer as well as for the participants.

6 DISCUSSION AND BROADER SIGNIFICANCE

In the account given in Section 4 attention is drawn to team members' apparent awareness that they are making little, if any, progress. Whilst they do not explicitly mention 'going round in circles' their comments show that each has a sense that they often return to an earlier point of departure *without new insights* about how to proceed. So it is not self-awareness that is at issue here. What is surprising is that the team do not use any strategies to try to fix things. For example, they don't use the repertoire from popularized design thinking (empathizing, problem defining, ideating/brainstorming, prototyping, and testing (IDEO, 2016)) to deal with problematic issues. It may be that knowing each other well and have worked together for some time inclines them to an informality which does not serve design task efficiency. Aside from each commenting, at different moments, that they are not making progress, they don't take formal measures to assess progress and decide how to proceed: there is little evidence of meeting management.

What design expertise does this team have for the design task they are tackling? They are employed as 'user involvement' designers and we hear them frequently hypothesize scenarios of user behavior (respondents' reactions to workshop activities). So in this there *is* evidence of the designers empathizing (e.g., Extract 1 029-032; Extract 2 091-093; and the whole of Extract 3) and of sketching scenarios (e.g., Extract 2 086). These *are* design thinking strategies. There is one point where Ewan experiments (prototypes) to test his idea that respondents will have differing perspectives on a particular concept ('*eco*'). But apart from these the practice of a range of design thinking skills is not in evidence. It is inappropriate to make any claims about the capabilities of the participants on the basis of examining their interaction in a single meeting, however the examination of this event draws attention to how the team's failure to make progress can be seen as a failure to impose discipline on their design task.

Framing of design tasks can lead to a course of design activity (design moves, reflection, further framing, reframing) and thence to design outcomes valued for effectiveness or for their innovation. In such cases, retrospectively, we can say the framing was good. Framing can also be regarded as poor retrospectively if the resulting design has shortcomings for some stakeholder(s), or if the design process was inefficient.

This study draws attention to how failure to regulate a design process can be interpreted as failure to operationalize framing effectively. Frame awareness, if we have it, gives us a way to confront our assumptions. In the design context, the concept of frames gives us a way to talk about how designers impose order on a design task, without which they confront under-constrained situations or paralyzing paradoxes. The account of meeting v5 given here in terms of frame failures exhibits both of these dilemmas as the team move around in a design space where too many possibilities are kept in play, yet everything they consider comes up against a constraint with which it is incompatible. Any frame has generative capacity, but it is a designer's skill to set *suitable* frames and *operationalize their capacity to impose order*. Using framing as a lens to interpret the meeting described here allows us to see that the opportunities the discipline of framing imposes contrast with what took place. The causes of what is observed is another matter: it might be team members' lack of design skills or lack of expertise in designing co-creation workshops specifically. Or there may have been external factors to which we have no access that led the team to make so little progress on this occasion.

Design professionals, when they are designing in some domain in which they have experience and expertise, have a repertoire of frames (Dorst, 2011), stylistic predilections (Tonkinwise, 2011) or, expressed more broadly, guiding principles (Lawson, 1994) acquired from that experience that allow them to shape designs through framing effectively. If we accept that this is the case, there remains a set of tantalizing open questions over the extent to which framing is a generic skill – a designerly way of approaching effectively (any) design situation, and the extent to which operationalizing framing successfully rests on the imposition of order derived from expertise in some particular design domain.

REFERENCES

Biskjaer, M. & Halskov, K. (2014). Decisive constraints as a creative resource in interaction design. *Digital Creativity*, 25:1, 27–61, Doi: 10.1080/14626268.2013.855239.

Brown, T. (2009). *Change by Design: How design thinking transforms organizations and inspires innovation*. New York, USA: Harper Collins.

Christensen, B. T. & Abildgaard, S. J. J. (2017). Inside the DTRS11 Dataset: Background, Content, and Methodological Choices. In: Christensen, B. T., Ball, L. J. & Halskov, K. (eds.). *Analysing Design Thinking: Studies of Cross-Cultural Co-Creation*. Leiden: CRC Press/Taylor & Francis.

Christiaans, H. & Dorst, K. (1992). Cognitive Models in Industrial Design Engineering: A Protocol Study. In Taylor, D.L. & Stauffer, D. A. (eds.) *Design Theory and Methodology*. New York: ASME.

Dong, A., Kliensmann, M. & Valkenburg, R. (2009) Affect-in-Cognition through the Language of Appraisals in J. McDonnell and P. Lloyd (eds.), *About Designing: Analysing Design Meetings*. Oxford: Taylor & Francis Group. 119–133.

Dorst, K. (2011). The Core of 'Design Thinking' and Its Application. *Design Studies* 32. 521–532.

Dorst, K. (2015). *Frame Innovation: Create New Thinking by Design*. Cambridge: MIT Press.

Dorst, K. & Cross, N. (2001). Creativity in the design process: co-evolution of problem-solution. *Design Studies*, 22, 425–437.

Glegg, G. (1971). *The selection of design*. Cambridge: CUP.

ten Have, P. (2007). *Doing conversation analysis*. Second ed. London: Sage.

IDEO. (2016) *Design Thinking Resources*. [Online] Available from: http://www.ideou.com/pages/design-thinking-resources. [Accessed February 21, 2017]

Lawson, B. (1994). *Design in Mind*. Oxford: Butterworth Architecture.

McDonnell, J. (2009). Collaborative Negotiation in Design: A Study of Design Conversations between Architect and Building Users. *CoDesign*, 5:1, 35–50, DOI: 10.1080/15710880802492862.

McDonnell, J. (2011). Impositions of order: A comparison between design and fine art practices. *Design Studies*, 32, 557–572, ISSN 0142-694X, Doi: 10.1016/j.destud.2011.07.003.

McDonnell, J. (2012). Accommodating disagreement: A study of effective design collaboration. *Design Studies*, 33, 44–63, ISSN 0142-694X, Doi: 10.1016/j.destud.2011.05.003.

Onarheim, B. & Biskjaer, M. (in press). Balancing Constraints and the Sweet Spot as Coming Topics for Creativity Research. In: Ball, L. (ed.) *Creativity in Design: understanding, capturing, supporting*.

Schön, D. (1983). *The Reflective Practitioner: How Professionals Think in Action*. New York: Basic Books.

Schön, D. & Rein, M. (1995). *Frame Reflection: Toward the Resolution of Intractable Policy Controversies*. New York, USA: Basic Books.

Stanford University (2016) *Our Way of Working*, Stanford University http://dschool.stanford.edu/our-point-of-view/ #design-thinking.

Stumpf, S. & McDonnell, J. (2002). Talking about Team Framing: Using Argumentation to Analyse and Support Experiential Learning in Early Design Episodes. *Design Studies*, 23, 5–23.

Tonkinwise, C. (2011). A Taste for Practices: Un-repressing Style in Design Thinking. *Design Studies*, 32, 533–545.

Valkenburg, R. & Dorst, K. (1998). The Reflective Practice of Design Teams. *Design Studies*, 19, 249–271.

Valkenburg, R. (2000). *The Reflective Practice in product design teams*. PhD Thesis. TU Delft.

Wiltschnig, S., Christensen, B.T. & Ball, L. (2013). Collaborative problem-solution co-evolution in creative design. *Design Studies*, 34, 515–542.

Chapter 21

Planning Spontaneity: A Case Study About Method Configuration

Koen van Turnhout, Jan Henk Annema, Judith van de Goor, Marjolein Jacobs & René Bakker

ABSTRACT

The methodological literature for generative forms of research such as co-creation is less mature than for some other types of research. This may be because we do not understand well enough the key issues in generative research and the means designers have to address them. In this paper we report on a study of a design team preparing a co-creation session. We focused on the means this team employs in releasing the tension between planning the study focus and making room for spontaneity. This tension forms a primary difficulty in ensuring the quality of the co-creation session. Using the theoretical lens of method configuration, we identified five means the designers employ to manage spontaneity and planning: participants, activities, content, group atmosphere and moderation. Each of these could be tuned towards enabling a more spontaneous or more planned co-creation session. They were planned in close coordination with each other. Moderation was used as a wildcard in this process. Apart from advancing our understanding of the way designers deal with planning and spontaneity, the study is also the first to extend the method configuration framework to generative research.

I INTRODUCTION

Design is both a planned and a spontaneous process. Many companies treat design as following a fairly generic preplanned process, but in practice, designers experience design as chaotic and fluent (Dorst, 2003; Hadfield, 2014). In the DTRS11 dataset (Christensen & Abildgaard, 2017) we were able to observe a European design team meticulously planning a co-creation session in a foreign country. The team appears to be aiming at a session with an informal, creative atmosphere, which should deliver rich insights in the values and desires of the participants. This has to be achieved with participants who aren't designers and who are from a different cultural background as the design team. The upcoming co-creation is thus characterized by many uncertainties, which, according to the design team, should be met with a flexible setup of the co-creation session. In the end, the session is planned in much detail, even up to the point where the team discussed how the food might prime the participants of the session. In the preparation of the co-creation session, planning is needed to ensure focus in the session and to make sure the team gains the desired input from participants. But the team also aims for spontaneity to ensure flexibility, creative input, and ownership of the participants. The resulting balancing act between these two extremes leads to the central question of the paper: "How does the design-team manage the tension between planning for focus and allowing for spontaneity to occur in the co-creation session?"

Our approach to answering this question has been to examine how discussions in the team about planning and spontaneity boil down to concrete decisions about the co-creation session. For this, the notion of *method configuration* forms a suitable theoretical lens. Woolrych, Hornbæk, Frøkjær, and Cockton (2011) introduce this term in a programmatic paper which critiques the way design research methodology is studied and taught in the academic community. They argue that in the practice of design, methods are seldom executed in their textbook form (as a recipe), but rather as loose assemblies of the elements that make up a method, such as 'participants' or 'tasks'. These elements are called *resources* by Woolrych *et al.* Practitioners configure and adapt method resources to fit the needs of each specific project. Following Woolrych *et al.*, a naturalistic study of design methodology should not focus on methods as 'indivisible wholes', but it should identify and study the individual resources, their interconnections and the influence of the project context on the method configuration. Taking the method resources as our unit of analysis allowed us to trace how discussions about planning and spontaneity unfold and how they materialize into concrete decisions for the upcoming co-creation session.

Despite the apparent relevance of the method configuration framework, it has not yet been applied to co-creation. Woolrych *et al.*'s paper has been influential and led to an increased attention to how evaluation methods are executed in practice, for example, in Følstad, Law, & Hornbæk (2012) and Lárusdóttir, Cajander, & Gulliksen (2014). It also provided support for the general concern about our understanding of interaction design methodology in practice, for example, in Lee (2013), Johnson *et al.* (2014), Gray (2016) and Smeenk, Tomico, and van Turnhout (2016). Despite this uptake, however, few have advanced the program as it was outlined by Woolrich *et al.* (2011) beyond their original focus on usability evaluation. This meant part of our work was to explore the applicability of the concepts of the framework, which are rooted in usability work, to generative research. Since the method configuration framework was intended to be generally applicable, in the first place, we felt applying it to co-creation could result in a valuable stress-test for the framework as an added benefit of our study.

This study was set up following a 'Field Reframing' research design pattern (Van Turnhout *et al.*, 2014). In a field reframing study, something in the field is studied from a novel theoretical point of view and its results are transformed to reusable theoretical insights. The paper is organized accordingly. In the next section we will expand our theoretical frame, which is the notion of method configuration, and we will elaborate on our approach to the data analysis. We will argue that the most important resources in the configuration of the co-creation session include 'participants', 'activity planning', 'group atmosphere', 'content management' and 'moderation'. For each of these we will highlight how they can be tuned towards planning or spontaneity. Moreover, we will highlight interdependencies with the planning of other resources.

2 ANALYTICAL FRAMEWORK AND APPROACH

2.1 Analytical framework

To study the tension between spontaneity and planning in the preparation of the co-creation session, we adopted the notion of method configuration as it was proposed by Woolrych *et al.* (2011). In this section we elaborate this framework. The central argument is that methods can be decomposed into a loose and diverse set of *resources*. For usability work, for example, Woolrych *et al.* identify resources related to participant recruitment, task selection, reporting formats, problem identification, knowledge of the application domain, problem classification, and so on. They can be combined into an approach which leads to a higher-order goal such as a usability assessment. How practitioners configure these sets of resources may depend on many contextual factors such as the design purpose, business context, project leadership and other practical constraints. A naturalistic understanding of a method (such as co-creation in our study), thus focuses on the resources. We need to know the resources employed, we need to understand the impact of each resource on the outcomes of a study as well as the way in which resources were combined. One could argue that a set of resources form an epistemic chain which is no stronger than the weakest link.

As the resources identified by Woolrych *et al.* (2011) form a varied set, they propose a classification of seven *resource categories*. First, *scoping* resources indicate the extent of a method's applicability in terms of the purpose and usage contexts of what is being evaluated. Second, *axiological* resources indicate the values underpinning a method. Third, *project management (process)* resources situate a method within an embracing development and collaboration context. Fourth, *expressive* resources communicate the output of a method, among others via specifications and reports. Fifth, *procedural* resources guide the use of a method, including partial automation through tools. Sixth, *instrumentation* resources collect issues and measures for evaluations. Seventh, *knowledge* resources underpin one or more of the previous resource classes. Table 21.1 enlists examples of these types of resources for the case of co-creation. This list of resource categories helped us to identify resources in the data. Of course, the team did not extensively discuss resources of all types in materials that we examined, so some resource types are missing from our analysis.

The DTRS11 dataset allowed us to see method configuration at work. A central focus in our analysis has been the special character of co-creation as a form of generative research (see Sanders & Stappers, 2012) leading to the potentially conflicting goals of stimulating participants into creative spontaneity and simultaneously guaranteeing the relevance of the outcomes of the session. We attempted to map out the most important resources that were under discussion by the design team and the choices the team made about each resource as they configured it. In particular, we looked at the different role each resource had to play in alleviating the tensions between planning and spontaneity. Finally, we looked at interconnections between the resources.

2.2 Approach

The core-research team consisted of four members, all of whom were staff of our university. Our analysis consisted of several phases. In the first phase we divided up the complete DTRS11 dataset and shared our first insights. An expert review meeting (as described in Van Turnhout *et al.*, 2013) was held with experts within our university. This exploration phase led to a sharper idea of what we wanted to research and a selection of data to analyze. We focused on the parts of the data where the design team actually engaged in designing the co-creation sessions: v02, v03, v04, v05, v09, and v10). In the second phase we immersed ourselves in the data (see Sanders & Stappers, 2012). In this phase all researchers independently watched one session and applied open coding, thus marking down salient moments in the video considering the research question. Next we watched each video with the whole team sharing and discussing interpretations of key moments. We raised novel questions and answered existing questions provisionally, but our aim was always to stay open to novel interpretations rather than reaching closure. Therefore, we did not decide on a definite list of resources in this phase. In the third phase, the interpretation phase, researchers individually noted down quotes, pointing out important moments of the video about the way in which the design team handled spontaneity and planning. We clustered these with affinity diagraming. This interpretation session led to the final list of resources as well as insights into how they were configured. Based on this session the material was revisited to answer the specific research question in this paper. We shared and discussed these in a second expert review meeting.

3   RESULTS

3.1   Setting the stage

At the start of the design of the co-creation sessions (v02) the design team discusses their expectations for the upcoming co-creation. In retrospect it turned out that the design team already talked about most of the resources that became subject to extensive discussion later on in this first meeting. As such we will use this meeting of the design team as a way to set the stage for our discussion as well.

The first resource is *participants*. Participants (at other times called respondents or people) are mentioned first as a source of uncertainty by Kenny, but also as something that can be controlled to some extent.

KENNY: *"Ehm of course really depending on the people that you work with. I guess that you need to somehow calibrate the users before you can calibrate the method."* (V02, 12)

Kenny expresses the need to fit the way of working to the participants of the co-creation, but also claims it is hard to predict who will come in. Later in the meeting Ewan expresses a wish for people who are ready 'to open up' and Abby points to the screener as a way to recruit the 'right' people. A later video (v03) shows the team discussing this screener. In the classification scheme of Woolrych *et al.* (2011) *participants* is an *instrumentation resource*, a resource supporting issues and measures for evaluation.

The second resource is *activities*. This is what Kenny in the previous quote refers to when he talks about 'calibrating the method' and later discussions in the first meeting include 'something with Lego' and prototyping tools as activities. In this paper we also place (structured) group discussions in the category of activities. During most planning sessions that follow the design team uses and adapts a draft activity plan. The activities resource is a *procedural resource,* which guides the use of a method.

The third resource is *group atmosphere.* While the group atmosphere can be seen as an emergent property, which is achieved by configuring other resources, such as activities and participants, it is still treated in the planning process as a separate resource. For example, an early discussion about the atmosphere centers around the question of whether a 'party with friends' is a good metaphor for the atmosphere that is considered ideal, aptly summarized by Ewan

EWAN: *And we're setting the scene, it's very artificial, super artificial. And stuff that you're not necessarily super interested in, it becomes very- but, then again, I think that is my analogy very often, to say "okay. Imagine these people are your friends, and you're doing something awesome together".* (V02, 45)

Here the team introduces the group atmosphere as something that is partly under their control and that they need to get 'right'. In later videos, discussions on atmosphere include the question of how to avoid hierarchy and mobile phone policy. We see the atmosphere resource as an *axiological* resource, which points out the values underpinning a method.

Ewan's quote also refers to something that is discussed at great length in two later videos, which we will refer to as *content. Content* is the fourth resource. Ewan refers to it as 'setting the scene' (repeating a remark by Abby) and (talking about) 'stuff'. Content contains, for example, the way participants need to be primed and what materials should be prepared to bring focus to the discussion. *Content* is in our view also an *instrumentation resource.*

The fifth, and last, resource is *moderation.* Although many discussions of the design team focus on the role of the moderator, we prefer the broader term moderation. We take it to include moderation by others than the dedicated moderators and also the preparation of the moderation through the construction of a moderation guide. Ewan mentions moderation only once in passing in the first meeting. But moderation turns out to play a central role in later discussions. Moderation is a resource which bridges instrumentation, axiology and procedure.

These five resources were not the only ones that were employed in the co-creation session. We chose to focus on these because they were subject of heated discussions in the dataset and because they turned out to play a role in the configuration of the tension between spontaneity and planning. Other research questions may have led to a different selection of resources. Also, not all resource *types* identified by Woolrych *et al.* were the subject of discussion in the DTRS11 dataset. Table 21.1 gives an impression.

Next we will discuss the selected resources one by one, focusing on the interrelations between resources and on the ways in which each of these can be tuned towards or away from spontaneity in the co-creation.

3.2   Participants

Participants are an important resource to any method involving users, making *recruiting* an important step to take. As mentioned, Kenny framed participants almost literally

Table 21.1 List of resource types.

Type of resource	Example of application in co-creation planning
Scoping resources	Out of the focus of the DTRS11 database
Axiological resources	*Group atmosphere*
Project management resources	For example: the involvement of stakeholders within the company with the project, not the focus of this paper.
Expressive resources	For example: the sticky notes and PowerPoints used for share back, not the focus of this paper. The sticky notes were studied in much detail by Dove *et al.* (2017)
Procedural resources	*Activity* planning, *Moderation,*
Instrumentation resources	*Participants, Content, Moderation*
Knowledge resources	Experiences with co-creation

as configurable in the first meeting, but participants were also seen as a source of uncertainty in this discussion: you have to deal with the people you get in. The participants resource is related to activities (as these need to fit the participants) and moderation (as the moderator needs to deal with the particularities of each participant). Also, group atmosphere is dependent on the participants, but this connection isn't discussed much in the dataset.

In one video we can watch the design team setting up the screener for the recruiter who is selecting the participants (see Kuniavsky, 2003, for a textbook example). Here the team expresses some demands for the participants such as representativeness (fitting in the client-profile), being power users (having a premium car) and being willing to take part and doing homework. The language barrier (can we find participants who speak English (?)) is an important topic for the group, but eventually it is decided an extra translator can solve this issue.

A striking observation is that, apart from this session about the screener, the particularities of the chosen participants are hardly ever discussed. Participants are treated as people who can deliver valuable comments and can be primed and questioned. But no references are made to the configured particularities of the respondents or the diversity of the participants that may enter the session. Once the screener has been made the abstracted 'group' is used as the way to talk about participants.

In summary, participants are seen as an important source of information who need to be carefully selected. However, despite Kenny's remark that participants need to be calibrated in combination with the methods, they do not play a central role in the further configuration of the co-creation session. To ensure focus, participants need the right background, and to ensure a creative atmosphere, participants need to be 'willing to open up'. Apart from this, participants are not used as a major means to dissolve the tension between planning and spontaneity. Literature on co-creation focusses on 'diversity' (Moller & Tellestrop, 2013; Schepers, Dreessen, & Huybrechts, 2014) as a major concern in participant selection, but this isn't discussed explicitly in the data.

3.3 Activities

Activities play a central role in the planning of the co-creation, as most of the time the design team makes use of, and adapts, an outline activity plan for the sessions. Planned activities include, for example, a warm up exercise, a brainstorm and a 'why-why-why' session. The second session also includes a role-play in which the participants are asked to play out a company (probably, best characterized as a design game; Brandt, 2006). Activities are planned in close conjunction with *content*. Taken together, activities and content form the backbone of the co-creation as it is planned. In the meetings of the dataset, the activity planning seems to be fairly uncontroversial, whereas the content, which we discuss further on, is a topic of much debate. An important role of the choice of activities is to ensure a good *group atmosphere*. The *participants* are discussed only in passing when deciding on activities, and always in an abstracted form (participants in general, not specific participants). One example where an activity is treated as a way to ensure spontaneity is the warm-up-exercise. Here the activity is to ensure a good creative atmosphere with 'designers' values', such as: there are no wrong answers. Similarly, after discussing concerns about the willingness of participants to participate and co-create, which they think might be worsened by a stronger sense of hierarchy in the target country, Ewan concludes:

> "And the point is that we need to make them forget all about this. And in the beginning it will be like that, but we are aware, and the awareness makes us strong and makes us design tools and exercises and the first half an hour is typically only spent on getting people to relax and trust each other: and then so on. Eh: even though they- it might seem for them that we already started on the co-creation part, and we kind of are but we're warming up, we're getting people on board." (v02, 169)

Using warm-up exercises (Sanders & Stappers, 2012) and discouraging hierarchy (Moller & Tollestrup, 2012) are also recommended in the literature as stimulants for group atmosphere.

Elsewhere the activity planning plays a role in ensuring focus. Here we see the team discussing 'when' the participants are invited to mention things that were not in the original 'pillars of good life'. Placing the activity first would invite the participants to think broadly, whereas placing it later in the program would ensure only 'important things' were mentioned, thus, bringing more focus. Abby states her preference:

> "So how- if:- I don't know why, I just have a feeling that it's better to introduce it later because then they have worked a little bit with this, and they would know if something is not fitting in to "these". Whereas, if it's in the first exercise then they can see it, and maybe it will be really hard for them to come up with a new area." (v04, 259)

In sum, the activity resource can be used to tune the session towards spontaneity or focus by ensuring a creative atmosphere and by the choice for 'deep' or 'broad' activities as well as by the order in which activities are placed.

3.4 Group atmosphere

A good *group atmosphere* is an important value for the team and a topic that they return to regularly in the design of the co-creation session. It gets configured indirectly, through the employment of other resources such as activities, content and moderation, but it is nevertheless something that is 'designed' by the team. *Activities* and *moderation* are the most important resources that support group atmosphere.

In order to ensure a spontaneous atmosphere, the group appears to strive for egalitarian values. They set up an exercise that stipulates it is 'ok to fail' and strive for a situation where everyone is on board. The design team appears to be afraid of the sensitivity towards hierarchy in the culture of the target country, which may spoil an equal contribution of all participants. This is remediated by enforcing a rule that participants cannot disclose their job titles and by instructing the moderators, eventually through the moderation guide, to present their selves as equally important from the beginning. Although the group atmosphere seems to be primarily targeted towards spontaneity some checks and balances are also put in place, in particular to keep the group manageable. At several instances the team makes references to the fact that a group is harder to manage than an individual. One issue is the occurrence of 'baby discussions', a term used by the team to refer to discussions between participants outside of the central conversation. Baby discussions are seen as a hindrance and need to be prevented with a 'one discussion at a time' rule. Suggested for example by Amanda:

AMANDA: *I think we need to have a ru'- eh: just on the practical side, a rule for this discussion, like one discussion at a time. (v09, 335)*

The rule is endorsed somewhat jokingly by Ewan in a 'fight club' analogy. Also, some activities are set up in such a way that groupthink can be prevented. Arguably the outcomes of a session could be impaired by groupthink while the group atmosphere could benefit.

It appears that the design team is most concerned with getting the group going and less concerned with an atmosphere that is too loose or creative. The team spends much time discussing group atmosphere, but is not clear how the moderator is informed about the specifics of this resource. There are no explicit remarks about atmosphere in the moderation guide.

3.5   Content (management)

With the resource *content* we refer to the team's efforts to influence what is talked about during the co-creation. The team can prime the participants, they can frame the discussion, they give guidance to the topics that are talked about and they can constrain the discussion. One example is the way the company's focus will be introduced through a number of 'pillars' and the way in which these will be phrased and introduced to the participants. Another example is the material that is selected to prime the topic of 'good life' to participants in the first session. The configuration of content is strongly related to that of activities, as the activities chosen can increase or decrease the focus of the conversation, and because the activities would make little sense without content. The configuration of content is also strongly related to moderation because the moderator has some power to steer the discussion during the co-creation session.

The tension field between planning and spontaneity is very tangible in discussions about content. If the team focuses the content too much the participants could resolve to socially desirable answers rather than bringing in their own perspectives. If the team focuses the content too little the team might go off-track into discussing irrelevant things for the company. For most materials, including the presentation of pre-defined pillars of good life, materials that should bring participants in the right mindset during lunch and even the food itself, the team wonders whether they bring sufficient focus and whether or not they prime too much. This becomes clear in a short exchange about putting up 'priming materials' during lunch, which should keep the participants 'in the bubble of the co-creation'.

EWAN: *yeah yeah, so we want them to be not talking about (.) some stuff that is too disconnected (.) 'cause then we need to work, to fish them in again*
KENNY: *yeah, and maybe we can:: put something on the table that: somehow makes them- forces them into- to::*

An idea that gets delimited only a few minutes after by Abby:

ABBY: *and i guess we- yeah and props. Whatever we can come up with. And i think we also: need to be careful not to prime them too much because (..) if we prime them too much this is just (.) what they will come up with. (v04, 105)*

Eventually the team decides the materials should be 'broad' or 'elusive'. At some points the designers try to make space for the participants' perspectives. One example is the 'mystery pillar' that can capture a theme which was not in the definition of 'good life' set out by the design team but that is deemed important. Voting sessions, which are planned at several times during the co-creation, are another way to give participants a way to set the agenda for the co-creation session. The team appears to strive for a balance between spontaneity and planning. In most discussions they play down materials that bring focus but they do not include much space for participants' perspectives either.

An interesting observation when discussing how the team deals with content is the role of the moderator. Discussions about content are sometimes relayed to discussing moderation, as the moderator can make room for spontaneity during the co-creation session, while preserving focus. In the next section we will discuss moderation in more detail.

3.6 Moderation

When it comes to releasing the tension between planning and spontaneity, moderation turns out to be an invaluable resource. Moderation is done by a dedicated moderator with support from team members who are present in the session. The design team also assembles a "moderator guide", which supports the moderator and reflects decisions about moderation. In the method-configuration process 'moderation' turns out to act as a central hub between all resources. The moderator is in control of introducing the activity plan to the participants; (s)he is charged with ensuring a good atmosphere and plays a crucial role in content management.

In his discussion of the role of plans in ongoing work situations, Bardram (1997) makes a distinction between plans as pre-hoc representations of works, as ad-hoc improvisations and as post-hoc reconstruction of work. As the moderator is capable of ad-hoc improvisations, whereas other resources need to be precompiled, moderation can soften the tension between planning and spontaneity on the spot, giving it a unique position in the configuration. However, getting such improvisations 'right' requires detailed background knowledge of the general objectives and those of each activity, in particular when it comes to content and group atmosphere. This becomes tangible when enlisting all the responsibilities of the moderator that are explicitly discussed during the meetings of the design team.

The moderator should, for example, guard the content by provoking participants, by asking follow-up questions and by making sure participants stay within the scope of the session. Some of these demands are summed up by Ewan in the preparation of the second co-creation.

EWAN: *Yeah. And we should- and that part, that's really: where we need Rose and Will to be:- own it. Own the direction, alright, because they will, you know, they should keep them within and give them provocation when needed and give them support when needed and so on. Help the river flowing. (v09, 330)*

Many of these aspirations for the moderation have been voiced earlier in more extensive discussion where moderator is presented as a gatekeeper of the playing field of the conversation and as a person who can encourage depth in the answers.

EWAN: *Yeah. So we- the moderator needs to always:, and we need to always, if they go out on a tangent and think only money or whatever, we need to draw them back into "this is what we know is important, this is what we agreed on". (v05, 39)*
EWAN: *I think it can be done in the moderation. That- that Rose or Will just ask "why did you not choose that instead?" or "why did you not choose that?" (v05, 114)*

The moderator also needs to 'make connections' between the different comments of the participants during the co-creation session. On top of his responsibility for the content, the moderator is also responsible for the group process and for supporting participants when they need it, while maintaining a single conversation focus (preventing subgroups with private discussions from emerging).

EWAN: *Absolutely. And this eh: puts a big pressure on the moderator, because, I mean, it is not an interview one to one, so it's- there will be seven respondents at the table, and if Rose or whoever is moderating, she really needs to dig in to all these eh: [emotions for example, "okay, oh, so what do you mean?"] (v02, 117)*

To be able to play these roles, the moderator is involved in the planning process through regular updates in which the team also takes on suggestions from the moderator. In this pre-hoc phase the moderator is consulted as a cultural expert. Later on, the moderator receives a moderator guide. The moderator guide contains an outline activity planning and some generic pointers about the content. Considering the detailed discussions the team had about the way in which the content needed to be managed and the role the moderator could play in this, we wondered how much of this background knowledge could actually be transferred to the moderator in the outline plan and during a couple

Table 21.2 Tuning each resource.

Resource	Tuning towards spontaneity	Tuning towards focus
Participants	Willingness to participate, open up, and doing homework, communication skills (English Language)	Choosing representative users, including power users
Activities	Warm up exercises to foster creative atmosphere, broad exercises. Using activities that allow participants to bring their own perspectives to the table	Narrow, focused exercises
Group Atmosphere	Promoting egalitarian and creative values trough exercises and prompts	Keeping discussions manageable
Content (management)	Using broad and elusive materials and prompts	Using exercises that go in depth
Moderation	Provoking participants, stimulating a creative atmosphere	Asking follow up questions, making sure participants stay within the scope of the session; having the moderator ask focused questions

of meetings with the design team. But from the DTRS11 materials it is hard to judge how the team managed this transfer.

In summary, the moderator can alleviate the tension between focus and spontaneity as his ability for ad hoc improvisations during the co-creation session allows him to navigate both ways. To be able to do this, it seems necessary for the moderator to have an intimate knowledge about the playing ground and about the desired outcomes. As such, the moderator needs to be much more than a neutral facilitator of the co-creation session.

3.7 Overview

Table 21.2 enlists for each resource which we just discussed how it can be configured towards either spontaneity or planning.

These resources are not configured one by one, but rather in a closely knit web of interwoven decisions. Figure 21.1 summarizes some of the interdependencies between the resources.

4 CONCLUSIONS AND DISCUSSION

In this paper we have discussed the planning of the co-creation session as a case study in method-configuration. The central theme of our analysis was how the resources employed by the design team could be configured towards planning and spontaneity and how they could alleviate the tension between these two conflicting goals. We have

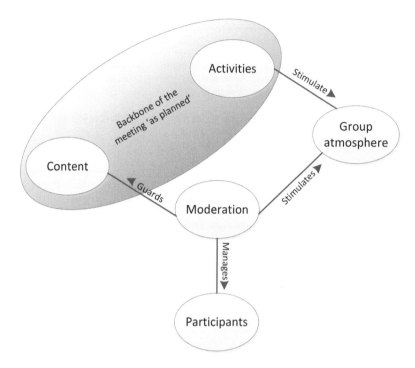

Figure 21.1 Interdependencies between the five resources. Resources affect each other in different ways. Depicted are (projected) relations during the meeting, more information about the interrelationships between the resources during the planning is in the text of this paper.

highlighted five resources which were central in the planning of the design team: *participants*, *activities*, *content*, *group atmosphere* and *moderation*. Each of these could be tuned towards both ends of the spectrum to some extent. Rather than configured one-by-one, the resources were jointly orchestrated throughout the meetings of the design team. Discussions about one resource were often relayed to other resources. For example, if the design team was unhappy with the amount of focus in an activity they would shift the discussion to a different resource such as the content. This was especially true for moderation, which seemed to play the role of a wildcard in the discussions, in particular when the design team needed to decide to leave space for more spontaneity.

The tension between planning and spontaneity played a very central role in the configuration of the co-creation. However, it did not become clear to us as observers on what grounds the decisions were taken and exactly how much space for spontaneity was left in the co-creation. The design team seemed to rely on personal experiences with co-creation sessions and tried to align their expectations based on those experiences (a process also observed and described by Zhang and Wakkary, 2014, and to some extent in Yliriskua *et al.*, 2017). We managed to identify and describe the decision-making process, but we did not find evidence of a guiding principle or a language in which different spontaneity-planning resolutions could be expressed. Such a language might support this decision making.

When we talk to designers who use generative research methods, we find the tension between planning and spontaneity to be a heart-felt concern. However, during this study we were unable to track down a focused body of literature addressing this issue. Methodological textbooks, such as Sanders and Stappers (2012) and Moller and Tollestrup (2013, ch8), treat the subject only in passing, and we found it difficult to locate an academic discourse that treated the topic spot on. This is a pity. We hope that our description of the way this team configured the spontaneity of their session can be a stepping-stone towards a deeper understanding of how generative methods can be tuned to be precisely spontaneous enough.

The notion of method configuration turned out to be a useful frame of reference in this study. Identifying resources helped us to trace the narratives about spontaneity and planning to concrete elements that *can* be configured. This clarified how the discussions of the team did boil down to concrete decisions about the co-creation. However, since method-configuration has not been applied to co-creation before, in this study we had a dual task. We needed to uncover the resources as well as the way they were configured.

Possibly we could have improved the process of identifying resources. In this sense Yliriskua *et al.* (2017) seem to offer a more solid way to identify those based on an interactionist view of the notion of a resource. But probably the best solution would be to have prior knowledge about common resources in co-creation. This could have strengthened our analysis of the configuration. Some evidence that our choice of resources made sense comes from comparing the resources with textbook instructions about co-creation. Sanders and Stappers (2012), for example, mention four of the five resources that we identified as 'topics' in their instruction: participants, activities, content (split into choosing focus and scope, and designing the toolkit) and moderation. Nevertheless, the broad selection of things that count as a resource in the method configuration framework might derail into long debates on how resources need to be demarcated, seriously reducing the analytic power of the approach.

The main argument of Woolrych *et al.* (2011) is that configuring a method is a much more complex and delicate process than following a method recipe, and this argument is well supported by our analysis. While most of the resources we found also structure textbook narratives, in the textbook they are described as abstracted steps to 'think about' or to 'do one after the other'. The in-depth and interdependent decision making about these resources that we have seen and described, aren't remotely captured by such a description. We do not doubt the value of method recipes for practitioners. But the mismatch between the simple abstraction of such recipes and the complexities involved in configuring a method in a real project is telling. It begs the question of how the latter can be supported in a better way. The creation of a generative method-design language in which the core configuration dilemmas play a central role is one possibility.

ACKNOWLEDGEMENTS

The authors would like thank Job Vogel, Tim Lecomte and Thea van der Geest for sharing their expertise with us during the study.

REFERENCES

Bardram, J. E. (1997). Plans as situated action: an activity theory approach to workflow systems. In: *Proceedings of the Fifth European Conference on Computer Supported Cooperative Work*. Springer Netherlands. pp. 17–32.

Beyer, H., & Holtzblatt, K. (1997). *Contextual design: defining customer-centered systems*. Elsevier.

Brandt, E. (2006, August). Designing exploratory design games: a framework for participation in participatory design?. In *Proceedings of the ninth conference on Participatory design: Expanding boundaries in design-Volume 1*. ACM. pp. 57–66.

Christensen, B. T. & Abildgaard, S. J. J. (2017). Inside the DTRS11 Dataset: Background, Content, and Methodological Choices. In: Christensen, B. T., Ball, L. J. & Halskov, K. (eds.) *Analysing Design Thinking: Studies of Cross-Cultural Co-Creation*. Leiden: CRC Press/Taylor & Francis.

Dorst, K. (2006). *Understanding design*. Bis Publishers.

Dove, G., Abildgaard, S. J., Biskjaer, M. M., Hansen, N. B., Christensen, B. T. & Halskov, K. (2017). Grouping Notes Through Nodes: The Functions of Post-It Notes in Design Team Cognition. In: Christensen, B. T., Ball, L. J. & Halskov, K. (eds.) *Analysing Design Thinking: Studies of Cross-Cultural Co-Creation*. Leiden: CRC Press/Taylor & Francis.

Følstad, A., Law, E., & Hornbæk, K. (2012, May). Analysis in practical usability evaluation: a survey study. In *Proceedings of the SIGCHI Conference on Human Factors in Computing Systems*. ACM. pp. 2127–2136.

Gray, C. M. (2016). It's More of a Mindset Than a Method: UX Practitioners' Conception of Design Methods. In: *Proceedings of the 2016 CHI Conference on Human Factors in Computing Systems* ACM. pp. 4044–4055.

Hadfield, J. (2014). Chaosmos: Spontaneity and order in the materials design process. In: *English Language Teaching Textbooks*. Palgrave Macmillan UK. pp. 320–359.

Johnson, M., Hyysalo, S., Mäkinen, S., Helminen, P., Savolainen, K., & Hakkarainen, L. (2014). From recipes to meals . . . and dietary regimes: method mixes as key emerging topic in human-centred design. In: *Proceedings of the 8th Nordic Conference on Human-Computer Interaction: Fun, Fast, Foundational*. ACM. pp. 343–352.

Kuniavsky, M. (2003). *Observing the user experience: a practitioner's guide to user research*. Morgan Kaufmann.

Lárusdóttir, M., Cajander, Å. & Gulliksen, J. (2014). Informal feedback rather than performance measurements–user-centred evaluation in Scrum projects. *Behaviour & Information Technology*, *33*(11), 1118–1135.

Lee, J. J. (2013). "Method-making as a method of designing." *Proceedings from the 5th Nordic Design Research Conference, Nordes 2013*.

Moller, L., & Tollestrup, C. (2012). *Creating Shared Understanding in Product Development Teams: How to 'Build the Beginning'*. Springer Science & Business Media.

Sanders, E. B. N., & Stappers, P. J. (2012). *Convivial design toolbox: Generative research for the front end of design*. Amsterdam: BIS.

Schepers, S. M., Dreessen, K. P., & Huybrechts, L. A. (2014, October). Hybridity in MAP-it: how moderating participatory design workshops is a balancing act between fun and foundations. In *Proceedings of the 8th Nordic Conference on Human-Computer Interaction: Fun, Fast, Foundational* ACM. pp. 371–380.

Smeenk, W., Tomico, O., Turnhout, K., van, (2016). A Systematic Analysis of Mixed Perspectives in Empathic Design: Not One Perspective Encompasses All. *International Journal of Design (IJDesign)*. Vol 10, no 2, pp. 31–48.

Turnhout, K. van, Hoppenbrouwers, S., Jacobs, P., Jeurens, J., Smeenk, W., & Bakker, R. (2013). Requirements from the Void: Experiences with 1: 10: 100. *Proceedings of CreaRE 2013, Essen, Germany.*

Turnhout, K. van, Bennis, A., Craenmehr, S., Holwerda, R., Jacobs, M., Niels, R., Zaad, L. Hoppenbrouwers, S., Lenior, D. & Bakker, R. (2014). Design patterns for mixed-method research in HCI. In *Proceedings of the 8th Nordic Conference on Human-Computer Interaction: Fun, Fast, Foundational* ACM. pp. 361–370.

Woolrych, A., Hornbæk, K., Frøkjær, E., & Cockton, G. (2011). Ingredients and meals rather than recipes: a proposal for research that does not treat usability evaluation methods as indivisible wholes. *International Journal of Human-Computer Interaction, 27*(10), 940–970.

Ylirisku, S., Revsbæk, L. & Buur, J. (2017). Resourcing of Experience in Co-Design. In: Christensen, B. T., Ball, L. J. & Halskov, K. (eds.) *Analysing Design Thinking: Studies of Cross-Cultural Co-Creation.* Leiden: CRC Press/Taylor & Francis.

Zhang, X., & Wakkary, R. (2014, June). Understanding the role of designers' personal experiences in interaction design practice. In *Proceedings of the 2014 conference on Designing interactive systems* (pp. 895–904). ACM.

Chapter 22

Problem Structuring as Co-Inquiry

Robin S. Adams, Richard Aleong, Molly Goldstein &
Freddy Solis

ABSTRACT

The purpose of this study was to investigate the social process of a group working together in an authentic work setting as individuals collaboratively map a problem space. We used a collaborative inquiry paradigm that mirrors a perspective of design as a social process of interacting object worlds as a lens for investigation. We (1) the dimensions of the problem space the group collaboratively renders; (2) the collaborative inquiry (co-inquiry) practices used; and (3) the ways by which the group renders coherence to the problem space. We found that the group traversed a multi-dimensional and integrated problem space linking human-centered perspectives with organizational perspectives for realizing human-centered products and services. We characterized co-inquiry practices as four modes of evoking ways of knowing (experiential, presentational, propositional, practical) and two modes of building coherence (intersubjectivity and critical subjectivity practices). Finally, through the analysis of a rich episode, we characterize rendering coherence of a problem space as an unfolding co-inquiry process. This study offers a new perspective on design as a co-inquiry process – a social process of building coherence to co-construct valid knowledge. By using the collaborative inquiry paradigm, this study also offers practical implications for design methodologies, facilitating design learning, and helping designers develop self-awareness as co-inquirers.

I MOTIVATION

This study is situated in a shared dataset (Christensen & Abildgaard, 2017). Rather than taking an a priori approach to the data, we reviewed all videos multiple times. We observed a group in an authentic work setting, designing and facilitating a series of "co-creation" workshops in China. There were also debriefing and synthesis sessions where workshop experiences were shared and discussed. Participants in these meetings included a core group of three user experience design professionals, three consultants with expertise about Chinese users and innovation strategies, and two Company stakeholders. Each seemed to bring their own perspective to these meetings – drawing from a wide assortment of personal experiences including experiences with the Company. The leader of the core team, Ewan, stated a project goal was to shape the delivery of a "concept package" that would be used within the Company to answer the question: *How might we evoke and capture the attention of the Active Urbanite so that we secure their emotional engagement and establish long-term Company brand/product/service commitments?* (Christensen & Abildgaard, 2017). This goal statement constitutes a design problem frame (Dorst, 2011) – it articulates a value to be achieved (secure long-term Company commitments) and a process for achieving that value (evoking and capturing the attention of the Active Urbanite).

Many consider problem framing as the essence of design (Cross, 2002; Lawson & Dorst, 2009; Schön, 1983), emphasizing the role of abductive reasoning (Dorst, 2011), co-evolution (Dorst & Cross, 2006), and iterative reflective conversations with a design situation (Adams *et al.*, 2002; Schön, 1993). For this project, we focus on problem framing as a process of structuring a problem space – a space of design intention configurations used to direct and react to solution development opportunities (Nelson & Stolterman, 2003; Schön, 1993). Diverse perspectives shape this space and are negotiated as a social process across "object worlds" that embody personal values, knowledge, and perspectives about good design practice (Bucciarelli, 1994). Object-world ways of seeing cannot capture the process of designing. They do not cover the "full breadth and depth of social context and historical setting" (p. 18), and mask "the uncertain and the unknown" (p. 177) in ways that show that "there is no single perspective that can control or manage the design process in object-world terms" (p. 124). Instead, designing is a social process that requires participants to negotiate object-world ways of seeing, working towards consensus to construct meaning. For this study, our research goal is to investigate *collaboratively structuring a problem space as a social process*.

2   COLLABORATIVE INQUIRY AS PARADIGM FOR UNDERSTANDING DESIGN AS A SOCIAL PROCESS

We use the collaborative inquiry paradigm (Heron & Reason, 1997; Kasl & Yorks, 2002) as a lens for characterizing design as a social process as *co-inquiry* and focus on how a group collaboratively renders coherent a problem space. While some in the design community use the "collaborative inquiry" to characterize design methodologies (such as participatory design), to date we have not identified empirical work that uses the collaborative inquiry paradigm to characterize design as a social process.

As shown in Figure 22.1, the collaborative inquiry paradigm mirrors Bucciarelli's (1994) perspective on design as a social process of interacting object-worlds. It assumes a participative worldview that emphasizes the way participation, reflection, and action are essential for meaningful inquiry into the dilemmas, questions, and problems that are a part of the human condition (Bray *et al.*, 2000). Here, participants engage in small groups to inquire into compelling questions of mutual interest for transformative and action-oriented outcomes (Heron & Reason, 1997; Kasl & Yorks, 2002). Each participant is a co-inquirer and co-subject, drawing on personal experience inside and outside the inquiry group to provide a collective pool of experience and insight for creating meaning (Bray *et al.*, 2000). Because group members hold diverse perspectives and take on different roles in the inquiry, they both reveal and engage with uncertainty as they seek coherence among the different perspectives in ways that may result in generating new knowledge. As such, this paradigm emphasizes the ways interactions

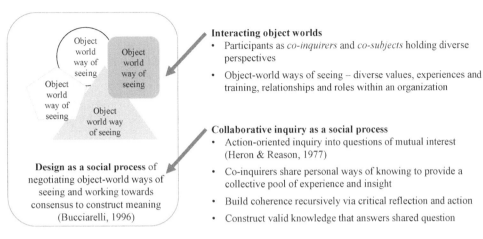

Interacting object worlds
- Participants as *co-inquirers* and *co-subjects* holding diverse perspectives
- Object-world ways of seeing – diverse values, experiences and training, relationships and roles within an organization

Collaborative inquiry as a social process
- Action-oriented inquiry into questions of mutual interest (Heron & Reason, 1977)
- Co-inquirers share personal ways of knowing to provide a collective pool of experience and insight
- Build coherence recursively via critical reflection and action
- Construct valid knowledge that answers shared question

Figure 22.1 Collaborative inquiry paradigm mirrors design as a social process of interacting object worlds.

among object-worlds reveal ambiguities and uncertainties. Through sharing experiential knowing situated in object-world ways of seeing and engaging in critical dialogue as a recursive process of reflection and action, the group works as a community of inquiry "towards a holistic construction of valid knowledge that answers a question of importance to them" (Bray *et al.*, 2000 p. 6). For this project, the knowledge that is constructed and critiqued for validity is the problem space.

2.1 Co-inquiry practices

Collaborative inquiry involves two intertwined co-inquiry practices: (1) ways of sharing individual experiential knowing in ways that make it accessible to group members so they can construct new knowledge as a group; and (2) ways to make meaning from collective experiences and judge its validity. As shown in Figure 22.2, ways of sharing individual experiential knowing involves evoking four ways of knowing that constitute the space of our subjectivity – experiential, presentational, propositional, and practical knowing (Heron & Reason, 1997; Kasl & Yorks, 2002). These modes of knowing are interdependent and combine to provide evidence for knowledge claims.

The collaborative inquiry paradigm emphasizes the primacy of subjective knowing. Rather than suppress subjective knowing, we accept that it is our experiential articulation of being in a world that provides grounding for all our knowing (Heron & Reason, 1997). We also accept that our subjective experience is open to distortions that can collude to limit understanding. But, how do we know what we know, that we are not deceiving ourselves? As shown in Figure 22.2, ways of collaboratively judging the validity of knowledge to build coherence in the group involves engaging in intersubjectivity and critical subjectivity practices (Heron & Reason, 1997). These coherence practices are interdependent but work together to build coherence – consummating each other and deepening the complementary ways they are grounded in each other.

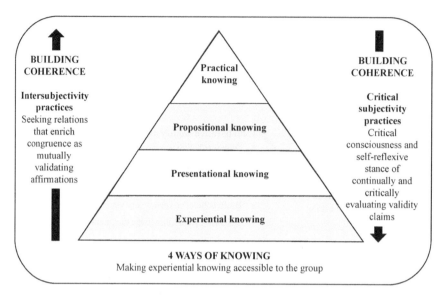

Figure 22.2 Collaborative inquiry paradigm (based on Heron & Reason, 1997, p. 282).

Table 22.1 describes these four ways of knowing and two building coherence practices, and provides connections to design thinking.

2.2   Method

We reviewed the DTRS11 technical report (Christensen & Abildgaard, 2017) and selected datafolders that offered diversity in: (1) rich interactions around collectively inquiring into the problem space; (2) practices for collectively traversing this space; (3) perspectives within the core team (Ewan, Abby, Kenny, David, Nina) with their training in human-centered methodologies, consultants with training in innovation strategies and the Chinese context (Amanda, Rose, Will), and Company stakeholders; and (4) ways of sharing and creating knowledge within the group, including the use of visual and material artifacts (e.g., analogies, gestures, embodied performances, Post-It notes). Table 22.2 summarizes the data used in terms of meeting goals, work products, data collection date, and participants (Core, Consultants, and Stakeholders).

Analysis was highly iterative with cycles of viewing the data and building a language for capturing observations, moving back and forth between the data and existing research or frameworks that might substantiate observations and identify implication pathways. We used the collaborative inquiry paradigm (Heron & Reason, 1997) as our guiding methodology, which also provided a language for sharing, discussing, and making sense of the data. We evaluated the ontological and epistemological underpinnings of this paradigm (Heron & Reason, 1997) and affirmed its relevancy and validity for pursuing research questions about the group's social process for collaboratively mapping a problem space. Working from this paradigm, we culled 11 episodes from the selected dataset that capture critical moments in mapping and traversing

Table 22.1   Collaborative inquiry: Four ways of knowing and two ways of building coherence.

Ways of knowing	**Experiential knowing:** Knowing through participative, empathic resonance grounded in direct encounter with an energy, entity, person, place or thing (Heron & Reason, 1997). It is knowledge experienced in context. *Connections to design: The situated nature of design problems (Dorst, 2004) shapes how designers frame problems (Dorst, 2011) and engage in reflective practice (Schön, 1993).*
	Presentational knowing: Knowing that emerges from our felt attunement with the world and is expressed through symbolic and spatiotemporal forms (Heron & Reason, 1997), metaphorical imagery (Bray et al., 2000), and material, musical, and verbal imagery (Kasl & Yorks, 2002). *Connections to design: Designers create and think through symbolic and material forms, and use non-verbal codes to translate abstract ideas into concrete objects and sketches that can embody and convey those ideas (Cross, 2006; Goldschmidt, 1991; Visser, 2006). These also serve as boundary objects for catalyzing communication, reinterpretation, and negotiation of ideas (Tverksy, 2011).*
	Propositional knowing: Knowing in conceptual terms ("knowing that"), expressed as descriptive reflective statements asserting or proposing a view of the world that are grounded in our experiential articulation of a world (Heron & Reason, 1997). The act of making assertions pro vides pathways inviting co-inquirers to validate truth claims (Kasl & Yorks, 2002). *Connections to design: In the context of design, designers draw upon broad and domain specific conceptual knowledge to set intentions, ideate, and evaluate performance.*
	Practical knowing: "Knowing how" to do something as demonstrated in performing an action (Heron & Reason, 1997). Practical knowing presupposes a conceptual grasp of principles (propositional knowing), presentational elegance, and experiential grounding – and brings these forms of knowing into fruition for purposeful action. *Connections to design: Practical knowing may be a form of synthetic sensemaking (Kolko, 2010) – an effort to integrate experiences into a rendering of the world that forges and prioritizes action.*
Building coherence	**Intersubjectivity practices:** Seeking out consummating relations between the four ways of knowing in ways that enrich congruence as mutually validating affirmations (Heron & Reason, 1997). Examples include affirmations and perspective taking (Duranti, 2010), amplifications and positive feedback loops (Davis & Sumara, 2008), and finding affinities such as resonating points of mutual recognition (Bray et al., 2000). Critical intersubjectivity involves challenging collective self-deception, and group think (Heron & Reason, 1997). *Connections to design: One mechanism for enacting design through language involves accumulating discourses where team members foster coherence by enabling semantic links at the same or higher level of abstraction (Dong, 2006).*
	Critical subjectivity practices: Accepting and attending to the subjectivity of experiential knowing with a critical consciousness and adopting a self-reflexive stance of critically evaluating the ground on which one is standing and divesting oneself of presuppositions upon which beliefs are built (Heron & Reason, 1997). Examples include asking for the basis of assertions, reasoning via data that is valid within the inquiry group, evaluating threats to validity, triangulating among claims to probe for evidence and alternative meaning, and challenging previously held perspectives (Bray et al., 2000; Heron & Reason, 1997; Kasl & Yorks, 2002). *Connections to design: Critical subjectivity practices resonate with a reflection-in-action stance, a process of inquiry under conditions of uncertainty that involves cycles of framing, experimenting (moving), evaluating, and reformulating (Schön 1993). It also resonates with a knowing-in-action stance in which the goal of inquiry is to surface critique and restructure knowing (refute, corred, abandon, or shift) in ways that supports theory building toward future action (Schön, 1993).*

Table 22.2 Summary of selected data for analysis.

Datafolder 1	Datafolder 8	Datafolder 14	Datafolder 16	Datafolder 18	Datafolder 20
Ewan's interview	Insights 1st Co-creation Workshop	Insights 2nd Co-creation Workshop	Discuss workshop insights (product-sales-story)	Clustering into opportunity areas	Debrief with Stakeholders
Work products: Project report	Work products: Post-it notes on boards, summary analysis slides	Work products: Post-it notes on boards	Work products: Post-it notes on boards	Work products: Post-it notes on boards, analysis reports, magic charts	Work products: Outcomes of "clustering" meeting as magic chart and post-it notes
Date: Oct 9	Date: Dec 2	Date: Dec 7	Date: Dec 8	Date: Dec 9	Date: Dec 9
Ewan	Core (Ewan, Abby, Kenny), Consultants (Amanda, Rose, Will)	Core, Consultants, Stakeholders	Core (Abby, Kenny), Consultants (Rose), Stakeholder	Core, Consultants (Amanda)	Core, Consultants, Stakeholders
	3 episodes	3 episodes	1 episode	3 episodes	1 episode

the problem space and provide richness for examining the group's social processes (see Table 22.2). These episodes are the focus of the analyses presented in the following sections.

3   RESULTS AND DISCUSSION

As shown in Figure 22.3, results are organized sequentially by three questions grounded in our motivation to understand design problem structuring as a social process. First, we characterize the context of inquiry as a multi-dimensional problem space as interacting layers of User, Concept Package, and Collaboration Bridge considerations. The group's inquiry and objects of discussion exist within and across this space. Next, we characterize the co-inquiry practices observed: personal sharing as evoking four ways of knowing and making it available to the group (*experiential, presentational, propositional, and practical*), and participating in *critical subjectivity* and *intersubjectivity practices* to build coherence around knowledge claims. These problem space dimensions and co-inquiry practices provide the language for answering our third question: How does the team (as co-inquirers) render coherence of the problem space? For this last section, we focus on a rich episode drawn from datafolder 20 (Christensen & Abildgaard, 2017) that involves all problem space dimensions and includes all inquiry group members.

In the following sections, we use capitals to signify specific problem space dimensions and italics to signify co-inquiry practices. We use a notation of ("v"datafolder, line number – i.e., (v8, 120)) to denote data sources. We also embed discussions in each section as a precursor for summarizing contributions and implications of the work.

3.1   A Map of the problem space

As shown in Figure 22.4, Ewan's description of the problem space traverses three dimensions: User Experience, Concept Package, and Collaboration Bridge.

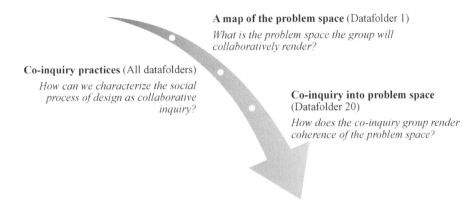

Figure 22.3 The research path from motivation to contributions through three questions.

Figure 22.4 Problem space dimensions (symbols are from project reports; see Christensen & Abildgaard, 2017).

The *User Experience* dimension signifies creating product experiences users would value and desire. This focuses on (1) user experiences, values, needs, and decision making and (2) strategies for eliciting and making sense of user values and needs. Ewan's description of the User Experience emphasizes "pure desirability...with desirability, feasibility and viability" (v1, 093), "a deep understanding of our Chinese users" (v1, 067), and differentiating brands "through experiences" (v8, 234) as compared to surface features of products such as color (v20, 103). He describes this dimension as central to his prior training and experience, as well as Abby's and Kenny's – "most of the tools we are using are from the design eh:: methodology, design thinking field...user centred approach is really important for us" (v1, 016). Ewan's emphasis on designing for meaning and experience indicate a perspective of human-centred design (as compared to user-centred design) (Krippendorff, 2006), and Ewan's language suggests human-centred design methodologies for finding synergy among desirability-feasibility-viability considerations (Brown, 2009; IDEO, 2011).

The *Concept Package* dimension signifies the key deliverable – a unified approach that "will outline and exemplify how a regional relevant holistic approach will increase the take rates and brand penetration in China...a merger of product, story and sales" (v1, 042). This focuses on: (1) integrating elements of the Concept Package (product portfolio – sales channels – communication stories); and (2) strategies for identifying and evaluating innovation opportunities.

Ewan describes the Concept Package as merging three different "streams" into an aligned innovation plan: "by a concept package we basically mean (..) ALL kind of what you need to (.) to make:: a product or service work...the product itself (.) ehm:: the communication needed and the: business stream of it" (v1, 009). This is not a solution to be implemented, but rather a conceptual "road map for how to execute this...and exemplify how a regional relevant holistic approach will increase the take rates and brand penetration into China" (v1, 037-042). Ewan's integrated perspective of the Concept Package was based on his prior ethnographic research with fourteen Chinese Active Urbanites where he discovered "that it is not product alone that is our issue, portfolio alone...it is clearly several things as well...a interconnected development bundle" (v1, 042). Creating the Concept Package involves identifying opportunity pathways from the reassembly of current Company products into a "regional reimagined product prototype" (v1, 035) that aligns users with "the core of what the Company is" to a "regional relevant story" pathway (v1, 036) that involves heavy investment in new services, sales channels and stories. This requires "cross functional collaboration...with different stakeholders" (v1, 042) to enable its handover to the Company and "push this further into the system" (v1, 009). Finally, Ewan's depiction of the Concept Package as a unified approach goes beyond a feasibility-desirability-viability framework (Brown, 2009) to emphasize the organizational context in which the new product will be developed. For example, the names of the Concept Package elements (product, sales, story) reference the organizational stakeholders associated with product development, sales, and marketing. As such, this dimension links to User Experience and Collaboration Bridge dimensions.

The notion of a unified package links to existing literature, particularly innovation strategy and typologies. For example, some of the strategies observed seem to target competition on the basis of experience differentiation, rather than price, on an emerging market (Eyring *et al.*, 2011) with a heavy emphasis on storytelling as a key influence on innovation efforts (Beckman & Barry, 2007, 2009, 2011). Others appear to aim for "blue ocean" innovations of new user conquests (Kim & Mauborgne, 2005). The team also appears to target multiple types of innovation (Solis & Sinfield, 2015) such as business model innovations (Crossan & Apaydin, 2010) and service innovations (Miles, 1993), and use innovation strategies similar to a "jobs to be done" approach for finding unsatisfied opportunities to address unmet needs (Christensen & Raynor, 2003; Anthony *et al.*, 2008a).

The *Collaboration Bridge* dimension signifies efforts to support the handover of the Concept Package to the organizational unit that will implement it, and to enable design thinking at the Company as a core value. Ewan describes this as a metaphorical bridge linking the "concept package stream" to an "innsytrings stream" which serves as both a "handover stream" so the Company is prepared to "take the package...to industrialization and execution" and an "infusion stream" to "get our concept package infused into the machinery. Which is the organization" (v1, 022). This dimension focuses on (1) being aware of organizational barriers and enablers as well as Company stakeholder experiences and values that may influence the Concept Package handover, and (2) strategies for enabling a User Experience mindset at the Company.

For Ewan, this dimension represents a "higher goal in what we are doing, it's not ONLY about using the best methods for the projects that we are doing individually, but it's also to (..) to showcase a whole mindset or way of thinking to the organization(..)" (v1, 097). This involves educating and enabling design thinking capability within the Company as embodied in the User Experience and Concept Package dimensions to "spread the message…teach us how to be user centred…design thinking in our everyday life" (1, 073) and "showcase(ing) a whole mindset or way of thinking to the organization" (v1, 097). Ewan's inclusion of the Collaboration Bridge within the problem space was also based on prior experience. One reason for focusing on the handover are "challenges in…reaching out of our own little silo" (v1, 038) that limits critical alignments within the Concept Package and between local and global units of the Company. Another reason is that "another team will carry it in to industrialization and execution" (v1, 022). Ewan's approach to overcome these challenges is to support stakeholders "to get ownership and a possibility to…somewhat shape what we're gonna do in the rest of the project" (v1, 015). Another strategy is anticipating and being proactive towards areas of resistance and support within the organization: "we have some people that we know will support us…(and those that) don't necessarily subscribe to the methods and the thinking that we do…there are many that does…we want them on board… pushing the: the message" (v1, 014).

Ewan's perspective of the Collaboration Bridge as enabling design thinking capability in the Company can be anchored in change management literature (Kotter, 2007; Luecke, 2003; ODR, 1993). For example, Ewan's engagement of Company stakeholders could be considered attempts to "paint the picture" to help decision makers see the potential impact of an opportunity and build cascading organizational sponsorship as partnerships for change (Adams & Krautkramer, 2016; ODR, 1993). Anticipating areas of resistance and support within the organization and proposing strategies to leverage synergy to change is a feature of change management strategies (Kotter & Schlesigner, 1979) and organizational literacy (Adams & Krautkramer, 2016).

3.1.1 A multi-dimensional and intertwined space of intentions

Frames and frame creation are core to design activity (Dorst, 2011) and a necessary condition for problem solving (Schön, 1993). Here, the problem structure is the space of design intention configurations that will be used to direct and react to solution development opportunities (Nelson & Stolterman, 2003; Schön, 1993). Mapping a problem space involves naming the features of the problem that will be attended to and framing the space of solutions that could be explored (Schön, 1993). As shown in Figure 22.5, Ewan ascribes a holistic space of design intentions by naming three dimensions: User Experience, Concept Package, and Collaboration Bridge. Each dimension shapes the space of possibilities of another dimension – e.g., considerations of user values are intertwined with considerations regarding what the organization values or could value in the future. The characteristics and importance of these dimensions are situated in Ewan's experiential knowing, and can also be understood through broader frameworks for human-centered design, innovation, and organizational change.

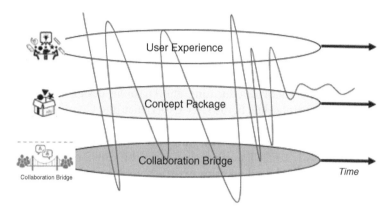

Figure 22.5 Problem space traversing over time as multi-dimensional and interconnected.

Table 22.3 Observed co-inquiry practices: Four ways of knowing and two practices for building coherence.

Collaborative inquiry paradigm		Observed co-inquiry practices
Ways of knowing	**Experiential knowing**	Sharing (offered or requested) personal experience, feelings, and knowledge – from relatively recent common experiences to prior personal experiences.
	Presentational knowing	Sharing and building on symbolic representations or enacting experience through expressive imagery – from the material to the verbal and performative.
	Propositional knowing	Proposing or making assertions about an aspect of the world – from depicting meaning to predicting or anticipating consequences.
	Practical knowing	Extracting insights across multiple ways of knowing to forge practical action – moving from knowing to doing.
Building coherence	**Intersubjectivity practices**	Invoking consensual affinities – from cumulative affirmations and amplifications to consummating points of resonance across ways of knowing.
	Critical subjectivity practices	Invoking (and self-invoking) validity questions to hold co-inquirers accountable for their reasoning – asking for the basis of assertions and critically evaluating threats to validity.

The process of mapping this problem space (depicted as the curved line in Figure 22.5) is a form of inquiry that sets in dialogue the ends to be achieved and the means for achieving them (Schön, 1993). In the next section, we characterize these inquiry practices.

3.2   Co-inquiry practices

Table 22.3 summarizes observed co-inquiry practices (right column) in reference to the collaborative inquiry paradigm (left column). The following paragraphs describe

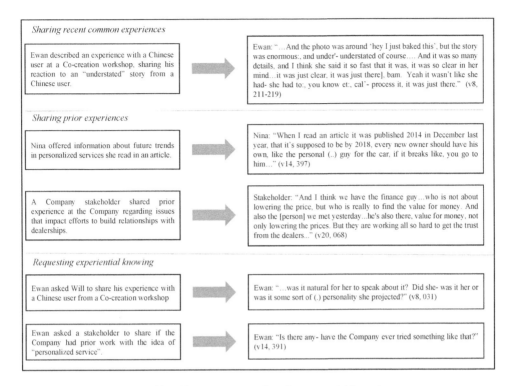

Figure 22.6 Illustrative examples of experiential knowing.

observed practices in terms of the ways co-inquirers made visible to each other their experiential knowing and the ways they sought to validate claims to build collaborative coherence (see Figure 22.2).

3.2.1 *Experiential knowing*

We observed experiential knowing as sharing personal experience, feelings, and knowledge – from recent common experiences to prior experiences (see Figure 22.6). Sometimes experiential knowing was offered; at other times requested. These practices were used to render dimensions of the problem space. For example, User Experience dimensions were evoked through sharing experiences from the Co-creation workshops. Experiential knowing regarding the Collaboration Bridge dimension often involved sharing experiences about Company values and perceived alignment with Concept Package opportunities. Other authors in this volume observed designers drawing on experiential knowing in ways that shape multi-cultural design work – such as cultural knowledge (Clemmensen *et al.*, 2017), resourcing stakeholders' previous knowledge (Ylirisku *et al.*, 2017), and relating observations to familiar contexts (Gray & Boling, 2017).

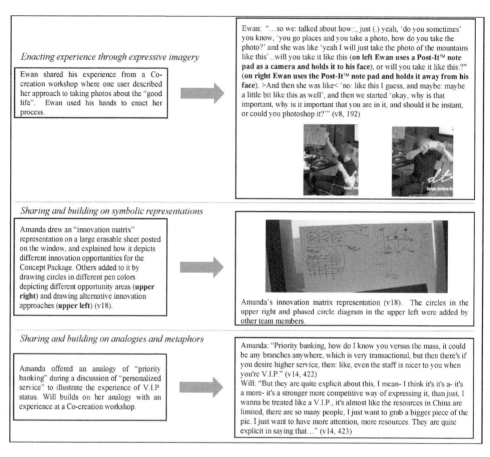

Figure 22.7 Illustrative examples of presentational knowing.

3.2.2 Presentational knowing

We observed presentational knowing as sharing and building on symbolic representations or enacting experience through expressive imagery (see Figure 22.7). Presentational knowing contributed to conversations across all problem space dimensions and was expressed through physical objects (such as Post-It notes organized on a board), analogies, and performances. Shared objects often evolved into collaborative objects. Others in this volume observed team members externalizing knowledge through Post-It notes (Dove *et al.*, 2017; Ylirisku *et al.*, 2017) and using visual and tangible materials to express their thoughts and rationales (Dhadphale 2017).

3.2.3 Propositional knowing

We observed propositional knowing as proposing or making assertions about an aspect of the world – from depicting meaning to predicting or anticipating consequences (see Figure 22.8). Observations were prevalent across the dataset, contributing to

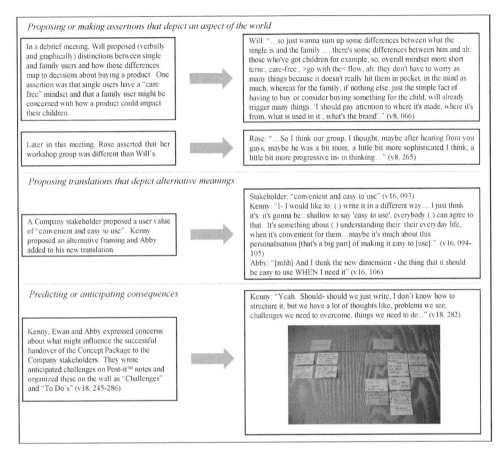

Figure 22.8 Illustrative examples of propositional knowing.

conversations across problem dimensions, and often co-occurring with other ways of knowing particularly presentational and practical knowing to build coherence within the group. Similarly, Dong & MacDonald (2017) observed how generating a new hypothesis about ambiguous conflicting observations can be a form of insight linking experience to assertion to validation. This process was described as semantic waves of bringing prior experiences into context, explaining surprising observations (abductive reasoning), and proving explanations from evidence (inductive reasoning). Others in this volume observed joint generation of insight (Jornet & Roth, 2017) and collective "teasing out" of differences among Chinese users (Ylirisku *et al.*, 2017).

3.2.4 Practical knowing

We observed practical knowing as extracting insights across multiple ways of knowing to forge practical action – moving from knowing to doing (see Figure 22.9). Observations of practical knowing were less prevalent, perhaps because of the ways practical

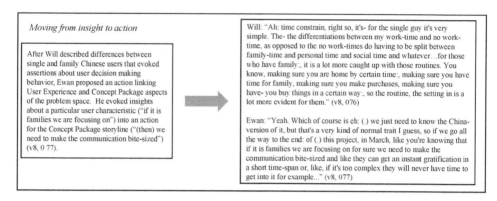

Figure 22.9 Illustrative examples of practical knowing.

knowing is often grounded in other ways of knowing. When it was observed, it often occurred as discrete moments in time signaling the end of a co-inquiry episode before a new topic was introduced. Other examples include the act of writing down an insight as a perceived indicator of value for future action (see also Ylirisku *et al.*, 2017).

3.2.5   Intersubjectivity practices

We observed intersubjectivity practices as invoking consensual affinities – from cumulative affirmations and amplifications to consummating points of resonance across experiential ways of knowing (see Figure 22.10). Cumulative affirmations had a "yes and" quality similar to improvisation techniques of accepting and then building upon an idea. Others had a quality of finishing each other's sentences or translating ideas into higher levels of abstraction (see Dong, 2006; Jornet & Roth, 2017). D'souza & Dastmalchi (2003) observed aggregations of ideas as the synthesis of ideas into a cohesive form as team members built on each other's statements. Ylirisku *et al.* (2017) observed a social process of building fragments of insights into collective small elaborations.

3.2.6   Critical subjectivity practices

We observed critical subjectivity practices as invoking validity questions to hold co-inquirers accountable for their reasoning – asking for the basis of assertions and critically evaluating threats to validity (see Figure 22.11). Most instances were associated with claims about user values and behaviors (User Experience). Similarly, Gray and Boling (2017) observed team members asking for clarification and explanation from translators for the purpose of validating statements about cultural understanding. Jornet and Roth (2017) observed aspects of collective reasoning as "making knowledge relevant and intelligible to others" such as joint generation of accumulating insights.

  In the next section, we use these co-inquiry practices to characterize in-depth an episode that illustrates how the team rendered coherent a multi-dimensional and intertwined problem space. We use the term "render" to emphasize the social process

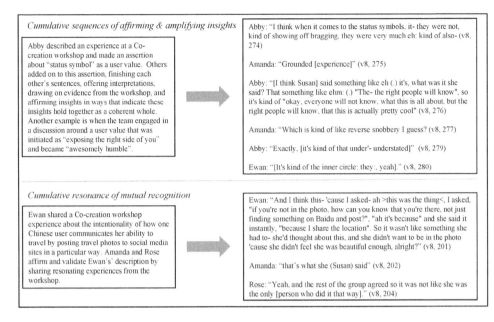

Figure 22.10 Illustrative examples of intersubjectivity practices.

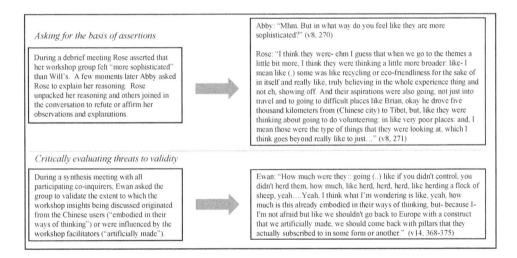

Figure 22.11 Illustrative examples of critical subjectivity practices.

of making individual experiential knowing visible and accessible to the group and working together to validate a coherent and shared perspective of the problem. The use of "rendering" also emphasizes how the problem space is emergent, fuzzy, and in flux as it unfolds through the ongoing co-inquiry (see also Jorney & Roth, 2017; Ylirisku *et al.*, 2017).

3.3   Rendering the problem space through co-inquiry

The team is sitting around a large conference table and everyone is facing Ewan, who is standing at the window delivering a debrief on insights from the day. Attached to the window are poster-sized notes generated during previous meetings, all of which represent different frameworks for integrating the product-sales-story Concept Package and identifying product development opportunity areas. Ewan asks the Company stakeholders, who participated as observers in the second Co-creation workshop and will eventually implement the Concept Package, to sit closer to him and they move closer. Everyone else is sitting on the opposite side of the table and interacts minimally. The conversation begins with Ewan using presentational knowing (pointing to the "innovation matrix" on the poster-sized notes on the window) to summarize current perspectives on the Concept Package, and propositional and practical knowing to *propose* and *pursue* the benefits for using the innovation matrix strategy to "dream big" and seek out "new conquest(s)". After Ewan finishes, Amanda asks: "Any thoughts, you guys, after hearing all this?" (v20, 037). A conversation (v20, 038-115) ensues.

This conversation is our focus for answering the question: How does the co-inquiry group render coherence of the problem space? We selected this episode because it invokes all three problem space dimensions, offers rich interactions associated with the Collaboration Bridge dimension, and provides an example of how new insights regarding Collaboration Bridge and Concept Package challenges reframed the problem space (integrating the dealer experience into the user experience). Our focus is on the co-inquiry dynamics of rendering the problem space as a process of building coherence. In the following figures, we follow this episode over seven chronological sub-episodes depicting: (1) the co-inquiry practices observed – evoking ways of knowing (experiential, presentational, propositional, practical) and invoking coherence through intersubjectivity and critical subjectivity practices; and (2) the problem space being rendered (User Experience, Concept Package, Collaboration Bridge). In each of the sub-episode figures, we offer an interpretive description and the corresponding evidence from the transcript as an abbreviated conversation. In each description, words in *italics* denote the specific instances of co-inquiry practices (see Table 22.3 and Figures 22.6 to 22.11). Words in capitals denote the dimensions of the problem space dimensions summarized in Figure 22.4.

3.3.1   Rendering concept package and collaboration bridge links through integration of user-company values

In the first sub-episode, Amanda invokes a *validity request* that sets in play an exchange of all four ways of knowing collectively among the co-inquirers. As illustrated in Figure 22.12, one stakeholder starts the dialogue, responding by engaging intersubjectivity practices and sharing experiential knowledge about the Company.

In this sub-episode, Ewan and Company stakeholders co-inquire into innovation and implementation issues that renders a link between the Concept Package and Collaboration Bridge dimensions. This link is a stakeholder affirmation of the Concept Package approach for developing opportunity areas, provided user values are aligned

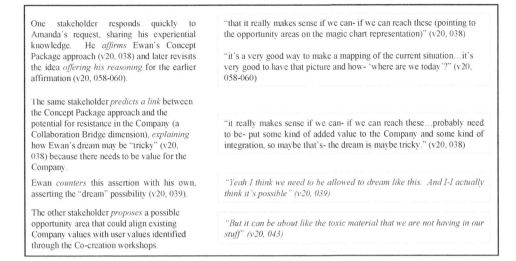

One stakeholder responds quickly to Amanda's request, sharing his experiential knowledge. He *affirms* Ewan's Concept Package approach (v20, 038) and later revisits the idea *offering his reasoning* for the earlier affirmation (v20, 058-060).	"that it really makes sense if we can- if we can reach these (pointing to the opportunity areas on the magic chart representation)" (v20, 038) "it's a very good way to make a mapping of the current situation...it's very good to have that picture and how- 'where are we today'?" (v20, 058-060)
The same stakeholder *predicts a link* between the Concept Package approach and the potential for resistance in the Company (a Collaboration Bridge dimension), *explaining* how Ewan's dream may be "tricky" (v20, 038) because there needs to be value for the Company.	"it really makes sense if we can- if we can reach these...probably need to be- put some kind of added value to the Company and some kind of integration, so maybe that's- the dream is maybe tricky." (v20, 038)
Ewan *counters* this assertion with his own, asserting the "dream" possibility (v20, 039).	*"Yeah I think we need to be allowed to dream like this. And I-I actually think it's possible" (v20, 039)*
The other stakeholder *proposes* a possible opportunity area that could align existing Company values with user values identified through the Co-creation workshops.	*"But it can be about like the toxic material that we are not having in our stuff" (v20, 043)*

Figure 22.12 Connecting Concept Package and Collaboration Bridge through seeking user-Company value integration.

with Company values. When Ewan negotiates for opportunity areas in a "dream big" category, one stakeholder affirms the possibility by offering an example of an integrated user-Company value story of not using toxic materials in Company products. From a co-inquiry practice lens, they are seeking coherence about user-Company value alignments through sharing experiential knowing (Company experiences), propositional knowing (predictions of resistance, propositions of opportunities), intersubjectivity practices (affirming and countering assertions), and critical subjectivity practices (reasoning).

3.3.2 Rendering an integrated problem space through "pushing the values story"

In the second sub-episode shown in Figure 22.13, Ewan follows up with an almost one-sided dialogue with the Company stakeholders, *making assertions* about User Experience principles (evoked through propositional knowing and presentational knowing) and *justifying* his emphasis on embodying user values in the final deliverable (invoking critical subjectivity practices).

Here, Ewan renders a link across all three problem space dimensions through propositional knowing (assertions), presentational knowing (analogy), and critical subjectivity practices (reasoning). He offers the Company stakeholders his user-centered mindset by invoking the coherence of a "values story" as "a key competitive advantage" that integrates the User Experience (identifies user values), Concept Package (embodies user values), and Collaboration Bridge (aligns user values with Company

Ewan *asserts* the importance of embodying user values (v20, 048) in the final deliverable, and justifies this by *explaining* how values, not surface features such as materials and cost, are critical for establishing product commitment. In the process, he invites stakeholders into a User Experience mindset, rendering a link to the Collaboration Bridge dimension.

> "And remember in the world of- that everything can be copied, we cannot- we cannot (.) it's not about the material, no sorry, it's not about like, how cheap it is, or how it looks like, 'cause that can really be copied. It needs to be the values: that it embodies. That cannot be copied." (v20, 048)

Ewan uses presentational knowing to *explain his reasoning* by offering the *analogy* of an air purifier (v20, 050-052). The stakeholders respond with affirmative statements of "Yeah" and "Mhm".

> "(e)veryone can make a really good air purifier...But can you make a good air purifier that is representing your value set?" (v20, 050-052)

Ewan continues to *explain his reasoning* by *asserting* the importance of the "values story" and connects all three problem space dimensions, (v20, 055), linking a need for authentic alignment between user and Company values as a key competitive advantage in developing new opportunity areas.

> "that's why we need to- we need to push the values story, and have good products supporting that. That is connected to: you know, the: authenticity to the source...even sustainability can be copied...we need to have a level above that, and that has to do with company culture: and everything needs to- to really be genuine" (v20, 055)

Figure 22.13 Integration of User Experience, Concept Package, and Collaboration Bridge through Company values.

values). The stakeholders seem to acknowledge and affirm his reasoning, although the nature of this affirmation is unclear.

3.3.3 Rendering concept package and collaboration bridge links through stakeholder experiences

In this sup-episode shown in Figure 22.14, a number of stakeholder experiences are invoked and shared among the co-inquirers that render a connection between the Concept Package and Collaboration Bridge.

Here, Ewan invokes a critical evaluation of potential Collaboration Bridge threats to the validity of the Concept Package strategy. After grounding threats to validity in the stakeholder's experiential knowing with the Company, the co-inquirers evoke propositional knowing, making assertions that affirm new insights about the dealers that sell Company accessories. This begins to render a new area of concern that links the consequences of dealer-Company relationships to user-dealer experiences.

3.3.4 Rendering concept package and collaboration bridge links about opportunity portfolios

In the sub-episode in Figure 22.15, the co-inquirers explore various opportunities for expanding product development. They move from propositional and experiential knowing to evoking and co-affirming practical actions that render connections between the Concept Package and Collaboration Bridge regarding short term and long term product development opportunities.

Ewan *requests experiential knowing* from the Company stakeholders on the challenges of adoption (v20, 063). This is a *request for validation* grounded in the stakeholder's understanding of the Company.	"What do you guys thinks-think are the biggest challenges for pushing something like this...a way to try to move away, or at least add much more (.) intangible value, to our products" (v20, 063)
One stakeholder *articulates* challenges from her own perspective and experiences with other stakeholders at the Company (v20, 066) that reveal Concept Package and Collaboration Bridge dimensions of the problem space. Two key issues include a potential for misalignment between the Chinese office and the central office (v20, 068), as well as across departments (v20, 068).	"...but they should not do something that goes against what we are doing, so we need to really be aligned and get all the departments needed, for instance communication." (v20, 066) "And I think we have the finance guy...who is not about lowering the price, but who is really to find the value for money... But they are working all so hard to get the trust from the dealers to the Company accessories." (v20, 068)
As a form of practical knowing, Ewan *extracts from this insight* and *asserts a consequence* (v20, 069) regarding the dealer challenge.	"So they're not even on customer level. They're on dealer level" (v20, 069)
The same stakeholder *builds on and amplifies* Ewan's assertion (v20, 070).	"They are to get the trust from the dealers." (v20, 070)

Figure 22.14 Connecting Concept Package and Collaboration Bridge through stakeholder experiences.

One Company stakeholder *proposes* Collaboration Bridge challenges he perceives within the Company. He emphasizes "the product side" (v20, 071) of the Concept Package and a need to expand the current portfolio.	"Really, the biggest challenge is really on the product side, to have that, something that you build up from." (v20, 071)
The same stakeholder *makes assertions* connecting this product-focused insight to the storytelling aspect of the Concept Package (v20, 071) and articulates a *practical action* to have both "short term" and "long term" approaches (v20, 073) to developing product opportunity areas.	"And then you have all these kind of ingredients about the communication, about the storytelling, but with the current program" (v20, 071) "...we need to have products really to- to- our new products...So, it's a two way. It's a short term eh action needed, but it's also on the long term." (v20, 073)
Ewan *affirms* and *builds on* this stakeholder's insights, *pointing* to the innovation matrix on the wall (presentational knowing) to connect this insight into different opportunity areas (v20, 076).	"And this is of course where this thing comes in (pointing to innovation matrix on wall), and we need to map those things, because we have something on the market, there will be something coming..." (v20, 076)

Figure 22.15 Connecting Concept Package and Collaboration Bridge through opportunity portfolios.

3.3.5 Rendering the boundaries of the concept package – the challenges with dealers

In the sub-episode in Figure 22.16, the team discusses challenges regarding the dealers who sell Company accessories. This renders visible implications for Collaboration Bridge and Concept Package dimensions because the issues speak to a challenge within

A stakeholder continues a previous conversation about dealer challenges, *sharing experience-based* details that *explains* how accessory dealers receive conflicting messages from the China-org and Company (v20, 077).	"...so then they get one message from there, and one message from the the Company and they have told the the Company China-org 'we are not allowed to sell your accessories'." (v20, 077)
Ewan *affirms* his understanding of the argument, and offers a "wild thinking" (v20, 080) *critical reflection question* about whether or not the core team is responsible for delivering the Concept Package to address this kind of challenge. Ewan *seeks coherence* regarding the boundaries of the Concept Package – does it include the dealer issue, or not?	"So, just to:- this is maybe a wild, wild thinking. (..) And that is, is it our responsibility or can we take on that responsibility with the dealers?" (v20, 080)
A Company stakeholder *affirms* a Collaboration Bridge boundary that includes dealer issues.	"...but we need to be aware of that this is boiling in the China-org, that we are supposed to get on board" (v20, 083)

Ewan *affirms* the assertion about the importance of dealer issues (v20, 086) and then *asserts* that the misalignment issue is a feature of the Collaboration Bridge dimension (between the China-org and the Company (v20, 089)) but not the Concept Package dimension of a "customer centered focus" (v20, 086).	"Yeah, it's just- I think it's important that we keep our focus eh customer- customer centered focus." (v20, 086) "...there's no doubt that dealership is a huge deal...a huge thing like eh misalignment within China-org in itself." (v20, 089)
A Company stakeholder adds personal insights regarding the dealer challenge by *sharing experience* with a "dealers dealers dealers" (v20, 090) emphasis in the Company and *proposing an action* of "maybe not have all the focus on the dealers" (v20, 090).	"But they are kind of (.) the challenge here might be that they are as- from top it has always been dealers dealers dealers dealers dealers period...focusing to get the dealers trust, of course, but, our challenge is to maybe not have all the focus on the dealers. And then we come from other part and make them realize about the rest." (v20, 090)
Ewan *affirms* her proposal, *re-asserting his "wild theory"* that the previous focus on dealer is "why we're here" (v20, 091) – that his team is involved in this project because "we haven't thought about 'what happens after dealer?'" (v20, 093).	"I think our wild theory is that it's- all the focus has been on the dealer dealer dealer, this is why we're here." (v20, 091) "we haven't thought about 'what happens after dealer?'" (v20, 093)

Figure 22.16 Identifying boundaries of the Concept Package through understanding stakeholders and challenges.

the Company but embodied within the product-sales-story features of the Concept Package.

Here, the group is co-inquiring into the overall scope of the problem space. They collectively and critically evaluate whether or not dealer challenges are a feature of the Collaboration Bridge or the Concept Package, or both. They seek consensual validity from each other grounding affirmations in experiential knowing and propositional knowing, and Ewan invokes a "wild thinking" critical subjectivity question for the group to reason through the role of dealers in the Concept Package. These intersubjectivity and critical subjectivity practices are intertwined. By the end of this segment, there appears to be some consensus that perhaps the issues with dealers is only relevant for the Collaboration Bridge dimension but may not be relevant for

Ewan offers a *thought experiment* to invoke critical reflection about the dealers. He *shares his reasoning* about the "zero value" (v20, 097) with respect to the Concept Package strategy.	"Okay let's say the dealer is perfect. Do we have a relevant product?" (v20, 095) "...the dealers are there, but dealers in itself has zero value. The customers that purchase from the dealer is the value." (v20, 097)
One stakeholder builds on Ewan's comments to *propose a reframing* of the Concept Package to include dealers and their role in communicating a story of user-Company alignment.	"as much as it (the dealers) is a challenge, it is also our strength....if we can explain it also like, put it, like structured way, so people understand, then they can kind of align with us easier" (v20, 098)

Ewan *affirms and amplifies* the stakeholder's insight evoking presentational knowing (a request to write down her insight) and invoking intersubjectivity practices (affirming her insight as embodying the project goals).	"I think what you're saying right now should be written down on a post-it, because...that is the value that this project will bring. Right there you kind of embodied it" (v20, 099)
Ewan captures this insight as a *practical action*, and *explains* how her insight embodies the meaning of the Concept Package in relation to the User Experience dimension, and *asserts* the importance of communicating this value within the Company via the Collaboration Bridge dimension.	"We will not come with that brand- 'this is the accessory, the cup holder that is also an ash tray'. It will not be that thing. It will be what you said right now. 'Look at this structure. Look at how you can see the products and the services and the story, and this is where the dealerships fits in, this is where the customer fits in, this is our gift. This is the value'" And, you know the more we can communicate that part, the better." (v20, 101)
Ewan then *grounds* the reasoning for his integrated approach in his *previous experiences*. He *explains* being "burned" (v20, 103) in the past when presenting a Concept Package and how people focus solely on the example in the package rather than the value of the Concept Package as an integrated framework. In this way, the Collaboration Bridge is rendered as a strategy not only for enabling a successful Concept Package handover but also enabling a mindset that values the integrated Concept Package idea.	"we have been burned so many times by coming with a suggestion-'this is an example of how it could be' and people see that example like 'okay, we gotta make that. This is how it is'. So instead of looking at these frameworks, they're looking at the little example we come with. And then they totally say 'yeah but I don't like it blue'. Like 'no. Don't like look at the colour. Look at how these things are working together'." (v20, 103)

Figure 22.17 Exploring the meaning of the Concept Package.

the Concept Package. At the same time, Ewan reintroduces a key aspect of the Collaboration Bridge – enabling a User Experience mindset at the Company through a customer-focused Concept Package strategy.

3.3.6 Rendering the meaning of the concept package as building value across dimensions

As illustrated in Figure 22.17, Ewan continues the discussion about dealers. He offers a thought experiment that evokes personal experience and propositions about the meaning of the Concept Package.

After Ewan invokes critical reflection questions this sub-episode becomes an example of co-inquirers finding coherence as evidenced by Ewan's request to capture a stakeholder's insight on a Post-It note. In the process, the group re-renders the Concept Package to include dealer issues, making a link between the Concept Package to the Collaboration Bridge. Intertwined intersubjectivity and critical subjectivity

Ewan *solicits validation* from the stakeholder's perspectives on the value and scope of the Concept Package deliverable.	"...do you think that that is: (.) or how do you feel by that as: a delivery? Like, that thing as a delivery?" (v20, 105)
One stakeholder *reaffirms* the value of the innovation matrix (represented on the wall) for mapping the different domains of customer opportunity areas. The second stakeholder responds with an *assertion* about the current state at the Company that *affirms* and *justifies* the value of the Concept Package approach because it aligns with Company values and fills a gap in understanding the Chinese customer.	"And to- to mapping the different eh, domain, yeah. (..) I think that's a good way" (v20, 108) "...yes, and it goes with the Company values as well, as you care about the customer, it is the (.) the thing for the Company and a core value as well, and of course we should see it from the customer point of view. And we have known so little about China, but now we have learned so much about China and the Chinese customer, and that we need to give to the others. So I think it's eh really really good." (v20 109)
The second stakeholder evokes practical knowing by assigning a *future activity* of enabling the Collaboration Bridge through effective communication of the Concept Package framework.	"Yes, absolutely. But if we get the framework and we can really explain it as well, in a structured way so everybody can follow that hasn't been involved, then it will hopefully easily (.) be aligned as well" (v20, 115)

Figure 22.18 Connecting the Concept Package and Collaboration Bridge through the innovation matrix framework.

practices are grounded and consummated through evoking all four ways of knowing (experiential, presentational, propositional, and practical).

3.3.7 Rendering concept package and collaboration bridge links – communicating the framework

In this sup-episode, the team is discussing the value of the innovation matrix framework in developing the Concept Package for the Company. As illustrated in Figure 22.18, the stakeholders identify benefits of alignment, insight, and communication. This effort to solicit feedback from stakeholders grounded in their Company experiences marks the end of the episode. The stakeholders confirm the value of the innovation matrix as a tool and the Concept Package as an integrated strategy, reaffirming links made between Concept Package and Collaboration Bridge dimensions. One stakeholder reaffirms the value of the innovation matrix in mapping product portfolio opportunities; the other stakeholder reaffirms the benefits of aligning user and Company values within the Concept Package. The episode ends with rendering a new link between Concept Package and Collaboration bridge in terms of effectively communicating the ideas embodied in the Concept Package within the Company to foster successful implementation.

3.4   Co-design of the problem space as co-inquiry

This analysis illustrates the social dynamics of rendering a problem space as a process of evoking experiential knowledge and seeking coherence among co-inquirers and their diverse ways of knowing. In other words, we observed *co-design of a problem space as a social process of co-inquiry.*

Through this episode, we show how a co-inquiry group collectively evoked experiential, presentational, propositional, and practical knowledge and engaged in

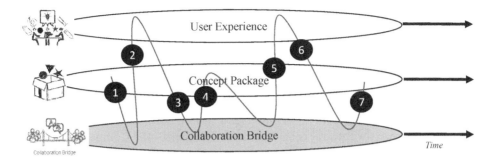

1. Rendering Concept Package and Collaboration Bridge links integrating user-Company values
2. Rendering an integrated problem space through "pushing the values story"
3. Rendering Concept Package and Collaboration Bridge links through stakeholder experiences
4. Rendering Concept Package and Collaboration Bridge links about opportunity portfolios
5. Rendering the boundaries of the Concept Package – the challenge with dealers
6. Rendering the meaning of the Concept Package as building value across dimensions
7. Rendering Concept Package and Communication Bridge links – communicating the framework

Figure 22.19 Rendering coherence of the problem space as an unfolding inquiry.

coherence practices to negotiate, interrogate, and create valid knowledge about a multi-dimensional problem space. Co-inquirers shared different object-world ways of knowing, made assertions and counter-assertions that encouraged dialogue and critical reflection, and extracted and reframed insights in productive ways. Sharing experiential knowing occurred through offering or requesting perspectives grounded in experience. Evoking presentational knowing was observed as using Post-Its, frameworks, and analogies as ways to make strategic knowledge available to the group. Propositional knowing was observed as making assertions (often grounded in experiential and presentational knowledge) and creating spaces for agreement and disagreement – spaces for working towards coherence about the problem.

Co-inquirers worked towards coherence through seeking consensual agreement and critically evaluating the validity and usefulness of assertions. Practical knowing was evident through building up connections across other modes of knowing and extracting those insights to direct future actions. All these practices are intertwined and interdependent, working together to build "coherence around new knowledge within the inquiry group" (Kasl & Yorks, 2002 p. 6) by grounding validity claims and building consummating relations among multiple ways of knowing (see Figure 22.2). While we present this insight for one episode, its essence was observed in all the selected 11 episodes from the Co-creation workshop debrief (v8) to the debrief with Company stakeholders (v20).

As shown in Figure 22.19, this episode also illustrates how *rendering coherence is an ongoing and dynamic process – an unfolding inquiry among co-inquirers* – rather than a static achievement. The co-inquirers contribute personal ways of knowing to a shared pool of experience and insight in ways that unveil elements of an integrated and comprehensive system of considerations linking human-centered design strategies (User

Experience), a unified innovation strategy (Concept Package), and change management strategies (Collaboration Bridge). As a system, these problem dimensions interact, and co-inquirers negotiate the boundaries of and relationships within this problem space in ways that seek integration of human-centered considerations with the ways organizations are prepared (or need to be enabled) to realize human-centered products and services.

Others in this volume provide complementary conceptual tools for rendering visible a multi-dimensional problem space, and enabling negotiation and meaning making across dimensions. Smulders and Dunne (2017) used the notion of "disciplina" as a tool to emphasize the integral role of intermediate users – the Company stakeholders as "the receiving actors" who will implement the Concept Package – to understand the goals, constraints or mindset of intermediate users and the broader context in which they operate. Similarly, Jornet and Roth (2017) observed the reframing of the problem to include the take-up of the Concept Package as a {Thinking | Feeling} unity. Fila and Hess (2017) used an empathy methodology to investigate challenges with designing for multiple users who may be in different problem spaces within the overall project. Lloyd and Oak focused on how problem space dimensions are in "value tension" and used this as a device to understand co-inquirer's use of narrative in resolving opposing values between Chinese users and the Company.

3.5 Contributions and implications

By using the collaborative inquiry paradigm, this study offers a new perspective on design as a social process that characterizes *co-design as co-inquiry* – a process in which co-inquirers share object-world ways of seeing the world (through evoking four ways of knowing) and build coherence among these object-world ways (through invoking critical subjectivity and intersubjectivity practices) to co-construct valid knowledge about a situation of shared interest. We have shown how these co-inquiry practices are observable and reveal mechanisms for understanding how one inquiry group collaboratively renders a problem space as an ongoing and dynamic emergent process. This problem space is multi-dimensional, integrating a human-centered perspective (the context of use) with an organizational perspective (the context of realization). As such, this study reveals a broader space of considerations for constructing and negotiating valid problem frames, a space that acknowledges how the broader organizational context shapes the problem frame into worlds of possibilities (and impossibilities).

By attending to the dynamics of negotiating among object-world ways, this view of design as co-inquiry offers an alternative framing of design as a social process of *negotiation* (Bucciarelli, 1994) to design as a social process of *building coherence*. This moves beyond negotiation as persuasion practices towards negotiation as invoking validity practices to build coherence and co-construct valid knowledge (Heron & Reason, 1997). Others in this volume use the concept of "coherence" in ways that resonate with this usage yet offer different insights. For example, McDonnell (2017) characterized observed failures in operationalizing a problem frame as a "lack of coherence". Menning et al. (2017) used coherence to indicate "relatedness"; however, their concern was a lack of critical intersubjectivity or group think that might foster fixation. Chan and Schunn (2017) measured semantic coherence as probability distributions of co-occurring topics to identify and cull for more detailed analysis "linguistically interesting" episodes from large unwieldy datasets.

By attending to the primacy of subjective knowing in collaborative inquiry, this view resonates with views on the nature of design as situated knowing and reflective practice (Dorst, 2006; Schön, 1993). Ylirisku *et al.* (2017) also took a situated lens to the data to investigate "resourcing" – how a group negotiates and signifies what is available and relevant to the team for co-designing. Resourcing takes on two roles: (1) making experience resources available to others; and (2) inviting further elaboration and resourcing of additional experiences. These roles align with collaborative inquiry (see Figure 22.2): (1) evoking and sharing of experiential knowing to provide a collective pool of experience and insight; and (2) building coherence recursively via critical reflection and action (Heron & Reason, 1997). Similar to the collaborative inquiry paradigm, resourcing locates the social process of design within a participatory methodology and emphasizes the primacy of experiential knowing as a resource that is continually shaped, re-contextualized, re-prioritized, and transformed in design projects over time (Bray *et al.*, 2000). Jornet and Roth (2017) approached the dataset through a design as inherently social and reflexive lens. They took a sociogenetic approach by focusing on the unity of social reasoning in design thinking that transcends individual manifestations to where "individuals and materials shape each other in mutually constitutive relations". This unity speaks to an entwined design thinking practice where thinking and communicating are different manifestations of the same practice (e.g., signified as {Thinking | Communicating}) and the outcome is emergent as a cumulative unfolding of communicative work. Mapping to the collaborative inquiry paradigm, we wonder if entwined practices of evoking ways of knowing and building coherence towards collective truth claims represent a unity (i.e., {Evoking knowledge | Building coherence}) where co-inquiry cannot be reduced to the sum of its parts but rather "transcends the different individual manifestations" (Jorney & Roth, 2017) as co-inquires engage in cumulative social reasoning.

An additional benefit of the collaborative inquiry paradigm is that it embodies a holistic research-to-practice continuum. As a research paradigm, collaborative inquiry guides practices for conducting human inquiry and developing knowledge as a social practice (Bray *et al.*, 2000). As a practice paradigm, the collaborative inquiry paradigm complements design methodologies such as participatory design that emphasize participants as co-inquirers, co-inquiry as action-oriented, and inquiry problems as human-centered. It is also a well-developed approach for facilitating adult and organizational learning (Kasl & Yorks, 2002). As a learning approach, co-inquirers share responsibility in the mutual pursuit of meaning, creating a learning structure that mirrors an ongoing experiential learning cycle arrived at through dialogue, reflection, and shared live experiences (Bray *et al.*, 2000). This can foster transformative learning in which co-inquirers become more cognizant of aspects of their life-world (Mezirow, 2000). For example, while co-inquirers engage in inquiry together, they do so for divergent reasons. Wrestling with this divergence can create disruptions that trigger critical dialogue and reflection on multiple meanings held within the inquiry group, and the realities those meanings represent (Bray *et al.*, 2000). This creates opportunities to collectively problematize those disruptions to make meaning and create a more comprehensive understanding of the world. In other words, designers (at any stage of expertise) can use these co-inquiry practices designers to monitor their design interactions with other co-inquirers – how they share experiential knowing, propose claims, critically reflect on those claims, and hold each other accountable in building validity and trustworthiness around claims.

REFERENCES

Adams, R.S. and Krautkramer, C. (2016). Harnessing the Elusive Expertise of Big Picture Thinkers. Proceedings of the University-Industry Interaction Conference, Amsterdam, June.

Brown, T. (2009). *Change by Design*. Harper Collins.

Bray, J.N., Lee, J., Smith, L.L., and Yorks, L. (2000). *Collaborative Inquiry in Practice: Action, Reflection, and Meaning Making*. Sage Publications: Thousand Oaks, CA.

Bucciarelli, L. L. (1994). *Designing engineers*. Cambridge, MA: MIT Press

Chan, J. & Schunn, C. D. (2017). A Computational Linguistic Approach to Modelling the Dynamics of Design Processes. In: Christensen, B. T., Ball, L. J. & Halskov, K. (eds.) *Analysing Design Thinking: Studies of Cross-Cultural Co-Creation*. Leiden: CRC Press/Taylor & Francis.

Christensen, B. T. & Abildgaard, S. J. J. (2017). Inside the DTRS11 Dataset: Background, Content, and Methodological Choices. In: Christensen, B. T., Ball, L. J. & Halskov, K. (eds.) *Analysing Design Thinking: Studies of Cross-Cultural Co-Creation*. Leiden: CRC Press/Taylor & Francis.

Clemmensen, T., Ranjan, A. & Bødker, M. (2017). How Cultural Knowledge Shapes Design Thinking. In: Christensen, B. T., Ball, L. J. & Halskov, K. (eds.) *Analysing Design Thinking: Studies of Cross-Cultural Co-Creation*. Leiden: CRC Press/Taylor & Francis.

Cross, N. (2006). *Designerly Ways of Knowing*. London: Springer-Verlag.

Davis, B. and Sumara, D. (2008). Complexity as a theory of education. *Transnational Curriculum Inquiry, 5*(2), pp. 33–44. URL: http://nitinat.library.ubc.ca/ojs/index.php/tci

Dhadphale, T. (2017). Situated Cultural Differences: A Tool for Analysing Cross-Cultural Co-Creation. In: Christensen, B. T., Ball, L. J. & Halskov, K. (eds.) Analysing Design Thinking: Studies of Cross-Cultural Co-Creation. Leiden: CRC Press/Taylor & Francis.

Dong, A. & MacDonald, E. (2017). From Observations to Insights: The Hilly Road to Value Creation. In: Christensen, B. T., Ball, L. J. & Halskov, K. (eds.) *Analysing Design Thinking: Studies of Cross-Cultural Co-Creation*. Leiden: CRC Press/Taylor & Francis.

Dong, A. (2006). The enactment of design through language. *Design Studies*, 28, pp. 5–21.

Dorst, K. and Cross, N. (2001). Creativity in the design process: Co-evolution of problem-solution. *Design Studies*, 22(5), pp. 425–437.

Dorst, K. (2006). Design Problems and Design Paradoxes. *Design Issues*, 22(3), pp. 4–17.

Dorst, K. (2011). The core of 'design thinking' and its application. *Design Studies* 32, pp. 521–532.

Dove, G., Abildgaard, S. J., Biskjaer, M. M., Hansen, N. B., Christensen, B. T. & Halskov, K. (2017). Grouping Notes Through Nodes: The Functions of Post-It Notes in Design Team Cognition. In: Christensen, B. T., Ball, L. J. & Halskov, K. (eds.) *Analysing Design Thinking: Studies of Cross-Cultural Co-Creation*. Leiden: CRC Press/Taylor & Francis.

D'souza, N. & Dastmalchi, M. (2017). "Comfy" Cars for the "Awesomely Humble": Exploring Slangs and Jargons in a Cross-Cultural Design Process. In: Christensen, B. T., Ball, L. J. & Halskov, K. (eds.) *Analysing Design Thinking: Studies of Cross-Cultural Co-Creation*. Leiden: CRC Press/Taylor & Francis.

Duranti, A. (2010). Husserl, intersubjectivity and anthropology. *Anthropological Theory*, 10(1), pp. 1–v20.

Goldschmidt, G. (1991). The Dialectics of Sketching. *Creativity Research Journal*, 4(2), pp. 123–143.

Gray, C. M. & Boling, E. (2017). Designers' Articulation and Activation of Instrumental Design Judgments in Cross-Cultural User Research. In: Christensen, B. T., Ball, L. J. & Halskov, K. (eds.) *Analysing Design Thinking: Studies of Cross-Cultural Co-Creation*. Leiden: CRC Press/Taylor & Francis.

Hess, J. L. & Fila, N. D. (2017). Empathy in Design: A Discourse Analysis of Industrial Co-Creation Practices. In: Christensen, B. T., Ball, L. J. & Halskov, K. (eds.) *Analysing Design Thinking: Studies of Cross-Cultural Co-Creation*. Leiden: CRC Press/Taylor & Francis.

Heron, J. and Reason, P. (1997). A Participatory Inquiry Paradigm. *Qualitative Inquiry, 3*(3), 274–294.

IDEO. (2011). Human Centered Design Toolkit: Atlas Books.

Jornet, A. & Roth, W. (2017). Design {Thinking | Communicating}: A Sociogenetic Approach to Reflective Practice in Collaborative Design. In: Christensen, B. T., Ball, L. J. & Halskov, K. (eds.) *Analysing Design Thinking: Studies of Cross-Cultural Co-Creation*. Leiden: CRC Press/Taylor & Francis.

Kasl, E. & Yorks, L. (2002). "Collaborative Inquiry for Adult Learning." *New Directions for Adult and Continuing Education, 2002*(94), 3–12. http://doi.org/10.1002/ace.54

Kolko, J. (2010). Abductive thinking and sensemaking: The drivers of design synthesis. *Design Issues, 26*(1), pp. 15–28.

Kotter, J. P. (1995). Leading change: Why transformation efforts fail. *Harvard Business Review*, March–April, pp. 59–67.

Krippendorff, K. (2006). *The Semantic Turn: A New Foundation for Design*. Boca Raton: Taylor and Francis.

Lawson, B. & Dorst, K. (2009). *Design Expertise*. Architectural Press: Boston.

Lloyd, P. & Oak, A. (2017). Cracking Open Co-Creation: Categorizations, Stories, Values. In: Christensen, B. T., Ball, L. J. & Halskov, K. (eds.) *Analysing Design Thinking: Studies of Cross-Cultural Co-Creation*. Leiden: CRC Press/Taylor & Francis.

Luecke, R. (2003). *Harvard Business Essentials*. Harvard Business School Publishing, Boston.

McDonnell, J. (2017). Design Roulette: A Close Examination of Collaborative Decision-Making in Design From the Perspective of Framing. In: Christensen, B. T., Ball, L. J. & Halskov, K. (eds.) *Analysing Design Thinking: Studies of Cross-Cultural Co-Creation*. Leiden: CRC Press/Taylor & Francis.

Menning, A., Grasnick, B. M., Ewald, B., Dobrigkeit, F., Schuessler, M. & Nicolai, C. (2017). Combining Computational and Human Analysis to Study Low Coherence in Design Conversations. In: Christensen, B. T., Ball, L. J. & Halskov, K. (eds.) *Analysing Design Thinking: Studies of Cross-Cultural Co-Creation*. Leiden: CRC Press/Taylor & Francis.

Mezirow, J. (1997). Transformative learning: Theory to practice. *New Directions for Adult and Continuing Education, 1997*(74), 5–12. doi:10.1002/ace.7401

Mezirow, J. (2000). Thinking Like an Adult. In: J. Mezirow and Associates, *Learning as Transformation*. San Francisco: Jossey-Bass.

Nelson, H. and Stolterman, E. (2003). *The Design Way: Intentional Change in an Unpredictable World*. New Jersey: Educational Technology Publications.

Schön, D. A. (1993). *The Reflective Practitioner: How Professionals Think in Action*. Basic Books, New York.

Solis, F. (2015). *Characterizing Enabling Innovations and Enabling Thinking*. Ph.D. Dissertation, Purdue University, West Lafayette, IN.

Solis, F. and Sinfield, J. (2015). Rethinking Innovation: Characterizing Dimensions of Impact. *Journal of Engineering Entrepreneurship, 6*(2), 83–96.

Smulders, F. & Dunne, D. (2017). Disciplina: A Missing Link for Cross Disciplinary Integration. In: Christensen, B. T., Ball, L. J. & Halskov, K. (eds.) *Analysing Design Thinking: Studies of Cross-Cultural Co-Creation*. Leiden: CRC Press/Taylor & Francis.

Tversky, B. (2011). Spatial thought, social thought. In: T.W. Schubert and A. Maas (eds.), *Spatial Dimensions of Social Thought* Germany; de Gruyter. pp. 17–38.

Visser, W. (2006). *The Cognitive Artifacts of Designing*. New Jersey: Lawrence Erlbaum.

Ylirisku, S., Revsbæk, L. & Buur, J. (2017). Resourcing of Experience in Co-Design. In: Christensen, B. T., Ball, L. J. & Halskov, K. (eds.) *Analysing Design Thinking: Studies of Cross-Cultural Co-Creation*. Leiden: CRC Press/Taylor & Francis.

Chapter 23

Designing the Constraints: Creation Exercises for Framing the Design Context

Claudia Eckert & Martin Stacey

ABSTRACT

Developing an understanding of a design problem by exploring the context in which the new product will be marketed and used is often a crucial part of the design process but has been little studied outside the fashion and textiles industries. A user experience design team in a European car company sought to understand the interests and values of potential Chinese customers by carrying out a co-creation exercise with a set of representative Chinese consumers, in order to understand how to design accessory products and services for them. This paper compares the co-creation exercise, which produced accounts of the consumers' values in verbal narrative form, to the constraint gathering research phase of artistic design processes, which typically produce sets of constraints, usable design features and desirable emergent properties to express the space of possible designs in visuospatial form as mood boards.

1 INRODUCTION: DESIGNING THE IDENTIFICATION OF CONSUMER VALUES

Designing a new car and bringing it to its multiple markets is one of the most complex engineering challenges, involving hundreds of thousands of people across the supply chain working together across the world. Many of the fundamental design decisions are taken by centralized design teams located in developed countries. However, the ultimate success of the car does not only depend on the technical quality of the car, but also on understanding these different markets and the needs and values of their customers. One of the great challenges is for design teams to understand the customers in these highly varied markets. It is not enough to analyze past sales: to enthuse the customers and open new markets the designers need to understand the values and the way of life of their target customers. The data analyzed in the book covers a series of co-creation exercises between a European user experience design team and a group of Chinese customers in which the designers are trying to elicit the values of their customers. In this paper we look at how designers in other design fields, notably fashion, go about understanding what their customers want to buy, to interpret the data as a value elicitation exercise.

The design team faced a completely open-ended task. They needed to come up with a product, feature or service which could be added to or sold in conjunction with an existing car, but targeted specifically at the Chinese market. To maximize the appeal to the target market segment in China, the design team wanted to understand the values that the potential customers have and find a way to align these to the brand values of the car company. To do this, the company's user experience team carried out what they termed a co-creation activity with a carefully selected group of Chinese consumers who fell into this market segment, where the customers started by talking about very broad themes, which were gradually narrowed to focus more on transport and cars. However, this did not involve any joint creating done together by the consumers and the European designers within the available dataset. Rather than asking the group directly about their values, or what features they would or would not like a car to include, the designers set up a number of exercises intended to reveal their values, culminating in the users designing and pitching a product which encapsulates "the good life" and gives them more "me-time". The planning of a sequence of activities for the participants in the study was itself a design process. To maximize the results from the two one-day workshops the team planned the activities to be carried out very carefully and provided detailed briefs that were issued to the various consumer groups. By removing the focus from cars, the design of the co-creation activity became even more open-ended.

In this paper we will argue that this co-creation process played a similar role in the overall design process to the search for sources of inspiration carried out by designers in artistic design domains, which at the same time frames what is an acceptable design for a particular market and also yields ideas for concrete products. In fashion and knitwear design the functional requirements for the product are well understood and largely taken for granted; the key is producing designs that fit the aesthetic preferences and self-understanding that the customers will have when the products are on sale. This involves understanding the space of acceptable designs within a particular fashion, and the functional and aesthetic relationships particular designs have both to complementary products and to competitor products (Eckert & Stacey, 2001, 2003a). This includes visual similarity and difference, aesthetic and cultural connotations, price and perceived quality. The aim is to produce products that target a particular market segment, not just by being fashionable but by being similar and different from comparable rival products in ways that correspond to how their customers see themselves and what they want to project, while maintaining the company's brand values. Similarly, the aim of the car company's user experience designers was to find ways to position the company's cars as distinctive from rival products in ways that both fit the company's longstanding brand values and appeal to the values and attitudes of Chinese consumers.

Fashion and knitwear designers gain this understanding of context by looking at other designs and a wide variety of other objects and images. The car company's user experience designers also looked far beyond cars by setting general themes like "well-being" and design exercise tasks for a group of potential users to gain an understanding of the values that drive their lives and purchasing decisions. The car experience designers were also open to concrete design suggestions for car accessories that they could use but that wasn't their main goal. In both cases designers want to gain an impression of the product and/or what it connotes for the customer, and suggest the space of possible designs. Artistic designers, like the knitwear designers we studied in the past, typically encapsulate their understanding in references to objects often expressed in mood boards, which can be appreciated visually. The user experience designers in this case study collected a set of stories and anecdotes, which encapsulated the values of the users, which might attract consumers to their company's cars.

Identifying the context for a product (through sources of inspiration or in this case through narratives of values) is an important part in the active constraint management that designers perform to formulate the right problems to solve. How they do this depends largely on the kind of problems they face: designers of open-ended problems engage in a constraint finding process in order to limit the search space of the open-ended task to something manageable (Stacey & Eckert, 2010). Here the designers designed the co-creation participants' activity by defining the exercises, and redirecting and narrowing their focus. Ewan, the head user experience designer, describes this as "we will go to a Chinese city. We will then – not as we normally do . embed ourselves in a very ethnographic way with the users for hours or days even. Now we will rent a venue and we will have a centralized point where we conduct co-creation workshops" (v1, 019). For the time the designers are investing this gives them considerably more control over the process. The outcomes of the co-creation exercise should contribute to constraining and focusing the design of the car company's products and services.

The paper outlines its method in Section 2, and in Section 3 briefly discusses the use of sources of inspiration and mood boards in artistic design, and the role of constraints in design processes. As the paper is part of a collection of papers covering the same co-creation activity the paper does not cover the co-creation literature in the expectation that other papers will cover this and that the workshop will reveal the antecedents of the approach taken by the car company's designers. Section 4 discusses the procedure followed by the designers as a repeated process of constraining the activities of the users and interpreting the results to meet the aims of the designers. Section 5 argues for this process being akin to the creation of a mood board, before conclusions are drawn in Section 6.

2   METHOD

The paper is based on the entire corpus of the transcripts (v1–v22). The authors listened together to the audio recordings of the sessions to gain an overview of the process and to relate the behaviour in the process to the authors' interests and experiences of design processes. In discussions about whether and how the co-creation activities corresponded to features of familiar design processes, they considered the idea that the activities were an exploration of the design context, thus that the purpose of the process was to understand the values and concerns of their potential customers, rather than to obtain concrete ideas for product features. One of the characteristics of early phases of artistic design processes is the use of strategies to reduce the search space through the introduction of constraints and quick decision making about which constraints to impose, which will constrain future steps. The result of this process is typically a coherent, but tacit, understanding of the range of acceptable designs within a fashion or a corporate style. Therefore the authors decided to look for points when the interaction designers constrained the process.

The first author then read through the transcripts highlighting points where the designers were setting constraints on their process and that of the users in the workshops. The highlighted constraints were then discussed with the second author to develop the argument of this paper.

Throughout the paper the user experience design team from Europe and China are collectively referred to as the designers and the participants in the workshops as users. In listening to the sessions the constant unchallenged reframing of the themes underlying the task of the users became apparent. The authors were immediately struck by the somewhat dismissive way the designers talked about the users as subjects to be studied rather than as colleagues, which signaled that they were not thinking of the user group as equals in a joint designing activity. This reminded the authors of knitwear designers keeping images and objects if they suit their purposes or throwing them away if they don't (Eckert & Stacey, 2003a; McGilp et al., 2016).

In the analysis we did not consider the dynamics within the design team and speak collectively of the 'the designers' unless we assign a quote to a particular individual. We choose not to differentiate between different team members, because there was little evidence of disagreement or different approaches amongst the group. The discussions of the protocols are intended as illustrative examples of the designers' behavior rather than a comprehensive analysis.

3   BACKGROUND: PROBLEM FRAMING IN OPEN-ENDED DESIGN

The DTRS 11 protocols cover some of the very early phases of an open-ended design task, where designers need to understand their customers, their markets, their competitors and the product they need to generate. Maybe unusually in this case study the designers are not even sure of the nature of the product or service they are aiming to define, as long as it is somehow associated with a car and can be sold in conjunction with a car or draw customers into buying a car. However, they are clear that they need to understand the way of life and values of their target customers.

3.1   Understanding the design space

The problem framing phase in open-ended design processes has three interrelated aspects: understanding the space of designs that could succeed in the market; finding sources of inspiration for these design, which also serve to frame the design space; and expressing the space of designs as potential product ideas.

3.1.1   A prepared mind

Much design research has concentrated on the process of designing from an idea for a product to the final design; however, the phase leading up to this process is less well understood. In some design processes developing a view of the problem to be solved, before anyone starts doing what would commonly be recognized as designing, plays a crucially important role. At this point designers need to understand the context in which their product will operate, the preferences and dislikes of their users or customers, and the offerings that competitors can bring to the market or have already brought to it. This is the stage when the space of possible designs is investigated.

Design spaces have been formally addressed in the generative design community (e.g., Woodbury *et al.*, 2000), which is interested in the range of designs that can be generated from an explicit set of rules. However, in most design fields the understanding of what is and is not a legitimate design within the space of possibilities is not explicit. One area where the development of the designers' understanding of the space of acceptable designs has been studied is fashion and textile design, where designers generate hundreds of designs every season and they are very open about their creative processes, which are largely centered around understanding the coming season through other objects that are on sale (Eckert & Stacey, 2003a). To identify the context of a coming season, fashion designers are looking at other garments in the same market, which allow them to tune their tacit perceptions of what is fashionable as well as setting explicit reference points for their own designs. They also look for themes that receive public attention and hit the mood of their times (Eckert & Stacey, 2003a). However, while designers are habitually on the lookout for ideas (Goldschmidt, 2015), understanding the coming season is not a random process, but typically a well-structured search through shops, magazines and online media, such as blogs (Eckert & Stacey, 2003a; Rocamora, 2011). Designers attend or follow the reports of fashion shows like London Fashion Week, which render 'visible the boundaries, relational positions, capital and habitus at play in the field' (Entwistle & Rocamora, 2006). One of the differences between fashion designers and the car company's designers in this dataset is that fashion designers rarely seek out interactions with their users, but search for objects that the consumers can buy. However, fashion designers are, of course, embedded in the culture of their consumers, whereas the problem for the car company's designers was that they weren't.

3.1.2 Sources of inspiration

Objects and images that trigger ideas play an important role in many design processes. In fashion and textiles design this is entirely overt. In some other domains this is less obvious or readily acknowledged, though precedents play a distinct role in the culture of architecture (see Goldschmidt, 1998). Sources of inspiration are at the same time material objects whose physical properties are important, and also manifestations of values and aesthetics. The same object can have multiple roles in defining a context and being the starting point for design ideas. In some cases sources of inspiration are adapted very literally where prominent features are developed in exactly the same way in the new design. However, they can also be the starting point for a chain of thought, either abstracting an idea that stands behind the design which is then instantiated with another design embodying the same abstract category; or simply as the starting point for loose association where designers might recognize similarities or just be reminded of something (Eckert & Stacey, 2003b). The sources of inspiration imply a space of possible designs. However, the space is very fluid and subjective as everybody interacting with it needs to interpret it in light of their own experience (Eckert & Stacey, 2000).

3.1.3 Expressing the design space

Rather than achieving an explicit understanding of the space of possible design and their properties, this understanding mainly remains tacit. The understanding of design spaces gained through looking for sources of inspiration are often expressed in mood boards, which are typically a range of images arranged in an aesthetically appealing way to give an impression of a future collection with an indication of the mood, colours, customers, details and aesthetics (Faerm, 2010) as well as the voice of the designer (Dabner, 2004). Lucero (2012) suggests that mood boards can play five main roles in the early stages of the design process: framing of both the problem setting and the problem solving activities; aligning divergent views of different designers and designers and clients; paradoxing, where designers can explore conflicting or contradicting ideas; abstracting enabling the designer to work on different levels of abstraction; and directing the future of the design process.

3.2   A prepared problem: framing and reframing

Framing the design problem is an important step in turning an open ended design task, with a potentially infinite design space, into a well-defined set of requirements. Designers actively construct their problems, when they begin and when they reach an impasse, not only to facilitate creative acts but to get the creative acts to yield the results they really want (Darke, 1979; Schön, 1983; Lloyd & Scott, 1995; see Cross, 2004a). Schön (1983) points out a creative design hypothesis "depends on a normative framing of the situation, a setting of some problems to be solved", which identifies pertinent aspects of the design out of the otherwise overwhelming complexity (Takeda *et al.*, 1999). Cross (2004b) found that what the three outstanding designers he studied had in common included exploring the problem space from a particular perspective in order to frame the problem in a way that stimulated and pre-structured the emergence of design concepts (Cross, 2004a). Sometimes designers need to see a design problem from another perspective (Kolko, 2010): this is reframing, which designers can pursue consciously as active construction of their problem or do unconsciously by being prompted by an external stimulus, like a source of inspiration or some ambiguity in the representations that they have generated (Schön & Wiggins, 1992).

3.3   Constraint management

Designers in many fields actively manage the constraints on the design that they use to focus their search for solutions (see Stacey & Eckert, 2010; Eckert & Stacey, 2014; Onarheim, 2012). We see constraint finding as a special form of problem framing. Another important aspect of problem framing is finding, reducing or reprioritizing constraints. This reduces the design space and focuses designer's attention to a particular part of the design space.

Research on creative idea generation, such as Finke's (1990) work on pre-inventive forms, where he encouraged people to imagine particular shapes, and then use them in creative tasks, indicates that tasks requiring imagination (but soluble in a wide variety of ways) are made easier by tight constraints that supply elements of solutions to be combined and adapted, and reduce the spaces of possible solutions (see Finke, Ward, & Smith, 1992). The nature of the constraints determine what people think the design problem *is*, as well as what appears to be a plausible part of a solution. The central role of the most salient constraints in guiding the conceptualization of the design problem and the generation of the key elements of the design has been well recognized by design researchers for a long time (see Darke, 1979).

Eckert and Stacey (2014) argue that constraints on design have three main sources:

- The problem that the design must solve or the need that the design must meet; this includes product requirements, manufacturing requirements, and constraints stemming from the strategic goals of the company.
- The process by which this is achieved.
- The emerging solution – since making certain decisions will rule out or restrict options for other later decisions.
- In some domains like software design, designers look systematically for constraints and requirements. However in artistic design processes designers have a habitual order of making decisions, for example selecting colour schemes, fabrics in tailoring or fonts in graphic design, which constrain the design space significantly (Stacey & Eckert, 2010). While some of the decisions are made after negotiations in meetings, others are the by-products of decisions or are made without people being aware of making these decisions.

4 DESIGNING THE CONSTRAINTS

While the user exercise is presented as a co-creation activity, its main goal is not to generate concrete ideas for accessories for cars, it is to gain a deeper understanding of the values that motivate car purchasing decisions in the Chinese market in the context of the values that drive Chinese society or as Ewan put it "the bleeding edge of our insights" (v18, 196). In v18 (006) Ken reflected on their exercise saying: "I think the nice thing about that is: it somehow reflects the way the Chinese or Asian people in general are thinking that they are not just that person. Everything that you do is always based on all of the people around you". Much of this is encapsulated in narrative.

The designers talked themselves about needing to "structure in some of the other constraints" (v10, 003) on the tasks the users carry out so as to shift their emphasis. They were conscious that they "can control [the activities] so much more" (v9, 114). However, they were mainly thinking about defining the activities for the users. In this section we will argue that the designers were also constraining the process through which the activities are designed and often made these decisions without much negotiation and concern for the effect on what the users could generate. One important element of this is reframing of the themes that the users are working with.

4.1 Selecting the participants

The first step of the designers was to define what a suitable group of users would look like, similar to designers in other fields deciding which shops or websites to look at for context and sources of inspiration.

Given the time constraints of 10 hours (v2, 64), they needed to use devices to get quickly at authentic insights. They opted for role play (v2, 102) and "slowly kind of trick them in to it, then suddenly they're building without even knowing" (v2, 109). They needed to guide and constrain the behaviour of the users very strongly. They wanted to "provoke" responses from the users (v4, 252). However, they were highly concerned that Chinese would behave differently from Americans (v2, 113) who are very open, so that the design of the workshop exercises and the selection of the people became critical. They wanted people who actively contribute to discussion rather than just agree with others (v2, 161) and are "in general extrovert" (v3, 142).

In spite of the constraints on the workshop they further specified the range of the target users from different disciplines (v3, 22) with different expertise (v3, 35), who were "average people" (v3, 81), who drove a "premium car" (v3, 81), but were not "too rich" (v3, 85) and were roughly from the same level of society (v3, 121). One of the areas of great concern to the designers was the language barrier between them and the users. Therefore they wanted to select people who are "lead users, they are comfortable both in English and Chinese" (v3, 341) or have lived abroad (v3, 280). The workshop had Chinese moderators, who could however not translate at the same time as running the workshop; therefore a long discussion ensued at the end of v3 about the ability to hire sufficient translators on the project budget.

4.2   Defining the activities

Prior to the activities covered by the protocols the team had taken a decision to run with co-creation workshops, which predefined and constrained the activities that they could carry out: fixed lengths of time, number of participants, workshop format. Within those constraints they needed to define activities. First, they decided on having at least some hands-on physical activities, which provided an authentic experience: "fidelity level, which is high enough for them to – to focus on what it actually is rather than: what they see, like, piece of paper or something, but focus on the concept instead" (v2, 10). Getting this balance right between anchoring the users with concrete designs and concrete problems and getting their reflection on issues to gain an insight into their values remained an issue throughout the process. They therefore saw the need "to somehow calibrate the users before you can calibrate the method" (v2, 10). They therefore decided on "going physical" (v2, 15). At the beginning of the exercise they anticipated outcomes on two levels: ideas for products and the communication with the consumers (v2, 54); however later they switched their attention mainly to the values that drive the users. Rather than obtaining a reasoned account of their user values or attitudes and needs for car usage, they want issues to "emerge" (v2, 57) in a way that "set their emotion on fire" (v2, 59) by putting them in the right mood. They want to see people's "gut reaction" (v3, 264).

The design activity was highly constrained by the format and the resources, but at the same time the pragmatic considerations about time, funding and the selection of people inevitably biased the results of the workshop, as the people were unlikely to be typical of the Chinese population or represent their values. However this was not discussed during the sessions. The designers didn't challenge each other during the session on the effects that the choices they made and constraints they set had on the results of the workshop or ultimately on the designs that would be derived from them.

4.3   Influencing the participants

The designers were very conscious that they had limited time with the users and therefore wanted to maximize what they can get out of them. They therefore planned the activities that the users were to carry out very carefully and set a tight schedule for each of the activities. After a number of warm up exercises like animal role play (v5, 495), they required the participants "to solve a problem ..., because this is the way the Chinese people think" (v5, 26). However, they saw a challenge that if the participants thought from a company perspective, they would think about money, rather than staying in their "own personal realm" (v5, 33). Rather than questioning whether setting them a company problem was the right way to approach their goal of obtaining values, they planned how to direct the participants and concluded "the companies [need to be] based on something a little bit more concrete than just the values and themes" (v5, 89). They wanted the users to pitch ideas at each other and then make investment decisions about them; however they were concerned that bad pitches by the users might obscure good ideas (v5, 219). The designers hoped the users' ideas will "align" with each other and to the themes (v5, 175; also in v5, 995).

The users were requested to write ideas on post-it notes and then stand up and present them (v5, 657). The designers also planned to mix up the groups to maximize the dialogue between different groups. The designers were interested in the "why they make different choices" (v5, 675), as the rationale would reveal the underlying values rather than remaining with the ideas as such. Essentially the designers were trying to direct the users into revealing what they are interested in without directly asking them the straight questions about what they would like to see as car accessories and what values they have that would influence the selection. The users provided sound bites and deeper explanations about their underlying values. The users were not given a chance to take ownership of the process or to make amendments to the running order; in that way they were passive in the process and produced ideas that could be picked up by the designers in a similar way to how objects in a shop are passive agents from which sources of inspiration are selected. The designers designed a process that exposed ideas and values, where at some level the ideas and the people who generate them were being treated are interchangeable.

4.4   Selecting the themes for discussion

In v7 the designers discussed the topics that emerged from the first session and the values underlying them. However, they knew very little about the participants and therefore used clichés to interpret them. In these interpretations the designers jumped from one topic to the next, without discussing or questioning these jumps. While systematically scrutinizing propositions by colleagues would not fit the free flowing spirit of the conversation, the designers rarely questioned each other.

This is exemplified by the discussion at the beginning of v7.

v7, 002   ...eh number one will be in terms of the body and the mind, eh: health. Right? So, physical and mental health, alright?

v7, 004   they were talking about eh body and mind eh: I mean several things of course some exercise, where to exercise, eh but they straight away started talking about things like relaxation

v7, 008   relax a little bit, when you get away from your daily grind and your stresses

v7, 009   ... the younger guy, who ... leads a slightly more disciplined life. ...He's not married, he's not. He talks about things like sleeping early:, going to bed by ten:, waking up really early by six:, you know, because your body starts to detox at eleven a clock.

v7, 015   my reading of him, I don't think he's very hard core in partying. Probably: it's only because he's (certainly?) younger, he's single:, so he's got more time for that kind of thing, but I think being a guy, he probably wants more of a party here
v7, 022   so he wants freedom
v7, 023   ... time and freedom

In this short session of less than 4 minutes the designers jumped from body and mind → physical and mental health → exercise → relaxation → sleep → freedom. While the session probably did cover these concepts, the way the conversation developed biased the way the designers set the topics for the next session on the values that they identify from the conversation. The discussion illustrates how the designers put their interpretation about the person into the development of the themes. The reference to the party illustrates how little this is founded on evidence about the people, since the person appears to have stated his desire for early nights and relaxing sleep, but as he is young and single they stated that he still might want "more of a party".

The designers started off with setting 'health' as a theme for the workshop, but quickly moved this on to 'good life'. In v4, (168) Abby rephrases the topics: "... to kind of force them a little bit to okay, what else could health be?'. Now they have to find that it could be: caring, it could be security, it could be all of this. Or the 'good life' what is the – I guess this is good life (...) [pointing at the white board] pillars, right?" Through this uncontested rephrasing, she shifted the discussion significantly and opened up significant new areas of discussion. As the term is broader it moved the discussion away from health. It was interpreted throughout the sessions in different ways, such as "me-time" (v8, 015, iterated in v9, 040), "maintenance" (v8, 038) for good health, "social time" (v8, 043) or work life balance (v8, 077). Later the theme morphed in "enjoying life" (v11, 001).

These themes are mentioned during the workshops by the participants, but in the designer's accounts of the sessions they selected particular themes and highlighted them. They constantly reframed the design problem given to the users, but described what they did as "rephrase" (v11, 001) or "tweak" (v11, 155).

Occasionally the discussion about themes became concrete discussion about cars. For example, in (v9, 28–59) they were discussing the car as a place to relax, giving the driver "a resting time from being a superhero" (v9,033). The designers briefly touched on the fact that the car company is also interested in "me-time" in the following dialogue.

v9, 040   And of course, in the company right now there's a lot of talk about me-time but in a very technical sense because it's like "what can you do while eh:-"
v9, 041   work, relax
v9, 043   [Yeah, work, relax, what can you do-], "when the car is driving itself, what do you do then?"
v9, 045   Ehm:. How can you stay un- not bored in the car:?
v9, 046   Eh: is it that way you think?
v9, 047   I'm just thinking of another analogy. It's like, you go to a café, because you want that social aspect, you want to be surrounded, that's why you're not at home in your study-room, right?

This very brief conversation was one of the few points when they touched both on company themes and on the technical context in which the accessory they are developing will eventually sit. It alluded to the future of self-driving cars or cars that at least will automate some core driving functions. While this might not have been the immediate concern of this group of designers the design they were developing might well eventually be used in autonomous vehicles. However instead of taking this point up, Amy responded with a non-technical analogy of a café as a private not social space.

4.5 Values

One of the aims of the designers was to identify the values that affect the users. They "want the customer to be able to find the connection of the- what choices they make in choosing the product that is according to their value" (v11, 438). They look for values that distinguish the car company's potential customers, therefore they want to avoid values that would be shared by all Chinese. In response to "taking care of our family" (v9, 015), Ewan, the leader of the group, responds: "[Oh, no no clearly], that's a core value, so then we can't use that one, we need to use something else, and that else is basically, you know, sorting your trash, or buying a product that is sustainable or, and that's the thing. Taking care of the family is eh: is I think too:- everyone is – everyone doing it so it doesn't add that effect, so we need to find something on the fringes" (v9, 016).

Some values were stated explicitly and briefly discussed like "status" in "let's take status as an example, for us when it comes to health and status. For me it is about letting my really eco-friends think that I'm eco-friendly. I am not really that into it, but it makes me feel cool" (v5, 906). Status is again discussed in (v11, 227). Others emerge out of wider discussions like "pride" (v8, 156), "quality of "experience" (v8, 222), "Chinese dream" (v8, 302), "comfort" (v14, 190), where the words often cropped up as a closure point to a particular discussion.

Other values are implied in the discussion of the users, as the following excerpt shows:

> "I think they were- ehm I guess that when we go to the themes a little bit more, I think they were thinking a little more broader: like – I mean like (.) some was like recycling or eco-friendliness for the sake of in itself and really like, truly believing in the whole experience thing and not eh, showing off. And their aspirations were also going, not just into travel and to going to difficult places like Brian, okay he drove five thousand kilometers from CHINESE CITY to Tibet, but, like they were thinking about going to do volunteering: in like very poor places: and, I mean those were the type of things that they were looking at, which I think goes beyond really like to just" (v8, 271).

Here the designers started with a general discussion of discussion explicitly stating "recycling" and "eco-friendliness", but referring indirectly to the status consciousness of the Chinese, which they had already identified in v5. The reference to Brian could potentially imply many values like "compassion", "willingness of help" or "guilt" over the oppression of Tibet, however these were left to be interpreted.

5   STORIES AS A MOOD BOARD OF VALUES

At the end of the exercise the designers have gained a quite detailed picture of the values and concerns of their Chinese user group, even though they still have few concrete ideas for products for the Chinese market. This understanding will be useful both for their design of accessories and services, but also for their technical design colleagues. Potentially it can be articulated in an explicit description of the values of the target market to provide an input to other development projects targeting the Chinese market.

However, at the end of the session this understanding has not yet been articulated or made entirely explicit. They were conscious that they would have to prioritize the values they have elicited (v18, 94). Their understanding of values was encapsulated in stories through which they had elicited the values or from which they had interpreted them, as in the example of the trip to Tibet in (v8, 227). There were many stories about families and the need to take care of them or the way people liked to spend their time, like the example of the young man who did or did not like partying. Like the objects that artistic designers use in their mood boards, these stories afford multiple interpretations and all the participants can draw their own inferences from them and use them in their own way, yet collectively the stories can support a shared understanding. However put together they give an impression of the affluent, time-poor, money-rich Chinese, who still maintain the importance of caring for their families.

Ultimately the aim was to position the car company's brand in China in a way that the explicit and implied brand values match those of the target customers, as well as understanding the requirements for a specific accessory they want to design. In this exercise they look at other brands. The following passage is revealing: "So, the: Nike Olympic is all about (.) this kind of performance basically. [Everyone gathers at and looks at laptop] Like one person (..) Hm: yeah. (..) Where is the: (..) like, the athletes. This is the: Chinese runner. So it's him:, it's the shoes:, it's the performance. And then we could have a series of: (..)" (v18, 25). It talked about Nike as a brand and what it stands for (performance) linked to the personal performance of the Chinese runner. For a running shoe the brand concepts and values would no doubt have been expressed as a visual mood board.

The designers in this exercise went through very similar processes of finding sources for stories, episodes and proxy product ideas to set the context for their own products. The process was constructed by the designers to provide revealing information. However as in the process of creating a mood board, the designers looked very widely and were quite unsystematic both in what they exposed themselves to and how they interpreted what they had found. They reframed the problems continuously by rephrasing and reinterpreting that appeared at times arbitrary.

As in the use of sources of inspiration in artistic design, the designers took an everything-goes attitude and collected and interpreted what they found. The details of any of stories seemed not to matter particularly. Individual designers generated an interpretation around the stories of the users. Other designers did not question each other's selection or interpretations of situations.

Like sources of inspiration in artistic design fields (Eckert & Stacey, 2003b), the information they obtained and the behavior they observed were beginnings of trains of thought. If the team liked the result the process was not questioned. However there is one critical difference between looking for sources of inspiration and putting a mood board together: a source of inspiration could also inform specific products rather than just establish the context. Here this role was split between the users stories and the ideas that they generated. In this exercise the designers were open to ideas that were provided by the users, even though they mainly wanted to elicit their values.

When looking for sources of inspiration designers deploy a mixture of systematic processes and complete serendipity. There was an element of serendipity in how the users in the car company's co-creation exercise responded to the tasks and how they interacted with each other. However the process was very tightly managed by the designers. It involved both setting constraints on the selection of the participants and deliberately constraining the activities of the users; this hugely biased the results without this bias being questioned.

Sources of inspiration are ephemeral: once they have fulfilled their purpose they are discarded. Similarly the designers in this case study made no effort to maintain or develop the personal narratives of the users across the different co-creation sessions. If the designers see the participants as generators of sources of inspiration it is not surprising that they dismiss their ideas lightly.

6 CONCLUSIONS

This paper considers the role the co-creation process covered by the DTRS11 protocols played in the design process for new accessories and services to accompany cars sold in the Chinese market. For the design team it was an information gathering activity that was part of the initial problem framing and constraint setting stage of design: part of the active construction of a problem to solve that is tightly specified enough to be tractable, from a much broader and looser initial problem. The co-creation activity served the role of identifying the values and behavior of the Chinese customers, which is an important part of understanding what would can sell in the Chinese market.

The problem framing activities of the car company's user experience designers shown in the protocols had striking similarities to the active problem construction performed by designers in artistic design domains like fashion and knitwear, where designers need to make preliminary decisions and add constraints to give themselves problems that are both relevant to their target markets and tight enough to facilitate imaginative thinking (Stacey & Eckert, 2010). Rather than bringing users to the designers in a workshop, in other design fields designers go out to collect sources of inspirations for their designs by looking at other object that customers use or could buy, that provide both usable ideas for individual designs and a sense for the space of acceptable designs within a fashion.

The car company's co-creation activity proved effective in revealing the values that drive the purchasing behavior of the car company's target customers. These were encapsulated by verbal quotes and narratives, in contrast to the use of mood boards in artistic design fields to express visual sources of inspiration and visuospatial ideas explicitly, leaving their cultural connotations and implications of values tacitly implied.

REFERENCES

Cross, N. (2004a). Expertise in design: an overview. *Design Studies*, 25, 427–441.
Cross, N. (2004b). Creative thinking by expert designers. *Journal of Design Research*, 4, 123–143.

Darke, J. (1979). The primary generator and the design process. *Design Studies*, 1, 36–44.

Eckert, C.M, & Stacey, M.K. (2000). Sources of inspiration: A language of design. *Design Studies*, 21, 523–538.

Eckert, C.M., & Stacey, M.K. (2001). Designing in the context of fashion – designing the fashion context. In: *Designing in Context: Proceedings of the 5th Design Thinking Research Symposium*. Delft, Netherlands: Delft University Press, 2001, pp. 113–129.

Eckert, C.M., & Stacey, M.K. (2003a). Sources of inspiration in industrial practice: The case of knitwear design. *Journal of Design Research*, 2003, 3(1).

Eckert, C.M., & Stacey, M.K. (2003b). Adaptation of sources of inspiration in knitwear design. *Creativity Research Journal*, 15, 355–384.

Eckert, C.M., & Stacey, M.K. (2014). Constraints and conditions: drivers for design processes. In: A. Chakrabarti & L.T.M. Blessing (Eds.) *An Anthology of Theories and Models of Design*. London: Springer. pp. 395–415.

Entwistle, J., & Rocamora, A. (2006). The field of fashion materialized: a study of London Fashion Week. *Sociology*, 40, 735–751.

Faerm, S. (2010). *Fashion Design Course. Principles, Practice and Techniques: The Ultimate Guide for Aspiring Fashion Designers*. London: Thames & Hudson.

Finke, R.A. (1990). *Creative Imagery: Discoveries and Inventions in Visualization*. Hillsdale, NJ: Erlbaum.

Finke, R.A., Ward, T.B., & Smith, S.M. (1992). *Creative Cognition: Theory, Research, and Applications*. Cambridge, MA: MIT Press.

Goldschmidt, G. (1998). Creative architectural design: reference versus precedence. *Journal of Architectural and Planning Research*, 258–270.

Goldschmidt, G. (2015). Ubiquitous serendipity: Potential visual design stimuli are everywhere. In J.S. Gero (ed.) *Studying visual and spatial reasoning for design creativity*. Dordrecht, Netherlands: Springer Netherlands. pp. 205–214.

Kolko, J. (2010). Abductive thinking and sensemaking: The drivers of design synthesis. *Design Issues*, 26, 15–28.

Lucero, A. (2012). Framing, aligning, paradoxing, abstracting, and directing: how design mood boards work. In *Proceedings of the Designing Interactive Systems Conference*. Newcastle, UK: ACM. pp. 438–447.

Lloyd, P., & Scott, P. (1995). Difference in similarity: interpreting the architectural design process. *Environment and Planning B: Planning and Design*, 22, 383–406.

McGilp, H., Eckert, C.M., & Earl, C.F. (2016). Don't Look Back: The Paradoxical Role of Recording in the Fashion Design Process. *Proceedings of DRS 2016, Design Research Society 50th Anniversary Conference*. Brighton, UK.

Onarheim, B. (2012). Creativity from constraints in engineering design: lessons learned at Coloplast. *Journal of Engineering Design*, 23, 323–336.

Rocamora, A. (2011). Personal fashion blogs: screens and mirrors in digital self-portraits. *Fashion Theory*, 15, 407–424.

Schön, D. A. (1984). Problems, frames and perspectives on designing. *Design Studies*, 5, 132–136.

Schön, D. A., & Wiggins, G. (1992). Kinds of seeing and their functions in designing. *Design Studies*, 13, 135–156.

Stacey, M.K., & Eckert, C.M. (2010). Reshaping the box: creative designing as constraint management. *International Journal of Product Development*, 11, 241–255.

Takeda, H., Tsumaya, A., & Tomiyama, T. (1999). Synthesis thought processes in design. In *Integration of process knowledge into design support systems: Proceedings of the 1999 CIRP International Design Seminar*, University of Twente, Enschede, The Netherlands. Dordrecht, Netherlands: Kluwer. pp. 249–258.

Woodbury, R., Datta, S., & Burrow, A. (2000). Erasure in design space exploration. In: *Artificial Intelligence in Design '00*. Worcester, MA: Springer Netherlands. pp. 521–543.

Part VI

Co-Creating With Users

Chapter 24

Cracking Open Co-Creation: Categorizations, Stories, Values

Peter Lloyd & Arlene Oak

ABSTRACT

In this chapter we look in detail at the relationship between values and storytelling in the design process. First, we consider the categorization of participants taking part in the co-creation sessions and the values that underlie selection. Second, we focus on how two different stories emerge: 'sexy commitment' and 'Mercedes guy'. We describe the function of stories within this co-creation process as holding value tension, encapsulating difference in a way that presents a trade-off experientially. A story in this sense can be mapped onto a bipolar structure through which to identify and categorize values that are potentially in conflict. We conclude by suggesting that the most effective stories, like that represented by the phrase 'sexy commitment', are flexible enough to bridge 'past particulars' – experiences and behaviours that the co-creation process reveals – with 'imagined particulars' – stories that place specific actors, objects and relations into an imagined context.

I INTRODUCTION

The recent resurgence of interest in the concept of framing as a way of both analysing and thinking about the design process has usefully focused our attention on structures that position the flow of discourse in design processes (Paton & Dorst, 2011; Umney *et al.*, 2014; Dorst, 2015; Jornet & Roth, 2017; Dong & MacDonald, 2017). Rather than looking in detail at specifically cognitive elements of understanding, as did the first DTRS common-data study (Cross *et al.*, 1996) and many DTRS papers since, framing and associated structures align much more readily with social and constructivist ways of looking at design behavior, and come closer to how design professionals conduct and describe their activity. The idea of storytelling (Lloyd, 2000; McClosky, 1996) is similar in many ways to that of framing, as recently defined, but in our view it is a more powerful analytical concept to work with as it implies a time-based operational logic; a sequence of actions by actors that make sense through a narrative arc being created or told. Our recent studies have looked at the interleaving of narrative in the presentation and performance of design (Lloyd & Oak, 2016) and the DTRS11 dataset provides an opportunity to develop the concept both in methodological and design process terms. Storytelling and narratives are widely studied in relation to creative practice (Oak, 2013b), with our approach being based around attention to 'small stories' (Bamberg & Georgakopoulou, 2008). A 'small story' approach avoids the analysis of, for example, interview-based life stories, and instead focuses on the brief narratives of conversational storytelling (Goodwin, 1984), and in particular on "how people actually use stories in everyday, mundane situations" (Bamberg & Georgakopoulou, 2008, p. 378).

Perhaps more than any other DTRS dataset, the DTRS11 data invite a critical reading; that is, not only an interpretation of what is happening but also an assessment about the value of what is happening and the outcomes that are produced (e.g., see Leggett, 2014). This is perhaps partly due to the story-like, episodic way in which these data are presented to the researcher (our previous studies have looked at televisual representations of design process (Lloyd & Oak, 2016) which bear some resemblance to how the DTRS11 data are presented) but also in how a critical distance between the subjects under study, and the researchers considering their behaviours, is created. As researchers who were not involved in data collection we can be fully disinterested and hence independently critical. The DTRS11 data thus provide a unique window into the current corporate design world and the practices labeled as 'design thinking' and 'co-creation'. We can look, for example, at the question of whether co-creation is simply a way of reinforcing existing social and cultural prejudices and predispositions, or whether more-or-less conscious biases might manifest as being, for example, gendered (Oak & Lloyd, 2016; Oak, 1998), or ethically problematic (Lloyd, 2009).

At the beginning of Session 1 of the DTRS11 data, Ewan, the main protagonist in the dataset, sets down the intention to: 'basically crack open co-creation in general and then maybe calibrate our expectations' (Line 5). Likewise this chapter seeks to critically examine aspects of the co-creation process and, in so doing, provide opportunities for readers to adjust their understandings of the norms, values, and categorizations that underpin co-creation and design thinking more generally. We first consider the selection of participants in the co-creation sessions along with the construction of contrasting values that results. We then look at a number of ideas around the concept of story, storytelling, and narrative as they apply to the dataset before focusing in detail on two particular stories that emerge: 'sexy commitment' which is also considered elsewhere in this volume (D'souza & Dastmalchi, 2017, pp. 15–17; Daly et al., 2017, p. 8; Dong & MacDonald, 2017, p. 10)[1] and 'Mercedes guy' considered elsewhere by Hess and Fila (2017, p. 8) and Chan and Schunn (2017, p. 10). We conclude that, in terms of design thinking expertise, stories help to negotiate what we term *value tension*, creating narratives that hold opposing values while suspending resolution.

2   APPROACHES TO ANALYSIS

The DTRS11 data are framed for analysis in a way that is more organised and episodic than previous DTRS datasets. The data are generally chronologically presented in 20 sessions that could be interpreted almost as a kind of design-practice television or internet 'reality' mini-series, with some sessions ending explicitly with a set-up for the next. Rather than concentrating on specific sessions of DTRS11 data, the paper takes a number of seams through the dataset and draws on two analytic approaches that centre on the analysis of discourse and language. Goffman's (1974) *Frame Analysis* describes how language is used to construct and 'frame' particular perceptions of reality and thus organize and structure experience. Membership Categorization Analysis (Housley & Fitzgerald, 2002; Sacks, 1972; Stokoe, 2012) explores how participants in social interaction assign and use social categories within talk, as they account for actions, and assess persons, places, products and situations.

There is clearly a lot that happens outside of the filmed material presented, so the editing of the data means that the researcher has to be alive to the multiple levels of potential narrative that are being offered, and also to the editorial decisions that have been made (e.g., choosing to provide some sections shown in speeded-up time-lapse, rather than 'real time'). While the dataset is undoubtedly showing 'real' design activities, it is nevertheless itself a complex text, framed through various presentation modes and featuring multiple narrations. Our approach to the data, then, was similar to previous work where we have looked at Television programmes of the design process (Lloyd & Oak 2016). In that work, and here, we looked through the entire dataset, highlighting areas of potential for ideas about what we thought was going on, seeking structures in the data where we had a sense that specific regularities or discourses were being enacted. We then looked for further corroboration and evidence as we developed our understanding of what was being presented before then thinking through mechanisms to explain what we observed. Although often appearing to emerge by chance, the ideas-through-talk that are produced in the sessions (that we term stories) are influenced by many different experiences and voices – supporting the idea of design being a multi-vocal practice (Bakhtin, 1981; Oak, 2013a, 2013b).

The selection of what the design team featuring in the DTRS11 dataset decide are salient features of a larger co-creation discussion foreshadows what we as researchers are doing with the dataset. In the same way that the episodes present fragments of a 'real' story, our analysis uses selective samples to construct a narrative around value construction and story emergence aiming to theorise some of that reality.

3 CONSIDERING AND CATEGORIZING THE PARTICIPANTS

In this section we look at how, in selecting the participants for the Co-Creation workshops, the users and audience for THE COMPANY's car accessories become actors in the construction of the stories that emerge and that we describe in following sections. The design team's attention to the potential users, and their recognition that the participants in the workshops must be thoughtfully selected and managed, makes up a considerable portion of the interaction that is recorded within the early phases of the DTRS11 data. For example, as the design team discuss their individual perceptions and background experiences of previous co-design activities in Session 3, they also relate various perceptions of how they will approach the participants. In the illustrative comments below, Kenny, Nina, Abby and Ewan (along with the DTRS11 representative, David) work together to plan the co-creation workshops:

14	K	I think we're gonna do some really good preparations here, but in the end, when we're sitting there with the people, then we really need to calibrate it against them
32	N	(We need tools to use) with the people, so we can get the information we need, but not in the direct way, like we cannot be asking them, like what do you want and stuff like that, we need to really make them play the role
45	E	We're setting the scene, it's very artificial, super artificial . . . but, then again, I think that is my analogy very often, to say "okay, imagine these people are your friends, and you're doing something awesome together".
56:	A	The communication is going to be really really interesting to see what kind of input we get, what stories will actually impact them and kind of set their emotions on fire.

Throughout Session 3 the team members recognize their agency in setting the scene in which strangers will interact and they discuss in some detail the possible use of prompts and tools. The team's sensitivity to the local social organization, that they will plan and help to enact through the workshop, is further indicated by their in-depth discussions of how to manage the workshop participants' self-and-other perceptions so that successful interaction can occur. The issue of how the workshop participants might view themselves and each other in ways that could impact the communicative competencies of individuals – and thereby impact upon the success of the workshop – arises at the beginning of Session 3. Here, the team members discuss a large number of 'screening' factors, but concentrate particularly on how to create balanced and egalitarian conditions that de-emphasize notions of social hierarchy and encourage communication. In effect, by aiming to have particular participants involved, the design team are helping to shape the future narrative that may unfold from the meeting. The design team talk about seeking people who are "not too high up the societal ranks in terms of hierarchy" (Line 24), "average people, no one too high up in the society" (Line 81). Ewan suggests that the users "need to be premium car users", but that they should also "be at the same level" (Line 82) with "some kind of alignment or common ground" (Kenny, Line 120) so that they are able to communicate effectively. The design team talks about how best to create this cohesion by, for instance, limiting the age range of participants, and/or including family members to increase the likelihood that participants will be comfortable speaking. The team then discusses whether they should be concerned to balance participant characteristics such as introversion and extroversion, levels of creativity, senses of humour, degrees of independence, and desire to actively participate in the workshop. Additionally, the design team consider requiring a dress code, Abby notes: "we could also kind of say to them 'okay, everyone needs to wear black' or white or something so that they don't differentiate themselves too much with fashion" (Line 228).

In their concern for managing the social categories associated with the future participants, and the experiences that the participants will have during the meeting, the design team are carefully shaping the workshop: managing its social organization and the potential psychological and emotional effects it will have. They are thus setting a scene and choosing the actors who will perform for them. For example, in Session 4 the designers recognize that, upon meeting at the workshop, participants may automatically place themselves into orderly social hierarchies, for example, through naming their occupations (e.g., CEO, housewife), or by attending to how people are attired. To disrupt such category-based affiliations and assumptions, and to create an alternative order, the team discuss having the workshop begin with participants each choosing two animals as avatars. These animals will then be joined together and each participant will explain to the others how the hybrid creature represents them and the particular values they wish to be aligned with. By choosing, combining, and then presenting animals (rather than forms of employment) as classifications of identity, the design team intends for the workshop members to have a particular experience of group order and cohesiveness: one that mimics the design team's (purportedly) 'flat hierarchy' (Background Interview, Line 11), as it builds on social equality and informality, rather than on income difference and public decorum.

The selection or 'screener' (v01, Line 24) for participants taking part carries with it a set of underlying values that the designers of the sessions wish to represent in the co-creation workshops, indeed values that arguably do not reflect Chinese culture more generally but are deliberately reflective of their own. In 'cracking open co-creation', early in Session 1, Ewan asks the group "maybe we can go one by one and just spend half a minute to talk about what we think it should be" (Line 5), which provides an opportunity for the designers to use their past experiences in creating a session that works on their cultural terms. Creating egalitarian conditions under which the workshop participants will be comfortable enough to share their experiences is clearly an important aim of the designers and can be seen as an expression of co-design's collaborative, participatory, user-centred approaches. Although the role of power within apparently collaborative processes has been questioned (Oak & Lloyd, 2016), the explicit aims of many co-designers are to encourage and empower consumers (Prahalad & Ramaswamy, 2004; Sanders & Stappers, 2008). The collaborative nature of co-creative practice also accords with Scandinavian design's underlying expectation that its designers and products should 'care' for their users in ways that align with, if only unevenly, political and cultural values of social responsibility" (Murphy, 2015). In the context of the DTRS11 data, the social and cultural values of 'care' and equality form part of the everyday norms of social life or 'norms-in-action' (Housley & Fitzgerald, 2009) that can be seen as performed by the design team as they work to choose participants (who represent the wider context of Chinese consumers) and create the workshop activities.

4 THE NARRATOR AND THE USE OF STORIES

The stories that are created and told by the designers, then, arise out of a very deliberate and rather exclusive set of social conditions, although it is Ewan who presents most prominently in all sessions of the DTRS11 dataset. He performs a number of functions but is essentially the person who controls the process as it unfolds and who comments most consistently on both how the process is going and the value of other's contributions. Ewan's role as the narrator and assessor is enhanced, certainly for the viewer/researcher, by him being the only person featured in the background and follow-up interviews, where he is able to describe, in a kind of 'once upon a time – happy ever after' temporal structure, the context and outcomes of the project. Ewan is thus placed in a position of authority by the Editors of the DTRS11 dataset (when compared to other project stakeholders such as Hans or Tiffany) since he is given the opportunity to narrate the story from his point of view. Ewan himself is clearly aware of the importance of creating and narrating 'stories', as he explicitly references 'stories' and 'storytelling' throughout the dataset, which means that he is both part of the story and the narrator of it. In this sense he somewhat resembles the TV presenter Kevin McCloud who in the popular series *Grand Designs* can be seen to arguably reinforce an heroic and male-centred version of the design process (Lloyd & Oak, 2016).

As well as articulating the need to create stories, Ewan is himself someone who self-consciously uses stories (in combination with analogies and metaphors) to communicate his thoughts to others. In Session 9 a story he tells takes on a moral disposition in a wider-ranging discussion about the idea of 'commitment' which has come up in the first Co-Creation Session (CC1). Commitment is not something that can be purchased by consumers, Ewan suggests with his story, but something that needs an opportunity to develop. The story Ewan tells derives from a parable from the Bible about courage:

> I'm not a religious man, but there's an amazing quote from the bible, which is this guy who asks for courage. He prays to God "can I get courage?" and he never gets the courage, he never gets courage at all. And then [he speaks to] a wise man [who says:] "what do you think? do you think God would just give you courage, or do you think he will give you the opportunity to be courageous?" I think that's what we need to do here, the opportunity to take commitment, instead of just giving them the free stuff. They need an opportunity to be courageous so they can shine through that in a way.
> (Line 18)

The structure of this story hinges around a re-framing of a need to become virtuous (in the case of the story, the need for courage). By using a story to suggest that you don't simply become virtuous by *fiat* (either through being given virtue or by purchasing it), but by being given an opportunity to develop virtue, Ewan is skilfully making a link to what a product (i.e., a car accessory) can represent to consumers. The story thus bridges two perceptions of value (as product or as opportunity) while holding their inter-relationship clear. The reasoning is rhetorical and suggestive, rather than being logical and conclusive, but it is the polarity of value the story presents that should be noted here.

5 STORY FINDING AND CONSTRUCTION

A particularly salient story that the group both search for and create is what is referred to as a 'region-relevant story', that is, a narrative that will underpin the design-related interpretations and decisions of both the designers and the potential users of the accessories. The explicit topic of the 'region-relevant story' continues through the 20 Sessions of the DTRS11 data. Figure 24.1 shows a graph of the occurrences of words

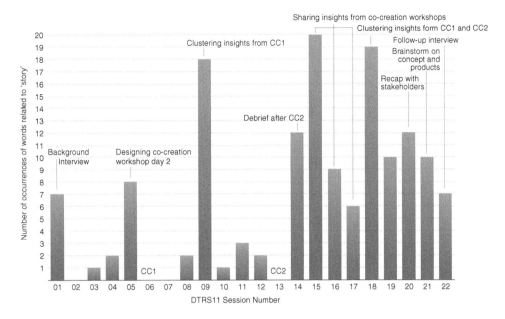

Figure 24.1 Number of occurrences of words related to 'story' in the DTRS11 dataset.

'story' and 'storytelling' for each session, including the two interviews with Ewan. In total there are 176 mentions of the two words although most mentions (105) occur subsequent to the second Co-Creation Session (CC2), which might suggest that this was a search that became more focused towards the end of Phase 2. Notable, however, is Session 9, with 17 mentions, and that appears to be the point at which the team develops consensus around the importance of having a 'story', perhaps at Ewan's direction, who mentions the word 'story' 12 times (70% of all mentions). To explore how the need for a 'region-relevant story' begins to be shaped by participants, Session 9 provides an example of a concept that emerges as an underpinning and recurring reference: 'sexy commitment'.

The phrase 'sexy commitment' is coined on Line 116 of Session 9, as the group seek out and begin to create a story that will combine both the idea of something that is seen as difficult but good (commitment) with the idea of something that is more indulgent (what they term 'me-time', and that becomes summarized by the word, 'sexy'). In the following excerpt Ewan begins by outlining the need for a story that can be summarized by this hybrid term:

110   E   It is a more complex thing, it is an eco-system of story, but we need a base story. It's just, I think just having this as one part of the story and having kind of the 'car-take responsibility of me' story, those need to go together [...] the whole thing around the me-time and stuff is an important part, but we just, as Kenny is saying, it is almost exhausted.

111   K   Yeah, but I also agree that it might be really interesting for accessories,

113	A	Exactly, because it's exhausted within the rest of the company. Or, not exhausted because it's going on right now so it is super relevant
114	K	Yeah, but I also think it's not necessarily, at least for me it's not super sexy. I think it's very easy, it goes in the practical way, that it's gonna solve some very-
115	E	Yeah yeah, calendars, and alarm clocks and like, whatever.
116	K	And maybe that's alright, but then we need to maybe spice it with the sexy commitment (*laughs*)
117	E	Yeah.
118	K	No not sexy commitment, but with the global awareness responsibility.
119	A	And the story around that 'okay, THE COMPANY actually takes care of you, now you have taken care of everyone else, now it's time for you to be pampered a little bit. This is your time'.
120	AM	Yeah.
121	K	So I think maybe this is more about how we tell the story.
122	A	Exactly. And then the products could absolutely live within there.

It is Kenny who introduces the idea, on Line 116, of making what Ewan refers to as 'the base story' – commitment to global sustainability – more palatable by 'spicing it up' and thus compensating for it not being 'super sexy'. Once coined, however, the group begins to identify what might be associated with the term[2].

Figure 24.2 shows that the phrase 'sexy commitment' appears over the course of the dataset a total of 30 times. The pattern of its appearance is also interesting as it

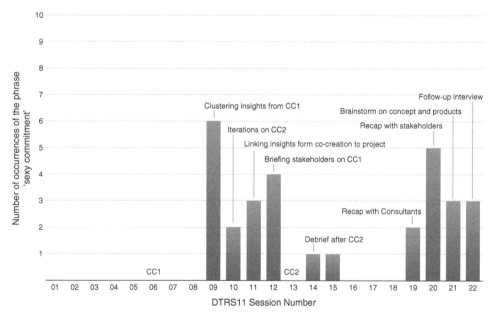

Figure 24.2 The origin and usage of the phrase 'sexy commitment' in the DTRS11 dataset.

seems to disappear as an idea or structuring story shortly after CC2 only to take off again in the final three sessions of the dataset (and also in the follow-up interview). The term comes about more by word association than by reference to common experience, but it provides a way for the group to engage with apparently contrasting ideas – sexiness and commitment – as they begin to shape an appropriate, region-relevant story that also has meaning for them, as designers.

The term 'sexy commitment' encapsulates for the group an ethical idea about what the project initially defined as 'the good life': "the different elements the Active Urbanite needs in his life to be able to have a good life" (DTRS11 Technical Report, p. 19, Unpublished report, note the male-gendered definition). As we will see, quite complex social ideas – politics, war, refugees – are used to develop the meaning of 'sexy commitment', but the good life, and especially the idea of 'being good' adds an explicitly moral tone. In the following excerpt, which comes from Session 11, Amanda and Ewan discuss the idea that acting for the common good is not something that can easily be talked or 'bragged' about in terms of the self, but can be rewarded in more subtle ways (and it is the idea that 'good' acts deserve reward that they think of as 'sexy'). In the excerpt Amanda begins by talking about the potential COMPANY car accessory user before Ewan tells a story about the 'good' act of a friend of his:

430	AM	It's about giving back, right? And that's what we want them to brag about.
431	E	Yeah. They're giving back to get something back. Hopefully. But it's the ultimate sacrifice. You give something without asking anything in return. But if you get something in return, it's really good.
432	AM	So I'm just thinking about the story where they show the location,
435	E	But that's a little disconnected 'cause that is about pure status symbol. Here we want them to help out at the school and yeah okay, so you show the location of that school in a photo. Like I have a good friend in Norway, and she's super first world, super Norwegian, tons of money, she quit her job to go- she was between jobs, and then she went down to: Greece, to help the refugees coming inland, which is a good thing to do. But she was, several times a day, posting images of her holding babies, saving babies. Why did she do it? I don't know. Did she do the whole thing to promote herself? It was done in a very understated way.
436	AM	Okay. So it's about finding that connection, right? that connection point.

Ewan develops the concept of 'sexy commitment' with a story, on Line 435, about a 'good friend' of his who 'quit her job' to go to Greece and help refugees arriving[3]. Describing this as 'a good thing' he then goes on to question her motivation for doing this because of the images she posted to social media with her 'holding babies, saving babies' (Line 45). He raises the possibility that his friend might have done this as an act of self-promotion rather (or perhaps additional to) a more straightforward self-less act of aid. Amanda understands that Ewan is sketching a polarity between the self that seeks reward, on the one hand, and the self-less, that seeks only to do good, on the other: she sums up his story by saying that 'sexy commitment', and the products that follow from it, are about finding the 'connection point' between the two poles (Line 436).

The example above is interesting not only for the explicit moral framing of the discussion – we are, after all, talking about a car-accessory – but also the way that Ewan draws in people outside of the project as a means to develop and justify the idea of 'sexy commitment'. The 'good friend' in the excerpt, as well as the refugees she helps, become, in effect, agents in the discussion by being what are termed 'implicated actors' (Latour, 1992; 2005). The fact that Ewan has cast doubt on his friend's motivation for acting in a morally good way, indeed has potentially shamed her as a person through doubting her altruism, without any possibility of her defending herself, could, however, be taken as itself a morally questionable (and in the context of a car accessory, again a slightly absurd) thing to do. It does, however, add another layer of meaning to what 'sexy commitment' is coming to symbolise for the group about 'the good life'; the idea that being good is both about doing good, but also being seen and recognised as doing good.

The phrase thus emerges and acts as an embodiment of a shared meaning for the group, becoming part of their vocabulary for talking about the psychological and cultural landscape where a future product might find purchase. However, in the three Sessions following CC2 (where it only appears on a Post-it note) it disappears altogether. It is only in Session 19 that it reappears, with the shared meaning evident when Ewan explains a tension as "a sexy commitment lives a little bit here" (Line 3). The phrase begins to reappear again for the group in the final three sessions of the dataset. In Session 20, the final session of the trip to China, Ewan summarises what the group have done for Tiffany and Hans, and sexy commitment is one of the themes he talks about, in effect telling a story about a story:

> And then this one, which is called "sexy commitment", and we have had this from the very first week, actually, where "sexy" is kind of what you get back, and "commitment" is what you sacrifice.
> (Line 12)

Neither Tiffany nor Hans ever use the term 'sexy commitment', perhaps because they have a more product-focussed role in THE COMPANY and perhaps also because they were not present when the concept emerged as meaningful within the group's own narrative of practice. Indeed, it is a phrase that perhaps represents the distance between Ewan's more conceptual 'co-creative' approach and the more conventional product development approach of Tiffany and Hans (see later discussion), a distance that is further reinforced by their late arrival in China. In Session 20 it is only Ewan and Kenny that use the phrase.

6   THE COMPANY STORY

While the designers plan a co-creation workshop that will be friendly and fun for the participants, the aim of the workshop is to accomplish particular tasks that are set within the context of THE COMPANY. The social and cultural values of care and equality may fit well within the designers' education and expectations of themselves as people who seek stories and help to shape narratives through co-operative, socially-engaged methodologies; nevertheless, the design team is also required to assist company profitability through developing new products that will engage consumers and maintain brand success. In the Background Interview Ewan tells the story of how the project came about through a 'random' meeting between two directors who pithily concluded: "we have low take rates in China, we need to do something about it" (Line 65). Therefore, throughout their activities and interactions, the designers must represent and so express both the 'norms-in-action' associated with the category 'being corporate' as well as those associated with the category of 'egalitarian sociability', a tension that is notably similar to the one they attempt to encompass with the term 'sexy commitment' a tension to which we will return later.

In the following excerpts of talk we see how the designers enact their affiliations with the more corporate, competitive environment that underpins and supports their activities. It is perhaps no surprise that the team's performance of these more explicitly competitive corporate values – expressed through phrases such as 'take rates', 'alignment', and 'sales peak' – occurs in settings where the group is talking with stakeholders from THE COMPANY (such as in Session 20). What is interesting in Session 20 is that a co-construction of meaning occurs where the socially responsible values that the design team have carefully constructed meet the values of the business environment they are operating in. This co-construction of social meaning shapes what happens within the locally-assembled group, and so has consequence for the collaboratively-constructed stories that are told and the nature of the emerging conception of what should result from the co-design process.

For instance, in the following sequences of talk from Session 20, below, Ewan first begins to outline the general areas that the design team have explored within the co-creation workshops, with some discussion of what the resulting, conceptual (not product-related) outcomes might be – for example, outcomes such as an allegiance to the 'sexy commitment' idea/story. A company stakeholder, Tiffany, is blunt in her response to Ewan, in which she questions (and implies a critique of) the design team's process in relation to THE COMPANY's financial interests (i.e., 'risks' and 'benefits'). Ewan's response is itself a question that challenges her budget-centred focus.

9	E	What's the risks of doing this? What are the benefits and so on?
10	T	And do we have the budget for it?
11	E	Yeah [laughter] or can we afford *not* to have the budget for it?

Soon after this exchange, however, Ewan's talk (Line 21) indicates his awareness of THE COMPANY's product-focused bottom line, as he references the creation of 'hero-products' that help to 'conquer' a market to ensure the COMPANY's growth. A recognition of the COMPANY's needs is further taken up as Ewan reiterates the need to 'conquest with an accessory' (Line 29). It is significant for our analysis that Ewan tells a story of how such a conquest might happen, with someone who might typically be categorized as a 'Mercedes guy' showing off his air purifier, a product not from Mercedes but from the COMPANY. The point of Ewan's brief narrative – a story told in the voice of the friend who is supposedly talking to the 'Mercedes guy' – is that this 'Mercedes guy' has associated himself with the 'good' qualities of the COMPANY (i.e., 'committed to the right quality') to the point that he has re-categorized himself (i.e., branded himself) through the use of the COMPANY's products.

| 21 | E | This is why we should have one or two hero-products that appeal to these people [. . .] how can we focus here to conquer these people? [. . .] We want to be in conquest, we want to grow. |
| 29 | E | And one of the things that we talked about is a scenario [. . .] was a conquest, a conquest with an accessory. There's a Mercedes guy, who has a friend over in his car, and he has a COMPANY approved air purifier. And the friend kind of asks 'wow I thought you were like a Mercedes guy' [. . .] and then the guy says 'yes I am a Mercedes guy, but above that I am a person that is committed to the right quality [. . .] and this air purifier [. . .] is really, really good' So they are then using THE COMPANY as a way to brand themselves'. |

What is the function of the story that Ewan tells in this excerpt? While recognizing that a business has to 'conquest' a market if it is to succeed, the telling of the story illustrates, in an imagined particular, the mechanism of success – how buying an accessory "makes people able to sample the THE COMPANY values, without making the full commitment of having a car" (Line 29). The story serves to negotiate and nuance the 'take rate' values of the business with the social responsibility values of the designers, decreasing what we might term *value tension*. Both the stakeholders and the designers can interpret the story within the dominant value system they relate to.

We can see how the story concerning the nature of the consumers' character shifts its value focus towards a business orientation in Session 21. Here, in the final sessions of the DTRS data, the designers talk with colleagues to help them come up with ideas for 'real' products. In the following extract from that meeting, Abby outlines for the colleagues what the workshop participants, as potential product users, value the most in their lives. Abby's comments transform the participants' multifaceted ruminations on 'the good life' and 'health' into a brief depiction that emphasizes individualism and acquisitiveness:

| 49 | A | They are super opportunistic, so [they want to] break free of their parents' way of living. They really go for every opportunity to earn extra money, to be able to differentiate themselves or build up their living standards. |

Later, in the same brainstorming session, Kenny reiterates the importance of THE COMPANY's product to the consumer's perceived and performed social status (Line 99). In Kenny's description, the product acts to identify knowledgeable and sophisticated consumers to each other, and so the product becomes more than an individual's fulfilled desire, it also becomes an agent of social capital.

| 99 | K | I think also the recognition that you get is also a confirmation that you are climbing up the social ladder because in order for them to recognize you they must have the same level of sophistication and intellect to understand the values that you have, so when they recognize you it is a confirmation that you are actually climbing up the social ladder at the same time. |

These brief descriptions of consumer behaviour widen the COMPANY's association with an underpinning narrative: not only are the potential consumer's choices associated with the product's demonstration of 'sexy commitment', they are also about the more overt performance of 'climbing up the social ladder'. The construction of the 'sexy commitment' story has not only served to hold *value tension* for the consumer (between 'sexiness' and 'commitment') but also for the designers in presenting a more acceptable business-focused logic to people whose task it is to come up with specific products.

The co-design process poses the implicit question 'is it possible to transpose knowledge gained of users through ethnographic and participatory experiences into viable products that will satisfy a company's needs?' Certainly the design team studied for DTRS11 are hopeful that they can succeed, but a range of actors (e.g., Tiffany and Hans) beyond the design team and the potential product users impact on the realization of the team's original aims – which were both to create concepts (if not actual products) and also to 'showcase' how co-creation's 'user involvement' approach could "work as change agents for the organization" (Background interview, Line 95). In this sense Ewan is telling (or maybe selling) a story of organizational change; of how a well-executed example of co-creation for a car accessory product can begin to influence the design process of the cars themselves. This shift in design thinking – from the shaping of products to the shaping of stories is what drives him. The mechanism of storytelling, and the *value tension* that it holds, is how he hopes to achieve this.

7 SUMMARY AND CONCLUSIONS

In this paper we've shown in detail how a very particular story, embodied by the phrase 'sexy commitment', emerges from the two co-creation sessions as a means of expressing contrasting values: commitment and duty on the one hand, indulgence and selfishness on the other. We showed how the phrase is taken up and used by the designers through subsequent sessions in the dataset as a means to classify and evaluate activity and ideas. Sexy commitment is not the only story that emerges by any means ('pockets of enjoyment' and 'sprinkle the king' are others we looked at) but its 'chance' emergence and consistent development marks it out as a significant contribution to the co-creation process of the DTRS11 dataset. The story that sexy commitment represents goes some way to meet the requirement for a 'region relevant story' in the project brief but the story arises out of a very controlled set of conditions; that is, the design of the co-creation workshops themselves. By carefully setting the criteria for the workshop participants (e.g., according to certain personality characteristics and professional roles) then prescribing the activities of the workshops to achieve particular social ends (a flat hierarchy in which everyone is equal) we saw how the sexy commitment story emerges over a more fundamental transmission of cultural values and is in a sense an expression of the *value tension* that results from the difference (crudely put) between the more egalitarian values of Scandinavian design and the more hierarchical values of Chinese culture[4]. The sexy commitment story, as other effective stories do, allow both value conditions to exist.

A further dimension of value is represented by the business and commercial aspects of THE COMPANY and this sets up another *value tension*, in this case between the 'socially responsible' team of designers involved in the co-creation process and THE COMPANY stakeholders, who are used to a more conventional product development and business environment focused on 'conquest', budgets, production plans, and dealerships. We showed how a story ('Mercedes guy') again allows both value conditions to exist and helps to preserve the implicit aim of Ewan in particular (and perhaps also the other designers) of changing the THE COMPANY's organizational culture around designing. Throughout the DTRS11 data, each designer indicates an ongoing awareness of the cultural and personal characteristics of the workshop participants. This seems particularly to be the case for Ewan (e.g., see his extensive discussion of the conflicts between the collective and the individual in Chinese society in Session 21, Line 2). However, as we also see above, all members of the design team adeptly translate this complex social information into simple phrases (e.g., 'sexy commitment', 'Mercedes guy') that are relevant to their corporate colleagues (particularly within the specific, local contexts of business meetings involving COMPANY representatives). The designers are able to skillfully notice, create, and tell contrasting stories that are underpinned by contrasting norms and categories of value.

We have described the function of stories within this co-creation process as holding *value tension* and the uniqueness and richness of the cross-cultural DTRS11 data has allowed us to develop this idea in detail. What stories seem particularly good at doing in a designing context is encapsulating difference in a way that presents a trade-off experientially. The stories we have identified, and we think also others in these data, have a bipolar structure that maintains a level of trade-off complexity but packages it in a simple and understandable way that others can easily relate to. Sometimes these stories come directly from someone's experience – Ewan's bible story, for example – but the most effective stories like that represented by the phrase 'sexy commitment' are flexible enough to bridge 'past particulars' – experiences and behaviours that the co-creation process reveals – with 'imagined particulars' – stories that place specific actors, objects and relations into an imagined context. A story in this sense provides a dynamic structure through which to identify and categorize value and ultimately provides a way of presenting an argument.

The design process, in the way that it bridges past and future, is a fertile ground for storytelling but the co-creation process that the DTRS11 data represents is particularly rich in the interleaving of stories in many forms. The expertise of the participants is notable here and contrasts with other accounts of design thinking expertise that focus around the activity of prototyping (e.g., in the DTRS7 data). It is particularly striking that this co-creation process is almost entirely text based with Post-it notes, and the structures of categorization and organization that they afford, central to the production of written and spoken text. This kind of design thinking appears less about the shaping of form through the construction of prototypes and more about the shaping of story through the selection of text and points the way to future analyses, and future practices of design thinking, particularly in the wider contexts in which design is now being discussed. Indeed it is relevant for the research process itself as practiced in this paper. Cracking open co-creation, then, can provide insight not only into new modes of design thinking but also new modes of research thinking.

NOTES

1 Interestingly, in the light of our opening sentence, Dong and MacDonald (2017) refer to the idea of sexy commitment as a 'frame' in the Schönian sense. Though it can be viewed as

such, in our view the use of the term in describing and allowing multiple narratives gives it a complexity that warrants the more dynamic term 'story'.

2 D'souza and Dastmalchi (2017, pp. 15–17) also analyse sections of text using the phrase 'sexy commitment', noting, as we do, its accumulation of meaning. They refer to the phrase as 'slang', but again (see footnote 1) we note that the phrase is used in more subtle and precise ways that we think of as story-like, belying the rather pejorative term 'slang'.

3 The 'wealthy-women-assisting-refugees' story is also nicely analysed by Dong and Macdonald (2017, p. 10) in terms of the semantics of 'status'.

4 D'souza and Dastmalchi (2017, pp. 15–17) note a slightly different value polarity when the Scandinavian Design Team develop the 'rational' meaning of commitment representing the Asian view and cohering around 'restraint' (commitment, environment, sustainability) to an 'embodied' meaning cohering around expressiveness (sexy, beautiful, tangible).

REFERENCES

Bakhtin, M. (1981). *The dialogic imagination: Four essays*. Austin and London: University of Texas Press.

Bamberg, M. & Georgakopoulou, A. (2008). Small stories as a new perspective in narrative and identity analysis. *Text & Talk*, 28(3), pp. 377–396.

Chan, J. & Schunn, C. D. (2017). A Computational Linguistic Approach to Modelling the Dynamics of Design Processes. In: Christensen, B. T., Ball, L. J. & Halskov, K. (eds.) *Analysing Design Thinking: Studies of Cross-Cultural Co-Creation*. Leiden: CRC Press/ Taylor & Francis.

Cross, N., Christiaans, H. & Dorst, K. (1996). *Analysing design activity*. Chichester: Wiley.

Daly, S., McKilligan, S., Murphy, L. & Ostrowski, A. (2017). Tracing Problem Evolution: Factors That Impact Design Problem Definition. In: Christensen, B. T., Ball, L. J. & Halskov, K. (eds.) *Analysing Design Thinking: Studies of Cross-Cultural Co-Creation*. Leiden: CRC Press/Taylor & Francis.

D'souza, N. & Dastmalchi, M. (2017). "Comfy" Cars for the "Awesomely Humble": Exploring Slangs and Jargons in a Cross-Cultural Design Process. In: Christensen, B. T., Ball, L. J. & Halskov, K. (eds.) *Analysing Design Thinking: Studies of Cross-Cultural Co-Creation*. Leiden: CRC Press/Taylor & Francis.

Dong, A. & MacDonald, E. (2017). From Observations to Insights: The Hilly Road to Value Creation. In: Christensen, B. T., Ball, L. J. & Halskov, K. (eds.) *Analysing Design Thinking: Studies of Cross-Cultural Co-Creation*. Leiden: CRC Press/Taylor & Francis.

Dorst, K. (2015). *Frame innovation: Create new thinking by design*. MIT Press.

Goffman, E. (1974). *Frame analysis: An essay on the organization of experience*. Harvard University Press.

Goodwin, C. (1984). Notes on story structure and the organization of participation. In: J.M. Atkinson and J. Heritage (eds.) *Structures of Social Action*, Cambridge: Cambridge University Press, pp. 225–246.

Haste, H. (2016). Jerome Bruner Obituar. *The Guardian*, 15th July. Available from: http:// tinyurl.com/zoulcaw [accessed July 25th 2016].

Hess, J. L. & Fila, N. D. (2017). Empathy in Design: A Discourse Analysis of Industrial Co-Creation Practices. In: Christensen, B. T., Ball, L. J. & Halskov, K. (eds.) *Analysing Design Thinking: Studies of Cross-Cultural Co-Creation*. Leiden: CRC Press/Taylor & Francis.

Housley, W. & Fitzgerald, R. (2002). The reconsidered model of membership categorization analysis. *Qualtitative Research*, 2, pp. 59–83.

Housley, W. & Fitzgerald, R. (2009). Membership categorization, culture and norms in action. *Discourse & Society*, 20(3), pp. 345–362.

Jornet, A. & Roth, W. (2017). Design {Thinking | Communicating}: A Sociogenetic Approach to Reflective Practice in Collaborative Design. In: Christensen, B. T., Ball, L. J. & Halskov, K. (eds.) *Analysing Design Thinking: Studies of Cross-Cultural Co-Creation*. Leiden: CRC Press/Taylor & Francis.

Latour, B. (1992). Where Are the Missing Masses? The Sociology of a Few Mundane Artifacts. In: W. E. Bijker & J. Law (eds.). *Shaping Technology/Building Society: Studies in Sociotechnical Change*, MIT Press, pp. 225–258.

Latour, B. (2005). *Reassembling the social: An introduction to Actor-Network Theory*. Oxford: Oxford University Press.

Leggett, W. (2014). The politics of behaviour change: Nudge, neoliberalism and the state. *Policy & Politics*, 42(1), 3–19.

Lloyd, P. (2000). Storytelling and the development of discourse in the engineering design process. *Design studies*, 21(4), 357–373.

Lloyd, P. (2002). Making a drama out of a process: how television represents designing. *Design Studies*, 23(2), 113–133.

Lloyd, P. (2009). Ethical imagination and design. *Design Studies*, 30(2), 154–168.

Lloyd, P. & Oak, A. (2016). Houses of straw: Grand Designs and the presentation of architectural design on television, *Design and Culture*, 8(2), pp. 155–180.

McCloskey, D. N. (1990). Storytelling in economics. In: Nash, C. (ed.) *Narrative in Culture. The Uses of Storytelling in the Sciences, Philosophy and Literature*. London: Routledge, pp. 5–22.

Murphy, K. (2015). *Swedish Design: An ethnography*, Ithica: Cornell University Press.

Oak, A. (1998). Assessment and understanding: An analysis of talk in the design education critique. In: S. Wertheim, A. Bailey, & M. Corston-Oliver (eds.) *Engendering Communication: Proceedings of the fifth Berkeley Women and Language Conference*, Berkeley: University of California, pp. 415–426.

Oak, A. (2009). Performing architecture: talking 'architect' and 'client' into being. *CoDesign*, 5(1), 51–63.

Oak, A. (2013a). *'As you said to me I said to them': Reported speech and the multi-vocal nature of collaborative design practice*. Design Studies, 3, pp. 34–56.

Oak, A. (2013b). Narratives in practice: The small and big stories of design. In: Sandino, L. & Partington, M. (eds.) *Oral History in the Visual Arts*, London: Bloomsbury, pp. 181–188.

Oak, A. & Lloyd, P. (2016). *'Throw one out that's problematic': performing authority and affiliation in design education*, CoDesign, 12(1–2), pp. 55–72.

Paton, B., & Dorst, K. (2011). Briefing and reframing: A situated practice. *Design Studies*, 32(6), 573–587.

Prahalad, C. & Ramaswamy, V. (2004). Co-creation experiences: The next practice in alue creation. *Journal of interactive marketing*, 18(3), 5–14.

Sacks, H. (1972). On the analyzability of stories by children. In: J. Gumperz & D. Hymes (eds.) *Directions in sociolinguistics: the ethnography of communication*. New York: Rinehart & Winston. pp. 325–345.

Umney, D., Lloyd, P. & Potter, S. (2014). *Political debate as design process: A frame analysis in Proceedings of DRS2014: Design's Big Debates*, Umea, Sweden, 16–19 June.

Sanders, E. & Stappers, P. J. (2008). *Co-creation and the new landscapes of design*, CoDesign, 4(1), 5–18.

Stokoe, E. (2012). Moving forward with membership categorization analysis: Methods for systematic analysis. *Discourse studies*, 14(3), 277–303.

Chapter 25

From Observations to Insights: The Hilly Road to Value Creation

Andy Dong & Erin MacDonald

ABSTRACT

Insights about complex and ambiguous environments can spot opportunities for new products and services. This paper develops a functional model of design insight by mapping verbalized statements associated with generative sensing onto a semantic scale established by the Legitimation Code Theory (LCT) dimension of Semantics. Analysis of discussions about co-creation workshops reveals that insights develop when observations are gradually represented in terms of abstract, general, and decontextualized features rather than their concrete, contextual, and incidental details or their abstract features alone. This study will show that knowledge-building associated with design insight entails a series of movements 'up and down' a semantic scale ranging from concrete details to decontextualized features. The decontextualization eventually reaches a limit, at which point a hypothesis is offered to explain the observations. The patterns of movements indicate that insight requires simultaneous decontextualization of evidence and observations into highly condensed meanings *and* their recontextualization into a new hypothesis.

I INTRODUCTION

Design is an act of value creation. At the root of a design-based mode of value creation is the construction of novel insights about ambiguous and complex existing environments (Kolko, 2010) as the basis for hypothesizing new goods that may have value (Dorst, 2011). Value includes profit, human experience (Shedroff, 2001), and ethics (Lloyd, 2009) among many possibilities. Design researchers tend to regard the initiation of value creation at the moment of insight, sometimes called the 'creative leap'. The moment during which an individual instantaneously appears to discover insight into a problem remains steeped in the lore of creativity. In contrast with the myth, research in creativity in design and the cognitive neuroscience of insight shows that the moment of insight occurs gradually rather than instantaneously and purposefully rather than serendipitously. Cross (1997) characterizes the "creative leap not so much as a leap across the chasm ... as it is building a bridge across the chasm between problem and solution." Architects report latent preparation through deep understanding of the design situation (Murty, 2007). Neuroimaging studies of moments of insight during problem-solving characterize the 'A-ha! Moment' as the culmination of a series of neural events rather than an independent, single neural process (Kounios & Beeman, 2009). Rather than singular moments of inspiration, reasoning through abduction toward an explanation to surprising observations or an intended value is generally regarded as the kernel of creativity in design (Dorst, 2011; Roozenburg, 1993).

Despite this recognition, knowledge-building associated with insight lacks a functional model – a structured representation of the necessary processes leading to insight. Where a model exists, this chasm-crossing is depicted simply as an arrow (Cross, 1997, Fig. 14) or as an 'unexpected discovery' contingent upon experience and tacit knowledge (Suwa, Gero, & Purcell, 2000). Moreover, studies in the adjacent area of inspiration in creativity tend to downplay knowledge-building. A review of research on inspiration (Vasconcelos & Crilly, 2016) noted that many studies ignored (or did not report on) dynamic knowledge-building during the task, despite the recognition of design as entailing the dynamic construction of knowledge (Visser, 2006) through the generalization of conceptual ideas (Suwa *et al.*, 2000). Cognitive research in entrepreneurial opportunity recognition has started to identify cognitive strategies such as inductive and deductive reasoning to generate profitable action possibilities (Cornelissen & Clarke, 2010) and heuristics such as counterfactual reasoning (Gaglio, 2004) but is similarly silent about knowledge-building associated with insight.

This paper addresses a gap in the study of the insight: knowledge-building associated with insight and its real effects on the fruitfulness of new design concepts that stem from the insight. First, this paper will propose an operational definition of insight by connecting insight to the hypothesis-setting characteristics of generative sensing (Dong, Garbuio, & Lovallo, 2016a, 2016b). The concept of generative sensing describes design as a process consisting of recursive cycles of logical reasoning during which design practitioners construct and test hypotheses. In this paper, explanatory hypothesis-making is considered as a type of insight. Second, the paper will illustrate knowledge-building associated with generative sensing by mapping statements expressed during generative sensing onto a semantic scale established by the Legitimation Code Theory (LCT) (Maton, 2013a) dimension of Semantics, or LCT(Semantics). Two LCT(Semantics) codes conceptualize the organizing principles underlying semantic structures: semantic gravity and semantic density (Maton & Doran, 2016a, 2016b). The semantic scale measures the complexity of knowledge based upon a continuum of strengths (or weaknesses) in semantic gravity and semantic density. The resulting semantic profile illustrates the way individuals create a knowledge structure from context specific and dependent observations, such as input from stakeholders in a co-creation design workshop, and decontextualized experiences and concepts having highly condensed meanings, such as social theories or design principles embodied in exemplars. Finally, based upon the analysis of discussions intended to generate design insights, the paper will propose a functional model of design insight.

2   THEORETICAL FRAMEWORKS

2.1   Insight

In psychology, insight is defined as the 'sudden' discovery of a solution to a problem despite repeated, prior failures (Bowden, Jung-Beeman, Fleck, & Kounios, 2005). According to current theory in psychology, insight requires a restructuring of the problem situation (Luo & Knoblich, 2007) including a re-interpretation or re-structuring of the problem (Ohlsson, 1992).

This research studies insight during the period in the design process in which the 'problem' is one of making sense of multiple and ambiguous or conflicting observations and information by inventing a hypothesis (Kolko, 2010). Individuals invent a hypothesis when present observations render current hypotheses as less likely to be true (Gettys & Fisher, 1979). Individuals are less likely to invent a hypothesis when they believe that current hypotheses are satisfactory (Garst, Kerr, Harris, & Sheppard, 2002). Therefore, to invent a new hypothesis requires individuals to relax their attachment to present hypotheses to see the problem in a new way, which is a general feature in functional models of insight. Current models of hypothesis generation in psychology explain the process as one of memory retrieval (Manning, Gettys, Nicewander, Fisher, & Mehle, 1980; Thomas, Dougherty, Sprenger, & Harbison, 2008) with no pathway for the invention of a new hypothesis not previously known to the individual or known in the context of the present observations. Some scholars of logical reasoning describe the former process of retrieving, or better-stated, selection, of a hypothesis from memory as selective abduction (Magnani, 2001). When it comes to the discovery of a scientific theory, wherein the current observations are anomalous per existing theory, the mere selection of a hypothesis from memory is no longer an appropriate description. Scholars of logical reasoning describe this type of reasoning as creative abduction (Magnani, 2001) or hypothetic inference to "discover causes" of observable phenomena (Niiniluoto, 1999; Peirce, 1932, 1998).

It could therefore be concluded that generating a new hypothesis to explain anomalous, ambiguous, or conflicting observations is a form of insight. To create the hypothesis, individuals must relax their present hypotheses, re-structure the elements of their observations, and detach themselves from prior experience to see the problem in a new way, all of which are elements of functional models of insight. In the data that will be presented, one of the participants, Will, describes hypothesis-as-insight succinctly (v11, 97): "I mean, usually when we write a concept or a proposition, right? We set it up with an insight. And the insight is usually a- some pain or tension point." In other words, the "pain or tension point" is a previously unobserved circumstance that requires a novel solution. The problems associated with insight in design are two-fold: first, to explain observations in new ways (Kolko, 2010), and second, to invent a product or service and associated set of solution principles to achieve an intended outcome (value) (Dorst, 2011). The first type of hypothesis is known as explanatory abduction whereas the second form is innovative abduction (Roozenburg, 1993). Both forms of abduction are central to the concept of generative sensing, a design strategy that describes the design process as a recursive process of generating and testing hypotheses, each of which are built upon the conclusions drawn from the previous hypothesis.

2.2 Generative sensing

Generative sensing describes a design approach 'through a design problem' consisting of recursive cycles of logical reasoning during which designers generate and test hypotheses until no further hypothesis could confirm or refute the realized design (Dong *et al.*, 2016a, 2016b). The cycle commences when individuals encounter a surprising observation, which might be an unusual data point or an intended value that cannot be satisfied by current solutions. The observation is explained by an abductive hypothesis. The hypothesis is tested through deductive or inductive reasoning and should include experimentation with prototypes. Finally, rather than accepting the conclusions of the deductive or inductive reasoning, the individual provides another abductive hypothesis to explain or undermine the conclusion, a hypothesis that can be tested through deductive or inductive reasoning, resulting in a continual testing and reinterpretation of the logic underlying a design problem and solution. The process continues recursively until no new hypothesis could confirm or refute the realized design. The realized design confirms the truth of the hypothesis. Building upon the concept of primary generators (Darke, 1979), generative sensing considers each hypothesis as a partial explanation only, an explanation that cannot address all the constraints and objectives. Designers address parts of the design problem and test propositions in a recursive manner. Crucially, generative sensing is not a trial and error process. New hypotheses explain, resolve, or challenge the evidence in favor of or against a design concept and are always grounded in the evaluation of the present design concept. Stated more loosely, generative sensing is not about discovering that a hypothesis (design solution) is wrong; therefore, a new hypothesis is generated. Rather, if the prior hypothesis were false (e.g., the prototype failed) the new hypothesis should propose a rule that would undermine the false conclusion.

One of the missing elements in the concept of generative sensing is the knowledge-building that occurs concurrently with the invention of the hypothesis. In other words, generative sensing assumes that the hypothesis is 'created' without regard to knowledge-building that underpins a hypothesis that better-explains the present observations. Knowledge-building based upon testing hypotheses and retrieving knowledge sources such as personal experience and external creative stimuli provides the mental preparation for the realization of the hypothesis. To perform this analysis, this research makes use of the Legitimation Code Theory dimension of Semantics.

2.3 Knowledge-building

Legitimation Code Theory (LCT) (Maton, 2013a) theorizes that fields of knowledge encode knowledge in semantic structures having underlying organizing principles. These organizing principles are conceptualized as semantic codes having strengths of *semantic gravity* (SG) and *semantic density* (SD). The strength of SG describes the degree to which meaning relates to its context. The meaning of a concept may have relatively stronger (+) or weaker (−) semantic gravity along a continuum. When the meaning is dependent upon its context, the concept has stronger semantic gravity and is denoted as SG+. When meaning is less dependent upon its context, the concept has weaker semantic gravity, denoted as SG−. To take a simple example from design, the 6-3-5 Method has stronger semantic gravity than the process of brainstorming which in turn embodies stronger semantic gravity than ideation. The strength of SD describes the extent to which the meaning of a concept is embodied in knowledge-oriented practices such as words, gestures, models, simulations, etc. When a concept has strong semantic density, denoted as SD+, the meaning is embodied within these practices and would be devoid of semantic strength if those practices were absent. In contrast, weaker semantic density (SD−) implies less condensation. The strength is not intrinsic to a concept *per se*. Again, to take a simple example from design theory, the semantic density characterizing the concept of 'function' is likely to be much stronger in a research publication in an engineering design journal than in its use in an engineering design class which in turn may be stronger than its use in a general interest technical magazine.

The codes SG and SD when taken together can establish a semantic scale to describe knowledge-building. The semantic scale ranges from weak semantic gravity and high semantic density (decontextualized and highly specific meanings) to strong semantic gravity and weak semantic density (concrete details and common meanings). Cumulative knowledge-building requires progressively reaching further up the semantic scale toward weaker semantic gravity and stronger semantic density (Maton, 2013b).

The next section describes the research methods used to describe knowledge-building during generative sensing that leads to the formation of insight. Based upon the analysis, a functional model of insight is proposed.

3   METHODS

To illustrate the dynamics of design insight, this paper analyses a design team's discussions about co-creation workshops. The dataset for the analyses come from the Design Thinking Research Symposium 2011 (DTRS11) dataset (Christensen & Abildgaard, 2017). A European car company assigned a team to design a product, a service, accessories, a communication package, or an event that can be used to promote the concept of "the good life" in a Chinese context. The design team convened two co-creation workshops with potential customers in China. The workshop conveners intended to develop insights into Chinese consumers and their interpretation of themes relevant to an automotive company as the basis for new product and service concepts. The value the company seeks is the emotional engagement of a Chinese premium consumer who is young and progressive.

The theoretical frameworks described previously provide the basis for the analysis of discussions about the workshop. Analysis of the transcripts commenced by finding statements of surprising observations, which is considered the start of generative sensing (Dong et al., 2016a). In this dataset, surprising observations were evident, even if idiosyncratic in expression. An example of an explicit statement is, (v07, 80) "It was interesting what he said". An example of a reference to a surprising observation is, (v15, 54) "that's definitely something". After identifying the surprising statement, the analysis continued by finding the abductive hypothesis that explains the surprising observation. Statements associated with generative sensing and statements supporting the hypothesis are mapped onto a semantic scale to illustrate the semantic profile of the knowledge-building. Semantic profiles in the same thematic area are connected across sessions to draw a more complete picture of knowledge-building associated with a specific insight. Further details on the tasks associated with each of these steps are described in the following.

3.1   Identifying abductive hypotheses

A key task of the analysis is finding instances of abductive reasoning and, in particular, the abductive hypothesis. Abductive reasoning occurs when individuals encounter a surprising observation or an intended value that is not currently satisfied by current knowledge. We follow rule-based guidelines for the identification of abductive reasoning in verbalizations (Dong *et al.*, 2016a; Dong *et al.*, 2015). Computational approaches based upon extracting adverbs of manner, referring to process and actions, can also be used to identify instances of abductive reasoning (Bedford *et al.*, 2017). An example of abductive reasoning from the session on clustering insights from the first co-creation workshop is:

> (v09, 239) if *you* actually also *buy these accessories* you have *become* even: *more,* eh: *responsible*
> The rule is: p → q: IF buy these accessories THEN responsible
> Alternatively, the rule could be stated as, "People who *buy these accessories* are *more responsible.*"

Once the abductive hypothesis is located, the generative sensing loops, that is, the transitions between deductive or inductive reasoning and abductive reasoning, are identified. These may occur both before and after the abductive reasoning statement. These statements are placed onto a semantic scale to create a semantic profile.

3.2 Mapping statements onto a semantic scale

The analysis of the position of statements on the semantic scale takes place on clauses, the combination of words into a short, coherent passages, and individual words. The following example demonstrates a clause with stronger semantic gravity than the next clause.

> SG+: (v07, 148) they are picking up or using ehm emotional indicators as self-expression:, as bragging rights:, as differentiation
> SG−: (v07, 63) Their kind of expression and mentality could also be a form of like status symbol.

The first clause has stronger semantic gravity than the second clause because it directly refers to an observation from the co-creation workshops. The second clause has a relatively weaker semantic gravity because it decontextualizes the workshop participants' "emotional indicators as self-expression ... as bragging rights ... as differentiation".

The second step analyzes the semantic density of a concept. Two codes characterize the strength of semantic density of words: *technical* and *everyday* (Maton & Doran, 2016b). These codes differentiate the extent to which the meaning of a word locates in specialized fields. Two codes types characterize the strength or semantic density of clauses: *connecting* and *augmenting*. By connecting, individuals relate concepts to established meanings, such as the schemes of classification associated with specialized domains of practice (Maton & Doran, 2016a). By augmenting, individuals add meanings without referring to any established systems of classification. In the previous example, the semantic density of 'self-expression' increases by connecting it to the concept of 'status symbol'. The following examples show examples of strengthening semantic density by connecting complex meanings with 'status' and weakening

$$\text{observations} \quad + \quad \frac{\text{explanatory}}{\text{abduction}} \quad = \quad \text{frame}$$

Figure 25.1 An explanatory abduction establishes the hypothesis for the frame.

semantic density by referring 'status' to everyday common meanings. Examples of connecting and augmenting from the dataset analyzed are:

Connecting (strengthening semantic density)
(v08, 269) consciousness about what kind of *social status symbols* work, what kind of *progressive attitudes* work or don't work
(v11, 232) And it's connected directly into *evolving status symbol*
Augmenting (weakening semantic density)
(v08, 197) It's *an understated way of like showing off*, that you don't have to
(v08, 276) it's kind of "okay, everyone will not know, what this is all about, but the right people will know, that this is actually pretty cool"

A statement is placed higher on the semantic scale, relative to prior statements, when the statement is more abstract, general, and contains decontextualized features. A statement is lower on the semantic scale when it is concrete, contextual, and contains incidental details. The mapping of statements onto the semantic scale proceeds until the participants move onto another topic.

3.3 Frame

The final step of the analysis connects hypotheses to the principle thematic areas that the workshop conveners established based upon their analysis of the workshop discussions and specific frames that emerged during their discussions. The purpose of drawing this connection is to understand whether their insights (hypotheses) turn out to be productive, that is, led to plausible products and services. To connect hypotheses to frames, the research builds upon the model of reasoning patterns established by Dorst (2011). In Dorst's model, value creation is based upon the hypothesis of a 'what' and a 'how' to create an intended value. We extend this model by explicitly linking the hypothesis to a particular frame. The frame is the conclusion of a hypothesis that explains the current observations. The frame is not arbitrary; an explanatory hypothesis leads to a frame as shown in Figure 25.1.

While the truth condition of the frame should ultimately be tested by market validation, the truth condition can be verified by the ability to generate an innovative hypothesis that explains a possible product and its solution principles, which are consistent with the frame. The existence of a solution is *ipso facto* evidence that confirms the validity of the frame. If no solution is possible that is consistent with the frame, then the hypothesis that led to the frame is likely invalid. The frame is the basis for the second abduction, the innovative abduction.

The innovative abduction, Figure 25.2, hypothesizes a product and its solution principles to create the intended value. The product and its solution principles remain consistent with the frame.

$$\text{frame} \quad + \quad \frac{\text{innovative}}{\text{abduction}} \quad = \quad \text{value}$$

Figure 25.2 An innovative abduction hypothesizes the product and its solution principles to create the intended value, consistent with the frame.

To illustrate this idea, consider the problem of road traffic. One possible frame for this problem is an economic one for which a plausible explanatory hypothesis for traffic avoidance is that drivers are utility maximizers. If the hypothesis is correct, and the corresponding economic frame, then if the intended value is to maximize utility, an innovative abduction could be any type of toll road ('what') in which the toll amount is based upon the time of day and level of congestion ('how'). Drivers can choose to use the toll road if the toll is less than or equal to the utility they derive from their use of the toll-road. Both the explanatory and innovative abduction may be false if in the implementation of the toll road, no driver uses it or even with a toll the congestion level fails to decrease. If so, the economic frame was not valid for traffic congestion. In this research, the ability of the participants to generate useful concepts from the insights will be used as a test of the validity of their insights.

3.4   Data in use

The analyses that follow report on how the design team generated insights within a set of themes (frames) that emerged from the co-creation workshops:

> *progress together* – working to accumulate wealth for family and community
> *conscious commitment* – being dedicated to values that acknowledge individuals as behaving responsibly; the dedication is reciprocated through social recognition that raises an individual's status
> *health as the enabler* – good health being fundamental to all other endeavors
> *in control and self-reliant* – being able to take of oneself and to control the controllable while sublimating those issues beyond individual control
> *endure and enjoy* – work hard now to enjoy the fruits of labor later in life

Most of the discussions associated with these themes occurred when the design team met to report, discuss, cluster, and reflect upon the workshops. Appendix 1 summarizes the product and service concepts generated for each of these themes. The following analyses will focus only on two themes: 'conscious commitment' and 'endure and enjoy. These two themes were chosen because of the richness of the conversation about these themes and a key difference between them: the team could develop design concepts for the 'conscious commitment' theme but struggled with the 'endure and enjoy' theme. The results will show important differences in the semantic profiles of knowledge-building. Based upon these differences, the paper draws some conclusions about knowledge-building for design insight and proposes a functional model of design insight. It should be noted that there were many other insights generated, such as about loyalty, smart living, and sense of achievement, which are not discussed in this paper due to space considerations.

4 FINDING

4.1 Insight into sustainability – from 'status' to 'conscious commitment'

This section illustrates a case from the dataset in which the participants progress between varying degrees of semantic gravity and semantic density as they exhibit generative sensing leading to new insight about 'status'. Their explanatory hypotheses about the origins of status in a Chinese context support the development of several product and service concepts. This theme also ties back to the car company's (v19, 133) "core traditional or primitive value. The sustainability story".

The development of this insight started in the debrief of the first co-design workshop and continued into their second insights meeting. The knowledge-building leading up to the first explanatory abduction expressing this insight began with the team discussing physical and mental health, one of the top-three high-level themes that the co-creation workshop participants voted as being the most important to them. They immediately bring the concept of physical and mental health into the context of the co-creation workshops (descend the semantic scale), mentioning various ways in which the workshop participants described relaxation and expressed concern for the environment, and thought about the effect of the environment on their health. By referring to the co-creation workshop data, the consultants strengthened the semantic gravity of the meaning of physical and mental health by bringing the concept closer to their direct observations from the workshop. This strengthening of semantic gravity is 'interrupted' by a set of *surprising observations*, later referenced to and summarized by Abby as:

Abby (v07, 80): Yeah. It was interesting what he said about that the (.) it's not about all the materialistic things, it's more about knowledge and like for sure that could be: not so rich, but very respected because of his position and eh knowledge eh

Abby references Amelia's anomalous observation about a participant's (v07, 57) "me making my own decision" sort of thing" as being about status rather than independence. Proceeding up the semantic scale (weaker semantic gravity), Will gives the explanatory abduction for these observations:

Will (v07, 63): And- but, I think *something we can think about* is eh this sort of expression, right? Eh and also for a more cosmopolitan, more progressive city like CHINESE CITY. *Their kind of expression and mentality could also be a form of like status symbol.*

Following Will's statement, Amanda recalls evidence (descends the semantic scale) from the workshop to support Will's hypothesis: (v07, 69) "Exactly. I really felt that also he- he kind of want to state it many times. *"No no, I don't take recommendations from my friends, I don't- no no, I (.) no one should decide, I want to decide this".*" Will generalizes the evidence with the statement that, (v07, 148) *"it's* [participants' statements] quite clearly indicating that *they are picking up or using ehm emotional indicators as self-expression}:, as bragging rights:, as differentiation}:,* you know."

This is the first time that the participants introduce 'status' as a concept into their observations that intangibles such as knowledge and behavior are a type of status.

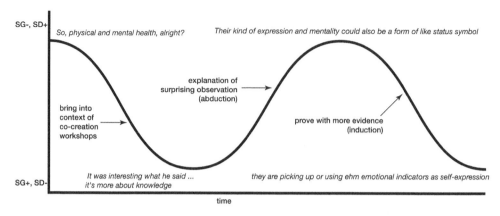

Figure 25.3   Semantic scale corresponding to the development of insight on 'status' leading to the theme of 'conscious commitment'. SG=semantic gravity (degree of dependence of meaning on context) and SD=semantic density (complexity of meaning).

As Will explains, (v07, 140) "*Notice that when she say it [good-life behaviors]*, she expresses it as, "*what other people want to do, but they: can't do*", *as opposed to* "what I want to do, but it's hard for me to do".[1]

'Status' becomes a frame for these observations. Following the framework shown in Figure 25.1, the frame is formed as:

observation	what other people want to do, but they: can't do
+	
explanatory abduction	their kind[s] of expression and mentality are a status symbol
=	
frame	status

To sum up, Figure 25.3 depicts the semantic wave from the data as the participants started from their observations to their hypothesis and subsequent strengthening of the semantic density of 'status' with reference to the workshop participants' statements. Italicized text represents actual quotes from the data.

The increasing semantic density of 'status' is worth noting. The design team recognizes that 'status' has both an everyday meaning (weaker semantic density) and a specialized meaning to them in the context of the co-creation workshops (stronger semantic density). As Lloyd and Oak (2017) point out, 'status' becomes entwined with other ideas including "a very sophisticated way of thinking" and "knowledge, awareness, time, investment". The team increases the semantic density of 'status' by Ewan connecting 'status' to the type of 'status symbol' embodied by a Rolex watch. Ewan's story about a wealthy woman in Scandinavia who travels to Greece to assist refugees while posting photos daily of her with babies supports the team's eventual view about conscious displays of status; his story augments 'status' with a story that has a character, motivation, and plot to lodge the concept into the consciousness of the team (Lloyd and Oak, 2017). At the peak of semantic density, 'status' redefines the concept of environmental sustainability. Amanda states, (v11, 409) "Yeah. So *it's not just about environment sustainability anymore*." Amanda concludes that 'status' becomes a capability (v11, 415) "to express their (.) to express their commitment". Amanda's statement references a key insight that Ewan made in a previous discussion through an explanatory abduction:

> (v09, 8) you don't prioritize your own needs to cater for others, but other people discover your commitment and sacrifice, and you get elevated to a higher status level.

In this statement is a hypothesis about the cause of 'status': People who have their commitment and sacrifice discovered are elevated to a higher status level. This hypothesis links together two key concepts: 'status' and 'commitment', leading participants to name this frame 'sexy commitment', later renamed to 'conscious commitment'. Daly *et al.* (2017) and Lloyd and Oak (2017) make the interpretation that 'sexy commitment' supplants the company's existing ideas about environmentalism and the 'good life': it is not enough simply to do good, individuals must also be seen and recognized as doing good.

4.2 Insight into the 'good life' – 'endure and enjoy'

This section shows a case from the dataset in which the participants dwell at a level of decontextualized meanings (weak semantic gravity). In this case, the participants will fail to reach productive insight as they will remain at a level of semantic gravity that is too weak because no one refers to the context, the co-creation workshops. As well, they exhibit stronger semantic density that packages up their observations into concepts that have meaning to individual members rather than the team. Therefore, they are unable to construct hypotheses that refer to observations. Their conceptualizations are generic such that they can apply to any situation and decontextualized from the co-creation data. Eventually, the consequence is difficulty in brainstorming concepts related to this theme because there was no knowledge-building.

According to the design team, the theme of 'endure and enjoy' is rooted to the Chinese work morale. In contrast to the Western world in which individuals want to enjoy their work, in China it is not important to enjoy work if there is the payoff later in life. Kenny explains this concept through reference to a prior engagement: (v11, 5) "The Chinese mentality of "endure now, benefit later" has become the recipe of success with aspirations to achieve a good life." In contrast to the teams' ability to reference the co-creation workshop context in relation to 'conscious commitment', with this concept, they continuously weaken semantic gravity while strengthening semantic density. They strengthen semantic density by connecting "endure now, benefit later" to (v11, 7) "concerns and challenges" to (v11, 10) "endure the uncontrollable" to (v11, 19) "break away from society and family constraints" to (v11, 10) "desire to retreat and escape" to (v11, 21) "manifestation of great autonomy". They do not appear to ascend or descend the semantic scale. Individuals added new connections rather than augmenting others' interpretations (weakening semantic density) or relating others' interpretations to the workshops (strengthening semantic gravity). Aptly, Kenny described the manifestation of new connections as "underlying subthemes" and noted with some optimism:

> (v11, 31) We kept like circling around. We- it was so easy to- to get in to good life. But then going from there into enjoying life, we kept circling around enjoying life with all the underlying sub-themes, just circling around it, but not really getting in to it. Eh: but I think I- I think we're happy about where we came now.

Some concepts never progress from an everyday meaning. The semantic density of 'endure and enjoy' remains at the level of an everyday meaning. The clause (v11, 19) "desire to retreat and escape from daily chores to accommodate me-time" is the only example in which the participants augment the meaning of 'enjoy' by relating it to an observation, the "cleaning robot quote", but this only confirms its everyday meaning. Given the lack of semantic complexity of these concepts, the participants do not create any new hypothesis to explain their observations. Their summary of this theme is, (v11, 21) "freedom of having pockets of enjoyment reflects a sense of achieve' – achievement towards a good life" – is simply a paraphrase of a statement made by Rose in their first debrief: (v8, 314) "the main goal actually I think it's still the enjoying life bit, that was what they really wanted, to have time, spending time with family, and so the choice is how to optimize time." The lack of deep insight is reflected in Will's statement, (v11, 23) "So that's (.) the result of eh: one and a half hour work, and I'm sorry it looks like so few words-".

4.3   Proposal for functional model of design insight

In the dataset analyzed, the design team developed insight into the theme of 'conscious commitment' but struggled to develop insights into 'endure and enjoy'. The difference in the semantic profiles of knowledge-building between these two themes is that for 'conscious commitment' the team traversed the semantic scale toward a novel hypothesis whereas for 'endure and enjoy' the team remained at weaker semantic gravity and weaker semantic density. The difference in patterns between the two cases suggests that insight occurs when individuals represent observations (or information) in terms of abstract, general, and decontextualized features (weak semantic gravity) that convey a novel central coherence (strong semantic density). As these abstractions become increasingly detached from the information, individuals establish the connection between the abstraction and the source information through hypotheses. In other words, the abstraction, which might be thought of as a new way of seeing the information, triggers the generation of a hypothesis, which if true, explains the cause(s) of the observations. The hypothesis becomes tested and further ones established through generative sensing until no new hypotheses can be generated that validate or invalidate the observations. Semantic structures characterized by varying degrees of strengths in terms of semantic gravity and semantic density operates as a structure upon which possible hypotheses can be grounded.

Based upon this observed difference, we propose a functional model of insight. Knowledge-building associated with the generation of the insight entails a series of movements 'up and down' a semantic scale ranging from concrete details to decontextualized features. Insights develop when observations are gradually represented in terms of abstract, general, and decontextualized features rather than their concrete, contextual, and incidental details or their abstract features alone. The decontextualization eventually reaches a limit, at which point an abductive hypothesis is offered to explain the observations. The patterns of movements across the semantic scale indicate that insight requires simultaneous decontextualization of evidence and observations into highly condensed meanings and their recontextualization into a new hypothesis.

In a follow-up interview, Ewan is proud of the team's insight into 'conscious commitment' and acknowledges the team (v22, 114) "getting stuck" in "the whole collectiveness versus- versus individual ... kind of network". 'Conscious commitment' is a re-contextualization of the car company's concept of sustainability in a Chinese context. "Conscious commitment" is about the dual effect of being dedicated to collectivism values including care for your family and other people. Value is about giving back to society, which is reciprocated through value recognition. As stated by David, (v21, 11:57) "Your investment there, the values that you emit will come back to you in the form of social recognition and social elevation." The person doing the recognition must have the same level of sophistication to recognize that the other individual is climbing up the social ladder. The theme of 'endure and enjoy' in contrast lacks a similar level of elaboration.

It should be noted that the analysis of the dataset occurred before the follow-up interview. The agreement between Ewan's statement on the themes with which the team succeeded and struggled and the team's trek across the hills and valleys of the semantic scale or along a high ridge line illustrates the importance of selecting and recontextualizing observations into hypotheses, which serve as the foundation of insights.

5   IMPLICATIONS FOR SUSTAINABLE DESIGN

We conclude the analysis by drawing some broader implications on the insights into sustainability derived from the co-creation workshops. Different cultures have different concepts of "sustainability" and value sustainability differently. The insights generated by the participants in the co-creation workshops suggest that China values sustainability as a status symbol and has only a vague interest in the details of how this sustainability is achieved. One study alone cannot support this broad generalization; however, if one assumes it applies more broadly, China has a potential advantage in adopting sustainable design over the United States (U.S.) and other Western countries. The following discussion is based on a broad application of this finding. Sustainable design, specifically eco-friendly or green design, faces a boundary in the U.S.: people are not willing to pay more for it, on the whole. In an extensive literature review, MacDonald and She (2015) find that it would be better if sustainability, as a feature, could compete with other features for which people are willing to pay more, such as luxury. In the absence of this willingness-to-pay, sustainable products must offer feature benefits that can compete with other non-sustainable products. These "triumph", or possibly "hero" as mentioned in the data set, products are not only better for the environment; they also have other feature advantages over competitors. Triumph products are more challenging to design, especially without added cost. Designing sustainable products for the Chinese market may not face this challenge, as the Chinese may not need triumphs, but rather as-good-as-competitor-products with the addition of superficial indications of sustainability. She and MacDonald (2013) created a design method for identifying "trigger" features that can be added to make products communicate thoughts of the sustainability through how they are perceived. This method could be used to add sustainable-communication features to aid in the communication of sustainability as a status symbol. The Western view of sustainable purchase as an altruistic action (MacDonald & She, 2015) is in conflict with the idea of sustainability being a status symbol. Also, when people view a purchase as altruistic, they expect to sacrifice on other features (quality, price) as part of their charitable action. This leads to the perception of sustainable products as less-good than non-sustainable competitors. Again, the Chinese have an advantage here, because without perceived altruism, the consumer can enjoy their status-purchase without guilt or perception of inferiority. Lastly, the study indicated that the Chinese consumer does not care much about the way sustainability was achieved. This allows designers and manufacturers to pursue sustainability in whichever way is most beneficial to them. However, as the number of products that claim sustainability increase, this flexibility on the part of the Chinese consumer will wane. The consumer will become more skeptical of claims and/or the prevalence of claims will significantly decrease their worth. If policy-makers in China get ahead of this trend, they can guide sustainability claims in a meaningful manner.

6   CONCLUSIONS

This paper described knowledge-building associated with the formation of design insight. The oft-valorized creative 'leap' is more a trek over semantic hills and across valleys, reaching toward higher levels of semantic complexity. At the nadir, the 'leap' is a hypothesis that explains a surprising observation. By mapping the semantic structures of generative sensing onto a semantic scale, and finding differences in semantic profiles between two representative cases, the paper proposed a functional model of insight. The model proposes that design insight entails the gradual representational detachment of observations away from their concrete features toward generalized principles until such time that a new hypothesis is necessary to explain the observations.

The analysis of co-inquiry by Adams *et al.* (2017) demonstrates an alternative approach to study knowledge-building toward insight. Using a collaborative inquiry framework (Heron & Reason, 1997), Adams *et al.* (2017) investigated the way that the participants created coherence about the problem space based upon knowledge brought into the co-creation process. In their analysis, co-inquiry is a knowledge-building process that entails finding a new relation between four ways of knowing (practical, propositional, presentational, and experiential) while attending to experiential knowledge in relation to critical frameworks. They identified insight as practical knowing, which is signaled by a tangible action that converts other forms of knowing into an externalized act. They found several instances of propositional[2] and practical knowing co-occurring, which is the same as our identification of insight as emerging from a hypothesis.

The approach taken by the design team to generate potential product or service concepts by developing insights rather than by searching for specifically-articulated user needs speaks to the value of *explaining* observations of individuals' behavior, experiences, and beliefs rather than simply reporting upon them. Specifically, the functional model of design insight proposes that design teams explain surprising observed behaviors through hypotheses that are not necessarily known to be empirically, scientifically, or logically true. That is, their hypotheses should be abductive rather than deductive or inductive. If the hypothesis could already be proven through established rules, then the hypothesis is not likely to be an insight. Actions should then be taken to test the hypothesis, which should include the introduction of a prototype. Under certain conditions of knowledge-building, those insights can lead to fruitful design concepts. For design strategists, this finding is important because it sheds light on the way that leaders should facilitate design insight workshops. Specifically, the model of design insight advises strategists to support the formation of design insight through several actions:

1   lead the team toward more detached representations of their observations of users by:

   a)   connecting their observations to other known concepts (e.g., psychological theories, social theories, design principles) and explanatory hypotheses
   b)   augmenting their observations with other concrete cases from their personal experiences

2   push the team toward the invention of a new way to explain their observations rather than accepting established explanations
3   discourage the proliferation of abstract representations of observations that are segmented from each other, that is, representations that fail to create explicit connections to prior representations.

ACKNOWLEDGMENTS

This research was supported under Australian Research Council's Discovery Projects funding scheme (project DP160102290).

NOTES

1   The original observation is, (v8, 22) "Susan [a co-creation workshop participant] is eh: a little bit more about looking good projecting eh certain kind of image, in front of other people. Not necessarily from the material stance, but she talked about things like- the comment that we really liked and we focused on is like she said something about "I want to do something that other people want to do, that they can't do". So it's what other people want, and want to achieve, not necessarily what- what I want to achieve, but I wanna show them that I can do what they can't do:"
2   Propositional knowledge in the form of "Proposing or making assertions that depict an aspect of the world" (Adams *et al.*, 2017) is an instance of abductive reasoning.

REFERENCES

Adams, R. S., Aleong, R., Goldstein, M. & Solis, F. (2017). Problem Structuring as Co-Inquiry. In: Christensen, B. T., Ball, L. J. & Halskov, K. (eds.) *Analysing Design Thinking: Studies of Cross-Cultural Co-Creation*. Leiden: CRC Press/Taylor & Francis.
Bedford, D. A. D., Arns, J. W. & Miller, K. (2017). Unpacking a Design Thinking Process with Discourse and Social Network Analysis. In: Christensen, B. T., Ball, L. J. & Halskov, K.

(eds.) *Analysing Design Thinking: Studies of Cross-Cultural Co-Creation*. Leiden: CRC Press/Taylor & Francis.

Bowden, E. M., Jung-Beeman, M., Fleck, J., & Kounios, J. (2005). New approaches to demystifying insight. In: *Trends in Cognitive Sciences, 9*(7), 322–328. doi: 10.1016/j.tics.2005.05.012

Christensen, B. T. & Abildgaard, S. J. J. (2017). Inside the DTRS11 Dataset: Background, Content, and Methodological Choices. In: Christensen, B. T., Ball, L. J. & Halskov, K. (eds.). *Analysing Design Thinking: Studies of Cross-Cultural Co-Creation*. Leiden: CRC Press/Taylor & Francis.

Cornelissen, J. P., & Clarke, J. S. (2010). Imagining and Rationalizing Opportunities: Inductive Reasoning and the Creation and Justification of New Ventures. *Academy of Management Review, 35*(4), 539–557.

Cross, N. (1997). Creativity in design: analyzing and modeling the creative leap. *Leonardo, 30*, 311–317.

Daly, S., McKilligan, S., Murphy, L. & Ostrowski, A. (2017). Tracing Problem Evolution: Factors That Impact Design Problem Definition. In: Christensen, B. T., Ball, L. J. & Halskov, K. (eds.) *Analysing Design Thinking: Studies of Cross-Cultural Co-Creation*. Leiden: CRC Press/Taylor & Francis.

Darke, J. (1979). The primary generator and the design process. *Design Studies, 1*(1), 36–44. doi: 10.1016/0142-694x(79)90027-9

Dong, A., Garbuio, M., & Lovallo, D. (2016a). Generative sensing in design evaluation. *Design Studies, 45, Part A*, 68–91. doi: 10.1016/j.destud.2016.01.003

Dong, A., Garbuio, M., & Lovallo, D. (2016b). Generative sensing: A design perspective on the microfoundations of sensing capabilities. *California Management Review, 58*(4), 97–117.

Dong, A., Lovallo, D., & Mounarath, R. (2015). The effect of abductive reasoning on concept selection decisions. *Design Studies, 37*, 37–58. doi: 10.1016/j.destud.2014.12.004

Dorst, K. (2011). The core of 'design thinking' and its application. *Design Studies, 32*(6), 521–532. doi: 10.1016/j.destud.2011.07.006

Gaglio, C. M. (2004). The Role of Mental Simulations and Counterfactual Thinking in the Opportunity Identification Process. *Entrepreneurship Theory and Practice, 28*(6), 533–552. doi: 10.1111/j.1540-6520.2004.00063.x

Garst, J., Kerr, N. L., Harris, S. E., & Sheppard, L. A. (2002). Satisficing in Hypothesis Generation. *The American Journal of Psychology, 115*(4), 475–500. doi: 10.2307/1423524

Gettys, C. F., & Fisher, S. D. (1979). Hypothesis plausibility and hypothesis generation. *Organizational Behavior and Human Performance, 24*(1), 93–110. doi: 10.1016/0030-5073(79)90018-7

Heron, J., & Reason, P. (1997). A Participatory Inquiry Paradigm. *Qualitative Inquiry, 3*(3), 274–294. doi: 10.1177/107780049700300302

Kolko, J. (2010). Abductive Thinking and Sensemaking: The Drivers of Design Synthesis. *Design Issues, 26*(1), 15–28. doi: 10.1162/desi.2010.26.1.15

Kounios, J., & Beeman, M. (2009). The Aha! Moment: The Cognitive Neuroscience of Insight. *Current Directions in Psychological Science, 18*(4), 210–216. doi: 10.1111/j.1467-8721.2009.01638.x

Lloyd, P. (2009). Ethical imagination and design. *Design Studies, 30*(2), 154–168. doi: 10.1016/j.destud.2008.12.004

Lloyd, P. & Oak, A. (2017). Cracking Open Co-Creation: Categorizations, Stories, Values. In: Christensen, B. T., Ball, L. J. & Halskov, K. (eds.) *Analysing Design Thinking: Studies of Cross-Cultural Co-Creation*. Leiden: CRC Press/Taylor & Francis.

Luo, J., & Knoblich, G. (2007). Studying insight problem solving with neuroscientific methods. *Methods, 42*(1), 77–86. doi: 10.1016/j.ymeth.2006.12.005

MacDonald, E. F., & She, J. (2015). Seven cognitive concepts for successful eco-design. *Journal of Cleaner Production, 92*, 23–36. doi: 10.1016/j.jclepro.2014.12.096

Magnani, L. (2001). *Abduction, reason, and science: processes of discovery and explanation.* New York: Kluwer Academic/Plenum Publishers.

Manning, C., Gettys, C., Nicewander, A., Fisher, S., & Mehle, T. O. M. (1980). Predicting individual differences in generation of hypotheses. *Psychological Reports, 47*(3f), 1199–1214. doi: 10.2466/pr0.1980.47.3f.1199

Maton, K. (2013a). *Knowledge and knowers: Towards a realist sociology of education.* London: Routledge.

Maton, K. (2013b). Making semantic waves: A key to cumulative knowledge-building. *Linguistics and Education, 24*(1), 8–22. doi: 10.1016/j.linged.2012.11.005

Maton, K., & Doran, Y. J. (2016a). Condensation: A translation device for revealing complexity of knowledge practices in discourse, part 2 – clausing and sequencing. *Onomázein.*

Maton, K., & Doran, Y. J. (2016b). Semantic density: A translation device for revealing complexity of knowledge practices in discourse, part 1 – wording. *Onomázein.*

Murty, P. (2007). Latent preparation. In: Y. Gang, Q. Zhou & D. Wei (Eds.) *CAADRIA 2007: Proceedings of the 12th International Conference on Computer-Aided Architectural Design Research in Asia: Digitization and Globalization.* Nanjing, China: Association for Computer-Aided Architectural Design Research in Asia. pp. 529–536.

Niiniluoto, I. (1999). Defending Abduction. *Philosophy of Science, 66*(Supplement. Proceedings of the 1998 Biennial Meetings of the Philosophy of Science Association. Part I: Contributed Papers), S436–S451.

Ohlsson, S. (1992). Information-processing explanations of insight and related phenomena. In M. T. Keane & K. J. Gilhooly (Eds.) *Advances in the psychology of thinking.* London: Harvester Wheatsheaf. pp. 1–44.

Peirce, C. S. (1932). *Collected Papers of Charles Sanders Peirce* (C. Hartshorne & P. Weiss Eds. Vol. II: Elements of Logic). Cambridge: Harvard University Press.

Peirce, C. S. (1998). *The Essential Peirce, Volume 2: Selected Philosophical Writings, 1893–1913* (N. Houser & C. Kloesel Eds.). Bloomington: Indiana University Press.

Roozenburg, N. (1993). On the pattern of reasoning in innovative design. *Design Studies, 14*(1), 4–18. doi: 10.1016/s0142-694x(05)80002-x

She, J., & MacDonald, E. (2013). Priming Designers to Communicate Sustainability. *Journal of Mechanical Design, 136*(1), 011001-011001. doi: 10.1115/1.4025488

Shedroff, N. (2001). *Experience design 1.* Indianapolis, IN: New Riders.

Suwa, M., Gero, J., & Purcell, T. (2000). Unexpected discoveries and S-invention of design requirements: important vehicles for a design process. *Design Studies, 21*(6), 539–567. doi: 10.1016/S0142-694X(99)00034-4

Thomas, R. P., Dougherty, M. R., Sprenger, A. M., & Harbison, J. I. (2008). Diagnostic hypothesis generation and human judgment. *Psychological Review, 115*(1), 155–185. doi: 10.1037/0033-295X.115.1.155

Vasconcelos, L. A., & Crilly, N. (2016). Inspiration and fixation: Questions, methods, findings, and challenges. *Design Studies, 42*, 1–32. doi: 10.1016/j.destud.2015.11.001

Visser, W. (2006). Designing as Construction of Representations: A Dynamic Viewpoint in Cognitive Design Research. *Human-Computer Interaction, 21*(1), 103–152. doi: 10.1207/s15327051hci2101_4

APPENDIX 1 BRAINSTORM

During the brainstorm session, the team generated a number of ideas based upon the themes. The following data shows the ideas they generated based upon a photograph of Post-it notes containing descriptions of the ideas and the video of the brainstorm session.

progress together

- spare parts for me and other people
- passing the item on adds value to that item
- durability for me and others: our accessories are made of durable materials and they can pass them on for generations. Sustainable in the sense that you can keep it. Other people in a timeline. Communicate durability in terms of family rather than it's been made out of something durable.
- soul of the car: small token which gathers info, bring to new car, hand to kids
- care for parents; target group buys accessories for their kids and parents
- parent/grandparent care just as childcare products; to take care of your parents, relevant products, nice materials
- family accessory subscription; you get the best + sustainable; get it delivered to your car, mounted

conscious commitment

- award e.g. "green tyres" which you 'earn' by choosing green parts for your car
- what is car company that BMW isn't, social behavior, sustainable
- buy this + donation goes to X
- car company to start academy where we teach environmental care, etc.
- "Eco-box" cleaning the air around the car; subtle indications that the car is cleaning
- rent out "parts" of your car: electricity, trunk, battery, roof rack, etc.
- cater for the local community, accessory pool, peer to peer equipment share
- we give 10 cars to a local community for them to use (for local good initiatives)
- locally produced "green" production "short travel parts"
- buy seats++ which are produced in your region of choice; money goes back to them
- refurbished accessories: sell back accessories to be refurbished
- green-colored tires. Maybe there is something connected with the engine. Maybe you could have a very nice color tint that communicates sustainability

health as the enabler

- sustainable "healthy" materials
- (popular) materials, cutting edge
- data collected while in the car about you to see if you're still healthy, tell you to slow down, you're tired, you should go to the doctor

in control and self-reliant

- weekend-getaway self-reliant kit (don't need hotels)

endure and enjoy

- weekend-getaway kit
- deliver cars with 2 bottles of Nordic water, apply to all interactions with car company "the extra treat enjoyment"
- team up with cool brand X
- "comfortable accessories" e.g. seat belt "puffs", heated child seat, etc.

Empathy in Design: A Discourse Analysis of Industrial Co-Creation Practices

Justin L. Hess & Nicholas D. Fila

ABSTRACT

Design researchers are increasingly promoting the development and utilization of empathy within design. However, there is a limited body of knowledge on how empathy operates in the world of design, particularly within professional contexts. In this study, we sought to understand how empathy manifests within an industrial design context by observing how a team of professional designers delivered and reflected on two co-creation workshops. We used discourse analysis as our guiding methodology, which directed our attention to micro-level discourses, or the specific words and phrases that designers used, along with macro-level Discourses, or the broader social, political, and environmental forces that deterred or encouraged designers to empathize with users.

We developed five themes from this analysis. First, the designers created and refined figured worlds, or user-assumptions, as a means to generalize from users to broader user groups and vice versa. Second, the designers utilized a series of empathic techniques to attempt to develop an empathically accurate understanding of the users. Third, designers developed personas from individual users to create idealized versions of their target user demographic. Fourth, through a telephone game of sorts, where the designers communicated their user-centric understanding to higher-ups in the company who in turn communicated with higher ups, the likelihood of holistic user understanding was potentially diminished. Lastly, the designers, constrained by corporate demands, faced the tension of "conquesting" clientele versus establishing meaningful concern for users. We describe these results and their implications for future design research and practice.

I INTRODUCTION

Empathy is a multifaceted phenomenon with cognitive, affective, and behavioral components (Davis, 1996; Kouprie & Sleeswijk Visser, 2009). Scholars have described at least eight distinct phenomena as empathy (Batson, 2009), ranging from an involuntary "mirror neuron" response (Iacoboni, 2009), a vicarious emotional response (Eisenberg & Fabes, 1990), a way of being (Rogers, 1975), to an advanced cognitive process that enables one to imagine and internalize the perspective of another (Batson, Early, & Salvarani, 1997). Further, empathy can be primed either automatically or voluntarily and separate empathy types may operate in isolation or concurrently. Thus, effectively empathizing with another is a complex and challenging process.

Leonard and Rayport (1997) were arguably the first scholars to describe an "empathic design" approach, and, similar to the wide-ranging conceptualizations of empathy, researchers have described this process in a variety of ways since then. For example, Mattelmäki, Vaajakallio, and Koskinen (2014) indicated the empathic design research focus has shifted from "elucidating experiences in an interpretive manner," to "co-design" or facilitating organizational and public collaboration, to "imagination" that goes beyond interpretation only (p. 69). More pointedly, Mattelmäki *et al.* (2014) suggested empathic design "focuses on everyday life experiences, and on individual desires, moods, and emotions in human activities, turning such experiences and emotions into inspirations" (p. 67). To achieve such an experience-to-inspiration translation, Postma *et al.* (2012) indicated that empathic designers must shift between participatory and expert mindsets. Specifically, they recognized four principles of empathic design: (a) "balancing rationality and emotions in building understanding of users' experiences," (b) making "empathic inferences about users and their possible futures," (c) "involving users as partners in NPD [new product development]," and (d) engaging "design team members as multi-disciplinary experts in performing user research" (p. 60).

Bridging the diverse understandings of "empathy" within empathic design literature, Kouprie and Sleeswijk Visser (2009) found three primary emphases: empathy as an intuitive but vague "quality" of design, empathy as a designer's "ability" and "willingness" to identify with others, and empathy as a design "technique" (p. 438f). As these findings indicate, the focus has seldom been on empathy as a psychological concept, but rather as a tool that may have utility at some point in the design process. This is likely because empathy is a complex phenomenon: it has affective and cognitive components, it is both a process and an internal disposition, and it has antecedents that will dictate its use or manifestation (Batson, 2009; Davis, 1996). Due to this complexity, it is difficult to specify *exactly* how empathy operates in design.

While "empathic design" emphasizes the development or utilization of empathy within design, any design process seems contingent upon anticipating the needs of users, and, hence, empathy. For example, Sanders and Stappers (2014) described co-design as the act of designers and non-designers "working together, using making as a way to make sense of the future" (p. 5). Here, designers vacillate between two mindsets: designing *with* users and designing *for* users. When moving between acts of inclusion and exclusion, designers' mindsets also vary. When "designing with" the mindset tends toward helping users, whereas when "designing for" the mindset tends toward provoking users. Further, when designing with, designers recognize users as "experts", whereas when designing for, the designers (the "experts") study users as subjects. Theoretically, the role of empathy shifts from an emphasis on affect when designing with (e.g., intuitively grasping users' emotions), to an emphasis on cognition when designing for (e.g., imagining the user perspective and trying to anticipate reactions to design decisions). This is not to suggest that affect or cognition ever disappear, but rather that their emphasis varies from one scope to the other.

Many scholars have posited that designers' development of empathic-related dispositions or tendencies will increase the likelihood that they will possess a participative mindset and thereby utilize empathy (Fila *et al.*, 2014; Kwok-leung Ho, Ma, & Lee,

2011; Postma *et al.*, 2012; Sanders & Stappers, 2014). However, some scholars have found that designers, particularly novice designers, tend towards a *designing for* mindset and, in the process, tend to devalue the emotional needs of users (Kwok-leung Ho *et al.*, 2011). Conversely, some scholars have explored how empathy manifests in varying design contexts. For example, students participating in a service-learning course demonstrated many instances of cognitive and affective empathy (Fila & Hess, 2015; Hess & Fila, 2016b) whereas students engaging in a non-immersive design task primarily utilized self-centric strategies that seemed to lack empathy (Fila *et al.*, 2016). Hence, one might posit that the manifestation of empathy in design will be partly due to designers' mindsets and partly due to the design context.

Sanders (2009) speculated that organizational culture was even more important than both individual mindsets and design contexts, as this culture dictates what behaviors and actions are appropriate and rewarded at a grander scale. More specifically, in Sanders' model, organizational values direct the mindsets employees ought to utilize when approaching design problems, which in turn directs appropriate methodologies, methods, and tools designers will use. Theoretically, then, the culture of the individual classroom, school, or the university will largely predict students' utilization of empathy for users. Similarly, in an industrial context, the company culture, including the explicit values of the company as well as unspoken norms, will largely predict designers' development of empathy or utilization of empathic techniques.

This study continues this line of work. Specifically, the DTRS dataset (Christensen & Abildgaard, 2017) allowed us to explore how professional designers tended to utilize and communicate empathy for users at individual and collective levels, along with how these designers' professional and social contexts influenced their empathic development and utilization. Furthermore, the context of the dataset is unique: the designer team's task is to design for a markedly distinct user-group from a foreign country. Hence, our findings also elucidate how empathy may develop cross-culturally within a professional design setting.

2 THEORETICAL FRAMEWORK

To guide our investigation, we utilized discourse analysis (Gee, 2011) to identify how empathy manifested within the verbal interactions of a European professional design team as they planned and reflected on two co-creation workshops featuring Chinese clientele. We grounded our exploration in an interdisciplinary conceptualization of empathy that integrates literature from neuroscience, social psychology, and design. Specifically, we conceptualized empathy as trying to understand another's thoughts and feelings, perhaps by experiencing an emotion through the internalization of another's state, and as acting upon that knowledge or feeling *as evident* through their verbal discourse (Batson, 2009; Hess & Fila, 2016a; Hoffman, 2000).

Discourse analysis was our guiding epistemological and methodological framework. We assumed that verbal discourses provided a valid means to explore and explain how empathy manifested within the world of design. We analyzed micro-level discourses, or the "specific words and phrases used" by the designers, to consider the macro-level Discourses, or "the reality evident/created by language and the sociopolitical influences on those meanings" (Baillie & Douglas, 2014, p. 4). At the micro-level, we explored how empathy manifested in the designers' conversations. At the macro-level, we considered broader influences that may have inhibited, catalyzed, or directed the designers' use of empathy. One individual's discourse both informs and is informed by their interactions with others and, theoretically, the macro-Discourse. Hence, we focused on the design team as a whole for our analysis, including their references to macro-level constraining forces.

With this framing in mind, our study addressed two **research questions** at the micro/macro levels:

- **Micro-level:** In what ways do professional designers develop empathy, as evident through their discourse, for a markedly distinct user group when utilizing a co-design approach?
- **Macro-level:** What overarching personal, corporate, and political structures inform the way professional designers communicate empathy when designing for a distinct user group?

3   RESEARCH METHODS

To guide our analysis, we utilized a modified version of Gee's (2011) discourse analysis methodology which incorporated elements of Braun and Clarke's (2006) thematic analysis approach. Specifically, we used a series of non-linear steps, each with extensive conversations between the authors to enhance the trustworthiness and quality of our investigation (Walther, Sochacka, & Kellam, 2013). During each step, we generated potential themes, which Gee (2011) described as hypotheses "about what we expect to find in further data or in a closer look at our original data" (p. 25). We continued to refine these themes throughout our investigation and through frequent dialogue between one another.

As a first step, we watched each video in the dataset to gain an understanding of the data's content and breadth. Throughout this process, we began noting where and how we potentially saw empathy in the team's discourse and we identified videos within which empathy for users appeared most common. The videos that were included in the next step were v07, v08, v09, v14, v18, v19, and v20 (Christensen & Abildgaard, 2017). To a large extent, this selection was triangulated by the semantic analysis conducted by Bedford *et al.* (2017) who found empathy (which they defined as "listening, sympathy, and understanding") to be largely present in videos where the team's debriefed post-workshop or shared insights (namely, v18, v14, v10).

Second, each week for two months, we selected one to two hours of video and identified, coded, or grouped excerpts. While we separately watched the videos, we identified discourses that were potentially empathetic and either created a new category or grouped these into an existing category underlying one of Gee's building tasks (described next). Next, through dialogue, we discussed what we coded, why, and then collaboratively refined our evolving narration of the themes. Third, we finalized our thematic selection by grouping codes within and across building tasks. As a final step, we formed narratives for each of these themes. These narratives demonstrate how we perceived empathy to be utilized, catalyzed, or constrained in the team's process.

Building tasks explain how one creates "reality" through language. We used building tasks to help us identify how empathy operated within the designers' worlds as they designed and reflected on a series of co-creation workshops. Gee described seven building tasks, each of which corresponds with questions (described below) that guided our analysis. Initially, we created codes and categorized these within each of the building tasks, but due to the time-intensiveness of this strategy, we started coding/grouping the discourses into their most pertinent category. In this sense, our analysis may more closely resemble thematic analysis rather than pure discourse analysis.

- **Significance**: How is this piece of language being used to make *empathizing with users* significant or not and in what ways?
- **Practices**: What practice is this piece of language being used to enact to get the team to empathize with users?
- **Identities**: With respect to users, what identities is this piece of language being used to enact? With respect to the self, how does this language help designers enact their identity?
- **Relationships**: What relationship is this piece of language seeking to enact with the users?
- **Politics**: What perspective on social goods is this piece of language communicating?
- **Connections**: How does this piece of language connect or disconnect users with other things? How does it make those connections relevant or irrelevant?
- **Sign systems**: How does this piece of language privilege/disprivilege specific sign systems?

4 RESULTS

In the following sections, we present five themes: *Figured Worlds*, *Seeking Empathic Accuracy*, *Persona-fication*, *The Telephone Game*, and *Conquest versus Concern*. These represent the most prominent discourses that we found in the designers' interactions. We present these discourses in order from micro to macro to first highlight individual or team-level discourses and then illuminate the overarching macro-discourses within which these micro-discourses were situated.

4.1 Figured worlds

The team commonly generated user-centric understanding by establishing this understanding with respect to pre-conceived assumptions, many of which the team assumed were universal. Such assumptions Gee (2011) described as figured worlds, or the "picture of a simplified world that captures what is taken to be typical or normal" (p. 71). The team relied upon figured worlds whenever they utilized the user context or categorical variables (e.g., "younger guy"; "he's not married"; "jogging club") as a means for understanding users or to justify their interpretations of user perspectives. Notably, the team often drew user assumptions through reference to the vehicles users drove. For example, while debriefing after the second co-creation workshop, the design team drew inferences from Yen (an oft-cited user whom they described as a "first mover") and his adoption of the Tesla (see Table 26.1).

Table 26.1 Extraction from debrief of co-creation workshop 2 (v14, 515–521).

R	… with the Tesla, at least for the first adopters, he [Yen] would have been […] in the initial group of people who have the Tesla. You gain that- that credibility just in that very small group […] they just feel good that they are already in the group, right?
E	And I think that is one of the reasons behind the success of Tesla, yeah (..) And it's like yeah, it's so: the right thing to do, eh:: because it is environmentally friendly and I look awesome, it is awesomely humble in a way, like (..) yeah, I'm doing this for the environment you can say, and people have to believe you, even though you can think you own thoughts in your head, but you can't argue against that.
AM	So being a first mover give them some intrinsic value kind of.
E	Mm definitely
R	And it- I guess it reflect something of them, to be the first-mover to begin with, right?
AM	Exactly
E	Ehh: I think it's very very interesting about, yeah, the whole- okay so maybe this guy he bought environmental things and he felt good about it, it was only for him. Still I think that he really wants to get repaid in one way or another, but maybe there is that he's- this is fueling some other parts with him, so it makes him proud, or more confident maybe. That is where he gets repaid. He walks into the room, he kind of own the room.

Table 26.2 Extraction from sharing insights from co-creation workshop 1 (v08, 038–039).

W	Traditional Chinese medicine is quite eh: a big thing, there is more for maintenance as opposed to: chronically eh: illness […] it's a little bit different from how: we think- or how they: will think of like western medicine, and we just- when you actually have a: big problem.
E	That's actually a very smart way of looking at it. Western medicine is after: you get sick, you get fixed up, here it's like "drink this tea:, do this thing and you won't: get sick".

In addition to the users' immediate contexts, the team generated user-assumptions via their socio-cultural contexts. Specifically, the team often attempted to enrich their understanding by juxtaposing eastern versus western norms. For example, the team used this dichotomization to understand user-attitudes about health. As shown in Table 26.2, Will described proactive (Western) versus reactive (Eastern) approaches to well-being, a notion that Ewan subsequently built upon.

4.2 Seeking empathic accuracy

While the team often relied on "insights" rooted in figured worlds, they also used empathic techniques to challenge, deny, and validate those assumptions. Specifically, the team constantly questioned their assumptions as they engaged in concept nego-tiation, seeking to enrich their emergent user-centric understanding. Such discourses, geared towards developing a holistic and accurate understanding of the user, permeated nearly all of the observed sessions, especially those in which the team was debriefing post-workshop or clustering insights. Throughout these dialogues, the team referenced assumptions (as described above), recalled prior interactions (with each other, the users, or prior research), negotiated user values, and structured their ideas with analytical

Table 26.3 Extraction from sharing insights from co-creation workshop 1 (v08, 191–204).

R	I really wanna share about what she [Susan] wanted to post on Instagram about her baking:
E	Eh yeah that was very good, so we: talked about how […] "do you sometimes […] go places and you take a photo, how do you take the photo?" and she was like "yeah I will just take the photo of the mountains like this" […] and then we started "okay, why is that important, why is it important that you are in it. […]
R	Oh sorry, can I interject first?
E	Oh, sorry.
R	But the thing was that she said didn't want to take a picture of her face, but she would share her location […]
E	[…], that was the understated:-
R	It's an understated way of like showing off, that you don't have to:-
E	And that was the thing, yeah not necessarily that you-
A	So it wasn't the selfie-thing, but just to have the: [location:]
R	[So people know] that you're there.
E	[…] I asked, "if you're not in the photo, how can you know that you're there […] and she said it instantly, "because I share the location". So it wasn't like something she had to- she'd thought about this, and she didn't want to be in the photo 'cause she didn't feel she was beautiful enough, alright?
AM	Yeah, that's what she said.
E	[…] then she added the location, perfect.
R	Yeah, and the rest of the group agreed so it was not like she was the only person…

tools (e.g., an innovation matrix, v18). Throughout their discourses, no information was off-limits. For example, the team continually juxtaposed their understanding with respect to successful Chinese marketing strategies outside of the automobile industry (e.g., Apple, Nike).

Many of these techniques resembled empathic techniques identified in previous studies (e.g., see Koskinen & Battarbee, 2003; Fila & Hess, 2015). For example, unsurprisingly, the team frequently reflected on their interactions with the users from the co-creation workshops. Through these reflections, the team negotiated their emergent user-centric understanding by prompting the users' own discourses, and, collectively, the team would add their interpretations of those discourses to the conversation. For example, Table 26.3 shows an interaction from the team sharing insights from the first co-creation workshop. Rose called to mind an interaction with Susan, a workshop participant. Ewan began detailing his understanding of Susan while Rose added nuance, Abby asked for clarification, Ewan expounded, and Amanda validated Ewan's addition. Eventually, the team appeared to reach a consensus.

4.3 Persona-fication

The team used numerous methods to convert their user-centric knowledge to actionable insights. One of the most prominent of these methods was creating user personas. This process of persona-fication was guided by the aforementioned themes: figured worlds and seeking empathic accuracy. In this context, a persona represents an idealized user framed by the loose collection of characteristics, perspectives, values, and contextual factors culled from the workshops. Hence, the process of persona-fication enabled the team to (a) methodically sort through the numerous individual user insights, (b) expand

Table 26.4 Extraction from debrief of co-creation workshop 2 (v14, 561–576).

E	What was the exact quote? Something you said, Abby, about unspoken…
AM	Superiority
E	Unspoken Superiority. Yeah.
A	That wasn't really what he told.
E	It is now!
A	It is now!
K	Because it's cool!
A	But it's all about the feeling inside.
E	Inside, yeah.
A	But of course he wants to – I mean also that type of guy is of course – he wants to kind of brag about that he makes all his own choices. So of course it is something that will say at some point, but he also feels that it's something that he shouldn't speak of.
E	Yeah
A	So he also knows that, okay, I also need to keep it a little bit to myself, but a guy like that, just, he seems like he–
E	Yeah, really we need that guy to buy THE COMPANY.

the focus beyond individual users to the broader target demographic, and (c) present a coherent story for company stakeholders outside the team.

As designers built personas, they seemed to become empathically distanced from authentic individuals. Hence, personas appeared to diminish empathic insights for more broadly drawn, abstract, and idealized user models. In part, this distance is inherent in the persona form, as a composite cannot represent a unified whole. Thus, the user-perspectives and values were often lost as the designers moved from the real user to the fictional one and as designers molded the personas to appeal to the target demographic. In doing so, the team continually considered the "story" that would allow them to infiltrate market niches. For example, the team built a persona based on a paraphrased quote, despite an acknowledgement that the persona might misrepresent user values. This misquote appeared to represent an attribute of the Company's ideal user (see Table 26.4).

Personas grew over time from many information sources. Hence, multi-project and cross-context references sometimes allowed the team to build more robust and broadly applicable conceptions of users, even while such discourses sometimes seemed to diminish the team's empathic understanding of the workshop participants. For example, when sharing insights from the first co-creation workshop, Will described a Scandinavian-specific persona that he then merged with a Chinese-specific one. Ewan used this discussion as a springboard to integrate non-user related insights, thereby applying his own idealizations upon this emergent persona (see Table 26.5).

Throughout the process, the team frequently considered how to present their findings to Tiffany and Hans from the Accessories Department. Personas became the team's method for presenting their findings. For example, the team created a "superhero" persona who represented the Company values (e.g., the Mercedes Guy, see Table 26.6).

4.4 The telephone game

Designers developed their empathic understanding of users with the knowledge that they would be distributing this understanding to others in the company. Hence, building and utilizing empathy was a collaborative process throughout the Company. While

Table 26.5 Extraction from sharing insights from co-creation workshop I (v08, 167–170).

W	I've detected in [Scandinavia], some moms will say things like "hey, I didn't care much about certain things when I was younger, but now I have a child, 'cause I'm- I'm interested and mindful about what kind of values I teach them now", so, it could include things like being environmentally conscious. "And that's the positive value I want to go in to our kid".
R	It was to get the kid to do it.
W	Yeah, and plus "I don't care. Maybe, deep down inside me I don't care, but because I wanna transmit the right values. […]
E	[…] I mean this is not breaking news, but like, you wanna try to have healthy stuff and environmentally friendly stuff (.) towards the human body because of like the very basic level, but also because of the value stuff. It becomes in the bottom and on the top. "I give the value, but I- or, I give a healthy orange because I don't want you to get sick, and I also give the value: of having a sustainable orange that I give to you because that is a good thing to grow up with".

Table 26.6 Extraction from clustering insights session (v18, 355–357).

E	I think that is connected to, if you take it all the way it could be a Mercedes Guy having a COMPANY air purifier in their car, for example.
AM	Exactly, yeah.
E	He didn't want the one from Mercedes. Maybe it works as well, but it's not made from materials that is sustainable or something like that. And he has the option to go to the other store and buy it, and that says something about him, like, "oh, I thought you were a Mercedes Guy?" "Yes, but over that, I am a responsible guy."

in China, during the co-creation sessions, outside consultants directly interacted with users. In turn, these consultants discussed users with the designers and joined the team in describing, negotiating, refuting, and accepting user insights. Hence, throughout the process, the team itself was "one-step removed" from the users. Second, in viewing the company as a whole, company personnel did not negotiate empathic understanding, but distributed their understanding from one sub-team to another. For example, the team communicated their insights to the accessories division (who were two-steps removed), who then interpreted and (possibly) shared the insights with other Company divisions (who were three or more steps removed). As these discussions worked their way up the company ladder, the discussers became increasingly distanced from the users, presenting an empathy telephone game of sorts.

One-step removed: While designers were present during the workshops, they relied on consultants to facilitate the workshops and interact with the users in their native language. The designers and consultants proceeded to discussed these workshops and develop insights over a series of "debriefing" and "clustering insights" meetings. During these meetings, especially in early meetings, the consultants adopted the role of Chinese-culture experts and shared their insights with the designers who generally accepted these insights. For example, during a debriefing session, Will led the meeting as Abby interjected, generally by accepting Will's claims (see Table 26.7).

Two-steps removed: After the initial debriefing meetings, team roles normalized and designers became more active in generating insights. As described in the Persona-fication section, when presenting insights to the Accessories Department, the

Table 26.7 Extraction from debrief after co-creation workshop 1 (v07, 140–149).

W	They ehm: (.) okay, yeah, the (.) the new experiencing new and different, I think: (.) the combination of […] have probably a lot to do with Yen. I think he's got more concrete examples about, you know, what you're do'- and even Susan she'll say that "ah, I wish I can do stuff that other people want to do, but they can't do". Notice that when she say it, she express it as "what other people want to do, but they: can't do", as opposed to "what I want to do, but it's hard for me to do".
A	Yeah yeah.
W	So that's also a bit of the, ehm, bragging rights,
A	Yeah, absolutely.
W	That she hope to have, but of course we didn't have status, she had less and much less opportunity.
A	Mhm.
AM	They can see it from her body language like, "I'll be like (.) like people look up to me".
A	Ah, mhm.
W	So it's quite interesting that from different expressions, eh: it's quite clearly indicating that they are picking up or using ehm emotional indicators as self-expression:, as bragging rights:, as differentiation:, you know.
A	Yeah.

Table 26.8 Extraction from recap with stakeholders (v20, 015–018).

E	Okay, then we have this one, which is kind of we're working with it, it's not like in a perfect shape […] And then you have this part here, which is- it started out, maybe it's a little difficult to see it now, but it started out that we are trying to make some sort of (.) ah we can say framework for how (.) the Chinese mindset is. […] And this is kind of where we stopped yesterday to try to: understand what this is. […] Eh: then up here, this is where we- what we didn't talk about yesterday and that is the different states- stages, or did we talk about this one? Or you wanna-
T	Very fast.
E	Real'- yeah, very fast. You want a recap of what we were talking about?
T	Don't you want him- to pull him up there dubi-dubi-doo?

team moved from direct user insights to fictionalized personas that represented opportunity areas for the company. As Amanda noted, "So you start with the stories, and the insights, to identify the needs, and then you come up with a concept or opportunity areas." In these meeting, nearly all traces of empathy evaporated as broad insights and a unified conception of a user demographic remained (e.g., as a persona). Even limited efforts to draw detailed portraits for the stakeholders were sometimes thwarted, as an interchange between Ewan and Tiffany during the stakeholder briefing demonstrates (see Table 26.8).

4.5   Conquest versus concern

Designers initiated user research with the espoused purpose of integrating users into the Company's decision-making system. Hence, Ewan utilized terms with strong empathic associations during the opening interview (e.g., co-creation, user integration).

He stressed his desire for the team to become Company "change agents," bridging the user-company disconnect, and promoting a Company-wide user focus. As he stated:

> ... *we have a higher goal in what we are doing, it's not ONLY about using the best methods for the projects that we are doing individually, but it's also to (..) to showcase a whole mindset or way of thinking to the organization.*

This user-centric mindset grounded the team's co-creation efforts to develop intimate knowledge of users, with the ultimate goal of developing a product that would improve users' lives. Given the lofty and altruistic vision presented by Ewan, we expected to find frequent empathic discourses throughout the videos, and in truth, there were many, as described in the seeking empathic accuracy section. Further, some of the discourses we found could be considered affective, such as when the team expressed concern for users feeling comfortable during workshops.

Yet, discourses regarding capital gain were commonplace throughout videos. Specifically, designers expressed a tension between (a) fostering intimate relationships with users and (b) developing a story to market to users and the demographic whom they represent. Hence, Ewan consistently referred to what he called "conquesting," or strategies for appealing to and exploiting users. While he pondered, "How can we focus here to conquest these people?" he expressed his underlying assumption, "We wanna be in conquest, we wanna grow." While conquest seems not to be Ewan's sole aim, it is a necessary goal for the team to exist and for the Company to thrive. Hence, the team abstracts and presents users in a persona form that meshes with the Company-imposing capitalistic objective (to design a marketable and profitable product), at the risk of sacrificing subtle user-centric knowledge.

As a result of this risk, Ewan carefully presented the design team's research to the Accessories Department managers. Slowly, these managers began to understand the team's findings, and perhaps began considering user perspectives. Eventually, Tiffany made the connection between the work the team has conducted, its importance, Company values pertaining to care for the customer, and the potential market niche that understanding the "customer" point of view enables (see Table 26.9). This singular dialogues exemplifies how the Company hierarchy can somewhat limit empathy for users, but also how the Company values can stand as a stark reminder to incorporate user perspective into the team's design process.

5   CLOSING DISCUSSION

Through this investigation we developed five themes. Three themes captured designers' micro-level discourses: (a) creation and refinement **figured worlds**, (b) empathic techniques utilized to develop **empathic accuracy**, and (c) **persona-fication**, or creations of idealized users that integrate user interactions, design constraints, and company needs. Two macro-level themes captured overarching Discourses that inhibited or promoted the team's empathic development or utilization: (d) a **telephone game**, where user understanding was generated then passed from one company representative to another, and (e) **conquest versus concern**, or the tension the team faced in empathizing with users versus manipulating those users for company success.

Table 26.9 Extraction from recap with stakeholders (v20, 103–109).

E	… we have been burned so many times by coming with a suggestion- "this is an example of how it could be", and people see that example like "okay, we gotta make that. This is how it is". So instead of looking at these frameworks, they're looking at the little example we come with. And then they totally say "yeah but I don't like it blue". Like "no. Don't like look at the color. Look at how these things are working together".
T	Mhm.
E	Ehm:. You- do you: (.) do you think that that is: (.) or how do you feel by that as: a delivery? Like, that thing as a delivery?
T	Our customer focused? You, or?
E	Or like this- these (.) these frameworks, these ways of looking at how to approach the customer.
H	And to- to mapping the different eh, domain, yeah. (..) I think that's a good way.
T	And I mean eh- yes, and it goes with the THE COMPANY values as well, as you care about the customer, it is the (.) the thing for THE COMPANY and a core value as well, and of course we should see it from the customer point of view. And we have known so little about China, but now we have learned so much about China and the Chinese customer, and that we need to give to the others.

In the introduction, we distinguished between *designing with* versus *designing for* users (Sanders & Stappers, 2014) noting the interplay of the two during empathic design (specifically, co-design). In our analysis, we only had access to snippets of co-creation sessions, which limited our ability to draw inferences on the team's acts of *designing with*. Hence, we focused on the team's co-inquiry process, specifically as they verbally reflected on their experiential knowledge developed during the co-creation sessions, presented their experiential knowledge of users to the team, and utilized this discourse to generate practical knowledge (Adams *et al.*, 2017). Importantly, by making this experiential knowledge explicit through verbal discourses, the team was able to collaboratively refine, validate, and prioritize their user-centric knowledge.

Yet, we could not help but imagine an empathic disconnect between not only the users and the final decision-makers (as indicated by the telephone game), but even the users and the design team. For example, the team's inability to speak the user's language forced them to rely upon translators. Incidentally, while this may have limited the team's experiential knowledge, it required that the translators be engaged throughout the process. This allowed the translators to continually add nuance to or challenge the team's emergent user-centric knowledge, particularly the Chinese-specific cultural assumptions the team verbalized in their discourses. Other researchers in this volume describe cross-cultural aspects in more detail (Clemmensen *et al.*, 2017; Gray & Boling, 2017).

Nonetheless, the Company's act of paying designers to travel to China indicated that the company was open to an empathic design approach. With the empathic knowledge in hand, in theory, the unit could become the "Collaboration Bridge" enabling the percolation of empathic knowledge throughout the organization (Adams *et al.*, 2017). Sanders (2009) described this top-down valuing as the most critical component for integrating empathy into any corporate environment. In alignment with her theory, Sanders created a pyramid structure with "culture" at its base, followed by mindsets,

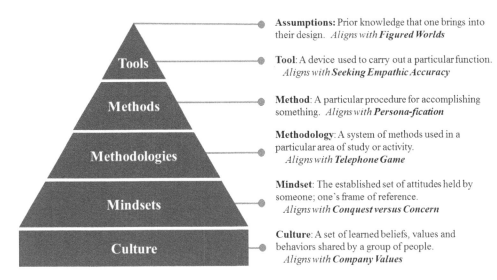

Figure 26.1   Empathic design contributors and constraints (adapted from Sanders, 2009).

methodologies, methods, and tools. An inspection of our themes in light of Sanders' theoretical structure conveys marked similarities (see Figure 26.1).

The base, "culture", incorporates elements of a theme that we excluded due to space limitations. This theme described the moderating influence of Company Values on designer techniques. As Tiffany from the accessories department reflected (see Table 26.9), the Company values included caring for the customer. This value explicitly allowed the manifestation of an employee mindset that prioritized customer well-being. Simultaneously, designers could not shed the corporate requirements of profit and growth, or what Ewan called conquesting. Hence, the team carried this tension with them throughout the entirety of the process, which explicitly manifested in discourses pertaining to fears that their process may not be accepted by superiors (Jornet & Roth, 2017).

This corporate culture, and its accompanying mindsets, supported and constrained the co-creation and telephone methodology utilized by the team. The team's methods and tools (the continually evolving components of the "problem space") were, in turn, supported and constrained by the overarching methodology or structure (Daly *et al.*, 2017). Specifically, the user-centric insights (or "empathic" knowledge) that the team generated and shared with the company were supported or constrained by the culture (e.g., company values), mindsets (e.g., conquest versus concern), methods they employed (e.g., persona-fication) and the tools they utilized (e.g., when seeking empathic accuracy). For example, the team used empathy to build a "Mercedes guy" persona who exemplified "sexy commitment" which (seemingly) met the ill-defined parameters of concern and conquest (Lloyd & Oak, 2017). Our addition to Sanders' model is what we called figured worlds, or the user-centric assumptions the team brought with and refined throughout their process. These included their constant invocation and interpretation of similarities and differences between Eastern versus

Western norms or values, as others in this volume also point out (Daly *et al.*, 2017; Gray & Boling, 2017).

Despite the apparent alignment between the company culture and the design team's approach, we cannot say with any certainty whether company executives have engaged with the Accessories Department. Postma *et al.* (2012) identified the unique challenge in large companies of determining who needs to empathize with clientele versus who needs to be informed of client-research. We posit that individuals who make decisions should interact with users, thereby generating their own experiential knowledge to reflect upon (Adams *et al.*, 2017), but we understand that organizational size limits this possibility. Nonetheless, we suggest that in moving forward with any solution the Company include the design team as evaluators. Ideally, in turn, the team would engage the translators who engage with users, thus re-orienting the telephone game as a two-way stream rather than the unidirectional stream as we have portrayed it. This, too, will potentially integrate empathy in "post-design" as Sanders and Stappers (2014) indicated it should.

ACKNOWLEDGMENTS

We would like to thank the DTRS11 organizing committee and the sponsors who made this study possible.

REFERENCES

Adams, R. S., Aleong, R., Goldstein, M., & Solis, F. (2017). Problem Structuring as Co-Inquiry. In: Christensen, B. T., Ball, L. J. & Halskov, K. (eds.) *Analysing Design Thinking: Studies of Cross-Cultural Co-Creation*. Leiden: CRC Press/Taylor & Francis.

Baillie, C., & Douglas, E. P. (2014). Confusions and conventions: Qualitative research in engineering education. *Journal of Engineering Education, 103*(1), 1–7.

Batson, C. D. (2009). These things called empathy: Eight related but distinct phenomenon. In: J. Decety & W. Ickes (eds.) *The Social Neuroscience of Empathy*. Cambridge, MA: MIT Press. pp. 3–15.

Batson, C. D., Early, S., & Salvarani, G. (1997). Perspective taking: Imagining how another feels versus imaging how you would feel. *Personality and Social Psychology Bulletin, 23*(7), 751–758.

Bedford, D. A. D., Arns, J. W., & Miller, K. (2017). Unpacking a Design Thinking Process with Discourse and Social Network Analysis. In: Christensen, B. T., Ball, L. J., & Halskov, K. (eds.) *Analysing Design Thinking: Studies of Cross-Cultural Co-Creation*. Leiden: CRC Press/Taylor & Francis.

Braun, V., & Clarke, V. (2006). Using thematic analysis in psychology. *Qualitative Research in Psychology, 3*(2), 77–101.

Christensen, B. T., & Abildgaard, S. J. J. (2017). Inside the DTRS11 Dataset: Background, Content, and Methodological Choices. In: Christensen, B. T., Ball, L. J., & Halskov, K. (eds.) *Analysing Design Thinking: Studies of Cross-Cultural Co-Creation*. Leiden: CRC Press/Taylor & Francis.

Clemmesen, T., Ranjan, A., & Bødker, M. (2017). How Cultural Knowledge Shapes Design Thinking. In: Christensen, B. T., Ball, L. J., & Halskov, K. (eds.) *Analysing Design Thinking: Studies of Cross-Cultural Co-Creation*. Leiden: CRC Press/Taylor & Francis.

Daly, S., McKilligan, S., Murphy, L., & Ostrowski, A. (2017). Tracing Problem Evolution: Factors That Impact Design Problem Definition. In: Christensen, B. T., Ball, L. J., & Halskov, K. (eds.) *Analysing Design Thinking: Studies of Cross-Cultural Co-Creation*. Leiden: CRC Press/Taylor & Francis.

Davis, M. H. (1996). *Empathy: A social psychological approach*. Boulder, CO: Westview Press.

Eisenberg, N., & Fabes, R. A. (1990). Empathy: Conceptualization, measurement, and relation to prosocial behavior. *Motivation and Emotion, 14*(2), 131–149.

Fila, N. D., & Hess, J. L. (2015). Exploring the role of empathy in a service-learning design project. In: R. S. Adams & J. A. Siddiqui (eds.) *Analyzing Design Review Conversations*. West Lafayette, IN: Purdue University Press. pp. 135–154

Fila, N. D., Hess, J. L., Hira, A., Joslyn, C. H., Tolbert, D., & Hynes, M. H. (2014). *Engineering for, with, and as people*. Paper presented at the IEEE Frontiers in Education Conference, Madrid, Spain.

Fila, N. D., Hess, J. L., Purzer, S., & Dringenberg, E. (2016). Engineering students' utilization of empathy during a non-immersive conceptual design task. *International Journal of Engineering Education, 32*(3B), 1336–1348.

Gee, J. P. (2011). *An introduction to critical discourse analysis in education: Theory and method* (3rd ed.). New York, NY: Routledge.

Gray, C. M., & Boling, E. (2017). Designers' Articulation and Activation of Instrumental Design Judgments in Cross-Cultural User Research. In: Christensen, B. T., Ball, L. J., & Halskov, K. (eds.) *Analysing Design Thinking: Studies of Cross-Cultural Co-Creation*. Leiden: CRC Press/Taylor & Francis.

Hess, J. L., & Fila, N. D. (2016a). *The development and growth of empathy among engineering students*. Paper presented at the American Society for Engineering Education Annual Conference, New Orleans, LA.

Hess, J. L., & Fila, N. D. (2016b). The manifestation of empathy within design: Findings from a service-learning course. *CoDesign: International Journal of CoCreation in Design and the Arts, 12*(1–2), 93–111.

Hoffman, M. L. (2000). *Empathy and moral development: Implications for caring and justice*. Cambridge, UK: Cambridge University Press.

Iacoboni, M. (2009). *Mirroring people: The science of empathy and how we connect with others*. New York: Picador.

Jornet, A., & Roth, W. (2017). Design {Thinking | Communicating}: A Sociogenetic Approach to Reflective Practice in Collaborative Design. In: Christensen, B. T., Ball, L. J., & Halskov, K. (eds.) *Analysing Design Thinking: Studies of Cross-Cultural Co-Creation*. Leiden: CRC Press/Taylor & Francis.

Koskinen, I., & Battarbee, K. (2003). Introduction to user experience in empathic design. In: I. Koskinen & K. Battarbee (eds.) *User Experience in Product Design*: IT Press. pp. 37–50

Kouprie, M., & Sleeswijk Visser, F. (2009). A framework for empathy in design: Stepping into and out of the user's life. *Journal of Engineering Design, 20*(5), 437–448.

Kwok-leung Ho, D., Ma, J., & Lee, Y. (2011). Empathy @ design research: A phenomenological study on young people experiencing participatory design for social inclusion. *CoDesign, 7*(2), 95–106.

Leonard, D., & Rayport, J. F. (1997). Spark innovation through empathic design. *Harvard Business Review, 75* (Nov.–Dec.), 102–113.

Lloyd, P., & Oak, A. (2017). Cracking Open Co-Creation: Categorizations, Stories, Values. In: Christensen, B. T., Ball, L. J., & Halskov, K. (eds.) *Analysing Design Thinking: Studies of Cross-Cultural Co-Creation*. Leiden: CRC Press/Taylor & Francis.

Mattelmäki, T., Vaajakallio, K., & Koskinen, I. (2014). What happened to empathic design? *Design Issues, 30*(1), 67–77.

Postma, C. E., Zwartkruis-Pelgrim, E., Daemen, E., & Du, J. (2012). Challenges of doing empathic design: Experiences from industry. *International Journal of Design, 6*(1), 59–70.

Rogers, C. R. (1975). Empathic: An unappreciated way of being. *The Counseling Psychologist, 5*(2), 1–10.

Sanders, L. (2009). Exploring co-creation on a large scale. In: P. J. Stappers & J. Szita (eds.) *Designing for, with, and from user experiences*. Delft: TU Delft. pp. 10–26.

Sanders, L., & Stappers, P. J. (2014). Probes, toolkits and prototypes: Three approaches to making in codesigning. *CoDesign, 10*(1), 5–14.

Walther, J., Sochacka, N. W., & Kellam, N. N. (2013). Quality in interpretive engineering education research: Reflections on an example study. *Journal of Engineering Education, 102*(4), 626–659.

Part VII

Design Iterations and Co-Evolution

Chapter 27

Information-Triggered Co-Evolution: A Combined Process Perspective

Phil Cash & Milene Gonçalves

ABSTRACT

Core elements of design work include the development of problem/solution understanding, as well as information and knowledge sharing activities. However, their interrelationships have been little explored. As such, this work aims to take the first steps towards a more integrated evaluation and description of the interaction between understanding and activity, based around co-evolutionary transition events; and start to answer the question: How can the link between co-evolution and activity be systematically characterized as a foundation for a more fundamental description of design activity? A protocol analysis is used to provide the basis for characterization of different types of co-evolutionary transition event. A number of distinct event types are described and significant differences in information use and team engagement are identified across transition events. Bringing these findings together, we propose a unitary model of the interaction between activity and understanding around co-evolutionary transition events. This has a number of implications for future theory building and testing in both design activity and wider design research.

I INTRODUCTION

The co-evolution model, formalized by Maher *et al.* (1996), describes how design work is underpinned by understanding development (Dorst & Cross, 2001). This model distinguishes between two conceptual spaces: the problem space, dealing with problem formulation, requirements, and constraints; and the solution space, dealing with possible solution structures. Designers iteratively transition between these spaces as they interpret the design brief and develop solutions, continuing until they are able to frame a problem-solution pair (Dorst & Cross, 2001; Schön, 1984). As such, this model positions transitions within and between spaces at the heart of design work (Maher *et al.*, 1996).

In parallel to this co-evolution of understanding, there are a number of manifest activities, which have also been described as characteristic of design (Cash *et al.*, 2015; Sim & Duffy, 2003). In particular, information (Robinson, 2010) and knowledge sharing (Dong, 2005) activities have both been highlighted as important. Further, Cash *et al.* (2015) describe the dynamic interaction between activities as a key element of design work. In this context, both information and knowledge sharing activities have been linked to understanding (Gibson, 2001; Kleinsmann *et al.*, 2012). For example, Gonçalves *et al.* (2014) highlight the effect of critical information stimuli, which can become inspirational, while Kleinsmann and Valkenburg (2008) focus on the gradual alignment of understanding and engagement fostered by effective sharing.

Connecting co-evolution and manifest activity through understanding, it is possible to identify transition events as key focal points connecting these phenomena (activity and co-evolution). For example, in their study of co-evolution, Dorst and Cross (2001) highlight how information can trigger transitions between and across spaces. However, there has been little further exploration of the link between these perspectives. This points to a critical question facing design research, which limits both theory building and empirical study: *How can the link between co-evolution and activity be systematically characterized as a foundation for a more fundamental description of design activity?* Design Activity describes the system of interacting activities, connecting information, knowledge, and object domains to underlying understanding development. For example, Gonçalves (2016) suggests that the use of information stimuli in design supports the transition between problem and solution spaces. However, it is currently difficult to describe how or why certain information stimuli are influential (Gonçalves *et al.*, 2014), or how design situations build up from these fundamental transitional 'units' (Cash *et al.* 2015). Further, there is a lack of clarity in the communication of theoretical frameworks describing the development of activity and understanding, hampering integration efforts (Dorst, 2008). As such, this work aims to take the first steps towards a more cohesive description of the interaction between understanding and activity, based around co-evolutionary transitions events.

This paper is structured as follows. Section 2 presents an overview of the current literature in co-evolution, information and knowledge sharing activity, before proposing a research framework linking these themes around transition events. Section 3 then outlines the research approach and analytical method, before the results are discussed in Section 4. Finally, a number of contributions are distilled in Section 5.

2 BACKGROUND

This section outlines the relevant components of co-evolution, information activity, and knowledge sharing, before framing their intersection via the research framework described in Section 2.4.

2.1 Co-evolution

Co-evolution constitutes a fundamental characterization of how knowledge structures develop over time with respect to a design task (Maher *et al.*, 1996). More specifically, it refers to how a designer's understanding of problem and solution iteratively mature over the course of a project. Although not directly dealing with design activity, the co-evolution model forms a fundamental driver for behavior in this context (Dorst & Cross, 2001). Specifically, interactions between cognition and manifest activity are key to behavioral processes (Bedny & Karwowski, 2004). This has been illustrated in the design domain by concepts, such as reflective practice (Schön, 1984). Using this conceptualization of behavior, co-evolutionary transitions can be framed as critical changes in understanding connected to specific activities. This connects models of design cognition and design activity via transition events (Aurisicchio *et al.*, 2010; Hult *et al.*, 2004).

The critical relationship between co-evolution and information activity has been highlighted by Dorst and Cross (2001), as well as authors more directly focused on various information activity processes, such as Gonçalves (2016) – around inspiration, Wiltschnig *et al.* (2013) – around team based practices, and Daly *et al.* (2017) – around co-creation practices and culture perceptions. Further, authors such as Aurisicchio *et al.* (2013) and Cash *et al.* (2015) have more generally highlighted information and knowledge sharing activities as interconnected elements of design work. Here, information processing explicitly links information acquisition, knowledge sharing, and understanding (Hult *et al.*, 2004; Simon, 1978).

2.2   Information and stimuli

Designers spend a considerable amount of time searching for information (Court *et al.* 1993), with their information needs changing substantially over the course of the design task (Hicks *et al.*, 2002; Gonçalves *et al.*, 2013). Here, information stimuli are particularly important during the early phases of the design process: enabling understanding and decomposition of the problem (Heisig *et al.*, 2010), and reducing uncertainty (Guo, 2011). Further, information stimuli support the generation and development of ideas (Eckert & Stacey, 2000; Gonçalves *et al.*, 2013), by acting as inspiration sources that can trigger ideation, reflection, or re-interpretation, for instance. Information stimuli can describe any information fragment, which has been interpreted by the designer and prompts a reaction to explore the problem or solution space (Sarkar & Chakrabarti, 2008; Gonçalves, 2016). These stimuli are considered inspirational when their use elicits creative ideas. However, certain stimuli can also trigger a dual-effect on design outcome and process (Cai *et al.* 2010; Perttula & Liikkanen, 2006). Instead of promoting a beneficial influence, information stimuli can also deter designers from exploring other solutions, as they become fixated on e.g. features or principles (e.g., Jansson & Smith, 1991). Thus, information stimuli can have a substantial impact on design work, as they are key to the expansion of the problem understanding (Eckert & Stacey, 2017) and exploration of the solution space (Dorst & Cross, 2011; Gonçalves, 2016).

2.3   Knowledge sharing and engagement

Information activity, besides supporting understanding development, encourages knowledge sharing between team members (Hult *et al.*, 2004; Lawson *et al.* 2009). Here, the gradual development of shared understanding within a team is built on the explanation of, reflection on, modification, and expression of agreement around knowledge structures (Dong, 2005). This activity has been identified as important not only for design understanding, but also for organizational issues, such as dissolving the barriers between different units (Kleinsmann & Valkenburg, 2008).

Knowledge sharing exchanges are characterized by the synthesis of information from different perspectives, in a process connecting context, rationale, strategy, and logical reasoning (Aurisicchio *et al.*, 2010). As such, there is an intimate connection between information use, knowledge sharing, and engagement within a team.

This again connects knowledge sharing to co-evolutionary transition events (Becattini *et al.*, 2014). Coupled with information activity, knowledge sharing, and iterative understanding development have been described in a number of empirical contexts, from organizational (Hult *et al.*, 2004) to design team (Aurisicchio *et al.*, 2010). As such, investigation of their intersection forms an important aspect of theory development in the design domain. Further, as with information activity, knowledge sharing manifests itself in different forms, such as by engagement and agreement expression (Ariff, Eris, & Badke-Schaub, 2013), physical interaction (Cash & Maier, 2016), and directed question asking (Deken *et al.*, 2012).

2.4 Research framework

Bringing together Sections 2.1–2.3, it is possible to characterize co-evolution as an understanding development process underpinning and connecting design activity. Further, reflecting on core models of activity (Bedny & Karwowski, 2004) and information processing (Hult *et al.*, 2004; Simon, 1978), it is possible to conceptually frame co-evolutionary transition events as key focal points where activity and understanding intersect (Yu *et al.*, 2015). This allows for a more fundamental description of design activity, which connects activity and cognitive perspectives around transition events over time. These events thus form the focus of this work, conceptually constituting a core transitional 'unit' of design work.

Building on this 'unitary' perspective each transition event must be characterized with respect to the three perspectives outlined in Sections 2.1–2.3: co-evolution, information activity, and knowledge sharing activity; as well as time. This conceptualization of transition events as intersections is illustrated in Figure 27.1, where progression in the x-axis represents time. Here, each event has distinct inputs/outputs with respect to each element, in line with systematic models of activity (Sim & Duffy, 2003). This allows for a generic description of each event as a unit within a wider process progression, as well as providing an explicit pictographic representing of design activity interactions.

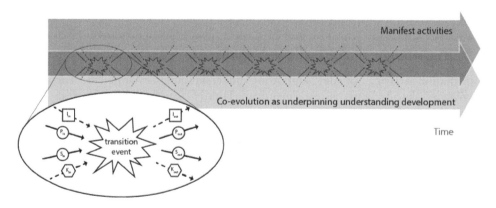

Figure 27.1 Co-evolutionary transitions as key focal points for process interaction.

Figure 27.1 connects an initial understanding state (Problem space P_{in}, Solution space S_{in}) together with an informational input (I_{in}) and knowledge sharing input, i.e., the people involved in the transition exchange (People$_{in}$). These are then transformed into outputs via the transition event. Collectively, these connect underlying problem/solution co-evolution and manifest information and knowledge sharing activities.

3 METHOD

Before starting the analysis, the analytical focus was first refined based on the different design situations available in the DTRS11 dataset: synthesis, co-creative workshops, and brainstorming. Selection at this stage used two criteria: first, that sufficient transition events were observable to allow for analysis of the different event types; second, that all three elements of the research framework could be observed (Figure 27.1). Based on these criteria, co-creative workshops and brainstorming situations were eliminated from consideration due to data limitations and unsuitability for the type of events researched here. This resulted in a focus on synthesis situations captured in the sharing, linking, and clustering sessions (v08, v09, v11, v15, v16, v17, and v18). Coding was then carried in two stages. First, all transcripts were coded in terms of co-evolutionary transition events. Second, each event was characterized in terms of its information and knowledge sharing inputs/outputs.

3.1 Co-evolution transitions

Co-evolutionary transitions were coded using a schema adapted from Becattini *et al.* (2014), which entailed four main transition types (Figure 27.2).

P–P (problem to problem space): Horizontal transitions across the problem space characterized by the decomposition of the problem into sub-problems, usually resulting in the refinement of the problem space. This type of transition includes the definition and/or clustering of (sub-) problems, formulation of requirements, definition of target

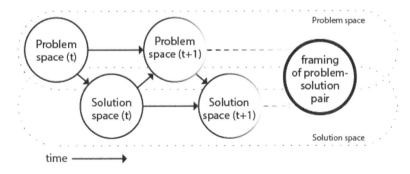

Figure 27.2 Co-evolutionary transitions across the problem and solution space.

groups, and company values. This type of transition can be seen in the following quote:

EWAN: *"They need to be something new, they want to stand out as well, they want to be like their parents or they want to go like 'how can we stand out? Ah, it's through minimalism; it's through experiences', and so on. But still of course with a touch of very shallow things like the container of the ingredients and the body and stuff like that."* (v05, 234).

In this excerpt, Ewan is decomposing the problem space. After the co-creation workshop (information input), the design team was able to formulate themes of exploration and to construct user profiles (information output), which enabled them to define and cluster different sub-problems.

S–P (solution to problem space): Diagonal transitions from the solution to the problem space include the reframing of the (sub-) problem, motivated by a solution. In this type of transition, a solution can trigger the identification of a new requirement, which demands a change of the problem definition and, thus, a reframing of the problem space. The following quote exemplifies this type of transition:

AMANDA: *"So they buy the domestic robot, because"*
EWAN: *"So they buy it for several reasons, they buy it because"*
AMANDA: *"To free up their time at home, so they can go out and contribute."*
EWAN: *"That could be one thing. Or, that's one thing, and, they also buy it because it is made of conflict free aluminum."* (v11, 461 to 464).

Here, the designers are using the hypothetical example of an existing idea – a domestic robot – to verify whether this pre-structure fulfills the requirements of the current problem definition. However, the designers do not consider the domestic robot as a valid solution, since it is an existing product.

P–S (problem to solution space): Diagonal transitions from the problem to the solution space indicate the exploration of solutions. These transitions use the problem (definition and/or requirements) to initiate and generate ideas, as illustrated in the following quote:

KENNY: *"the modular thing, I think- is it more about having different levels or a way to differentiate from the basic to premium - so if you buy something premium, you want to be sure this is premium compared to something else".* (v15, 375).

In this quote, Kenny has transitioned from an earlier insight, which arose from the comments of the co-creation session participants, to a possible solution direction (modularity to distinguish premium products). Therefore, what was previously an insight that helped to redefine the problem space became an exploration focus in the solution space.

S–S (solution to solution space): Horizontal transitions across the solution space are characterized by the synthesis and/or development of a specific solution. Transitions of this type have (parts of) a solution formulated earlier as starting point, which can result

Table 27.1 Summary of transition type coding based on Becattini *et al.* (2014).

Transition types	Description
P–P	Refinement and synthesis of the problem space
P–S	Generation of ideas from the problem space
S–P	Bridging of solution and problem space
S–S	Refinement and synthesis of the solution space

in commitment to specific solutions. These use a prior solution to initiate and generate ideas, thus changing the solution space. An example of this transition is observable in the following excerpt, which follows from the previous quote:

ROSE: *"I think what they meant was because this was a new product, a new service, so you have the modular to- for way- so one is to let people test, try it out a little bit so that you don't have to spend all your money that one".*
KENNY: *"so you don't need to buy the premium version first?"*
ROSE: *"the other one- yes yeah, then I think is was the thing about the modular was also because then you can differentiate your target customers, so you have who want everything and then you can get people to buy the stuff."* (v15, 376 to 378)

These transition types and their basic description are summarized in Table 27.1.

Transition events were first coded for video 09 by both authors in order to provide an inter-coder reliability check. This resulted in an initial percentage agreement of 60% using the schema from Becattini *et al.* (2014). This initial value was found to be low due to ambiguity in the definition of the problem/solution space division. Based on this, the schema was revised to that outlined above and a second round of coding resulted in 100% agreement. As video 09 was used for training purposes it was excluded from further analysis. Subsequently, all remaining videos were coded for transition events by the second author. Once this was complete, the first author coded the information and knowledge sharing manifest indicators as outlined in the following section. Using this procedure, both authors evaluated each transition event, based on the video 09 training described above.

3.2 Coding information use and knowledge sharing across transition events

Due to the exploratory nature of this research, a manifest approach was adopted to the coding of information and knowledge sharing activity indicators (Cash & Snider, 2014) i.e. no interpretation was applied.

In the information domain, the coding focused on the interaction with information stimuli, as manifest indicators of explicit connection to external information (McAlpine, Hicks, & Tiryakioglu, 2011). Manifest indicators here describe observable references to stimuli, which can be objectively identified, such as specifically pointing to a post-it (Cash & Snider, 2004). Here the intention was to characterize the type of information input/output, as well as the modality of the information representations used. For example, information input could entail insights from the co-creation participants or references to other brands, for instance:

AMANDA: *"Yeah. But I think stories like for example ah: what's his name, Yen, he said 'Yeah I started with an iPhone, I switched to Samsung, I'm now using a Huawei'. Huawei technology now is almost on par with the Western you know like, it's cheaper, value for money, same features, similar capability. So, it's tangible stories like this I think is very useful. And we should actually think more of this type of examples to put on the deck."* (v08, 129).

On the other hand, information output consisted of ideas or notes produced, usually represented on post-its. For the purposes of analysis, each information input/output was considered of equal value. More specifically, whenever a participant mentioned an information source, it would count as a single input/output, receiving the value of 1 for the purpose of counting the amount of manifest indicators.

Similarly, in the knowledge sharing domain, the coding focused on manifest indicators of engagement between participants (Ariff *et al.*, 2013), i.e. objectively observable by the designers' physical expression of agreement or interaction with the transition leaders. An example of such a manifest indicator of engagement is apparent in the following quote from several designers, where they are discussing how a possible solution structure aligns with the Company's values:

ABBY: *"and that's again all about what's THE COMPANY saying in the [values]".*
TIFFANY: *"mhh [Yes, yes,] it really is mmh"*
ABBY: *"It fits to THE COMPANY"*
KENNY: *"Yeah".* (v16, 122–125).

Here, the intention was to characterize the level of engagement input/output. For the purposes of analysis, each engagement expression was again considered of equal value (receiving the value of 1), even if some designers nodded multiple times, for instance.

These transition indicators and their basic description are summarized in Table 27.2.

Table 27.2 Summary of information and knowledge sharing manifest indicators coding.

Information	Manifest indicators
Type	The explicit description of the information being invoked e.g. participant inputs, information about participant background, reference to external brands, product or services,
Mode	The physical modality of the information being invoked e.g. post-it, poster, computer based website, slides

Knowledge sharing	Manifest indicators
Engagement	Engagement with the transition event discussion e.g. nodding, "yeah", "hum", gesturing support

3.3 Analysis

Analysis was carried out in two stages. First, every transition event was identified and characterized in terms of type (Table 27.1). Second, events were aggregated by type and characterized with respect to the activity (both information or knowledge sharing activities), in order to distill out common characteristics (Table 27.2). Thus, transition events form the basic unit of analysis. Significance values comparing input/output wcrc calculated using a two-tailed students t-test for within population samples (Walker, 2010). Using this approach, events were aggregated from across all the videos. Event level aggregation is appropriate for two main reasons. First, problem/solution co-evolution theory is general, suggesting that individual events should be distinct irrespective of the situation in which they occur (Dorst & Cross, 2001). Second, differences in situation level activity (as might be observed across videos) are driven by the dynamic interaction between multiple events over time as illustrated by analytical approaches such as Linkography (Goldschmidt, 1990).

In addition, a number of statistical robustness checks were undertaken to ensure that bias was not introduced based on the distribution of transition events across videos. First, all results and identified trends were evaluated at the individual video level. This analysis showed that no video displayed deviant behavior in terms of trend direction within individual event types or in terms of differences across event types. Second, results from each video were comparatively benchmarked in order to identify any significant differences in behavior. Here, the only differences were: change in engagement was lower for P–P events comparing v08 and v18; change in information was lower for P–S events comparing v15 and v18. However, these differences were in magnitude only and did not contradict any of the reported trends described in the following sections. Based on these two robustness checks, coupled with the theoretical appropriateness of aggregation at the event level, it can be concluded that the selected tests and reported results are robust within the context of this work.

4 RESULTS AND DISCUSSION

From the six videos used in the analysis (v08, v11, v15, v16, v17, and v18), 227 distinct transition events were identified, distributed across the four transition types (Table 27.1), as illustrated in Figure 27.3. Substantially more P–P transitions were found overall, and in each individual video. P–P transitions deal with framing and synthesis of the problem space. As such, their dominance within the dataset is aligned with prior theoretical and empirical literature, where problem framing is characteristic of the early phases of the design process (Dorst & Cross, 2001; Yu *et al.*, 2015). Thus, the relatively lower number of S–S transitions is also in line with theoretical expectations. Considering that a number of fellow researchers, who investigated the same DTRS11 dataset, also found many instances of co-evolution transitions or shifts (Shroyer *et al.*, 2017; Eckert & Stacey, 2017; Daly *et al.*, 2017), it is possible to argue for the validity of our results. From this, it is possible to conclude that our findings provide a meaningful foundation for further analysis of transition events in this context. As the dataset did not include further phases of the design process (i.e. the crystallization of a final problem-solution pair) it is possible to assume that S–S transitions would become more frequent with the development and refinement of the final product.

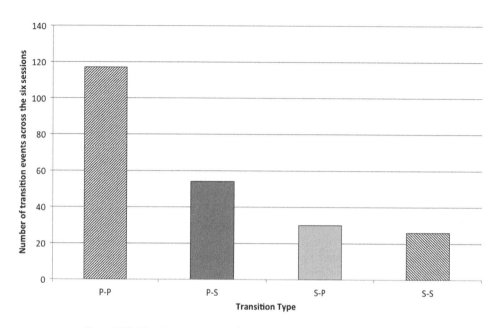

Figure 27.3 Transition events with respect to the four transition types.

4.1 Information use across transition events

In this section, we first present the results of the analysis of information stimuli input/output, across all transition events. With information stimuli, we refer to e.g. the production of new post-its/slides with ideas or insights (output) or reference to comments made by the co-creation participants (input). There was a generally significant reduction of information stimuli input/output across each transition event ($p < 0.001$), as illustrated in Figure 27.4. Specifically, designers consumed more information stimuli than they produced. This indicates that information was synthesized across each transition event, independently of its type (Table 27.1). For example, the designers commonly mentioned multiple prior notes, ideas or co-creation insights at the start of a transition, but only occasionally produced new information stimuli at the end of an event. Thus, there was a net reduction in information stimuli across the transition event.

Decomposing the results by type (Table 27.1), P–P, P–S, and S–P transitions showed significant reductions across time ($p < 0.001$), while S–S showed no change ($p = 0.279$). This synthesis of information stimuli across P–P, P–S, and S–P transitions aligns with prior literature, where information is consumed and integrated during design (Robinson, 2010; Wasiak *et al.*, 2010). This indicates that, as the project proceeds, the problem becomes more defined, which enables the synthesis of information and, eventually, the development of a problem-solution pair at the end of a project. Further, the lack of distinct synthesis and higher information stimuli output from S–S transitions also aligns with prior theory. S–S transition events are characterized by the extension and development of solution ideas, thus, they have been associated with the creation of new sketches or notes (McAlpine *et al.*, 2011).

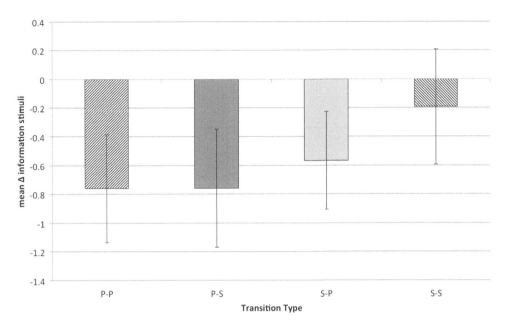

Figure 27.4 Difference between information input and output across transition events.

Reflecting on the observed sessions, more information stimuli were produced than those possible to identify during the transition events. For example, the designers verbally mentioned many directions and ideas throughout the session, which were only recorded at the end of the session. As these asynchronous notes could not be directly linked to specific transition events, they have not been considered here. Further investigation is needed to explore which events spark asynchronously produced information stimuli, because this is likely related to the perceived meaningfulness of a transition and thus related to concepts such as inspirational information (Gonçalves *et al.*, 2014). More specifically, certain information sources could lead to a transition between problem and solution spaces, thus becoming inspirational (Gonçalves, 2016).

It was also possible to evaluate the difference between transition types with respect to information use. This is detailed in Table 27.3, which shows the difference in quantity of information input/output across each transition type as well as their comparative significance. Here, S–S transitions were found to be significantly different in terms of information input/output when compared to P–P and P–S transitions. Specifically, the ratio between information consumed and produced was more balanced in S–S transitions. This indicates that, in P–P and P–S, much more information was consumed than it was produced. Designers referred to many sources during these transitions but did not produce as much new information. Further, the use of information in P–P and P–S transitions was nearly identical.

Two insights were evident when decomposing these differences in information stimuli use across transition events. First, despite the significant difference in quantity of information input/output across P–P and S–S transitions, the distribution of

Table 27.3 Difference in quantity of information use across transition types (*p < .05, **p < .01, ***p < .001).

	P–P	P–S	S–P	S–S
P–P	–	0.001	0.194	0.568***
P–S		–	0.193	0.567**
S–P			–	0.374
S–S				–

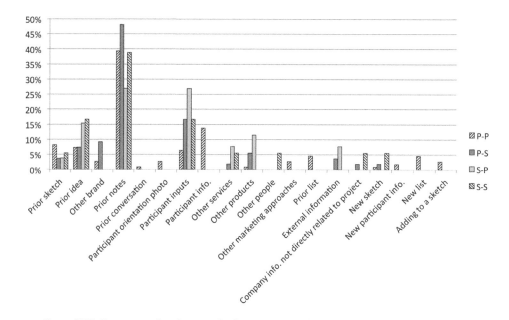

Figure 27.5 Percentage distribution of information stimulus types used in transition events.

information stimuli used was similar. Specifically, the distribution of stimuli types used in P–P and S–S transitions was significantly ($p < 0.05$) positively correlated ($r = 0.818$). This indicates that similar types of information stimuli were used in both P–P and S–S transitions, such as prior notes or sketches, and other marketing approaches (Figure 27.5). Both when defining the problem (P–P) and when synthetizing the solution space (S–S), designers seem to use the same stimuli type. Second, despite some correlation between the other transition types, there were no significant similarities/differences evident in the data. The overall distribution of information inputs is illustrated in Figure 27.5.

As with the information stimulus type, mode (Table 27.2) was closely related across all the transition types. As the designers were working together in creative sessions, most information stimuli were verbally communicated, however, post-it notes and the magic chart were also heavily used. Overall distribution of information stimulus mode is illustrated in Figure 27.6.

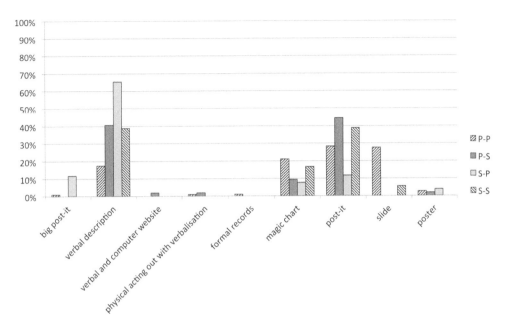

Figure 27.6 Percentage distribution of information stimulus mode used in transition events.

4.2 Knowledge sharing across transition events

Comparing the transition events with respect to engagement, there was a generally significant increase between input and output of engagement manifests ($p < 0.001$), i.e. transitions formed moments of positive team engagement. Decomposing by transition type, P–P, P–S, and S–P transitions showed a highly significant increase ($p < 0.001$), and S–S showed a significant increase ($p < 0.01$), as illustrated in Figure 27.7. As with the information results, these findings are in line with expectations from prior literature, where discussions and knowledge sharing events are key to engagement and sharing within a team (Ariff *et al.*, 2013; Deken *et al.*, 2012).

In contrast to the information results, there were no significant differences between the transition types in terms of change in quantity of engagement across the event (Table 27.4). All transition types were associated with a positive increase in engagement within the team. Thus, knowledge sharing by engagement triggers the occurrence of transitions events.

As no significant differences were found between transition types with respect to engagement, these findings were not further decomposed. This lack of differentiation is partially attributable to the generic nature of engagement through knowledge sharing activity (Ariff *et al.*, 2013; Dong, 2005), as well as the limited team sizes found in the observed sessions. In most cases, the maximum team size was five or less individuals, and thus change in engagement was naturally limited. It is possible that in larger teams where side conversations often split groups (Swaab *et al.*, 2008), the potential for higher differentiation in terms of quantity of engagement across transitions would be greater.

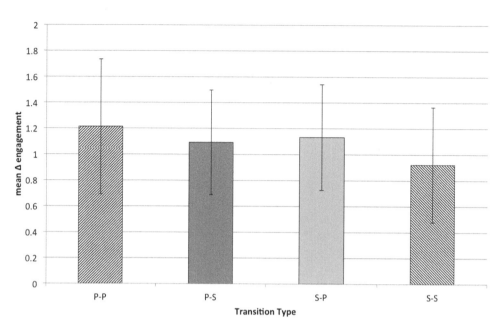

Figure 27.7 Difference between engagement input and output across transition events.

Table 27.4 Difference in quantity of engagement across transition types
($*p < 0.05$, $**p < 0.01$, $***p < 0.001$).

	P–P	P–S	S–P	S–S
P–P	–	0.121	0.080	0.291
P–S		–	0.040	0.170
S–P			–	0.210
S–S				–

4.3 The building blocks of design

Bringing together the findings discussed in Section 4.1, it is evident that there are a number of significant differences across transition events in terms of both information and knowledge sharing activities. Further, the coding of transitions themselves points to simultaneous changes in problem/solution understanding. These results provide empirical support for prior theoretical descriptions of understanding/activity interactions bridging transition events (Dorst & Cross, 2001). Thus, the results from this work empirically support co-evolution as a core descriptive model of design understanding development (Dorst & Cross, 2001; Maher *et al.*, 1996) and explicitly connect co-evolution transitions to distinct activities, such as information and knowledge sharing.

However, the fact that the transition types cannot be fully differentiated with respect to the examined manifest indicators points to the need for a more complete system of characterization; both in terms of identifying major perspectives on design

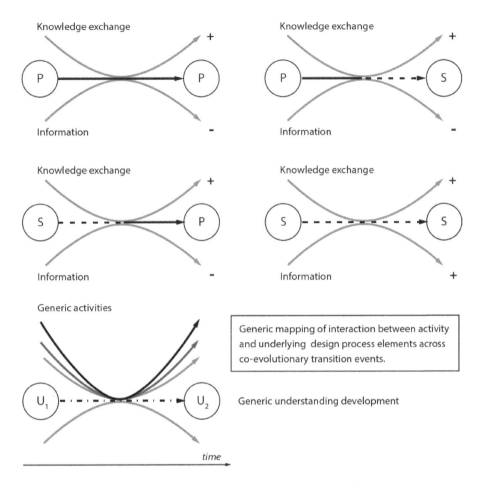

Figure 27.8 Developing pictorial representations describing the behavior of transition events connecting knowledge exchange and information activities. The bottom left figure illustrates the potential for connecting additional activity perspectives to the current framework.

activity (Cash *et al.*, 2015), and also in terms of integrating other aspects of relevant understanding (Hult *et al.*, 2004). Combining these descriptions from both activity and cognitive perspectives allows for the development of a more fundamental unit of analysis to describe design activity, as highlighted by the findings from this work.

A unitary perspective on design activity comes as a direct response to the need for a clear and a coherent conceptual framework that is able to support the communication and integration of theoretical descriptions of design activity and understanding (Dorst, 2008). This is important because prior works have typically formulated activity or understanding specific frameworks that make consistent representation and integration difficult across theories. As such, characterizing transition events, considering both problem/solution and activity, using a common visual language lays the foundation for consistent representation of theory and potential integration efforts in the design activity literature. This conceptualization is illustrated in Figure 27.8, which provides an initial pictorial representation of the interaction between co-evolution and activity. This provides a common means for integrating representations of design activity as well as the elaboration of relationships between cognition and activity. Such conceptualization can be useful to complement past findings on co-evolution, for instance, on different design environments or design domains (Yu *et al.*, 2015). Further, this can be complemented with, for example, the type of object interaction (Gero & Kannengiesser, 2004), in order to further characterize a design activity 'unit' over time (Section 2.4).

Utilizing this common unitary perspective opens the door for evaluation and characterization of design activity at a fundamental level by systematically linking understanding and activity. Describing situations in terms of their base transitional units provides two key opportunities for design research. First, it is possible to more systematically characterize a situation at the unit level (Cash *et al.*, 2015), rather than gestalt characterization at the whole situation level using protocol analysis, for example. Considering that the situational level is generally considered highly complex and difficult to effectively describe in terms of current design theory (Love, 2000), synthetizing design activity into unitary elements allows for a reduction of complexity, both theoretically and empirically. Second, theory at the unit level is more prevalent in design (Dorst & Cross, 2001; Gero & Kannengiesser, 2004; Yu *et al.*, 2015). Therefore, providing a platform for integrative evaluation in this context allows for more effective hypothesis generation and theory testing. Both of these elements are important for future theory building and empirical work in the design cognition and design activity domains. However, a number of questions remain to be answered in this context. First, it is not clear from the current literature what activities should be included in a core model of design activity (Cash *et al.*, 2015), that is, what additional perspectives should be integrated in the bottom left part of Figure 27.8. Second, further work is needed to describe other aspects of understanding development that should be included in a holistic model of design, such as the work on organizational barriers as distinct drivers for knowledge sharing activity (Kleinsmann & Valkenburg, 2008). Specifically, it is unclear what elements make up generic understanding development in Figure 27.8. Finally, there are numerous questions surrounding how design activity is affected by situational context and agential factors (Gallivan & Srite, 2005). These questions could be addressed by building quantitatively validated characterizations of the different transitional units. Further, the development of a pictorial language would provide an easier to implement and understand framework for the integration of theory and empirical data in design.

5   IMPLICATIONS AND LIMITATIONS

5.1   Implications

The identification and quantitative characterization of transition events as key focal points for interaction suggests the possibility of developing a more fundamental model of design work, that combines understanding, cognitive processes, and design activity. This specifically extends prior works that have typically focused on individual perspectives or interactions between cognition and single activity types. In this study, only information and knowledge sharing activities were considered. However, the significance of the results points to the value of further investigation in this area. As such, this work provides directions for further theory building and testing in the design activity domain.

The identification of significant interactions across transition events points to implications for design researchers seeking to empirically characterize design activity at a detailed level. In particular, by considering a unitary perspective, it is possible to test the characteristics that differentiate transitional units, allowing for the empirical interrogation of a number of design models that have only previously been described at the situational level. Further, this leads to key empirical hypotheses that could be tested in this context, validating basic assumptions of design transitions and their interaction with e.g. information activity.

Finally, the proposal of a transition event-based characterization of design activity (Figure 27.8) sets the stage for more transparent and easily integrated research in this domain. This also suggests the value of developing a common pictorial language for representing the interaction between activity and understanding.

5.2 Limitations

The range of design situations considered was constrained by the dataset. In particular, it limited quantitative characterization of transition events to situations where P–P and P–S transitions were more frequent (synthesis of insights from co-creation sessions). However, our results were significant and point to the value of further investigation across situation types.

The number of transition events found in the data restricted fine grain analysis. As the number of factors increases, the number of events required for quantitative analysis also significantly increases. As such, despite the broad significance of the results reported here, further data is needed if more fine grained decomposition is desired. For instance, this could enable the identification of specific differences in information stimulus type or interactions between elements.

Finally, this work has focused on examining transition events as key focal points where activity and understanding intersect. However, it was not possible to examine how events connected and influenced each other over time. In this context, Linkographic (Goldschmidt, 1990) or network based (Cash et al., 2015) approaches could provide valuable insights, building on the unitary characterization described in this work.

6 CONCLUSION

The aim of this work was to take the first steps towards a more integrated description of the interaction between understanding and activity, based around co-evolutionary transition events. Furthermore, we intended to start to answer the question: *How can the link between co-evolution and activity be systematically characterized as a foundation for a more fundamental description of design activity?* A protocol-based approach was used to identify transition events before they were quantitatively analyzed in terms of their relationship with information and knowledge sharing activities. This revealed a number of significant interactions across transition events in terms of both information and knowledge sharing. Further, it was possible to empirically differentiate transition event types via their differing interaction with these activities.

These results were brought together in a unitary model of design activity/ understanding interaction based on co-evolutionary transition events. This has a number of significant implications for future theory building and testing, particularly in the design activity domain.

Finally, the need for more extensive characterization of the different design activity elements, underlying understanding drivers, and nature of transition events were all highlighted as areas for further work.

REFERENCES

Ariff, N. S. N. A., Eris, O., & Badke-Schaub, P. (2013). How Designers Express Agreement. *5th IASDR Conference* Tokyo, Japan. pp. 1–10.

Aurisicchio, M., Bracewell, R., & Wallace, K. (2010). Understanding how the information requests of aerospace engineering designers influence information-seeking behaviour. *Journal of Engineering Design*, 21(6), 707–730.

Aurisicchio, M., Bracewell, R., & Wallace, K. M. (2013). Characterising the information requests of aerospace engineering designers. *Research in Engineering Design*, 24(1), 43–63.

Becattini, N., Cascini, G., & Rotini, F. (2014). An OTSM-TRIZ Based Framework towards the Computer-Aided Identification of Cognitive Processes in Design Protocols. *Design Computing and Cognition DCC'14*. pp. 103–122.

Bedny, G. Z., & Karwowski, W. (2004). Activity theory as a basis for the study of work. *Ergonomics*, 47(2), 134–153.

Cai, H., Do, E. Y.-L., & Zimring, C. M. (2010). Extended linkography and distance graph in design evaluation: an empirical study of the dual effects of inspiration sources in creative design. *Design Studies*, 31(2), 146–168.

Cash, P., Hicks, B., & Culley, S. (2015). Activity Theory as a means for multi-scale analysis of the engineering design process: A protocol study of design in practice. *Design Studies*, 38 (May), 1–32.

Cash, P., Hicks, B., Culley, S., & Adlam, T. (2015). A foundational observation method for studying design situations. *Journal of Engineering Design*, 26(7–9), 187–219.

Cash, P., & Maier, A. (2016). Prototyping with your hands: the many roles of gesture in the communication of design concepts. *Journal of Engineering Design*, 27(1–3), 118–145.

Cash, P., & Snider, C. (2014). Investigating design: A comparison of manifest and latent approaches. *Design Studies*, 35(5), 441–472.

Cash, P., Stankovic, T., & Storga, M. (2014). Using visual information analysis to explore complex patterns in the activity of designers. *Design Studies*, 35(1), 1–28.

Court, A., Culley, S., & McMahon, C. (1993). The information requirements of engineering designers. *Proceedings of the International Conference of Engineering Design*, 1993.

Daly, S., McKilligan, S., Murphy, L. & Ostrowski, A. (2017) Tracing Problem Evolution: Factors That Impact Design Problem Definition. In: Christensen, B. T., Ball, L. J. & Halskov, K. (eds.) *Analysing Design Thinking: Studies of Cross-Cultural Co-Creation*. Leiden: CRC Press/Taylor & Francis.

Deken, F., Kleinsmann, M., Aurisicchio, M., Lauche, K., & Bracewell, R. (2012). Tapping into past design experiences: Knowledge sharing and creation during novice-expert design consultations. *Research in Engineering Design*, 23(3), 203–218.

Dong, A. (2005). The latent semantic approach to studying design team communication. *Design Studies*, 26(5), 445–461.

Dorst, K., & Cross, N. (2001). Creativity in the design process: Co-evolution of problem-solution. *Design Studies*, 22(5), 425–437.

Dorst, K. (2008). Design research: a revolution-waiting-to-happen. *Design Studies*, 29(1), 4–11.

Eckert, C. M., & Stacey, M. K. (2000). Sources of inspiration: A language of design. *Design Studies*, 21(5), 523–538.

Eckert, C. & Stacey, M. (2017). Designing the Constraints: Creation Exercises for Framing the Design Context. Christensen, B. T., Ball, L. J. & Halskov, K. (eds.). *Analysing Design Thinking: Studies of Cross-Cultural Co-Creation*. Leiden: CRC Press/Taylor & Francis.

Gallivan, M., & Srite, M. (2005). Information technology and culture: Identifying fragmentary and holistic perspectives of culture. *Information and Organization*, 15(4), 295–338. JOUR.

Gero, J. S., & Kannengiesser, U. (2004). The situated function–behaviour–structure framework. *Design Studies*, 25(4), 373–391.

Gibson, C. B. (2001). From knowledge accumulation to accommodation: Cycles of collective cognition in work groups. *Journal of Organizational Behavior*, 22(2), 121–134.

Goldschmidt, G. (1990). Linkography: assessing design productivity. In: *Cyberbetics and System'90, Proceedings of the Tenth European Meeting on Cybernetics and Systems Research*. JOUR, World Scientific, Singapore. pp. 291–298.

Gonçalves, M. (2016). *Decoding designers' inspiration process*. TU Delft.

Gonçalves, M., Cardoso, C., & Badke-schaub, P. (2013). Through the looking glass of inspiration: Case studies on inspirational search processes of novice designers. *Proceedings of the international Association of Societies of Design Research, IASDR* 2013, Tokyo, Japan.

Gonçalves, M., Cardoso, C., & Badke-Schaub, P. (2013). Inspiration peak: exploring the semantic distance between design problem and textual inspirational stimuli. *International Journal of Design Creativity and Innovation*, 1(June 2013), 215–232.

Gonçalves, M., Cardoso, C., & Badke-schaub, P. (2014). What inspires designers? Preferences on inspirational approaches during idea generation. *Design Studies*, 35(1), 29–53.

Guo, B. (2011). The scope of external information-seeking under uncertainty: An individual-level study. *International Journal of Information Management*, 31(2), 137–148.

Heisig, P., Caldwell, N. H. M., Grebici, K., & Clarkson, P. J. (2010). Exploring knowledge and information needs in engineering from the past and for the future – Results from a survey. *Design Studies*, 31, 499–532.

Hicks, B. J., Culley, S. J., Allen, R. D., & Mullineux, G. (2002). A framework for the requirements of capturing, storing and reusing information and knowledge in engineering design. *International Journal of Information Management*, 22(4), 263–280. JOUR.

Hult, G. T. M., Ketchen, D. J., & Slater, S. F. (2004). Information processing, knowledge development, and strategic supply chain performance. *Academy of Management Journal*, 47(2), 241–253. JOUR.

Jansson, D. G., & Smith, S. M. (1991). Design fixation. *Design Studies*, 12(1), 3–11.

Kleinsmann, M., Deken, F., Dong, A., & Lauche, K. (2012). Development of design collaboration skills. *Journal of Engineering Design*, 23(7), 485–506.

Kleinsmann, M., & Valkenburg, R. (2008). Barriers and enablers for creating shared understanding in co-design projects. *Design Studies*, 29(4), 369–386.

Lawson, B., Petersen, K. J., Cousins, P. D., & Handfield, R. B. (2009). Knowledge Sharing in Interorganizational Product Development Teams: The Effect of Formal and Informal Socialization Mechanisms. *Journal of Product Innovation Management*, 26(2), 156–172.

Love, T. (2000). Philosophy of design: A meta-theoretical structure for design theory. *Design Studies*, 21(3), 293–313.

Maher, M. L., Poon, J., & Boulanger, S. (1996). Formalising Design Exploration as Co-Evolution. In J. S. Gero & F. Sudweeks (eds.) *Advances in Formal Design Methods for CAD*. pp. 3–30.

McAlpine, H., Hicks, B. J., & Tiryakioglu, C. (2011). The digital divide: Investigating the personal information management practices of engineers. *ICED 11 International Conference on Engineering Design*. JOUR, Technical university of Denmark. pp. 31–42.

Perttula, M., & Liikkanen, L. (2006). Exposure effects in design idea generation: Unconscious conformity or a product of sampling probability? *NordDesign 2006*. Reykjavik, Iceland. pp. 1–14.

Robinson, M. A. (2010). An empirical analysis of engineers' information behaviours. *Journal of the American Society for Information Science and Technology, 61*(4), 640–658.

Sarkar, P., & Chakrabarti, A. (2008). The effect of representation of triggers on design outcomes. *Ai Edam, 22*(2), 101–116.

Schön, D. A. (1984). *The reflective practitioner: How professionals think in action*. BOOK, New York, USA: Harper Torchbooks.

Shroyer, K., Turns, J., Lovins, T., Cardella, M., & Atman, C. J. (2017). Team Idea Generation in the Wild: A View from Four Timescales. In: Christensen, B. T., Ball, L. J. & Halskov, K. (eds.) *Analysing Design Thinking: Studies of Cross-Cultural Co-Creation*. Leiden: CRC Press/ Taylor & Francis.

Sim, S. K., & Duffy, A. H. B. (2003). Towards an ontology of generic engineering design activities. *Research in Engineering Design, 14*(4), 200–223.

Simon, H. A. (1978). Information-processing theory of human problem solving. *Handbook of Learning and Cognitive Processes, 5*, 271–295.

Swaab, R. I., Phillips, K. W., Diermeier, D., & Husted Medvec, V. (2008). The pros and cons of dyadic side conversations in small groups. *Small Group Research, 39*(3), 372–390.

Walker, I. (2010). *Research methods and statistics*. BOOK, New York, USA: Palgrave Macmillan.

Wasiak, J., Hicks, B. J., Newnes, L., Dong, A., & Burrow, L. (2010). Understanding engineering email: the development of a taxonomy for identifying and classifying engineering work. *Research in Engineering Design, 21*(1), 43–64.

Wiltschnig, S., Christensen, B. T., & Ball, L. J. (2013). Collaborative problem-solution co-evolution in creative design. *Design Studies, 34*(5), 515–542.

Yu, R., Gu, N., Ostwald, M., & Gero, J. (2015). Empirical support for problem-solution co-evolution in a parametric design environment. *AIEdam, 25*(1): 33–44.

Chapter 28

Team Idea Generation in the Wild: A View From Four Timescales

Kathryn Shroyer, Jennifer Turns, Terri Lovins, Monica Cardella & Cynthia J. Atman

ABSTRACT

The generation of novel ideas is an important and difficult part of the creative design process. Much of the research on idea generation (or ideation) has focused on for-malizing techniques to support idea generation and characterizing the effectiveness of the techniques as measured by quantity, quality, and creativity of ideas. Less is known about idea generation "in the wild," particularly idea generation across different timescales (i.e., idea generation across a multi-month project alongside idea generation in a period as small as minutes). We present a qualitative case study of a professional design team's use of idea generation with analyses at four timescales. The "share back" that occurs during structured idea generation emerges as an interesting phenomenon, and we provide a detailed analysis of the work done by the team during the "share back" of ideas.

I INTRODUCTION

Idea generation, as part of the creative design process, can be understood as an activity where designers consider multiple potential solutions to a given problem. A growing body of research investigates idea generation as a critical part of the design process. While idea generation is important, designers of all levels experience limitations in generating many diverse ideas (Linsey *et al.*, 2010; Sio *et al.*, 2015; Vasconcelos & Crilly, 2016). In response to these difficulties, a number of techniques for generating many diverse ideas have been developed and utilized in areas such as design, business, and engineering. Much of the current research on idea generation is conducted in lab settings and is concerned with the effectiveness of these techniques as defined by quantity, quality, and originality of ideas generated.

In the current scholarship on idea generation in design, there is room for more studies of idea generation in the wild, that is, idea generation as it takes place within a design team's process. Further, there is a need for studies of idea generation at different timescales, paralleling Lemke's argument of the need to explore learning across timescales (Lemke, 2001). The previously mentioned work is primarily done at the timescale of an "idea generation session". Questions about idea generation over the course of months, or the work of idea generation done during a few minutes, are less represented in the research. Thus, there is an opportunity for both research on idea generation in naturalistic settings and research on idea generation that crosses timescales. This paper takes up such an opportunity by focusing on the case of a professional design team working on a project over a multiple month time span. Four qualitative analyses are presented that highlight features of naturalist idea generation at different timescales (also referred to as levels in this paper): (1) twenty videos capturing segments of design meetings; (2) three formal idea generation sessions within these meetings; (3) the "share back" of ideas within one idea generation session; and (4) the discussion of one idea within that "share back".

2   BACKGROUND

Idea generation is a critical part of the creative design process and a part that may be related to the quality of the final design solution. For example, early phases of the design process, including idea generation, have been shown to have the highest impact on the quality and manufacturing costs of the final design (Römer, Weißhahn, & Lindemann, 2001).

While idea generation is important, designers of all levels experience limitations in generating many diverse ideas (Linsey et al., 2010; Sio et al., 2015; Vasconcelos & Crilly, 2016; Bruseberg & McDonagh-Philp, 2002). One reason for this is a cognitive difficulty, termed *fixation*, where designers develop an early attachment to their initial idea and stop considering alternatives (Ullman et al., 1988). Another reason, also a form of fixation, is an inability to break away from known examples or solutions (Linsey et al., 2010).

In response to these difficulties, a number of techniques for generating many diverse ideas have been developed in areas such as design, psychology, business, and engineering. These include brainstorming (Osborn, 1953), morphological analysis (Zwicky, 1969; Allen, 1962), synectics (Gordon, 1961), brainwriting (Geschka et al., 1976), nominal group technique (Van de Ven & Delbecq, 1974), and affinity diagraming (Mizuno, 1988). Brainstorming is most widely known and one of the earliest formalized idea generation techniques (Isaksen, 1998). As a result, brainstorming is often used to describe the idea generation process as a whole, but more specifically refers to a technique developed by Osborn (1953) as a means of increasing the creative productivity of groups.

Much of the current research on idea generation techniques is concerned with the effectiveness of the technique as defined by quantity, quality, and originality of the idea. A subset of this research explores the nature of idea generation, covering topics such as the effects of timing and time constraints (Tseng et al., 2008; Liikkanen et al., 2009), the use of examples (Pettula & Liikkanen, 2006), and the role of representation type such as text versus graphical (McKoy et al., 2001). Other research focuses on tools and techniques to support idea generation and methods by which to evaluate them (Dorta et al., 2009; Jonson, 2005; Bilda et al., 2006; Nelson et al., 2009; Shah et al., 2003; Hernandez et al., 2010; Linsey et al., 2011; Daly et al., 2012). A common theme in several of these studies is an emphasis on generating a large number of ideas or potential design solutions, with the understanding that a greater number of ideas will lead to a better design solution. Some research, however, challenges this notion, suggesting that there are diminishing returns for additional ideas after a point (Reinig & Briggs, 2008) and that the quality of idea generation is more important than the quantity of ideas generated (Reinig & Briggs, 2013).

Much of the work to date has been conducted in lab settings. As a result, we know very little about when and how designers employ these techniques, or other idea generation techniques, in their design process. More broadly, what do we know of designers in practice and their approaches to idea generation? The few studies addressing this topic are very general, for example, mention of particular idea generation techniques in interviews of professional designers (Herring, 2009).

3   APPROACH

The opportunity to explore naturalistically occurring idea generation was made possible by a specific video dataset collected and provided by DTRS11 (Christensen & Abildgaard, 2017). Our approach to studying idea generation was grounded in this dataset and thus emergent, resulting multiple levels of analysis across different timescales. Below, we briefly characterize the dataset and high-level features of our analyses, leaving the detailed methods to be presented alongside the findings.

3.1   Dataset

This study analyzes a shared dataset, which follows the User Involvement design team of an international car company from October 2015 to January 2016 as they engage in Phase II of a design project in collaboration with the company's accessories department. The design team's aim was to develop a concept package for premium car users that would increase "take rates" (v21, 026) of car accessories in the Chinese market by better allowing the company's brand values to "shine through" (v21, 026). The data captures the team's development, implementation and debrief of a two day co-creation workshop with Chinese lead users to understand cultural values that can inform the company's approach to developing and selling accessories in China. While there are a number of individuals involved in the design activities, the core design team is made up of three user experience researchers: Ewan, Abby and Kenny. Additional participants involved in the project include an intern, two stakeholders, three consultants, Chinese lead users, four colleagues brought in to assist with the generation of ideas, and a participant observer involved in collecting data *in situ*.

The dataset, which documents some of the work of this team, includes two interviews and 20 videos capturing interactions among the design team during their meetings. The videos are each 30 to 90 minutes in length, and represent interactions at different points in the project. In addition to the videos, we had access to transcriptions of the videos and photos of some of the artifacts generated by the team during their interactions.

3.2   Data analysis

Our qualitative analysis began with the broad question: Given the documented work of the design team, how does the team appear to generate ideas? The question was

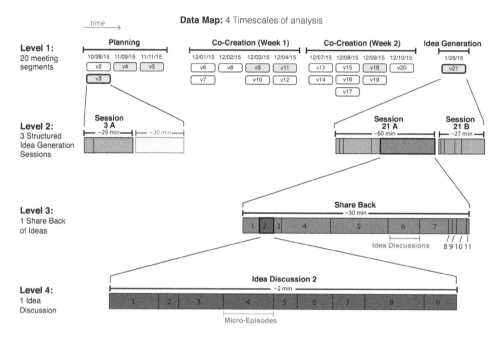

Figure 28.1 This figure illustrates (top to bottom) the four emergent timescales of analysis. Level 1 contains 20 videos (boxes) of design team activity on 12 days (columns) within a three-month period. Videos are grouped into 4 phases with shaded boxes indicating videos containing idea generation. Level 2 depicts the three formal idea generation sessions occurring in v3 and v21, their relative durations, and subcomponents (vertical lines). Level 3 depicts the share back of ideas in session 21 A subdivided into 11 idea discussions. Level 4 depicts the two-minute discussion of one idea within the share back subdivided into 9 micro-episodes.

approached in an exploratory nature, grounded in the data and guided by the following questions: who is generating ideas, what is the topic of the idea generation, when does the idea generation take place, what is the structure of the idea generation, and what techniques or tools does the team employ to generate ideas? This approach surfaced characteristics of the team's idea generation that prompted subsequent questions and a closer look at certain parts of the team's work. These emergent characteristics led to an analysis of different amounts of data across four different timescales (from three months to two minutes): (1) twenty videos capturing segments of design meetings; (2) three formal idea generation sessions within these meetings; (3) the "share back" of ideas within one idea generation session; and (4) the discussion of one idea within that "share back". The data selected for each level of analysis are detailed in Figure 28.1. This approach is resonant with Teixeira *et al.*'s (2017) approach, where they considered four timescales described in Goldschmidt's (2014) book: session, activity, episode and move. Teixeira *et al.*'s first timescale is analogous to our Level 2. We include a short methods section alongside each level of analysis in the work that follows.

What does quality mean for our analysis? Using a metaphor of continuous quality control, Walther *et al.* (2013) argue that the quality of qualitative research is affected continuously by decisions and actions related to the making and analyzing of data. In this case, we were not responsible for the collection of the data, but we are responsible for our choices concerning what data to address how to analyze the data. Shenton (2004) also provides guidance on how to think about issues of trustworthiness in qualitative research. Drawing on foundational ideas about what contributes to the quality of qualitative research, Shenton identifies specific strategies that a researcher can adopt to improve the trustworthiness of their work. In the work that follows, the notions of "providing a thick description" and "providing an audit trail" are central to the quality of the work.

4   LEVEL I: IDEA GENERATION ACROSS THE DATASET

In our initial analysis, we focus our attention on a large portion of the team's design process (videos of twenty design meetings spanning a three month period). Given the documented work of the design team in these twenty videos, how does the team appear to generate ideas?

4.1   Methods

Analysis of the dataset began by utilizing the 20 videos of design team meetings (v2–v21) as the initial scope of analysis. Two researchers familiarized themselves with the whole dataset by chronologically listening to the audio of the 20 design meeting videos (each lasting between 20 and 100 minutes and totaling over 15 hours). One researcher did a second pass, including viewing videos of the design meetings, and identified the videos where idea generation seemed prominent. A second author confirmed the existence of idea generation in these videos.

4.2   Findings

At this broad level, we see the design team planning a two-day co-creation workshop, running the workshop in China, and later conducting an idea generation session back home with the insights they gathered in China. The design team engages in generation and transformation of ideas throughout the dataset. A map of the data at this broad level is illustrated in Figure 28.2.

Idea generation was found to be present in at least seven sessions (v03, v04, v05, v09, v11, v18, and v21), indicated by the shaded boxes in Figure 28.2 and described here:

- Video 3: The design team, intern, and participant observer conduct a formal idea generation session related to choosing workshop participants. This is followed by a discussion about language and translation of the workshop where idea generation occurs as part of the discussion.
- Video 4: The design team and participant observer discuss the activities planned for the first day of the co-creation workshop. Idea generation occurs as part of the discussion.

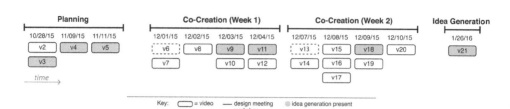

Figure 28.2 This figure chronologically maps (left to right) the 20 videos analyzed. Each block represents a video, solid outlines representing design meetings and dashed outlines representing workshops. The grey shaded boxes indicate videos found to contain idea generation. Videos are dated with each column representing a single day and grouped into four phases depicting the team's planning, their two weeks in China, and the final idea generation session.

– Video 5: The design team and participant observer discuss the activities planned for the second day of a co-creation workshop. Idea generation occurs as part of the discussion.
– Video 9: The design team and invited consultants discuss outcomes from the first day of the co-creation workshop and begin to adjust plans for the second day of the co-creation workshop. Idea generation occurs as part of their planning.
– Video 11: The design team and invited consultants discuss changes to the second day of the co-creation workshop. Idea generation occurs as part of their planning.
– Video 18: The design team and a consultant cluster insights from the co-creation workshop. Idea generation occurs as part of their thinking about how to incorporate the insights into products and services.
– Video 21: The design team members, plus two pairs of colleagues unfamiliar with the project, engage in formal idea generation sessions.

At this broad level, we see that while the team generates ideas throughout the dataset, there are moments where they explicitly set aside chunks of time dedicated to idea generation. We refer to these segments as formal idea generation sessions. The team engages in formal idea generation sessions in three prominent instances: the first part of video 3 (3A), the first half of video 21 (21A) and again in the second half of video 21 (21B). Outside of these three formal periods, members of the design team generate ideas as part of meeting activities not explicitly dedicated to idea generation.

5   LEVEL 2: FORMAL IDEA GENERATION WITHIN MEETINGS

Looking at all 20 videos of the design team meetings revealed multiple instances where the team engaged in formal idea generation sessions. The encounter of such sessions provided a chance to explore formal idea generation in a naturalistic case study. If we focus our attention more narrowly on exemplars of such formal idea generation, how does the team appear to generate ideas?

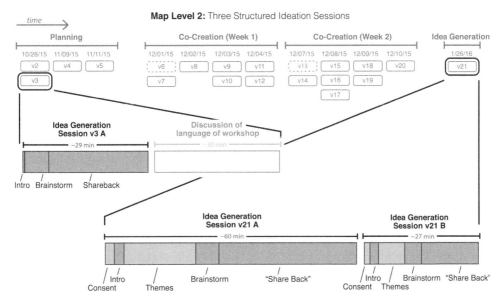

Figure 28.3 This figure maps the second level of analysis – three formal idea generation sessions further divided into various activities. The relationship of the three sessions to the overall dataset is seen in the top exploded view. Each session contains three common activities, an introduction, individual brainstorming and a "share back" of ideas (dark grey boxes). The sessions in video 21 also contain consent to videotape and an introduction to project themes activities (light grey boxes).

5.1 Methods

At this level, researchers selected three formal idea generation sessions as the scope of their analysis. Researchers identified three instances in which formal idea generation was particularly prominent (the first part of video 3 and two distinct periods in video 21). These were identified by the design team's explicit introduction of a topic and acknowledgement that they were going to dedicate time to generating ideas. One researcher qualitatively analyzed these formal idea generation sessions with attention to who was involved, the content of the discussion, and the actions that were taking place. Graphical representations of the sessions were created to support the analysis. These graphical representations are shown in Figure 28.3 and discussed below.

5.2 Findings

At a broad level, the three formal idea generation sessions selected for analysis differ in a number of ways: who is participating, the topic of idea generation, the duration and when in the larger design process they take place. Despite these differences, we notice a common set of activities that emerges in all three sessions. Figure 28.3 provides a graphical depiction of these three idea generation sessions. All sessions contain three main activities: introduction of the topic of idea generation, followed by individual brainstorming, and concluding with a "share back" of ideas. During idea generation sessions in 21A and 21B, there are two additional activities: gaining consent to video record and introducing the participants to the project and relevant cultural themes. These additional activities are likely a result of the fact that colleagues, new to the project, have been brought in to generate ideas, whereas in video 3 the idea generators were already familiar with the project.

We provide an overview of session 21A as an example of the structure of these formal idea generation sessions. The design team characterizes this as "a brainstorming session" (v21, 033) to generate "accessories related ideas" (v21, 033) that could help "brand values [of the company] ... shine through in China" (v21, 026). Two colleagues, Steve and Paul, who until this point were uninvolved with the project, are brought in to provide a fresh perspective and help "look at new angles" (v21, 003).

During this session (A), the design team explains and receives consent to video record the session. Next the team provides an introduction orienting Paul and Steve to the general format and goals of the session, which are quantity and breadth of ideas related to accessories, defined broadly as "products", "services", and "communication" (v21, 033). Next, the design team gives a short description of their project and shares the results of their China co-creation workshop, including cultural themes and subthemes. During the description of the themes, Paul and Steve are given Post-it notes and instructed to document ideas as they think of them. After introducing the themes, Ewan facilitates a five-minute "individual" brainstorm during which the colleagues continue to generate ideas. The rest of the session (about 30 minutes) is spent "sharing back" many, but not all, of Paul and Steve's ideas. Paul and Steve read ideas off their Post-it notes one by one. The team then spends various amounts of time discussing the ideas during which time they place the Post-it notes on the whiteboard near similar ideas.

We notice several interesting features when zooming in on the formal idea generation sessions. While there are some differences across the three sessions, the design team employs "brainstorming" and "share back" as major idea generation strategies in all three sessions (the quotation marks indicate the terms used by design team members).

We notice purposeful elements in the structure of the idea generation session implemented to aid in the generation of novel ideas. The structure of sessions 21A and 21B involves explicitly soliciting ideas from outsiders by bringing in people new to the project to generate ideas. Ewan explicitly acknowledges this during the introduction in 21A, telling the colleagues that the team has "been kind of institutionalized" by the project and has brought them in to "look at new angles" (v21, 003). Another strategy built into the structure of the sessions is the idea of individual idea generation versus group interaction. Each session has explicit time set aside for first generating ideas individually, then sharing them and building off of others' ideas.

Display and visual externalization is seen as another strategy the design team employs in the session. Pictures and text descriptions of each of the themes described in the introduction are displayed on the wall of the room where the session takes place. These visuals are gestured to during the introduction and parts of the "share back". The team also uses Post-it notes as a visual tool that seems to serve many purposes. Ideas generated are written on Post-it notes. When ideas are shared, these notes are put on a whiteboard in the room where everyone can see them. Here they are moved around and similar ideas are grouped together.

We also note strategies mentioned by the design team during the introduction that manifest as rules or guidelines for the session. One such rule is mass production and a focus on quantity over quality. During all three sessions, we see mention of quantity of ideas as a goal. For example in session 21B Ewan states "quantity is totally fine, doesn't have to be quality, just start with the obvious" (v21, 001). We also see breadth of ideas as a guideline mentioned several times in session 21A by different members of the design team. For example "whatever ideas, accessories related ideas, you have" (v21, 033). Abby even suggests that if the participants cannot think of ideas related to health, they can draw on experiences from "their own everyday life" (v21, 146).

6 LEVEL 3: "SHARE BACK" WITHIN A FORMAL IDEA GENERATION SESSION

Looking at the three formal idea generation sessions (v3, v21A, and v21B) revealed a common structure in which "brainstorming" was used to generate ideas and a "share back" was used to discuss those ideas and start to form clusters of related ideas. The work of the "share back" emerged as interesting in light of the fact that it is under-studied in literature and more was going on than just reporting back ideas. If we focus our attention more narrowly on the share back activity, how does the team appear to generate ideas?

6.1 Methods

The goal of this analysis was to provide a thick description of the "share back" of one collection of ideas during one formal idea generation session. At this level, researchers took the 32-minute "share back" session that begins video 21A to be the scope of analysis. The start of the "share back" begins when Ewan asks "should we start sharing a little bit, maybe other stuff will come up?" (v21, 199) and ends when Paul and Steve hand Ewan their remaining Post-it notes and leave the room. Transcripts, video, and a photo of Post-it notes arranged on the whiteboard were used to qualitatively analyze this "share back". One researcher focused on recreating the sequence with which ideas from the "individual brainstorm" (i.e., ideas represented on Post-it notes) were discussed by the group and placed on the whiteboard. The sequence is depicted chronologically in Table 28.1, and visually in Figure 28.4. Figure 28.4 also embeds the chronology with the imposition of numbers showing the sequence by which Post-it notes were added to the whiteboard. The creation of the chronology was supported by linguistic cues in the transcript (e.g., "okay I'll start with an obvious one" (v21, 207) and "eh did you have something else?" (v21, 400)). The creation of the visual depiction (i.e., the map of the Post-it notes) involved three passes through the video in order to make note of when the Post-it-note was written, what was written on it, who wrote it, and it's movement on the board (if any).

6.2 Findings

Here we begin to see not just structures of idea generation in time, but also the design team's use of space to organize ideas. Table 28.1 provides the chronological depiction

Table 28.1 The table provides a summary of the 11 discussions that occur in the "share back" activity of the formal idea generation session in video 21A. The first column numbers the discussions chronologically. The second column shows who authored the idea. The third column summarizes what was written on the Post-it note. The fourth column shows the duration (time) in minutes and seconds. The final two columns show which Post-its are used during the 11 discussions and when the Post-it was created (during the brainstorm or "share back").

		Level 3: 11 Discussions in a share back session			
Discussion	Author	Idea	Duration (m:ss)	Post-it from brainstorm	Post-it from share back
1	P	Cleaning the air around the car	2:20	P1	A1
2	P	Renting out parts of the car	1:56	P1	A2, E1
3	P	Famous table tennis player	1:09	P1	
4	S	Durability through time	6:41	S1, S2	A3, A4, E2
5	S	Local production	8:10	S3, S4	P2
6	S	Accessory subscription service	3:17	S5, P3	S6
7	S	Elder care	3:58	S7	
8	P	Company academy	0:32	P4	
9	P	Always get something extra	0:38	P5	
10	P	Teaming up with local brand	1:24	P6	
11	P	Comfy accessories	0:22	P6	

A = Abby E = Ewan K = Kenny P = Paul S = Steve

of the "share back" of session 21A; the depiction of Post-it note placement on the whiteboard is provided in Figure 28.4.

After individually generating ideas, Paul and Steve "share back" their ideas one at a time. Facilitated by Ewan, each person shares an idea and puts the Post-it note on the whiteboard. If there are similar ideas, they are put on the whiteboard at that time. The ideas are then discussed together. Additional Post-it notes are constructed during the "share back" to capture new information and are added to the board. Sometimes members of the design team do this, other times the design team asks the Paul or Steve to document new ideas or additional details. The "share back" activity, in this instance, lasts about 32 minutes and covers 11 idea discussions summarized in Table 28.1.

One of the most interesting features of the idea discussions is that they evolve and expand the idea into an exploration of the bounds and possibilities of the idea. Many of Paul and Steve's ideas are presented, not as one specific implementable solution, but as a constrained space of possible solutions. We see that the team discusses many options for ideas related to the Post-it note idea, not one specific idea. For example, during discussion 1, Paul lists a few different things his idea could be. He mentions a box "that you slap on your car, or maybe it's, you don't see the box maybe you see like a small post, or badge, or something" (v21 215). There are also instances when the design team explicitly acknowledges that there are multiple paths or subcomponents of an idea. For example, during discussion 4, Ewan says, "it has the two path thing, so you can recycle it or down cycle it into other parts, or is it the same object, …Yeah that's two interesting approaches" (v21, 350).

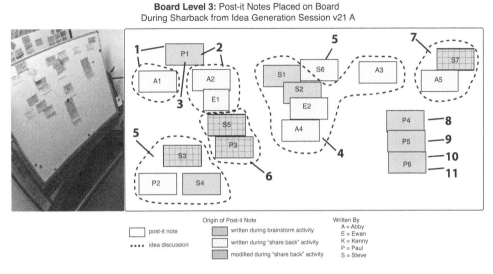

Figure 28.4   Figure shows a photo (left) and illustration (right) of the clustering of Post-it notes on the whiteboard by the design team during the "share back". Color indicates the origin of the notes (brainstorming, "share back", or brainstorming but modified during "share back". Numbers indicate the 11 discussions (detailed in Table 28.1) and which post it notes were placed on the board during this time. Labels indicate the team member who wrote down the post it and the order in which Post-its they were used during the discussion.

At the previous level we noted the use of display and visual externalization. Here we see in more detail that the team uses Post-it notes as a means of capturing ideas. Post-it notes from the individual brainstorm activity are used to represent and display ideas generated by Paul and Steve. Each Post-it note is intended to represent one idea that can then be displayed and moved around on the shared whiteboard to represent it's relationship to other idea Post-it notes. The importance of one note representing an idea is revealed when Paul has written multiple ideas on one Post-it note. This prompts several discussions on Post-it note logistics. Post-it notes are modified and new Post-it notes are added to capture new thoughts as ideas are discussed.

7   LEVEL 4: THE DISCUSSION OF ONE IDEA WITHIN "SHARE BACK"

Looking at the 32-minute "share back" activity in video 21 enabled us to notice the prolonged discussion around ideas documented on Post-it notes. Further, these prolonged discussions are potentially interesting because such discussions occasionally seem playful and disorganized, which raises a question about the role of the discussion. The opportunity to notice the prolonged discussion gave rise to the final level of analysis presented in this paper. If we focus our attention more narrowly on the discussion that occurs during the "share back" of one idea, how does the team appear to generate ideas?

7.1 Methods

At this level, the researchers took Discussion 2, the two-minute conversation about Paul's idea to "rent out parts of your car" (v21, 258), from the "share back" activity in video 21A to be the scope of analysis. The discussion is defined to begin when Paul remarks "and then I had a very vague, but – rent out 'parts' of your car" (v21, 258), and ends when Paul moves onto his next idea "then I have the last thing on this note" (v21, 304). This idea was chosen because the duration of its discussion was commensurate with that of other discussions during the "share back". The goal of this analysis was to provide a thick description of the "share back" of one idea. The content of the discussion of this one idea was analyzed in detail through a systems approach based on morphological analysis. Open coding of the content of the conversation was used to define *features* (different aspects of the idea) and *options* (choices within a feature). These features and options were observed through time and with attention to who was talking.

After the features were identified, the presence or absence of these six features was coded using a cleaned up version of the transcript. A transcript of this two-minute segment can be found in Appendix 1. Then, the first researcher coded the transcript, to begin to document the patterns that we describe below. The coding was passed along to a second researcher for confirmation. This confirmation phase led to the identification of minor disagreements resolved by discussion. The resulting coding in Figure 28.5 represents the agreement.

7.2 Findings

The discussion of Paul's idea begins with what Paul has written out on his Post-it note: "rent out parts of your car. e.g. battery, trunk, …" Through the two-minute discussion, the group expands his idea with more detail. By the end of the discussion, we see that this one idea could be renting out the trunk, the front, the electricity, or the inside of the car as a local community service (not for the world but for the car owner's neighborhood). We see that the trunk can be used for storage or delivery. We see that electricity could be a source of clean energy for the community to use rather than coal energy from a wall plug. Electricity can be used to charge phones, but not for hours – just enough to make a call. This electricity could be accessed via an induction charger on the car or even come through the trunk.

This elaborated sense of what "renting out parts of the car" might mean is the product of a variety of interactions among the group members. The structure of these interactions in time is detailed in nine micro-episodes found in the Appendix 1.

By applying a morphological analysis approach to the conversation around one idea, an idea substructure of *features* and *options* was revealed. This substructure is detailed in Figure 28.5. Our analysis identified six main features – *who, what action, what object, why, when,* and *type of accessory* – and a number of different options for each feature. Appendix 1 shows the transcript of the discussion coded for each feature (i.e., if the talk captured in a row of the table has something to do with the feature represented in the column of the table, then the intersecting cell is shaded). Analysis across each column made it possible for us to see the range of options associated with a feature.

Level 4: Ideaspace

Features	Option 1	Option 2	Option 3	Option 4	Option 5	Option 6	Option 7
WHO	others	not the world	your neighborhood	the mafia	the kindergarten	people cleaning cars	homeless guys
WHAT ACTION	Rent out	use	not drive				
WHAT OBJECT	car	parts of car	trunk	electricity	front	inside	
WHY	storage	practice	sleep	clean energy	charging phone	delivery	Community Service
WHEN	while owner not using car	while others use parts of car	temporarily				
TYPE OF ACCESSORY	service						

270 Kenny   "So the mafia they can rent the trunk for the people they kill
272 Kenny   just uh temporarily"

Figure 28.5 Example of the features and options proposed by the team during the second discussion around the idea "renting out parts of your car". The joke suggesting that the mafia can temporarily rent out the trunk for the people they kill is visualized as a connection of different options for five different features.

Figure 28.5 represents the collective set of features and options created by the group during the two-minute "share back". Because the figure shows the space of features and options associated with the idea, we refer to the table as an ideaspace. This collective set of features is created by different members contributions to the conversation. For example, Kenny's joke contribution "So the mafia can rent out the trunk for the people they kill" can be seen in the shaded boxes of Figure 28.5 as a connection of five different features with specific options.

Through the close examination of the discussion of an idea, we notice several strategies the design team employs to create the space of features and options, referred to as the ideaspace, that expands and enriches the idea as captured on the Post-it note.

The features and options of the ideaspace are drawn from the content of the discussion. Paul, the author of the idea, as well as Steve and the design team all contribute examples of what the idea can be. Each example provides a stitching together of options for certain features. Not all features are present in each example. Some examples make clear what features go together, others tack on features in a way that makes them only loosely connected. An example configuration with the mafia joke was can be seen in Figure 28.5.

Individuals use a variety of different levels of detail when expanding the idea. Throughout the conversation there is gradual movement from a broad level of detail to a closer level of detail with recaps throughout that jump back up to broad levels of detail.

This can be seen for example by tracing one feature of the ideaspace. The *who*, for example, begins as defined by Paul as "others" (v21, 262). During the joke section Kenny, Ewan, Paul, and Steve define who these others might be in more detail "the mafia", "the kindergarten", "people cleaning cars", "and homeless people" (v21, 270, 273, 275 and 276). Later, Ewan refocuses to a broader level of detail "not for the world, but for your neighborhood" (v21, 284).

Another action the group engages in is a bounding of the space by mentioning what the idea is not. This strategy is used several times in the discussion of this idea.

For example, Paul frames the length of time someone can use the electricity from the car to charge their phone:"yeah like maybe you can't like stand for an hour, but maybe you can get just a bit of juice so you can make a phone call." (v21, 289). In this example, Paul defines the length of time. There is not an exact number of seconds that one should necessarily charge a phone, but giving an example of what it is and what it is not gives the general range of acceptable durations that the electricity could be rented out.

Ewan also uses this technique when rearticulating *who* will be using the service. He explains "it's not for the world necessarily, but it is for your neighborhood" (v21, 284). Rather than just mentioning exactly who the service is for, Ewan gives an example of who it is not for. Again this seems to serve the purpose of better defining the outside boundaries of the ideaspace.

8   DISCUSSION

In this paper, we have presented a qualitative case study of a professional design team's engagement in idea generation with analyses at four timescales. Below, we comment on the findings at each timescale.

8.1   Idea generation across the dataset

Looking across the entire dataset (i.e., the 20 videos capturing interactions among team members at various points in their process), we saw that the design team does employ formal idea generation at some points while idea generation emerges naturally at other points. It would be interesting to better understand why the team chose formal idea generation approaches at the points they did. Daly *et al.* (2017) offer a complementary analysis in their chapter through their focus on tracing the evolution of the design problem across the dataset. The shifts in the problem space lead to shifts in solutions considered by the design team.

8.2   Formal idea generation within meetings

At the level of the idea generation session, we saw familiar elements including explicit allusions to "brainstorming," the writing of individual ideas on Post-it notes, and the aggregation and clustering of these notes on a shared public space (see Dove *et al.*, 2017, for more discussion of the functions of the Post-it notes in the design team's process). We were drawn to the work the design team did during the aggregation of the notes on the shared public space; a type of work they referred to at least once as "share back." As we moved forward in our analysis, we leveraged their term and identified "share back" as worthy of additional investigation.

8.3 "Share back" within a formal idea generation session

The work during "share back" was found to be a combination of affinity mapping (or clustering) of ideas and discussion of those ideas. At the beginning of the "share back", Ewan noted that "maybe other stuff will come up" (v21, 199) during the "share back". However, no one made it clear that the work would involve representing the ideas on Post-it notes and clustering them. Although the work clearly involved clustering the notes (and moving them around), this was not articulated. The "share back" as used by the team was transformational and expanded the ideas. Although some additional Post-it notes were generated during "share back" (and some existing Post-it notes were annotated), such generation and annotation does not happen uniformly. In other words, during the "share back", the team is using Post-it notes as a tool, but this tool no longer seems to fit their process. The discussion seems to go far beyond what is represented on the Post-it notes.

8.4 The discussion of one idea within "share back"

Our analysis of the team's brainstorming and "share back" in the second part of video 21 brings to light an interesting tension between the team's espoused procedure and their enacted practice. The procedure, as described by the design team during the introduction to the session, gives privilege to ideas as individually generated units to be created, captured on Post-it notes, described by the originator, and organized on the whiteboard. However, what we see upon closer examination of the discussion of one of these ideas is a collectively generated space made up of many partially complete idea fragments. While is it mentioned that "maybe other stuff will come up" (v21, 199) during the "share back"; the fragmented, collective, and dynamic nature of this other "stuff" is not explicitly acknowledged.

In this paper we used the term ideaspace to describe this phenomenon. A detailed analysis of the discussion around the idea "renting out parts of your car" helped to make clear the multitude of features and options named during the discussion. The discussion resembles the group reacting to Paul's somewhat vague "idea" much in the way they might react to a mini design brief. Rather than passively listening as Paul explained his idea, individuals offered new content. The content was, however, not delivered in cohesive tidy units. It was instead fragmented (individuals only presented pieces of ideas), in various levels of detail, overlapping (individuals borrowed and added what was being offered), and scattered (in terms of the order in which the discussion progresses). Instead of one cohesive idea, what the discussion created is a space full of possible fragments of ideas that together describe what "renting out parts of the car" might mean. Eckert and Stacey (2017) provide a complementary discussion of what we call an "ideaspace" in their chapter as they compare the creation exercises used by Ewan and his team to the artistic design processes used by fashion and knitwear designers, as these "artistic designers" use "mood boards" to express the space of possible designs and fragments of ideas are collected as a range of images capturing different aspects of possible design solutions.

Our analysis underscored the significant work that was being done during the "share back" portion of the brainstorming session. Looking back, these results are interesting because what seems relatively coherent now that it has been analyzed was hard work to see. This causes us to pause and think about what made it hard to see, whether the participants could see it, and whether others in a similar situation might be able to see it. Given the work required of us to bring the ideaspace into focus, it is interesting to consider whether the design team itself could see it. While we do not know if the team could see it, we can note that the group was diligent (in their talk) about getting "the ideas" on Post-it notes and getting the Post-it notes on the whiteboard, but was not similarly diligent about documenting the features and options that we ultimately framed as the ideaspace.

9   CONCLUSIONS

Through a qualitative analysis focusing on four timescales of analysis, we drew attention to structures, techniques, and tools used for idea generation. We noted that the design team engaged in formal idea generation at least three times, and looking at these three formal idea generation sessions, we see the design team use an individual brainstorming technique to generate ideas followed by a "share back" (discussion and affinity diagraming) to expand the ideas and generate additional ideas. Looking at the "share back", we see the team employ a variety of heuristics to share and expand idea generation. Using a morphological chart to represent the expansion of one idea during "share back", we brought to the fore how the team, knowingly or not, expanded the idea through a collection of features and options. This study begins to provide a picture of how designers use formal idea generation techniques *in situ* and brings to light the importance and need to further study the "share back" discussion as a mechanism for expanding ideas.

A key idea that frequently surfaced during the analysis, was how difficult ideas are to pin down. The mix of incomplete and varying levels of detail makes ideas difficult to trace. Using a chart to break ideas down into their features and options could be a productive approach, not just for generating ideas, but for documenting them and tracing them through time. This is useful for both researchers, and as a documentation and reflection artifact for designers. What might we learn by putting this in people's hands as a documentation tool? Or, as a means to teach students?

The decision to highlight the work of the design team at four different timescales emerged from interaction with the dataset. While the dataset is exceptionally comprehensive, it clearly leaves out a great deal of the design team's work on the project over time. For example, there were likely other team meetings that were not captured in the dataset, as well as each team member's individual thinking and doing. With additional data, we might have showcased additional levels, or different examples, of the current levels of analysis. However, such additions would not take away from the illustrative value of what we have presented here.

ACKNOWLEDGEMENTS

We would like to acknowledge the DTRS11 team for the amazing amount of work it took to pull this dataset together. We would also like to thank the individuals and organizations that we studied. Additionally, we would like to thank the Consortium to Promote Reflection in Engineering Education (CPREE) and the Center for Engineering Learning & Teaching (CELT) for helping to support this work.

REFERENCES

Allen, M. (1962). *Morphological creativity: The miracle of your hidden brain power; a practical guide to the utilization of your creative potential.* Englewood Cliffs, N.J.: Prentice-Hall.

Bilda, Z., Gero, J., & Purcell, T. (2006). To sketch or not to sketch? That is the question. *Design Studies,* 27(5), 587–613.

Bruseberg, & Mcdonagh-Philp. (2002). Focus groups to support the industrial/product designer: A review based on current literature and designers' feedback. *Applied Ergonomics, 33*(1), 27–38.

Christensen, B. T. & Abildgaard, S. J. J. (2017). Inside the DTRS11 Dataset: Background, Content, and Methodological Choices. In: Christensen, B. T., Ball, L. J. & Halskov, K. (eds.). *Analysing Design Thinking: Studies of Cross-Cultural Co-Creation.* Leiden: CRC Press/Taylor & Francis.

Daly, S. R., Yilmaz, S., Christian, J. L., Seifert, C. M., & Gonzalez, R. (2012). Design heuristics in engineering concept generation. *Journal of Engineering Education,* 101(4), 601.

Daly, S., McKilligan, S., Murphy, L., & Ostrowski, A. (2017) Tracing Problem Evolution: Factors That Impact Design Problem Definition. In: Christensen, B. T., Ball, L. J. & Halskov, K. (eds.) *Analysing Design Thinking: Studies of Cross-Cultural Co-Creation.* Leiden: CRC Press/Taylor & Francis.

Dorta, T., Lesage, A., & Pérez, E. (2009). Design tools and collaborative idea generation. Joining Languages, Cultures and Visions-CAADFutures 2009, *Proceedings of the 13th International CAAD Futures Conference,* 65–79.

Dove, G., Abildgaard, S. J., Biskjaer, M. M., Hansen, N. B., Christensen, B. T., & Halskov, K. (2017) Grouping Notes Through Nodes: The Functions of Post-it Notes in Design Team Cognition. In: Christensen, B. T., Ball, L. J. & Halskov, K. (eds.) *Analysing Design Thinking: Studies of Cross-Cultural Co-Creation.* Leiden: CRC Press/Taylor & Francis.

Eckert, C., & Stacey, M. (2017) Designing the Constraints: Creation Exercises for Framing the Design Context. In: Christensen, B. T., Ball, L. J. & Halskov, K. (eds.) *Analysing Design Thinking: Studies of Cross-Cultural Co-Creation.* Leiden: CRC Press/Taylor & Francis.

Geschka, H., Schaude, G., & Schlicksupp, H. (1976). Modern Techniques for Solving Problems. *International Studies of Management & Organization,* 6(4), 45–63.

Gordon, W. (1961). *Synectics, the development of creative capacity.* (1st ed.). New York: Harper.

Herring, S. R., Jones, B. R., & Bailey, B. P. (2009). Idea generation techniques among creative professionals. In *System Sciences, 2009. HICSS'09. 42nd Hawaii International Conference on* IEEE. pp. 1–10.

Hernandez, N., Shah, J., & Smith, S. (2010). Understanding design idea generation mechanisms through multilevel aligned empirical studies. *Design Studies,* 31(4), 382–410.

Isaksen, S. G. (1998). *A review of brainstorming research: Six critical issues for inquiry.* Creative Research Unit, Creative Problem Solving Group-Buffalo.

Jonson, B. (2005). Design idea generation: The conceptual sketch in the digital age. *Design Studies,* 26(6), 613–624.

Lemke, J. L. (2001). The long and the short of it: Comments on multiple timescale studies of human activity. *The Journal of the Learning Sciences*, 10(1–2), 17–26.

Liikkanen, L. A., Björklund, T. M., Hämäläinen, M. P., & Koskinen, M. (2009). Time constraints in design idea generation. DS 58-9: *Proceedings of ICED 09, the 17th International Conference on Engineering Design*, 9, 81–90.

Linsey, J., Clauss, E., Kurtoglu, T., Murphy, J., Wood, K., & Markman, A. (2011). An Experimental Study of Group Idea Generation Techniques: Understanding the Roles of Idea Representation and Viewing Methods. *Journal Of Mechanical Design*, 2011 Mar, Vol. 133(3).

Linsey, J., Tseng, I., Fu, K., Cagan, C., Wood, & Schunn. (2010). A study of design fixation, its mitigation and perception in engineering design faculty. *Journal of Mechanical Design, Transactions of the ASME*, 132(4), 0410031-04100312.

Mizuno, S. (1988). *Management for quality improvement: The seven new QC tools*. Cambridge, Mass.: Productivity Press.

McKoy, F. L., Vargas-Hernández, N. D., Summers, J. J., & Shah, J. (2001). Influence of design representation on effectiveness of idea generation. *Proceedings of the ASME Design Engineering Technical Conference*, 4, 39–48.

Nelson, B., Wilson, J., Rosen, D., & Yen, J. (2009). Refined metrics for measuring idea generation effectiveness. *Design Studies*, 30(6), 737–743.

Osborn, A. (1953). *Applied imagination; principles and procedures of creative thinking*. New York: Scribner.

Perttula, M. K., & Liikkanen, L. A. (2006). Exposure effects in design idea generation: Unconscious conformity or a product of sampling probability? *Proceedings of NordDesign 2006 Conference*, 42–55.

Reinig, B. A., & Briggs, R. O. (2008). On the relationship between idea-quantity and idea-quality during idea generation. *Group Decision and Negotiation*, 17(5), 403–420.

Reinig, B. A., & Briggs, R. O. (2013). Putting quality first in idea generation research. *Group Decision and Negotiation*, 22(5), 943–973.

Römer, Pache, Weißhahn, Lindemann, & Hacker. (2001). Effort-saving product representations in design–results of a questionnaire survey. *Design Studies*, 22(6), 473–491.

Shah, J. J., Vargas-Hernandez, N. M., & Smith, S. (2003). Metrics for measuring idea generation effectiveness. *Design Studies*, 24(2), 111–134.

Shenton, A. K. (2004). Strategies for ensuring trustworthiness in qualitative research projects. *Education for information*, 22(2), 63–75.

Sio, U. N., Kotovsky, K., & Cagan, J. (2015). Fixation or inspiration? A meta-analytic review of the role of examples on design processes. *Design Studies*, 39, 70–99.

Teixeira, C., Shafieyoun, Z., de la Rosa, J. A., Cai, J., Li, H., Xu, X., & Chen, X. (2017) Structures of Time in Design Thinking. In: Christensen, B. T., Ball, L. J. & Halskov, K. (eds.) *Analysing Design Thinking: Studies of Cross-Cultural Co-Creation*. Leiden: CRC Press/Taylor & Francis.

Tseng, I., Moss, J., Cagan, J., & Kotovsky, K. (2008). The role of timing and analogical similarity in the stimulation of idea generation in design. *Design Studies*, 29(3), 203–221.

Van de Ven, A. H., & Delbecq, A. L. (1974). The effectiveness of nominal, Delphi, and interacting group decision making processes. *Academy of management Journal*, 17(4), 605–621.

Walther, J., Sochacka, N. W., & Kellam, N. N. (2013). Quality in interpretive engineering education research: Reflections on an example study. *Journal of Engineering Education*, 102(4), 626–659.

Vasconcelos, & Crilly. (2016). Inspiration and fixation: Questions, methods, findings, and challenges. *Design Studies*, 42, 1–32.

Zwicky, F. (1969). *Discovery, invention, research through the morphological approach*. 1st American ed. New York: Macmillan.

APPENDIX I "RENT OUT PARTS OF YOUR CAR," MICRO-EPISODES

Here we describe the content of this section of session #21 by briefly describing nine micro-episodes that occur within this particular two minutes of time.

- **Micro-episode 1:** (v21, 258–262) Paul begins with an introduction during which he introduces his idea generally – calling it out as "vague" (v21, 258)- and gives two more specific possible implementations of the idea – renting out electricity and trunk space.
- **Micro-episode 2:** (v21, 263–269) Next, Ewan recaps or clarifies Paul's idea by restating it- "oh that's a good idea, so parts of your car." (v21, 263)
- **Micro-episode 3:** (v21, 270–277) Next, Kenny begins a string of jokes that fill in detailed alternative implementations of renting out different parts of cars to different audiences for different reasons. Abby does not participate; she is at the table documenting ideas on Post-it notes.
- **Micro-episode 4:** (v21, 278–281) Following the joke, Paul refocuses the conversation, giving more detail about the renting out electricity implementation with a justification that the car can provide clean energy. During this explanation, Kenny interjects – combining the electricity and trunk concepts with, "and you could even combine it with a trunk" (v21, 280).
- **Micro-episode 5:** (v21, 282–284) Ewan again recaps, bringing into focus the goal of the accessory as a "community service" (v21, 282). Following this the group again enters a period of co-developing specifics of the idea.
- **Micro-episode 6:** (v21, 285–289) Steve, Ewan, and Paul continue to develop the electricity implementation by bringing in more details about how the power can be used to charge a phone with an induction charger, but just for a short period of time.
- **Micro-episode 7:** (v21, 291–292) Next Paul wonders if this idea meets the criteria of being an accessory, and is assured it's a service type of accessory.
- **Micro-episode 8** (v21, 293–300): Abby and Kenny follow up by mentioning one of the company's current services, "in car delivery" (v21, 293), that China is requesting.
- **Micro-episode 9:** (v21, 301–302) The conversation ends with Abby and Kenny relating this current service back to Paul's idea, "so it could be renting a trunk just ... pick up stuff in any random trunk (v21, 301).

APPENDIX 2 "RENT OUT PARTS OF YOUR CAR," TRANSCRIPT CODING

The discussion surrounding the "rent out parts of your car" idea and our coding of the discussion to draw attention to the features of the ideaspace. In order to save space, we have omitted turn-taking utterances that were exclusively filler words (e.g., um) or affirmative words (e.g., okay). Because these turns were not assigned codes in our coding scheme, the omission in the figure does not alter what is needed to understand our explanation of the results.

Level 4: Coded Transcript

Micro-Episodes	Line #	Initial	Transcript Segments	WHO	ACTION	OBJECT	WHY	WHEN	TYPE OF ACCESSORY
1. Intro	258	P	and then I had a very vague, but - rent out "parts" of your car		■				
	260	P	so while you don't use it.					■	
			It can be electricity,			■			
			because you probably have an electric car or something.						
	262	P	It can be using the trunk space, whatever.			■			
			But basically letting others utilize your car while it's parked	■		■			
2. Recap	263	E	oh that's a good idea, so parts of your car			■			
	266	P	not using the car as such, but			■			
3. Jokes	270	K	so the mafia they can rent the trunk for the people they kill	■					
	272	K	just uh temporarily					■	
	273	E	while the kindergarten use the front	■					
	275	P	or people cleaning cars can practice on your car for example	■					
	276	S	homeless guys can sleep in the car	■					
4. More Details	278	P	It started more with the electricity thing,			■			
			like renting out the-		■				
			it's basically a battery, if it's on a electric car,			■			
			it's just standing there so-					■	
			from your car			■			
	280	K	and you could even combine it with a trunk			■			
	281	P	rather than coal energy from a wall plug			■			
5. Recap	282	E	community hub						■
	284	E	it's not for the world necessary,	■					
			but it is for your neighborhood	■					
6. More Details	285	S	so you have like a spot on your car where you can place your phone			■			
	287	S	where you could charge				■		
	288	E	like a yeah- like an induction- charger			■			
	289	P	yeah like maybe you can't like stand for an hour					■	
			but maybe you can get just a bit of juice so you can make a phone call				■		
7. Accessory?	291	P	I don't know it's an accessory,						■
	292	E	but eh: the [service part can be an accessory]						■
			[the service part of course]						
8. Example of current service	293	A	and they have this eh: (..) delivery, what do they call - [in car delivery]				■		
	294	K	[in car delivery]						
	295	A	which is something that they are driving apparently		■				
	296	K	yeah they are driving and actually they said		■				
	297	P	roaming delivery or what?			■			
	298	K	yeah and they said that China is actually eh: requesting [that] service					■	
	300	K	say it's very likely that I guess it will come there-						
9. Relate to current service	301	A	so it could be renting a trunk		■				
			just - (.) picking up stuff in any random trunk			■			
	302	K	yeah and electricity could come through						

Initial Key:
A = Abby, K = Kenny, E= Ewan, N= Nina, S= Steven, P = Paul

Chapter 29

Structures of Time in Design Thinking

Carlos Teixeira, Zhabiz Shafieyoun, Juan Alfonso de la Rosa,
Jun Cai, Honghai Li, Xing Xu & Xu Chen

ABSTRACT

The significant increase in both the scale through which designers collaborate and in the complexity of the problem-solution space they confront makes design thinking a lengthier and more fragmented process than it used to be, challenging commonly accepted notions of timeframes associated with shaping a well-integrated solution. Part of the challenge is because design thinking activities have a variety of unique structures that optimize concept generation through a combination of timeframes. However, planning and managing time for concept generation through design thinking is based on past experiences and guesswork, because no time structures and principles have been developed so far, leaving designers with no guidelines for planning concept generation activities. Through literature review and analysis of an industry case study, this study proposes a model of how to structure time in design thinking processes in order to optimize time and resource allocation, improving productivity in collaborative and complex projects of concept generation through design thinking. Therefore, the aim of this study is to develop a comprehensive model of timeframes in design thinking.

1 INTRODUCTION

As design is moving from an individual task to team collaboration (Chiu, 2002; Stempfle & Badke-Schaub, 2002) and from shaping simple forms in products to solving complex problems (Dorst, 2011), the design process is becoming more and more fragmented and the designer's task more and more complicated to shape an integrated solution for an ill-defined problem (Cross, 2001). Therefore, planning effective team-based design thinking processes for concept generation in a complex space of innovation has become a very challenging task for managers of design projects. However, project planners lack structures and models for defining the optimal allocation of time and resources in design thinking processes.

This study investigates the structures of time in design thinking. It attempts to answer the question of 'how time is structured in design thinking processes?'. The findings from this study lead to the proposal of a classification of timeframes in design thinking, where timeframe is defined as a set period of time made up by a unique pattern of design activities, in which certain routines are expected to occur. Through literature review this study looks at timeframes from past studies in design thinking and compares them to the timeframes used by the designers in the industry case study. At a later stage of this study, after all papers were presented and published during the DTRS11 event, the findings from our study were also compared to and supported by another study of the same case by Shroyer et al. (2017), which also focused on time throughout the idea generation process and identified different timescales similar to our study. As a result, our study proposes timeframes in design thinking in order to equip designers with concepts and models for optimizing time and resource allocation when planning and managing concept generation projects.

2   METHODOLOGY

This study is based on compiling key concepts and findings from past studies and articles published in the journal *Design Studies* about timeframes and their structures in design thinking, and applying these concepts and findings to an analysis of the planning of the design thinking process from the case study provided by the organizers of DTRS11 (Christensen & Abildgaard, 2017), which concerns a real industry project executed by professional designers. Therefore, we would like to identify 'What are the structures that designers use to plan timeframes for concept generation activities?' and 'Are there any structures for planning and managing timeframes that can help designers optimize concept generation flows and improve productivity in design thinking?' As suggested by Tsenn et al. (2014), most of what we know about timeframes in design thinking is based on lab experiments, therefore there is a lack of knowledge regarding what we could learn if we study timeframes based on real industry case studies. The aim of this approach is to bring together the findings from these two different contexts (lab experiments and an industry case study) and develop a more comprehensive, complete, realistic, and evidence-based model of the overall structure of timeframes in the design thinking process. The analysis of articles published in *Design Studies* examined studies that had time and its structures (timeframes) as a key variable in laboratory experiments for understanding design thinking, ideation flows and concept generation activities. An additional study was conducted to analyze past studies that used Protocol Analysis and Linkography to examine the design thinking process. In her recent published book on Linkography, Goldschmidt (2014) consolidates the findings from multiple past studies. The literature review focused on identifying key structures of time that model timeframes at a macro scale, defining the total length of the process and its stages, and at the micro scale, defining blocks of time and its step-by-step activities. Within each category, structures were compared in terms of their properties in order to cluster similarities and to map the range of different structures of time in design thinking that are already known by researchers in the design field.

The analysis of the field case study examined multiple sources from the dataset provided by the organizers of DTRS11. It focused on interpreting how the design team planned and managed time and its structures throughout the whole process and at specific moments of co-creation activities. At the macro scale the documents examined detailed the overall project plan and its phases (Christensen & Abildgaard, 2017) and the audio and transcripts of an interview with the project leader (v01, Background Interview, 2015). At the micro scale the examination focused on the detailed agenda, schedule and guidelines of two, five-hour co-creation workshops

(Christensen & Abildgaard, 2017) as well as the videos (v02, v03, v04) and transcripts from the planning meetings of these workshops. Similar to the literature review, structures of time were categorized as either a macro or micro scale timeframe. Within each category, structures were compared in terms of their properties in order to cluster similarities and to map the range of different structures that professional designers use to plan and manage design thinking.

Timeframes at both the macro and micro scales identified in the literature review and field case study were integrated into a model based on cross-referencing findings from multiple studies. The goal is to develop a model for understanding structures of time in design thinking, consequently providing principles and guidelines for planning and managing timeframes in concept generation processes more effectively.

3   LITERATURE REVIEW

Starting from the question 'how time is structured in design thinking processes?' this study investigated articles in the journal *Design Studies*, searching for descriptions of unique patterns of design activities depicted in terms of a set period of time. An additional study was conducted to analyze the findings from the recently published book on Linkography by Goldschmidt (2014), in which the author consolidates findings from multiple past studies. The literature review focused on identifying key structures of time that model timeframes.

The literature on timeframes in design thinking is embedded and fragmented within studies of design projects, processes, and tasks. While time is just a tangential aspect of the studies, this dataset provides fragmented discoveries that if compiled together can be a reliable and revealing source for understanding the structure of time in design thinking. Studies in design thinking reveal that there are basically two levels of analysis, the macro level and micro level, through which professionals and researchers categorize timeframes.

When a project is measured in its entirety, time is measured at the macro scale. These studies characterize time in relation to design as a project. Bashir and Thomson (2001) studied design in the form of projects and argue that design projects are similar to any other type of project, requiring planning, coordination, control and management of activities through time. Kleinsmann and Valkenburg (2008) investigated design projects in terms of how infrastructural conditions for design activities work as interfaces that serve as enablers or barriers for the creation of shared understanding, which is the expected outcome of a collaborative design process. Austin *et al.* (2001) focused on the conceptual phase of design projects, defining it as a vibrant, creative and dynamic period. They propose a framework to model the iterative nature of design activities in this phase, mapping the exchange of information, opinions, and ideas between team members.

Additionally, studies from Baker and van der Hoek (2010) codified designers' activities in terms of sessions, in which designers focus on exploring solutions for very narrow problem spaces. Designers then routinely repeat those narrow ideation sessions reconsidering previously developed ideas in new contexts, transforming sessions into design cycles. These cycles serve the purpose of generating a large number of ideas, helping designers understand a problem and envision ways to solve it. These studies reveal the common understanding that projects, phases, and sessions are the key concepts that define structures of time in design thinking. They characterize concept generation as a process with a beginning and an end, defined as a project divided into phases of intense and focused design thinking sessions, in which design activities and thinking are performed.

When only a small section of the project is observed, then time is defined at the micro scale. These studies address time in relation to design as tasks and activities. In the book Linkography, Goldsmith (2014) describes the step-by-step process of how designers do concept generation by focusing on a very small design space during a very short and intense period of time. The author defines design activities as intense and focused sequences of moves performed by designers constructing episodes in which they explore promising associations to resolve a very narrow aspect of the overall problem. In addition, Busby (2001) studied error in design tasks in terms of distributed cognition and heuristics that happen through multiple types of interconnected interactions over full-scale design processes. It highlights the interdependence of asynchronous design tasks, revealing the critical role of tasks in shaping non-linear design processes.

Goldsmith (2014) provides many contributions to the understanding of time and iteration in design thinking. First, she provides a taxonomy for breaking down at multiple levels the sequence of developments involved in design thinking at the micro scale, as illustrated in Figure 29.1. It is possible to identify four different levels. The first one consists of a Session (Level 4), which is usually the length of a concept generation period (Goldsmith, 2014, Chapter 3). At the macro scale, design projects tend to have multiple concept generation sessions for iterating solutions. Sessions result from a sequence of design activities (Goldsmith, 2014, Chapter 2), therefore Level 3 is composed of well-defined and self-contained activities. At Level 2, activities are interpreted as permutations of episodes involving designers trying to find and explore promising associations to resolve a very narrow aspect of the overall problem (Goldsmith, 2014, Chapter 5). Level 1 is the lowest level and is composed of *moves*, which are actions that ignite a follow-up action or reaction (Goldsmith, 2014, Chapter 4).

These studies reveal the common understanding that a design session can be defined as a sequence of activities or tasks. Activities are structured as intense and focused sequences of moves performed by designers constructing episodes in which they explore promising associations to resolve a very narrow aspect of the overall problem.

Alternatively, some other studies look at time in relation to design as a process. This interpretation privileges the understanding that design thinking is best represented as a flow or process. Chiu (2002) framed design as a collaborative process that requires coordination of tasks and information flows, especially in complex and large projects. The author claims that structured organization can facilitate communication in design processes affecting the success of a design project. Studies by Tsenn *et al.* (2014) reveal two central structures to the concept generation process. They describe the concept generation period as the moment when designers develop associations that frame and reframe the problem and its solutions, and the incubation period that serves to break designers' fixation on redundant concepts, expanding the exploration of the problem-solution space. They argue that the sequencing of these two periods of concept generation and incubation during the idea generation process can impact the quantity

Session					
Activity 1		Activity 2		Activity 3	
Episode 1	Episode 2	Episode 3	Episode 4	Episode 5	Episode 6
M1 · M2 · M3	M4 · M5 · M6	M7 · M8 · M9	M10 · M11 · M12	M13 · M14 · M15	M16 · M17 · M18

Figure 29.1 Four levels of time structures at the micro scale of design thinking.

and quality of ideas, and the size of the solution space. Gale and Kovacs (1996) propose that the best way of reproducing full-scale, real-world designing, defined as the period from the brief to the final design, is to reveal the 'logic of design' manifested through designers' train of thought that results from imagining and reasoning. Oxman (2002) proposes a definition of the phenomenon of emergence in design as a process that is guided and anticipated through perceptual and cognitive components playing a critical role in how reasoning emerges in design processes. Carmel-Gilfilen and Portillo (2010) identified different trajectories of concept generation in design thinking according to designers' thinking abilities. Dualistic thinkers develop a linear path toward defining a design solution, while multiplistic thinkers adopt a multifaceted approach, thinking holistically about the solution space. Dorta *et al.* (2007) proposes the notion of design flow, through which designers engage with tools to augment their capacity to ideate. The concept of design flow interprets the ideation process as a fluid and uninterrupted interaction between designers and tools during the idea generation of a design solution.

These studies reveal the common understanding that process and flow are the key concepts that define structures of time in design thinking. They characterize concept generation as a process that connects problem framing and design solutions. Nelson *et al.* (2009) investigated idea generation in the design process, and focused on the specific moment when designers are exploring the possibility space. Idea generation is defined as the moment in the design process between problem definition and the development of a final solution. Therefore, the design process can be defined as a continuum that goes from the beginning to the end of the project, but it is primarily acted out during the intense periods of design thinking activities and sessions. These two timeframes, activities and sessions, are noticed both in studies that define design as a project (macro level), as well as a task (micro level). Therefore, activities and sessions can be characterized as happening at the meso level of design thinking. Meso level analysis is generally applied to indicate how the macro and micro levels connect, highlighting unique organizational structures that results from models and practice. Here, meso level is applied to indicate that concept generation as a process happens at the intersection of the macro and micro levels.

Table 29.1 summarizes the levels and units of analysis through which structures of time in design thinking are defined in the literature, providing a classification scheme to map different types of timeframes and to study how designers plan concept generation

Table 29.1 Categories of analysis of timeframes in design thinking.

Levels of analysis	Unit of analysis
Macro	Project
	Phase
Meso	Session
	Activity
Micro	Episode
	Move

projects by planning and managing time, design activities, and resource allocation in design thinking processes.

4   FIELD STUDY

The project led by the design team was estimated to have a total duration of 4 months (Oct 2015–Jan 2016) divided into seven phases, comprised of intensive insight generation and co-creation design activities connected to inbound and outbound streams of concept development to generate continuous feedback loops through the involvement of the multiple stakeholders from the project. Using the classification scheme developed through literature review, the dataset from the field study was analyzed to study how designers plan concept generation projects by planning and managing time, design activities, and resource allocation in design thinking processes, and especially what timeframes they use for planning design thinking projects.

In an interview about the project background (Audio 1: Background Interview, 2015), the project team leader explains how he envisioned the project as an ideation flow to be generated through the integration and orchestration of design activities, stakeholders' participation, and concept development streams over a period of time. In the interview, the team leader outlines the principles and structures underpinning his macro scale planning of the project timeframe. While some of his rationale is explicitly explained in his answers, some can only be revealed by interpreting the proposed project plan. Therefore, this study used not only the interview and the project plan (Christensen & Abildgaard, 2017) to interpret how timeframes of design activities were established, but also analyzed the videos and transcripts from v02, v03, and v04 concerning the planning meetings of the two co-creation workshops.

The analysis of this material indicates that timeframes were based on three underlying principles. First, they should maximize the allocation of limited resources through time and across multiple locations. Second, they should combine concept development and validation. Third, they should keep all stakeholders 'up to speed' with the concepts in development. These principles demonstrate an overall concern to plan and manage time in the most effective way possible. In order to address these principles, timeframes of design activities were structured to happen both sequentially and simultaneously,

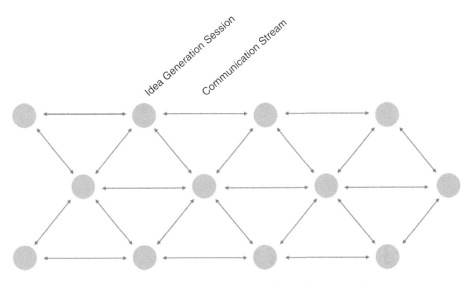

Figure 29.2 Macro structures of time from the project plan.

distributed across multiple locations, and connected through a network of streams. The team leader defined the structure of time for the design activities in terms of sessions, phases, and stages, characterizing timeframes as intense and focused moments of co-creation and team de-briefing, functioning as nodes connected together forming a collaborative concept development network. While the model can be interpreted as a linear process, the team leader emphasizes in his description that workshops are centralized points of co-creation, feeding each other and networking concept development with multiple stakeholders. As nodes, these moments have multiple inbound and outbound streams for sharing information before and after the intensive workshops, creating non-linear and non-sequential feedback loops and ideation flows.

The two co-creation workshops realized in a Chinese city on December 2015 with Chinese users (v06; v13) exemplify this structure of intense ideation moments as nodes in a fragmented and dispersed concept development network. They were planned by the design team (Christensen & Abildgaard, 2017) to involve customers in the co-creation of concepts. Each session had a total duration of five hours and they happened five days apart. In-between these sessions the design team worked on framing the problem-solution space, debriefing insights from co-creation sessions, and sharing the findings with the multiple stakeholders involved in the project. After the workshops were planned at the macro scale as critical nodes in the concept development network, the design team shifted the scale of their planning to define the design activities to be organized during the two five-hour workshops. It shifted from a node in a network (macro level) to a step-by-step five-hour sequence of design activities (micro level).

The workshops were divided basically in two main blocks: immersion and ideation. Immersion consisted of building common ground and framing the problem-solution space in collaboration with external participants. Ideation consisted of activities for

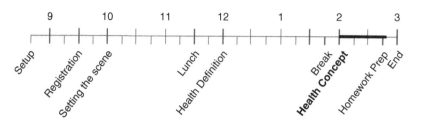

Figure 29.3 Time structure in co-creation workshop day 1.

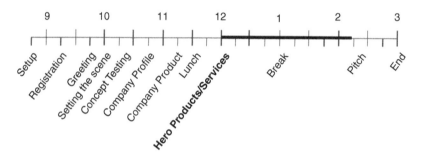

Figure 29.4 Time structure in co-creation workshop day 2.

exploring alternative solutions to narrowly framed problem spaces. Figures 29.3 and 29.4 represent the timeline of the activities that started at 8:45am and ended at 3pm (see Moderation Guide CC1, CC2, 2015; v02; v03; v04 in Christensen & Abildgaard, 2017).

The analysis of the structure of time from these workshops reveals that ideation activities that were focused on key themes (marked in red in the diagrams) were considered the *epicenters* of the overall concept generation sessions. Therefore, all other activities (marked in grey in the diagrams) were planned as providing a supporting role to ideation.

Figure 29.6 illustrates a combined timeframe of the activities from both workshops. Seen as a sequence of activities, iterations emerge as a pattern where each workshop is an iteration of the overall concept development. Each iteration consists of a lengthy period of immersion to build a common ground, followed by an intensive and compact moment of ideation to explore concepts in a given problem-solution space, trailed by an extended period of incubation to break designers' fixation on redundant concepts, expanding the exploration of the problem-solution space before initiating the next concept development iteration. In total the two workshops consisted of ten hours of design activities, from which seven hours were dedicated to immersion, and three hours to ideation, having five days of incubation separating the two iterations.

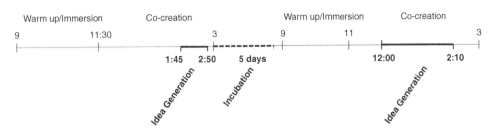

Figure 29.5 Overall activities timeframe of co-creation workshops day 1 and 2.

5 DISCUSSION

Analysis of the planning meetings from the industry case study indicates that concept generation sessions, which are critical nodes in the design thinking process, were planned from the macro to the micro scale, with designers first allocating broad strokes of time to collaborative design activities. It could be identified that designers decided to plan two five-hour co-creation workshops (concept generation sessions) at different days. At this scale, an hour was the main unit of time. Then, the five-hour session was used as a placeholder to structure a sequence of collaborative activities. At this scale, activities were planned in periods of 15, 30, 45, and 60 minutes. Last, designers worked on defining the step-by-step episodes of each activity. When planning the step-by-step of episodes within an activity, designers used rehearsal to explore and experiment how the ideation flow would happen. The biggest challenge of the design team when planning the ideation flows was to organize activities that were effective in helping participants break fixation, consequently exploring new associations of ideas and concepts. Therefore, they used rehearsal as a technique to model the ideation flow to be generated through their facilitation. Timeframes of activities were defined by putting themselves in the flow, rehearsing questions, prompts, primers, exchanges, deviations, and confusions, therefore anticipating through experience and sensing through intuition if the timeframe of the step-by-step activity is properly structured. Designers envisioned the concept generation process as sequences of activities and iterations triggered by prompts (questions, primers, tools) to stimulate participants to break fixation and explore new associations. However, to avoid unintended redundancies, time constraints were used to limit the unnecessary continuous iteration of an activity that already achieved its goals. Quantifiable time units are seldom articulated in the planning of the step-by-step activities. At the episode level no measurement of time was used, but they were structured as very brief tasks expected to happen in few minutes or fractions of a minute (seconds). The sense of correctness is defined by sensing if the flow would 'fit' in the pre-established timeframes allocated to each session. Therefore, each session was custom made and discussions on planning activities were based on past experience, not on any model or structure for time planning considering the uniqueness of concept generation through design thinking.

Through the categorization scheme developed by means of the literature review of timeframes in design thinking and the mapping of timeframes from the field study, it is possible to develop a description of how time is structured in design thinking. At the macro level design thinking consists of basically four different types of timeframes: flow, intensive, stream, and feedback loop. Flow is the recurrence of timeframes as series of repetitions irrigated by communication streams, providing the opportunity for comprehensive and novel examination of the problem-solution space, consequently shaping well-integrated solutions (Dorta *et al.*, 2007; Chiu, 2002; Kleinsmann & Valkenburg, 2008). Intensive relates to the conceptual phases of design projects in which information, ideas, and points-of-views are exchanged and co-created by team members of a project (Austin *et al.*, 2001). Stream is a continuous flow of inbound and outbound communication between periods of ideation and incubation (Chiu, 2002). Feedback loop is defined as communication streams that happen as loops, reshaping information, ideas, and opinions. At the micro level design thinking is primarily enacted in the ideation period, and consists of sequential moves intended to explore alternative possibilities to frame a problem-solution space and its solution (Oxman, 2002; Goldsmith, 2014). A sequence of moves that generate insights creates episodes that are focused on resolving a particular issue, by cross-examining, weaving or reaffirming a specific aspect of a concept (Goldsmith, 2014; Chandrasekera & D'souza, 2013).

At the meso level design thinking is primarily enacted through the concept generation process. Immersion is the period of framing the problem-solution space, solving a narrow aspect of the problem, and defining well-integrated solutions (Gale & Kovacs, 1996). Incubation is the period of putting ideas to rest in order to break fixation from a given solution, providing the time to open up new associations that can lead to new or improved solutions (Tsenn *et al.*, 2014). According to Tsenn *et al.* (2014, p. 502), a concept generation process is effective when it lasts until the concept generation rate slows substantially, but just before the participants become fatigued, losing the capacity to generate non-redundant ideas. Figure 29.6 illustrates this process. According to Tsenn *et al.* (2014, p. 502), the initial phase of the concept generation process is usually fast-paced and intense because designers tend to use familiar ideas, which are easily accessible. As familiar and easy to access ideas become scarce over time, the pace of idea generation tends to slow down and extends over longer periods of time because designers have to explore unfamiliar concepts and remotely connected associations. In this second phase, as uncommonness of ideas and remoteness of associations increase, while the pace of concept generation slows down as time passes, solutions increasingly become more creative and unique. Therefore, studies (Tsenn, 2014, p. 503) suggest that it is in the later phase of the concept generation process that novel and high quality solutions are developed.

Based on Tsenn *et al.*'s study (2014, p. 502), while the rate of idea generation never drops to zero, which means that designers never stop developing new concepts, over time there is a significant decrease in their effectiveness to generate non-redundant concepts, suggesting that extended timeframes won't result in more and better ideas. This is because designers' fixation on concepts previously developed blocks their capacity to explore new associations, consequently making ineffective any additional time spent on making more associations and exploring new concepts.

As an antidote to break designers' fixation, Tsenn *et al.* (2014) developed experiments to better understand the role that an incubation period can have in the creative process. An incubation period is a time in the concept generation process that happens

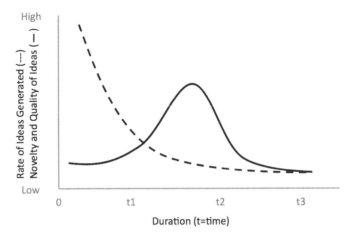

Figure 29.6 Concept generation timeline based on Tsenn *et al.*

at the point when there is a significant decrease in the development of non-redundant concepts, and designers stop actively working on the problem. The incubation period enables designers to break fixation, effecting designers' capacity to develop new insights. It is the period that it takes for designers to stop consciously working on the design problem, consequently reducing their fixation on familiar solutions, so they can work on expanding and exploring the solution space of a problem, until their concepts start to become redundant again. Therefore, creative processes that include incubation periods are more likely to develop greater quantity, quality, novelty and variety of solutions than processes without incubation periods (Tsenn *et al.*, 2014, p. 507). They break fixation, opening up new associations from concepts developed in previous idea-generation periods.

The results from this analysis, combined with the evidence collected from the field study, and a comparison to the findings from the study of the same case by Shroyer *et al.* (2017), produce a model of timeframes for design thinking in collaborative and complex projects. It consists of a multi-level approach to structures of time in design thinking. At the macro scale, concept generation through design thinking in collaborative and complex projects is centered around ideation sessions, followed by incubation, identified by Shroyer *et al.* as "idea generation across the dataset". At the micro scale, concept generation results from ideation sessions, defined by Shroyer *et al.* as "formal idea generation within meetings", composed of iterative design activities that start with fast-paced and intense association of ideas to explore, expand, and cross-exam multiple aspects of the problem-solution space. Shroyer *et al.* defined such timeframes as periods of "discussion of one idea within share back". Then, ideation sessions transition at the later stage to less intense associations enabling designers to spend longer periods of time exploring unfamiliar concepts and remotely connected associations, developing more creative and unique solutions until they become fixated on a concept. In Shroyer *et al.*'s study, these timeframes were interpreted as "share back within a formal idea generation session". In this context, ideation sessions function as

Table 29.2 Categories of timeframes in design thinking.

Level of analysis	Unit of analysis	Dimension of analysis	Timeframe
Macro	Project	Month	Flow Intensive
	Phase	Week	Stream Feedback Loop
Meso	Session	Day	Immersion Incubation
	Activity	Hour	Framing Iteration Fixation
Micro	Episode	Minute	Ideation
	Move	Second	Insights

nodes connected together forming a collaborative and iterative concept development network. As nodes, these sessions have multiple inbound and outbound streams for sharing information before and after, creating streams of communication and ideation flows.

6   CONCLUSION

As illustrated by the industry case study, significant increase in both the scale through which designers collaborate and in the complexity of the problem-solution space they confront makes design thinking a lengthier and more fragmented process than it used to be, challenging commonly accepted notions of timeframes associated with shaping a well-integrated solution for an ill-defined problem. Consequently, planning and managing the flows through which concepts are generated has become a major challenge in design thinking, requiring new models and structures of time management for design activities in collaborative and complex projects. The findings from this study indicate that timeframes in design thinking for professional practice should be planned and managed to maximize the allocation of limited resources through time and across multiple locations, while keeping all stakeholders 'up to speed' with the concepts in development. These principles identified through the field study demonstrate an overall concern to plan and manage time in design thinking processes in the most effective way possible. Therefore, the classification of different configurations through which time and design activities can be structured and the understanding of their overall function in design thinking processes can aid designers and design teams in finding ways to optimize time and resource allocation in concept generation projects. In addition, these research findings can be used as the starting point for exploring and testing different sequencing of timeframes and searching for optimal configurations of design activities in relation to different circumstances. Moreover, it can serve as the basis for investigating the possibility of measuring the optimal average, maximum, and minimum duration of timeframes based on collecting evidence from multiple industry case studies. This future research could reveal time standards for design thinking, transitioning the planning of concept generation projects from intuition and estimation to a more standardized practice supported by evidence from field studies, not just lab experiments.

REFERENCES

Austin, S., Steele, J., Macmillan, S., Kirby, P., & Spence, R. (2001). Mapping the conceptual design activity of interdisciplinary teams. *Design Studies, 22*(3), 211–232.

Baker, A. & van der Hoek, A. (2010). Ideas, subjects, and cycles as lenses for understanding the software design process. *Design Studies, 31*(6), 590–613.

Bashir, H. A. & Thomson, V. (2001). Models for estimating design effort and time. *Design Studies, 22*(2), 141–155.

Busby, J.S. (2001). Error and distributed cognition in design. *Design Studies, 22*(3), 233–254.

Carmel-Gilfilen, C. & Portillo, M. (2010). Developmental trajectories in design thinking: an examination of criteria. *Design Studies, 31*(1), 74–91.

Chandrasekera, T., Vo, N. & and D'souza, N. (2013). The effect of subliminal suggestions on Sudden Moments of Inspiration (SMI) in the design process. *Design Studies, 34*(2), 193–215.

Chiu, M. (2002). An organizational view of design communication in design collaboration. *Design Studies, 23*(2), 187–210.

Christensen, B. T. & Abildgaard, S. J. J. (2017). Inside the DTRS11 Dataset: Background, Content, and Methodological Choices. In: Christensen, B. T., Ball, L. J. & Halskov, K. (eds.). *Analysing Design Thinking: Studies of Cross-Cultural Co-Creation.* Leiden: CRC Press/Taylor & Francis.

Cross, N. (2001). Design cognition: results from protocol and other empirical studies of design activity. In: Eastman, C., Newstatter, W. and McCraken, M. (eds.) Design knowing and learning: cognition in design education. Oxford, UK: Elsevier, pp. 79–103.

Dorst, K. (2011). The core of 'design thinking' and its implication. *Design Studies, 32*(6), 521–532.

Dorta, T. (2009). Design flow and ideation. *International Journal of Architectural Computing,* 6(3), 299–316.

Dorta, T., Perez, E. & Lesage, A. (2008). The ideation gap: hybrid tools, design flow and practice. *Design Studies, 29*(2), 121–141.

Galle, P. & Kovács, L. B. (1996). Replication protocol analysis: a method for the study of real-world design thinking. *Design Studies, 17*(2), 181–200.

Goldschmidt, G. (2014). *Linkography: unfolding the design process* [Kindle Edition]. The MIT Press eBooks.

Kleinsmann, M. & Valkenburg, R. (2008). Barriers and enablers for creating shared understanding in co-design projects. *Design Studies, 29*(4), 369–386.

Nelson, B. A., Yen, J., Wilson, J. O. & Rosen, D. (2009). Refined metrics for measuring ideation effectiveness. *Design Studies, 30*(6), 737–743.

Oxman, R. (2002). The thinking eye: visual re-cognition in design emergence. *Design Studies,* 23(2), 135–164.

Shroyer, K., Turns, J., Lovins, T., Cardella, M. & Atman, C. J. (2017). Team Idea Generation in the Wild: A View from Four Timescales. In: Christensen, B. T., Ball, L. J. & Halskov, K. (eds.) *Analysing Design Thinking: Studies of Cross-Cultural Co-Creation.* Leiden: CRC Press/Taylor & Francis.

Stempfle, J. & Badke-Schaub, P. (2002). Thinking in design teams – an analysis of team communication. *Design Studies, 23*(5), 473–496.

Tsenn, J., Atilola, O., McAdams, D. A., & Linsey, J. S. (2014). The effects of time and incubation on design concept generation. *Design Studies, 35*(5), 500–526.

Chapter 30

Tracing Problem Evolution: Factors That Impact Design Problem Definition

*Shanna Daly, Seda McKilligan, Laura Murphy &
Anastasia Ostrowski*

ABSTRACT

Design problems evolve throughout many typical design processes. Little research has focused on the extent to which design problems evolve and the role that various factors play in this evolution. In this research, we drew from data gathered for DTRS11 that traced a design team's process as they progressed from the end of a large-scale user data-gathering phase to a deeper understanding of the problems to be addressed. Analysis revealed evidence of three factors that impacted the way the design problem was defined: the structure of the co-creation sessions; cultural perceptions and norms of the team and the users; and user data and its translation by the team. Understanding factors that guide the definition of the design problem can support designers in expanding their awareness in design decision making and problem solving because they can be more reflective and explicit about how and why their understanding of the design problem changes and more intentional about exploring the design problem space.

1 INTRODUCTION

The act of designing involves "searching in a hypothetical space of many possible ideas." This includes exploration of both the problem space and solution space, where the problem space is a hypothetical space representing differing ways to understand the problem, and the solution space contains all possible solutions to the determined problem. Designers must define the size and scope of these spaces, and this definition evolves throughout a design process (Cross, 2004; Cross & Roozenburg, 1992; Dorst & Cross, 2001; Hybs & Gero, 1992). A variety of factors may influence this evolution of the problem space, including data gathered, user feedback, testing, team preferences, and solution ideas (Goel & Pirolli, 1989, 1992; MacLean, Young, Bellotti, & Moran, 1991). The continuous and iterative impact of solutions on the definition of the problem, and vice versa, is known by the term problem-solution co-evolution (Dorst, 2011; Dorst & Cross, 2001; Maher, Poon, & Boulanger, 1996; Maher & Tang, 2003; Wiltschnig, Christensen, & Ball, 2013). While there is a body of work on the way problems and solutions shift in response to one another (Dorst & Cross, 2001), there is limited work on what and how other factors drive or limit changes in designers' understanding.

Thus, our research focused on understanding ways a design problem was defined throughout a design project based on knowledge gathered from lead users in co-creation sessions and design team conversations on those sessions. Specifically, we investigated how the design problem space was decomposed, framed, and structured throughout a team's process and what factors impacted the definition of the problem.

2 BACKGROUND

Defining a design problem includes understanding a design need, background to the need, and solution requirements and constraints (Dieter & Schmidt, 2013; Dym, Little, & Orwin 2013; Fogler, LeBlanc, & Rizzo, 2014; Ulrich & Eppinger, 1995; Yock *et al.*, 2015), which requires synthesis of multiple data sources, including prior knowledge, engineering principles, and contextual information (Howard, Culley, & Dekoninck, 2008; Restrepo & Christiaans, 2004; Simon, 1973). As information is processed and relevant elements are better understood, a redefined design problem is structured. This evolution of the "real" problem can be understood as the exploration of a problem space, which we define as a hypothetical space that represents all potential understandings of a design problem, grounded in Newell and Simon's theoretical description of design spaces (Newell & Simon, 1972; Simon, 1978). This exploration is described as iterations of defining, framing, structuring, and scoping the problem. It does not only dominate the initial steps of the design process, but also re-occurs throughout as designers make decisions about constraints and hone in on certain areas of the space (Brophy, 2001; Cross, 2004; Goel & Pirolli, 1992; King & Sivaloganathan, 1999). At various times throughout a design process, decisions are made and certain aspects of the design problem are considered well understood and "set." Sometimes these "set" aspects of the design problem hold, and other times, they are later iterated upon once again. No matter the phase in a design problem, the current understanding of the problem impacts the direction of the team in solving the problem as well as the possibilities considered for design solutions.

Both internal and external factors shape the ways designers view and navigate problem spaces, and multiple aspects impact design process direction as a whole (e.g., Lopez-Mesa & Thompson, 2006). These factors include: the use of human-centered design tools and approaches, including co-creation sessions and design ethnography techniques; the characteristics of each person on the design team, including their demographics, disciplinary perspectives, and prior experiences; and the environment, including the general atmosphere, company norms and priorities, and team dynamics (Bucciarelli, 1996; Kelley, 2001; Lopez-Mesa & Thompson, 2006; Mohedas, Daly, & Sienko, 2014; Mohedas, Sabet-Sarvestani, Daly, & Sienko, 2015; Rhodes, 1961; Salvador, Bell, & Anderson, 1999; Wiltschnig *et al.*, 2013). Additionally, potential solutions to the version of the design problem that is relevant at the time have also been shown to have an influence on the way a design problem is understood (Dorst & Cross, 2001; Maher *et al.*, 1996).

Co-creation is a familiar term in management and marketing research, where it is defined as the joint, collaborative, peer-like process of producing new value, both materially and symbolically (Galvagno & Dalli, 2014), in which the active participation of customers and end users is enabled through multiple interaction channels.

Similarly, design researchers use the term design participation to define a user's integration in the decision-making process (Cross, 1972, 1995), allowing users to build on and provide recommendations based on their own experiences (Bødker, 1996; Sanders & Stappers, 2008; Stappers & Visser, 2007). Prahalad and Ramaswamy (2000, 2004, 2013) adopted the term participatory design for the business community and introduced co-creation as an environment where consumers can have active dialogue and co-construct personalized experiences.

Past work on design cognition has been solution-focused, with little attention being given to facilitating the exploration of problems (Simon, 1995; Studer, Yilmaz, Daly, & Seifert, 2016) and with less therefore being known about the factors that influence the way a design problem evolves and how these factors function. For these reasons our research was guided by the following research questions:

- How does the design problem evolve for a design team as they progress through the design process?
- What factors impact the definition of the design problem?

3   RESEARCH METHODS

3.1   Participants and setting

The dataset included a professional design team's activities during their design process (Christensen & Abildgaard, 2017). The team's goal was to develop a concept package of accessories for premium car users in the Asian market as well as to exemplify how a regional relevant holistic approach would increase brand penetration in Asia.

The design team included eight designers: two from the accessories department, three from the user involvement department, and three external design consultants. A stated company goal of bringing designers from two different departments together was to explore and understand users' behaviors and values through the application of user research methods. The three designers in the user involvement department worked full-time on the project and had expertise in communication and multimedia design. The two designers in the accessories department specialized in car accessories and thus collaborated with the core design team to make decisions on the project's direction and its relation to other stakeholders and its overall implementation. Three external design consultants had expertise in Asian markets and assisted the design team in planning and facilitating the co-creation sessions, as well as translating the language and cultural diversities and traditions. The team members' names (as pseudonyms), their roles in the project, and their nationalities are provided in Table 30.1.

3.2   Data analysis

We identified a subset of the full data based on the design team's discussions on aspects of the problem: co-creation workshop sessions 1 (CC1) and 2 (CC2), insights from both workshops, recaps with consultants and stakeholders, and the design team's brainstorming session on potential concepts and products to pursue (v06, v07, v08, v09, v11, v13, v15, v16, v17, v18, v19, v20, v21), totaling 543 minutes of video recording.

Table 30.1 Project participants.

	Team member	Role	Nationality
Core Design Team	Ewan	Team Leader/ UX Researcher/Designer	Western
	Abby	UX Researcher/UX Design Specialist	Western
	Kenny	Tech Support/UX Researcher/UX Prototype Engineer	Western and Asian
	Nina	Intern	Western
	David	DTRS11 Observer	Western
External Design Consultants	Rose	Researcher/Cultural Translator and Moderator/Design Thinking Expert	Asian
	Amanda	Design Researcher/Consultant/ Design Thinking Expert	Asian
	Will	Market Researcher/Consultant/ Cultural Translator and Moderator	Asian

Across the dataset two researchers separately characterized the current version of the design problem (based on how it was represented by a member of the design team) throughout the team's work. The two researchers met regularly with each other and with the other two researchers on our team throughout this process to discuss what we noticed across the dataset and to define structures to help us reliably represent patterns we noticed in the data.

We defined aspects of the design problem definition to include design needs, relevant background to the needs, and solution requirements and constraints, in alignment with descriptions of design problems from design texts (Dieter & Schmidt, 2013; Dym, Little, & Orwin 2013; Fogler, LeBlanc, & Rizzo, 2014; Ulrich & Eppinger, 1995; Yock *et al.*, 2015). In our analysis, we focused on identifying "shifts" in the design problem. A shift in the design problem was considered to represent exploration of the problem space, and included defining and/or changing focus, priorities, emphases, scope, and boundaries.

As an example of a shift, during a team discussion on requirements for car accessories, the team enlarged the scope of the design problem (representing a shift in the design need) to include redefining and branding the company values:

> *"We kind of want to change [the company] values, and for this project it's kind of possible... I'm just thinking we need to find some way where we can actually ... [still have the kind of normal value] [the company] values, but also have these ... extra values ..."* [v9, 237].

We recognized that not all team members likely had the same conceptualization of the problem as the individual speaking, however, our analysis focused on the most recently presented version of the problem so then we could also track what factors may have prompted this new version. Based on this approach, three people expressing three

Table 30.2 Factors that impacted shifts in the problem space.

Factor in problem evolution	Definition of factor
Structure of co-creation sessions	How the two co-creation sessions were planned and executed by the design team
Cultural perceptions and norms	How the design team perceived and interpreted their own culture compared to other cultures
User data translation	How the design team incorporated data collected from users

different ideas about some aspect of the problem one after the other was recorded as three different shifts in the problem space. The majority of shifts in the problem space were related to solution requirements and constraints.

As we identified shifts in the problem, we also described reasons suggested by the data that the design problem was understood in its current form. Using an inductive coding approach (Creswell, 2013; Patton, 2002), we grouped the reasons into categories, which we called "factors impacting problem exploration." Factors were defined as reasons the design team viewed the problem in a particular way. Factors could prompt a shift in the understanding of an aspect of the design problem or limit the extent to which aspects of the problem space were further explored. Thus, a factor could prompt a revision to the design problem or set a boundary to how the problem was understood. We did not evaluate either impact of the factor as good or bad, as there are times in design when ideas should be explored as well as times when choices should be made and parameters are set (Brophy, 2001; Cross, 2001; Guilford, 1984).

Through multiple rounds of discussions and reviewing the data, we iterated on our list of factors and definitions of those factors. We ultimately identified three factors that explained many of the shifts in the team's understanding of the problem space.

4 FINDINGS

Three key factors emerged in driving the evolution of the problem space: (1) structure of the co-creation sessions; (2) cultural perceptions and norms; and (3) user data translation. These factors are defined in Table 30.2 and described in the following subsections with regards to their impacts on the evolution of the problem. We included several examples of how each factor caused shifts in the ways the team understood design needs, relevant background, and requirements and constraints in the problem space.

4.1 Structure of the co-creation sessions

The design team's approach in developing the co-creation sessions impacted how the design problem was perceived and shaped. This was most evident in the themes on which the design team focused in the sessions. The team developed the co-creation structure in two phases, the initial creation of both sessions, and additional development of CC2 after CC1. During the initial creation of the sessions, the team designated 7 pillars to guide the discussion: environment/sustainability; wellbeing; health;

social versus individual; comfort and convenience; safety and security; and evolving status. CC1 was structured around the participants confirming and/or rejecting the current pillars and/or creating alternative pillars:

> *"I'm thinking hopefully they will not because hopefully so maybe they're there to validate the themes that we have actually selected"* [v5, 131].

These pillars represented the team's understanding of the problem space at the time. The team discussed the impact that these pillars might have on the data they yielded. For example:

> *"I think we also need to be careful not to prime them too much because… if we prime them too much this is just … what they will come up with."* [v4,104]
> *"So if they see a picture of a happy family then they will say 'a family!' … is really important to me."* [v4, 106]
> *"We- we should not tell them [prompt them to think of other ideas], because if it's not really important if it's just like a random thought they have, then they'll forget it if it's not important. But if it is something that they really feel "why is this not here?", and they keep thinking the same thing, for sure they'll remember it when we reach 'this' point, and then they get the opportunity in the end to unload the-"* [v4, 263]
> *"[But I think] that Mia she referred to these as assumptions, and I think that's totally right now, in the- in this way. And they are assumptions, and that's fine, and we need to [verify and validate that assumptions are true or- or not:, in the end]."* [v4, 586]

This structural decision likely impacted how and to what extent the problem space evolved. When participants were asked to define "good life," they were presented with the seven pillars and the team planned to limit their discussion to these topics. After discussing these pillars, the team then asked the participants to identify anything else that was missing. In their earlier conversations, they said that often the answer to this type of question was "no":

> *"typically a question where people … even though they thought of something, they typically will say, "no, no, nothing." "But if they don't have anything, do we want to provoke anything?"* [v4, 458].

The team decided to focus the conversation on "good life" to the seven pillars. For example, the team did not want the conversation to surround money and luxury:

> *So that we have narrowed the "good life" down a little bit, when we start, so that they don't come in and say 'okay, what is good life to you?' 'It's money and being able to spend everything on luxury."* [v4, 234]

The team made the decision to start with the pillars knowing that this would shape their problem space, but decided it was the best decision to optimize the time they had with users and where they were in the design process.

"Yeah, and we want them to gravatize around something. And we want that something to- since we are limited on time we can't start from everything, we need to start from something." [v4, 240]

"And we- since we've done the research before, we think that these pillars will represent what they will gravatize around, but we don't know the specifics of it." [v4, 242]

CC2 was structured around the top-voted theme from the CC1, "Freedom/Enjoy life." This theme was the over-arching goal for the CC2 participants to keep in mind for the day's activities. The design team prompted CC2 participants to create a company centered around '(Evolving) Status Symbol' and 'Health/Well-being'. The goal was to create a 'fictional' company with its own product-line, purpose, mission, values, and culture. Participants were also tasked with developing a preliminary product that would be manufactured by their company. Four constraints were placed on the participants in the activity: safety/security; comfort and convenience; environment/sustainability; and 'sexy commitment'.

The comfort and convenience constraint was introduced by the team but never brought up by the participants. The team incorporated comfort and convenience as customer priorities important in understanding the problem. The only time participants discussed comfort and convenience was when it was introduced by the team. [v14, 206].

A: *"Yeah... I think the convenience here, to be focused even more, is more about- it's not just a pointless add-on, it has to be a seamlessly integrated... kind of convenience"*

E: *"Yeah, mmm"*

A: *"And Yen also said that he didn't want to choose between comfort and convenience"*

E: *"Okay, so he- no compromise"*

A: *"Yeah but Rose forced him"*

The design team structured the CC sessions specifically by grounding them in the seven pillars identified from Phase 1 work. Thus, Phase 1 guided the CC sessions, and the CC session structures facilitated certain priorities in the problem space to remain present. This statement is not intended to be an evaluative one, that is, it's neither "good" nor "bad" that the team made this decision, but rather, this choice had an impact of the exploration of the problem space throughout the rest of their process.

4.2 Cultural perceptions and norms

Cultural perceptions and norms also played a crucial role in shaping the problem. There were cultural differences between the design team (Western culture) and the lead-users (East Asian culture). East Asian cultures traditionally view the world as the sum of many parts and emphasize how parts fit into the whole enabling it to function, while Western cultures focus on objects (Gautam & Blessing, 2007; Kuhnen *et al.*, 2001; Nisbett, 2003). Western culture is rooted in individualism, while East Asian culture is rooted in collectivism. These Western cultural characteristics contributed to how

East Asian culture was understood throughout the design process and represented by the team in the final deliverable. Cultural perceptions and norms impacted the design team's understanding of multiple aspects in the problem space; here we highlight how it shaped the team's understanding of solution requirements such as the environment, role of status, and user freedom.

Environment was initially defined by the team as 'green' environmentalism, which involves protecting and improving the health of the natural environment. This was challenged by participants' holistic perspective of the environment as including additional elements in their understanding of environment beyond the natural environment, such as political and societal spheres, and also included the relationship of these components to one another. For example, one of the participants referred back to her experience of teaching recycling to her child and that the goal was not to do it for others to see it.:

> "... they also talked about environment, but in the big picture environment so like, political stability for example: and, ehm yeah, overall societal progression, yeah that was when "good life" went together with society, and all four of them surprisingly talked about that, like, once there is-, they really saw it as like the whole society moving up together and having a more harmonious ..." [v8, 294]

The team questioned why the users did not talk about environment in terms of solving environmental issues like the air pollution. While the team noticed the difference, they did not immediately shift their understanding of the problem. The team's Western cultural background conflicted with this alternative view, and the team maintained a priority on green environmentalism:

> "Food, water, and air... Maybe ASIAN CITY [is] not as bad, but generally the environment and what you consume. I think that consumption, environment is pre that." [v7, 89].

This demonstrated that the team believed that environment was before the actual consumption with the product and theorized that Asian cities are more focused on the environment at the point of consumption. As the team's work continued, the team leader indicated several times that the team's environmental understanding was a Western view:

> "A lot of Asian people are not one hundred percent embracing... 'I should recycle, or I should do the right thing' ... because not everyone is doing it" [v21, 12].

Collectivism was evident in the team's description of the environment feature of the problem space when discussing the scope of environment according to the users:

> "... we talk about environment and like this clean eco through- all the way, and then there's a bit about protection that you just mentioned, it's not just- they're doing it for its own sake but in return you know you'll always benefit, you and your future generations ..." [v19, 150].

While the team ultimately made this shift in their understanding of the environment, this also caused another shift in the problem space, as they incorporated their 'green' definition of environment into the competitive theme of the *status* feature. The team used the term 'sexy commitment' to describe the status potential through 'green' environmentalism, translated as people see what you are doing and your values, therefore, you gain status in society because these actions and values make you attractable to others:

"It's not just- they're doing it for its own sake but in return you know you'll always benefit, you and your future generations, and also they're sexy compliment, and the leadership, and the role-modeling as well." [v19, 21].

Green environmentalism as perceived in the Western culture was incorporated into the team's understanding of the status requirement in the problem space. Therefore, when speaking about status, the team continued to emphasize caring for the environment in the status definition, which is contradictory of how the Asian culture views the environment and how status is manifested in Asian culture. The problem space ultimately included both the team's Western view of environment and their understanding of the Asian environment definition.

The team's understanding of the design requirement of freedom also evolved due to cultural perspectives and norms. Early in the process, the team's Western individualist cultural background and their interpretation of the collectivist culture drove the freedom requirement to represent a release from responsibilities, including family:

"Taking a long drive at night, on the express way for an hour ... he's had a lot to do in the time, like him and the lady who's got a kid, I think they're very concerned about, you know "where's my freedom? I do not have autonomy anymore. So that all the things are- you know, what will be my control?" [v7, 28].

The team interpreted participant comments from the first co-creation session as the participants wanting to feel in control and make their own decisions, and not wanting to think about family:

"The other part was the family thing which Amanda mentioned... I think some subconscious they want some freedom or way from (..) some liberation from, but I ... think the way we written it, may have been kind of extreme, or may not be something that is (INAUDIBLE) correct to say – Yeah, it sounds more like a burden" [v14, 291–292].

However, this contradicts cultural norms in the collectivist society. The team seemed to adapt the Western value of freedom to the Asian culture, defining freedom in the problem space as participants wanting to break free of family constraints:

"... of course they don't see that you need to really break free from old bonds whether it of family or- ... Shackles! The shackles of family" [v14, 295–296].

While the participants emphasized family and diligence as values, the team pushed the problem space more to align with Western norms of freedom:

"We totally respect about family and it is about enduring ... but it is also about enjoying and living life to fullest" [v11, 243] and *"he likes to feel that he's in control ... so he say 'don't tell me what to do, don't try to sell me stuff. I will decide what I want' so, there's a lot more: of, his own decision making ... and not wanting to feel like he has been pressured into making some kind of commercial decision or purchase decision, or don't want other people to influence his choice, he likes to find time, you know at home, well either at home or out, you know, at night, because that's the only time he can find some escape from his child, from his wife, and everything else, so he spoke a little about, you know, going out on a cruise in a car for an hour, on highway."* [v8, 4]

To push against the team's emphasis on this Western value of *"enjoying and living life to the fullest,"* one of the external consultants stated the team's perception had the *"connotation of just whole enjoyment"* [v11, 259], which is not traditional of Asian culture. This prompted a shift in defining freedom as "pockets of enjoyment." This maintained priority in the problem space on freedom, but in a way that accommodated Asian values:

"Becomes a manifestation of great autonomy. The freedom-" maybe it should say *'that freedom of having pockets of enjoyment reflects a sense of achieve'-achievement towards a good life"* [11,21].

When the team discussed hard work and the Asian people's endurance, the idea of freedom as "pockets of enjoyment" was identified as an accurate assumption. Diligence is emphasized in the Asian culture, where diligence refers to the amount of effort people put toward reaching their goals, which requires self-respect and knowledge and emphasizes an individual's resilience toward challenges and difficult contexts (Zhang, 2011). Different views of the value of work and family affect how the team perceived the information and how it was incorporated into the problem space.

The way the team discussed the definition of freedom sometimes shifted back towards this individual freedom. At various points when these shifts were happening in the problem space, one of the external consultants directed emphasis back on Asian values of life, for example:

"I think some subconscious they want some freedom or some way of liberation but I don't think the way we have written it, may have been kind of extreme ... not correct way to say" [v14, 291].

The emphasis of freedom returns to the ideas of *"pockets of enjoyment"* after his interjection. [v14, 274]. Throughout the conversations, he helped shape this definition of freedom to be more in line with East Asian culture. The team felt solid in this final definition of freedom:

"It's good to... emphasize these pockets there, because what we kind of focused on in ASIAN CITY FROM PHASE 1 was the "okay, endure now", and that's from until you're fifty, and there is not these small pockets at all, it's just endurance. Hard work, and then: it's freedom and happy life." [v11, 142].

Throughout the process, the interpretation of the Asian people and their freedom expression changed as the team better understood the collectivist nature. Although there were references to individualistic freedom at times, the team supported "pockets of enjoyment" as a final definition of freedom.

Culture and perception of norms had a significant effect on how the problem space was articulated, investigated and redefined. These perceptions especially impacted the team's understanding of three solution requirements in problem space: environment; status; and freedom. The design team's Western background focused more on the individualistic values, whereas participants' Asian background facilitated their interpretation of these three requirements from a collectivist perspective.

4.3   User data and its translation by the team

The way the team translated user data also impacted the problem space. For example, the team seemed to struggle to translate what participants said about *status*, as the data from the participants' contrast the way the team shaped the problem space in response. On the other hand, *trust* is an example where the translation of the user data aligned well with what the participants said in the co-creation sessions, which prompted a major shift in the team's understanding of the problem.

The emphasis of *status* in the problem space by the team was not reflected in the data from participants' discussions in the CC sessions. Status was frequently discussed and emphasized; however, the user data did not reflect that status was a priority for participants. For them, status was not a motivating factor for their actions, for example, one of them discussed his volunteer experiences in a poor area of Tibet and described his motivation for doing this as due to his belief in the holistic experience, not to show off or brag about it:

> "*I think they were thinking a little more broader like recycling or eco-friendliness for the sake of in itself and really like, truly believing in the whole experience thing and not showing off. And their aspirations were also going, not just into travel and to going to difficult places like Brian, okay he drove five thousand kilometers from ASIAN CITY to Tibet, but, like they were thinking about going to do volunteering: in like very poor places: and, I mean those were the type of things that they were looking at, which I think goes beyond really like to just- ...*" [v7, 271].

In translating this to define status in the problem space, the team created a narrative for why customers do things, proposing that customers take action in order to gain social status for themselves. This narrative created by the design team contradicts the initial testimony of the CC participants.

> "*I'm sacrificing something over here, to gain something over here... I'm using my free time, which I worked too hard to get, to help out in a school over there, so that people will see me, hopefully see me, understand what I'm doing and place me in society here, which I wanna be.*' [v11, 367].

In the problem space, these insights on status turn into a product requirement where the final product or service must be "*brag-able*" [v11, 385]. Later in the design process, this interpretation of status as a user motivation manifests through the idea of gamification as a key aspect to include. For example, the team suggested green environment as one thing that could be braggable to achieve status:

> *"You could compete if you … have a product like who are environmentally friendly drivers for instance and then you can share it and show it"* [v16, 191].

While status seems to be a more prominent driver as the team's work continues, the data from the co-creation session do not emphasize status or competition as motivating factors which suggests another transformation of the problem space.

Another design requirement emphasized was trust. Trust manifested itself as the conflict between the perceptions of customers of local versus foreign companies. Participants described this as believing that foreign companies produce high quality products, but at the same time them desiring a company to be producing locally:

> *"I need to have your company where your customers are"* [v15, 172].
> *"And also- then the other one started to say that it was also important that it was actually a local brand, a place in ASIAN CITY if the products were made for ASIAN CITY"* [v14, 135].

The team's response was that based on prior experience with low quality local goods an Asian customer may trust foreign quality better, but that he/she may not connect with or understand the foreign values. This leaves a market opportunity for the team to translate those foreign values to values that local customers can understand, potentially through a popular figure mediating this translation:

> *"… maybe a foreign brand has some values, but it's difficult to understand that value, because they're foreign values"* [v14, 156].
> *"How can we find the local counterpart of those values"* [v14, 156].

This understanding led the team to redefine aspects of the problem space and discussing the need to build customer trust through, for example, local manufacturing while also producing foreign quality. The team's translation of the user data was that in order to be successful in the Asian market, the customers have to both trust the quality of the product as well as trust in the values of the company. The team then explored redefining what "manufacturing locally" actually means to the customer in order to take advantage of the existing company structure:

> *"The other brands are not doing as well when they produce locally, but if you can differentiate by what type of manufacturing locally really means, then there' … trust and assurance of the final product, that it's not gonna lower the quality"* [v17, 154].

Ultimately, when recording the translation from participant data to the design decisions made by the team, we uncovered instances of alignment with user data that were also in our dataset as well as instances where the user data were not present or did not align with the evolved understanding of the solution requirements in the problem space. The user data impacted the process as did the team's translation of these data into their design problem descriptions.

5   DISCUSSION

Our analysis demonstrated how the problem space evolved through the co-creation sessions and the team's interpretations of the conversations. The team's understanding of the design problem and solution requirements shifted throughout discussions in their process, and as they got input from lead users. For example, the team's understanding of needing to address environment in their goals to get into the Asian market had large shifts from a green/sustainability definition to one that focused on a more holistic notion of environment. At the same time, this original definition of environment was incorporated into another criterion, the need to address the status of the potential user. The evolution of the problem space described in our work is consistent with Adams *et al.*'s (2017) discussion of co-inquiry as a framework for mapping changes in the team's understanding of the design problem.

We were also able to document evolutions in the team's understanding of the seven pillars they used to articulate the problem space. Within this documentation, we sometimes saw pathways that were linear and building on prior understandings, and other times, saw tensions between multiple views of the problem. This documentation highlights the iterative and complex nature of a design problem and how a team comes to understand the real needs that need to be addressed. Past research supports this finding as design experts simultaneously and iteratively explore a problem while searching for solutions (Dorst & Cross, 2001), and the problems are constantly envisaged, posed, formulated and created through a great deal of effort spent in restructuring to reach solutions (Getzels & Csikszentmihalyi, 1976; Restrepo & Christiaans, 2004).

Our analysis also revealed three specific factors that had a demonstrative impact on the evolution of the problem space: the structure of the co-creation sessions; cultural perceptions and norms; and user data translation. The team's decision to ground the co-creation sessions in the seven identified pillars developed from Phase 1 impacted the breadth of the discussion that likely ensued during the sessions. This is known in the psychology literature as "priming" (Bargh, Chen, & Burrows, 1996; Levin, Schneider, & Gaeth, 1998; Tversky & Kahneman, 1981) Priming can both preclude and promote certain discussion (Klein, 1993, 1997; Ramser, 1993).

Cultural perceptions and norms of the design team and the co-creation participants played a critical role in shaping the problem space. The different perspectives of the two groups meant that the design team had to understand their own norms, of which they might not readily be aware, and develop a deep enough understanding of another culture's realm to be able to interpret what co-creation participants were saying. Other literature has documented that knowledge of the culture influences solution characteristics (Felgen *et al.*, 2004); in this work, we additionally highlighted impacts it can have on problem characteristics. Designers are often given the role of a cultural interpreter (Kimbell, 2011), collecting insights from different cultures and translating this information. Understanding these cultural variations has been shown to have significant impact on creating relevant products and solving the right problems (Gautam & Blessing, 2007). In this study, differences in Asian and Western cultures were challenging for the team to understand, and they spent much of their discussions navigating how their own culture influenced their understandings of the problem. This navigation played a role in how they approached their design process, consistent with how Clemmensen *et al.* (2017) describe design thinking as a "culturally situated practice." Other work has demonstrated dramatic differences in the nature of Asian and European thought processes (Nisbett, 2003). According to Nisbett's findings, Asians are more holistic in the way they perceive the world whereas Westerners approach it from an analytic perspective. Asians focus on the context and situation and the relationships among objects and people. On the other hand, Westerners try to categorize the objects with labels they create in their perceptions, and treat the world as static and governed by rules. Asians' are more interdependent, considering self as a part of a larger whole, while Westerns thinking more independently, seeing self a unitary free agent. These values and tensions were evident throughout the evolution of the features of the problem space.

User data gathered also informed modifications to the problem space. In our comparisons, at times these data translated directly from what participants said, and at other times, these data and their translation seemed to be in conflict. Literature has documented struggles in this translation of user data to solution requirements (Mohedas, Daly, & Sienko, 2014; Mohedas, Sabet-Sarvestani, Daly, & Sienko, 2015). However, in the translation of user data to design decisions, designers also incorporate their own lenses and expertise (Cross, 2004; Daly, 2008), thus everything a user says is not automatically incorporated or incorporated at all into a design decision. The designer has a broader perspective on the problem, the market, and other aspects that will impact the success of the design. In cases where the data were not directly translated, we do not know why this was so from the data we had. For example, the definition of environment as being focused on sustainability was not lost in the problem statement, but incorporated into status. Perhaps the team had insights on the way the auto industry is moving and regulations of the industry to know this had to be maintained as part of the problem even if users were not emphasizing it.

The factors of Cultural Perceptions and Norms and User Data Translation are likely related. However, as discussed by Gray and Boling (2017), in many cases cross-cultural design practices and user-centred design practices have not yet converged to a common approach. Translating user data is by nature, cross-cultural, so long as the designer is not also the user (Cooper, 2004). While cultural perceptions and norms were explored as part of an explanation for why there were inconsistencies in what users said in the sessions and the ways it was translated to design decisions by the team, there were likely other reasons for why there was not a one-to-one match in what users said and how the team defined the problem, including the expertise of the team in knowing what would make the project successful, company priorities, and the ways the team leveraged user-centred design tools.

Several implications for research and pedagogy emerged from our work. One research implication is to explore additional factors that might impact problem space exploration in a diverse range of problem contexts. A related pedagogical implication is that design instructors can be more explicit on how and when to explore a problem space, perhaps even using examples from studies such as these representing how what might seem like a simple choice or an easy interpretation has implications for the way the design problem is understood. In many cases, designers take the design problem as it is rather than fully iterating on it (Cross, 2001). Specifically, intentional instructions on the need to explore the problem and how various factors may be influencing an understanding of the problem can support more reflective practice.

As discussed by Valgeirsdottir and Onarheim (2017), process awareness can support better understandings of the problem space.

6   CONCLUSIONS

From our analysis of the evolution of a design problem throughout a team's work, we revealed three factors that prompted shifts in the design team's understanding of the real needs and solution requirements. The problem space evolved due to the nature of the co-creation sessions, including their setup and structure, designers' perceptions of other cultures, and their interpretation of the data gathered from the users. These results highlight areas in which a design team can focus, allowing them to be more aware of the ways their problem understanding is being impacted, and strategize different choices or interpretations they might consider as they define their current understanding of the design problem. While much emphasis has been given to exploring solutions in design to achieve quality design outcomes, exploring multiple perspectives from which to view the design problem can also significantly support design success. Recognizing ways these perspectives are shaped is an important step to leveraging them intentionally in a design process.

REFERENCES

Brown, T. (2009). *Change by Design: How Design Thinking Transforms Organizations and Inspires Innovation.* London: HarperCollins Publishers.

Adams, R. S., Aleong, R., Goldstein, M. & Solis, F. (2017). Problem Structuring as Co-Inquiry. In: Christensen, B. T., Ball, L. J. & Halskov, K. (eds.) *Analysing Design Thinking: Studies of Cross-Cultural Co-Creation.* Leiden: CRC Press/Taylor & Francis.

Bargh, J. A., Chen, M., & Burrows, L. (1996). Automaticity of social behavior: Direct effects of trait construct and stereotype activation on action. *Journal of personality and social psychology, 71*(2), 230.

Bødker, S. (1996). Creating conditions for participation: Conflicts and resources in systems development. *Human-computer interaction, 11*(3), 215–236.

Brophy, D. R. (2001). Comparing the attributes, activities, and performance of divergent, convergent, and combination thinkers. *Creativity Research Journal, 13*(3&4), 439–455.

Bucciarelli, L. L. (1996). *Designing Engineers.* Cambridge: MIT Press.

Christensen, B. T. & Abildgaard, S. J. J. (2017). Inside the DTRS11 Dataset: Background, Content, and Methodological Choices. In: Christensen, B. T., Ball, L. J. & Halskov, K. (eds.) *Analysing Design Thinking: Studies of Cross-Cultural Co-Creation.* Leiden: CRC Press/Taylor & Francis.

Clemmesen, T., Ranjan, A. & Bødker, M. (2017). How Cultural Knowledge Shapes Design Thinking. In: Christensen, B. T., Ball, L. J. & Halskov, K. (eds.) *Analysing Design Thinking: Studies of Cross-Cultural Co-Creation.* Leiden: CRC Press/Taylor & Francis.

Cooper, A. (2004). *The inmates are running the asylum: Why high-tech products drive us crazy and how to restore the sanity.* Indianapolis. I Sams.

Creswell, J. W. (2013). *Research Design: Qualitative, quantitative, and mixed methods approaches* (4th ed.). Thousand Oaks, CA: Sage Publications.

Cross, N. (1972). *Design participation.* Paper presented at the International Conference of Design Research Society, London, UK.

Cross, N. (1995). Discovering design ability. In: R. Buchanan & V. Margolin (eds.) *Discovering design: Explorations in design studies*). Chicaho, IL: University of Chicago Press. pp. 105–120.

Cross, N. (2001). Design cognition: Results from protocol and other empirical studies of design activity. In: Eastman, C., Newstatter, W. and McCracken, M. (eds.) *Design knowing and learning: cognition in design education*. Oxford, UK: Elsevier, pp. 79–103.

Cross, N. (2004). Expertise in design: an overview. *Design Studies, 25*(5), 427–441.

Cross, N., & Roozenburg, N. (1992). Modelling the design process in engineering and in architecture. *Journal of Engineering Design, 3*(4), 325–337.

Daly, S. R. (2008). *Design across disciplines*. (PhD Dissertation), Purdue University, West Lafayette, IN.

Dieter, G. E., & Schmidt, L. C. (2013). *Engineering design*(Vol. 3). New York: McGraw-Hill.

Dorst, K. (2011). The core of 'design thinking' and its application. *Design Studies, 32*(6), 521–532.

Dorst, K. H., & Cross, N. (2001). Creativity in the design process: co-evolution of problem-solution. *Design Studies, 22*(5), 425–437.

Dym, C.L., Little, P., & Orwin, E. (2013). *Engineering Design: A Project-based Introduction*. Hoboken, NJ: John Wiley & Sons.

Felgen, L., Grieb, J., Lindemann, U., Pulm, U., Chakrabarti, A., & Vijaykumar, G. (2004). *The impact of cultural aspects on the design process*. Paper presented at the International Design Conference, Dubrovnik, Croatia.

Fogler, H.S., LeBlanc, S., & Rizzo, B. (2014). *Strategies for Creative Problem Solving*. Upper Saddle River, NJ: Prentice Hall.

Galvagno, M., & Dalli, D. (2014). Theory of value co-creation: A systematic literature review. *Managing Service Quality: An International Journal, 24*(6), 643–683. doi:10.1108/MSQ-09-2013-0187

Gautam, V., & Blessing, L. (2007). *Cultural influences on the design process*. Paper presented at the International Conference on Engineering Design, ICED'07, Paris, France.

Getzels, J. W., & Csikszentmihalyi, M. (1976). *The creative vision: A longitudinal study of problem finding in art*. New York, NY: Wiley.

Goel, V., & Pirolli, P. (1989). Motivating the notion of generic design within information processing theory: The design problem space. In: *AI Magazine, Spring*, 18–36.

Goel, V., & Pirolli, P. (1992). The structure of design problem spaces. *Cognitive Science, 16*(3), 395–429.

Gray, C. M. & Boling, E. (2017). Designers' Articulation and Activation of Instrumental Design Judgments in Cross-Cultural User Research. In: Christensen, B. T., Ball, L. J. & Halskov, K. (eds.) *Analysing Design Thinking: Studies of Cross-Cultural Co-Creation*. Leiden: CRC Press/Taylor & Francis.

Guilford, J. P. (1984). Varieties of divergent production. *Journal of Creative Behavior, 18*(1), 1–10.

Howard, T. J., Culley, S. J., & Dekoninck, E. (2008). Describing the creative design process by the integration of engineering design and cognitive psychology literature. *Design Studies, 29*(2), 160–180.

Hybs, I., & Gero, J. S. (1992). An evolutionary process model of design. *Design Studies, 13*(3), 273–290.

Kelley, T. (2001). *The art of innovation: Lessons in creativity from IDEO, America's leading design firm*. New York, New York: Doubleday.

Kimbell, L. (2011). Rethinking design thinking: Part I. *Design and Culture, 3*(3), 285–306.

King, A. M., & Sivaloganathan, S. (1999). Development of a methodology for concept selection in flexible design strategies. *Journal of Engineering Design, 10*(4), 329–349.

Klein, G. (1993). A recognition primed decision (RPD) model of rapid decision making. In: J. O. G. Klein, R. Calderwood, & C. E. Zsambok (eds.) *Decision making in action: Models and methods*. Cambridge, MA: MIT Press. pp. 205–218.

Klein, G. (1997). The recognition-primed decision (RPD) model: Looking back, looking forward. *Naturalistic decision making*, 285–292.

Kuhnen, U., Hannover, B., Roeder, U., Shah, A. A., Schubert, B., Upmeyer, A., & Zakaria, S. (2001). Cross-cultural variations in identifying embedded figures: Comparisons from the United States, Germany, Russia, and Malaysia. *Journal of Cross-Cultural Psychology, 32*, 366–372.

Levin, I. P., Schneider, S. L., & Gaeth, G. J. (1998). All frames are not created equal: A typology and critical analysis of framing effects. *Organizational behavior and human decision processes, 76*(2), 149–188.

Lopez-Mesa, B., & Thompson, G. (2006). On the significance of cognitive style and the selection of appropriate design methods. *Journal of Engineering Design, 17*(4), 371–386.

MacLean, A., Young, R. M., Bellotti, V. M. E., & Moran, T. P. (1991). Questions, options, and criteria: Elements of design space analysis. *Human-computer interaction, 6*(3–4), 201–250.

Maher, M. L., Poon, J., & Boulanger, S. (1996). Formalising design exploration as co-evolution: a combined gene approach. In: J. S. Gero & F. Sudweeks (eds.) *Advances in formal design methods for CAD*. London, UK: Chapman and Hall.

Maher, M. L., & Tang, H. (2003). Co-evolution as a computational and cognitive model of design. *Research in Engineering Design, 14*, 47–63.

Mohedas, I., Daly, S. R., & Sienko, K. (2014). Design ethnogpraphy in capstone design: Investigating student use and perceptions. *International Journal of Engineering Education, 30*(4), 888–900.

Mohedas, I., Sabet-Sarvestani, A., Daly, S. R., & Sienko, K. (2015). *Applying design ethnography to product evaluation: A case example of a medical device in a low-resource setting*. Paper presented at the International Conference on Engineering Design, Rome, Italy.

Newell, A., & Simon, H. A. (1972). *Human problem solving*. Englewood, NJ: Prentice-Hall.

Nisbett, R. E. (2003). *The geography of thought: How Asians and Westerners think differently ... and why*. New York, NY: The Free Press.

Patton, M. Q. (2002). *Qualitative evaluation and research methods (3rd ed.)*. Thousand Oaks, CA: Sage Publications, Inc.

Prahalad, C. K., & Ramaswamy, V. (2000). Co-opting customer competence. In: *Harvard Business Review, 78*(1), 79–90.

Prahalad, C. K., & Ramaswamy, V. (2004). Co-creation experiences: The next practice in value creation. *Journal of Interactive Marketing, 18*(3), 5–14.

Prahalad, C. K., & Ramaswamy, V. (2013). *The future of competition: Co-creating unique value with customers*. Boston, MA: Harvard Business Press.

Ramser, P. (1993). *Review of decision making in action: Models and methods*: American Psychological Association.

Restrepo, J., & Christiaans, H. H. C. M. (2004). Problem structuring and information access in design. *Journal of Design Research, 4*(2), 1551–1569.

Rhodes, M. (1961). An analysis of creativity. *Phi Delta Kappa, 42*, 305–310.

Salvador, T., Bell, G., & Anderson, K. (1999). Design ethnography. *Design Management Journal, 10*(4), 35–41.

Sanders, E. B.-N. (2006). Design serving people. *Cumulus working papers Copenhagen, 15*(05), 28–33.

Sanders, E. B.-N., & Stappers, P. J. (2008). Co-creation and the new landscapes of design. *Co-design, 4*(1), 5–18.

Simon, H. A. (1973). The structure of ill structured problems. *Artificial Intelligence, 4*, 181–201.

Simon, H. A. (1978). Information-processing theory of human problem solving. In: W. K. Estes (ed.) *Handbook of Learning and Cognitive Processes*. Hillsdale, New Jersey: Lawrence Erlbaum Associates.

Simon, H. A. (1995). Problem forming, problem finding and problem solving in design. In: A. Collen & W. W. Gasparski (eds.) *Design and systems: General applications of methodology*.

Stappers, P. J., & Visser, F. S. (2007). *Bringing participatory techniques to industrial design engineers*. Paper presented at the International Conference on Engineering and Product Design Education, Newcastle, UK.

Studer, J. A., Yilmaz, S., Daly, S. R., & Seifert, C. M. (2016). *Cognitive heuristics in defining engineering design problems*. Paper presented at the ASME 2016 International Design Engineering Technical Conferences (IDETC); 13th International Conference on Design Education (DEC), Charlotte, NC.

Tversky, A., & Kahneman, D. (1981). The framing of decisions and the psychology of choice. *Science*, 211, 453–458.

Ulrich, K. & Eppinger, S. (1995). *Product Design and Development*. New York, NY: McGraw-Hill.

Valgeirsdottir, D. & Onarheim, B. (2017). Metacognition in Creativity: Process Awareness Used to Facilitate the Creative Process. In: Christensen, B. T., Ball, L. J. & Halskov, K. (eds.) *Analysing Design Thinking: Studies of Cross-Cultural Co-Creation*. Leiden: CRC Press/ Taylor & Francis.

Wiltschnig, S., Christensen, B. T., & Ball, L. J. (2013). Collaborative problem-solution co-evolution in creative design. *Design Studies, 34*(5), 515–542.

Yock, P.G., Zenios, S., Makower, J., Brinton, T.J., Kumar, U.N., Kurihara, C.Q., Denend, L., Krummel, T.M. & Watkins, F.J. (2015). *Biodesign*. Cambridge University Press.

Zhang, L. F. (2011). Hardiness and the big five personality traits among Chinese university students. *Learning and Individual Differences, 21*(1), 109–113.

List of Contributors

Sille Julie J. Abildgaard studied Sociology and Psychology of Language at The University of Copenhagen, where her research focused on the use of multimodal communicative resources in collaborative work. Currently she works as a Research Assistant at Copenhagen Business School, Denmark on the CIBIS project, exploring creativity in blended interaction spaces.

Robin S. Adams is an Associate Professor in Engineering Education at Purdue University and holds a PhD in Education with degrees in Materials Science and Mechanical Engineering, and was a Senior Design Engineer in semiconductor packaging. She researches cross-disciplinary ways of thinking, acting and being; design learning; and education transformation.

Ömer Akin has been a faculty member at Carnegie Mellon University since 1977. He earned his Ph.D. from the graduate programs of the Department of Architecture in 1979 and has been conducting research on the design process and architectural ethics since then.

Richard J. Aleong is a Ph.D. student in the School of Engineering Education at Purdue University. His research interest focuses on integration as a creative process in the design of learning systems and educational development. He earned his M.A.Sc. and B.Sc.E. in Mechanical Engineering from Queen's University, Kingston, Canada.

Jan Henk Annema is lecturer in the field of communication and multimedia design at the HAN University of Applied Sciences, Arnhem, The Netherlands. He has a background in cognitive psychology and interaction design. He has worked as a researcher in human-computer interaction and now works mainly as a lecturer.

Dr. Jennifer Weil Arns is an Associate Professor at the School of Library and Information Science at the University of South Carolina. Her research focuses on cultural heritage informatics, public management, and the construction of knowledge in public and private settings.

Cynthia J. Atman is a Professor in the Department of Human Centered Design & Engineering at the University of Washington and Director of the Center for Engineering Learning & Teaching (CELT). Her research focuses on engineering design learning, considering context in engineering design, and the use of reflection to support learning.

Olaitan Awomolo is a PhD Candidate in the School of Architecture and Department of Civil and Environmental Engineering at Carnegie Mellon University.

Rene Bakker is Professor Networked Applications at the HAN University of Applied Sciences, Arnhem, The Netherlands. His research is directed at innovative internet applications in culture, energy, health and well-being.

Linden J. Ball is Professor of Cognitive Psychology and Dean of the School of Psychology at the University of Central Lancashire, Preston, UK. His research examines the role of metacognition in thinking, reasoning and problem solving using laboratory-based tasks as well as in real-world domains such as creative design.

Denise Bedford is Adjunct Professor, Georgetown University Communication Culture and Technology; adjunct faculty at Schulich School of Business, York University, Visiting Scholar University of Coventry, a Distinguished Practitioner/Virtual Fellow with U.S. Department of State. Senior Information Officer, World Bank (retired) and Goodyear Professor of Knowledge Management, Kent State University (retired).

Michael Mose Biskjaer is an assistant professor at Aarhus University. Building on interaction design, philosophy, aesthetics, and creativity research, his work applies an interdisciplinary humanistic perspective on central theoretical themes in individual and collaborative creative processes in art, design, and innovation. He currently studies how analog and digital constraints may improve creative performance.

Elizabeth Boling is Professor of Instructional Systems Technology at Indiana University, Bloomington, Indiana, USA. Her research focuses on design knowledge as it used by practicing designers and developed in design students.

Jacob Buur is Professor of User-Centred Design and Director of SDU Design Research at University of Southern Denmark. His main research contribution is an understanding of how video and physical objects can scaffold co-design, in particular in manufacturing industries.

Mads Bødker is an associate professor at the Department of IT Management, Copenhagen Business School. He holds a master degree in Film/Media Studies, and a Ph.D. in Human-Computer Interaction, IT University Copenhagen. His work combines human-computer interaction design, user experience and design research with philosophy, social theory and anthropology.

Jun Cai is Professor, Academy of Arts & Design, Tsinghua University, Director at Design Management Research Lab. His research area includes Design strategy and innovation based on user lifestyle development, knowledge transformation and structure change of design innovation and design management.

Monica E. Cardella is the Director of the INSPIRE Research Institute for Pre-College Engineering and an Associate Professor of Engineering Education at Purdue University. She is interested in how practicing professionals, undergraduate students and young children engage in design – and what the similarities and differences are amongst these groups.

Philip Cash's core research focus is on better understanding and describing design activity by integrating perspectives from Design, Cognition, and Management theory. Philip also has a keen interest in research methods and scientific development in design research.

Joel Chan is a Postdoctoral Research Fellow in the Human-Computer Interaction Institute at Carnegie Mellon University. His research explores new forms of innovative design at the intersection of computing and collective creativity, with a special focus on understanding and supporting the socio-cognitive basis of effective collective creativity systems.

Chih-Chun Chen is a Research Associate in the Design Practice Group in the Cambridge Engineering Design Centre. With a background in Experimental Psychology, Philosophy and Computer Science, her current research focuses on effectively communicating Complexity concepts to designs, engineers and technologists.

Xu Chen is Ph.D. Candidate in Academy of Arts and Design and Design Management Research Lab of Tsinghua University, China PR. He also is associate professor in School of Innovation Design of Guangzhou Academy of Fine Arts. His research area includes design history and design management.

Bo T. Christensen is Professor w/ special responsibilities in Creative Cognition at Copenhagen Business School. A cognitive psychologist by training, his works include both ethnographic studies of design practice and experimental studies of design cognition. His theoretical focus is on creative cognitive processes such as analogy, simulation and incubation.

Torkil Clemmensen is a Professor MSO at Department of IT Management, Copenhagen Business School, Denmark. His interest is in Human-Computer Interaction, in particular psychology as a science of design. The focus of his research is on cultural psychological perspectives on usability and user experience.

Nathan Crilly leads the Design Practice research group at the Cambridge Engineering Design Centre. His research interests lie in design, creativity and communication. He and his research group employ an interdisciplinary approach to studying the development of products, the properties they exhibit and how people respond to them.

Nigel Cross is Emeritus Professor of Design Studies at The Open University, UK. He is a leading international figure in design education and research. His current research is in creative cognition in design, based on studies of expert and exceptional designers. This work has been published in many journal articles, and in his books *Analysing Design Activity, Designerly Ways of Knowing* and *Design Thinking: Understanding How Designers Think and Work*.

Shanna R. Daly is Assistant Professor of Mechanical Engineering and Co-Director of the Center for Socially Engaged Design at the University of Michigan, Ann Arbor. Her research focuses on design innovations through divergent and convergent thinking as well as through deep needs and community assessments using design ethnography.

Mohammad Reza Dastmalchi is a Ph.D. student at department of Architectural Studies, University of Missouri. His research is focused on collaborative design teams, design cognition, and digital media. He has practiced interior design and taught at different art and design programs in United States.

Tejas Dhadphale is an Assistant Professor in Industrial Design at Iowa State University. He is currently developing a methodological framework for designers and researchers to integrate cultural aspects into the design process.

Franziska Dobrigkeit is a research associate at the Hasso-Plattner-Institute in Potsdam, Germany and member of the HPI-Stanford Design Thinking Research Program. Means to seamlessly integrate design thinking activities into agile software development is the focus of her research.

Andy Dong is Professor of Engineering Innovation at the University of Sydney, Sydney, Australia. His research aims to discover knowledge structures underlying design processes and objects of design. His research theorizes that the structure of design knowledge has real effects, such as the productivity of design or the progress potential of a new product.

Graham Dove is an interaction design researcher in the Center for Participatory IT at Aarhus University, Denmark. He holds a PhD in Human-Computer Interaction Design from City University London, where he was a member of the Centre for Creativity in Professional Practice. His research focuses on understanding design creativity, and developing designerly approaches to working with personal data and intelligent systems.

Newton D'souza is an Associate Professor of Interior Architecture at the College of Communication, Architecture & Arts at Florida International University, Miami. He has worked as a practicing architect and his research and teaching have focused on design cognition, creativity, human factors for design, environment-behaviour and emerging media environments.

David Dunne is Professor and Director, MBA Programs at the Peter B. Gustavson School of Business, University of Victoria. His research focuses on the application of design-led innovation in organizations in the private, public and non-profit sectors.

Claudia Eckert is professor of design at the Open University. Her research focuses on understanding and supporting the design processes of the complex products, and investigating underlying theoretical concepts.

Benedikt Ewald is a research associate at the Hasso-Plattner-Institutein Potsdam, Germany and a PhD candidate with the Department of Industrial/Organizational and Social Psychology of the Technical University of Braunschweig. He researches design team interactions within the HPI-Stanford Design Thinking Research Program with a special focus on the design thinking process as a mediator for innovation team interactions.

Nicholas D. Fila is a PhD Candidate in the School of Engineering Education at Purdue University. He earned a M.S. in Electrical and Computer Engineering and a B.S. in Electrical Engineering at the University of Illinois at Urbana-Champaign.

His current research focuses on the role and manifestation of innovation and empathy in engineering design, particularly among engineering students.

Molly H. Goldstein is a PhD student in the School of Engineering Education at Purdue University with a research focus on characterizing trade-off behaviors in novice designers. She earned her B.S. in General Engineering (Systems Engineering and Design) and M.S. in Systems and Entrepreneurial Engineering from the University of Illinois.

Milene Gonçalves is an assistant professor at the Delft University of Technology. A designer by training with a PhD on Design creativity from TU Delft, she identifies herself as a Design Researcher, with an emphasis on creativity, inspiration and cognition.

Judith van de Goor is lecturer Industrial Product Design at the HAN University of Applied Sciences, Arnhem, The Netherlands. Her main interest are social design, design for sustainability, design research and design methodology.

Bastien Grasnick is a master student in IT Systems Engineering at the Hasso Plattner Institute in Potsdam, Germany and research assistant in the HPI-Stanford Design Thinking Research Program. He is focusing on natural language processing and the analysis of design team interaction.

Colin M. Gray is an Assistant Professor in the Department of Computer Graphics Technology at Purdue University, West Lafayette, Indiana, USA. His research focuses on the role of student experience in informing a critical design pedagogy, and the ways in which the pedagogy and underlying studio environment inform the development of design ability.

Kim Halskov is Professor in Interaction Design at The Department of Digital Design and Information Studies at Aarhus University, Denmark. His research area includes innovation processes, design processes and methods. Current projects include the CIBIS project, which explores creativity in co-design.

Nicolai Brodersen Hansen is a Postdoc on the Creativity in Blended Interaction Spaces-project at Aarhus University. He holds a PhD in Interaction Design from Aarhus University, his doctoral work investigating the role of materials in Participatory Design processes. His research focuses on understanding how tools and materials shape design thinking as well as support multidisciplinary collaborative design processes.

Justin L. Hess is the Assistant Director of the STEM Education Innovation and Research Institute at Indiana University Purdue University Indianapolis. His research interests include exploring the functional role of empathy within engineering and design; designing STEM ethics curricula; and evaluating students' learning in the spaces of design, ethics, and sustainability.

Javaneh Jabbariarfaei graduated from Carnegie Mellon University with a Masters in Architecture Engineering Construction Management.

Marjolein Jacobs is lecturer and researcher in the field of communication and multimedia design at the HAN University of Applied Sciences, Arnhem, The

Netherlands. Her main areas of research interest are the visual arts and the interplay between art & technology, design research and research methodology.

Matilde Bisballe Jensen is PhD Student in Eng. Design and Innovation at the Department of Mechanical and Industrial Engineering at the Norwegian University of Science and Technology (NTNU). She investigates prototypes roles in early stage product development, experimenting with materials effect on the prototyping out-put as well as teaching larger companies to become more prototype-driven.

Alfredo Jornet is a postdoctoral researcher in the Department of Education at the University of Oslo, Norway. His research examines thinking and learning in and across formal (educational) and informal settings, including professional design practice, with a focus on the social and cultural roots of affect and cognition in creativity.

Honghai Li is Ph.D. Candidate in Academy of Arts and Design and Design Management Research Lab of Tsinghua University, China PR. He also is Lecturer in Industrial Design Dept. of Beijing Information Science and Technology University. His research area includes practical design research and design driven innovation.

Peter Lloyd is a Professor of Design in the School of Architecture and Design at the University of Brighton and Associate Editor for the Journal 'Design Studies'. He teaches in the areas of design methods, design thinking, and design ethics and his research looks at all aspects of the design process. You can read his blog at https://iprofessdesign.wordpress.com

Terri Lovins is a research scientist in the Center for Engineering Learning & Teaching (CELT) and holds a Master of Science degree from the Department of Human Centered Design and Engineering at the University of Washington.

Erin MacDonald is Assistant Professor of Mechanical Engineering at the Stanford University, Stanford, USA. Her research informs product design through quantified choice to represent how people will prefer and react to cognitively complex product purchase, use, and involvement decisions such as those associated with sustainable products and technologies.

Janet McDonnell is Professor of Design Studies and Head of Research at Central Saint Martins, University of the Arts London. Her research explores design processes and the nature of design expertise; published work includes fine-grained analyses of spoken interaction which draw out the skilled practices that enable creative collaboration.

Seda McKilligan is Associate Professor of Industrial Design at Iowa State University, USA. Her research focuses on investigations of creativity and cognition in interdisciplinary design teams, design ideation process, and integration of design thinking as a transformative skill in engineering education.

Axel Menning is a research associate at the Hasso-Plattner-Institute in Potsdam, Germany and member of the HPI-Stanford Design Thinking Research Program. Within the project Visual Diagnostics for Design Thinking Teams, he investigates the topical course of design conversations.

Karen Miller is a Doctoral Candidate at the University of South Carolina School of Library & Information Science. Blending her background in business and legal studies with her current cognate concentration in statistics, she is interested in applying machine learning algorithms and social network analysis to identify patterns and new knowledge in large datasets.

Ella Miron-Spektor is an Assistant Professor in Organizational Psychology at The Faculty of Industrial Engineering and Management at the Technion – Israel Institute of Technology. Her research areas include creativity and innovation, paradox, organizational learning, culture and emotions.

Laura Murphy is a senior in Mechanical Engineering at the University of Michigan, Ann Arbor. She is the co-founder and CEO of the disability design company ADAPT and her research interests lie in understanding in vivo strategies for front-end design.

Maria Adriana Neroni is a Research Associate at the Design Practice Group in the Cambridge Engineering Design Centre. With a background in Experimental Psychology, Cognitive Neuroscience and Education, her current research focuses on developing novel experimental paradigms to investigate creativity in engineering design.

Claudia Nicolai is academic director at the Hasso Plattner Institute – School of Design Thinking in Potsdam, Germany and a principal investigator in the HPI-Stanford Design Thinking Research Program. She studied Business Administration, Economics and Social Science and received her Ph.D. in Strategic Marketing.

Arlene Oak researches and teaches in the area of design studies and material culture in the Department of Human Ecology at the University of Alberta. Her research interests are particularly focused on how language (especially conversation) relates to the creation and mediation of the material world.

Balder Onarheim is Associate Professor in Creativity and part of the research group Innovation, Design and Entrepreneurship at Department of Management Engineering, Technical University of Denmark. His expertise lies within a neurobiological understanding of creativity, and methods to use this understanding to improve individuals' capabilities in creative problem solving.

Anastasia K. Ostrowski is a graduate student in biomedical engineering at the University of Michigan, Ann Arbor. Her research examines idea generation tools, biomedical engineering design, and factors that impact idea generation in engineering education.

Susannah B. F. Paletz is an Associate Research Scientist at the University of Maryland Center for Advanced Study of Language in College Park, Maryland. She combines her fundamental social psychology background and her applied experience to examine organizational, social, and cultural influences on team dynamics and cognition.

Apara Ranjan is a postdoctoral researcher at Copenhagen Business School. She holds a Ph.D. in psychology and interdisciplinary studies from the University of British

Columbia, Canada, and an MA in Cognitive Science from University of Allahabad, India. She enjoys researching and teaching usability engineering, psychology of humor, organizational creativity and psychology of creativity.

Line Revsbæk is Assistant Professor in Participatory Innovation at SDU Design Research, University of Southern Denmark. As an organizational psychologist by training and working from process philosophies and complexity theory perspectives, her research at SDU Design focuses on the social dynamics of innovation and co-design.

Juan de la Rosa is an associate professor in the graphic design department at Universidad Nacional de Colombia. His current research interest is related to the notion of design knowledge and how we can learn from the objects that do not-exist-yet and the future systems they could define.

Wolff-Michael Roth is Lansdowne Professor of Applied Cognitive Science in the Faculty of Education at the University of Victoria, Canada. He investigates knowing and learning across the lifespan in formal educational, workplace, and leisure settings. His work crosses disciplines, theoretical frameworks, and research methods.

Christian D. Schunn is Professor of Psychology, Intelligent Systems, and Learning Sciences and Policy at the University of Pittsburgh. He studies expert engineering and science teams. From that work he builds innovative technology-supported STEM curricula for secondary students, and then studies factors that influence student and teacher learning and engagement.

Nairiti Singh graduated from Carnegie Mellon University with a Masters in Architecture Engineering Construction Management. Her Bachelor's degree is in Civil Engineering from India. At CMU she was Research Assistant under Prof Ömer Akin where she worked on the DTRS11 data.

Zhabiz Shafieyoun is a Research Scholar at IIT Institute of Design. She holds a PhD in Design at Politecnico di Milano. Her background is Industrial Design and her PhD thesis was in Emotional Design at Healthcare Centers. Her research area includes Design for Emotion, Healthcare Design, Design Thinking and relation between Mathematic and Design.

Kathryn E. Shroyer is a graduate research assistant in the Center for Engineering Learning & Teaching (CELT) and a PhD student in the Department of Human Centered Design & Engineering at the University of Washington.

Frido Smulders is Associate Professor in Design, Innovation and Entrepreneurship at the Department of Industrial Design Engineering, Delft University of Technology. In his collaboration with industry he aims to develop practice relevant theoretical notions on the socio-interactive dimension of design enabled innovative processes.

Freddy Solis is a former postdoctoral research associate at Purdue University with a research focus on innovation. He earned his Ph.D. in Civil Engineering, MBA, and M.S. in Civil Engineering degrees from Purdue, as well as a B.S. in Civil Engineering from the Universidad Autonoma de Yucatan in Mexico.

Martin Stacey is senior lecturer in computer science at De Montfort University. His research focuses on design thinking and interdisciplinary understanding of design processes.

Martin Steinert is Professor of Eng. Design and Innovation at the Department of Mechanical and Industrial Engineering at the Norwegian University of Science and Technology (NTNU). His research focus lies on Fuzzy Front End paradigms in general and especially on critical HSI. Projects are multidisciplinary (ME/CS/EE/Neuro- and Cognitive Sc.), connected with industry or academic units.

Arlouwe Sumer is an Industrial-Organizational Psychology practitioner from the University of Maryland, USA, and recent M.S. graduate of Carlos Albizu University, Miami, USA. His past research includes enterprise human resource management solutions, cognitive processes in leadership, and communication and behavior in the workplace.

Carlos Teixeira is an Associate Professor and PhD Program Coordinator at IIT-Institute of Design with research interests in the areas of design strategy, open innovation, and sustainable solutions. He teaches graduate courses and advises doctoral students on the strategic use of design capabilities in complex spaces of innovation.

Koen van Turnhout is senior lecturer and researcher interaction design at HAN University of Applied Sciences, Arnhem, The Netherlands. His research is about research and design methodology and education. He is involved in several design research projects in the health and cultural sector.

Jennifer Turns is a Professor in the Department of Human Centered Design & Engineering at the University of Washington. Her research focuses on the role of reflection in undergraduate engineering education, educator professional development, and graduate student preparation.

Dagny Valgeirsdottir is a PhD fellow and part of the research group Innovation, Design and Entrepreneurship at Department of Management Engineering, Technical University of Denmark. Her research focuses on ways to enhance individual creativity by optimizing creativity training through the application of metacognitive approaches.

Rianne Valkenburg is (part-time) Professor Designerly Innovation at The Hague University of Applied Sciences and Value Producer at LightHouse/expertise in smart lighting & smart cities @ TU/e. Her research focusses on design thinking for innovation, her projects aim to create better quality of life for citizens through sustainable smart city solutions.

Luis A. Vasconcelos is a PhD Student in the Design Practice Group at the Cambridge Engineering Design Centre. With a background in design research and methodology, his research explores new paths in design inspiration and fixation studies. He is interested in the design process at both macro and micro levels.

Andreas Wulvik is PhD Student in Eng. Design and Innovation at the Department of Mechanical and Industrial Engineering at the Norwegian University of Science and Technology (NTNU). His research interests are how to use sensors and computational tools to measure human behaviour, both in Human-Human and Human-Machine interactions.

Xing Xu is Ph.D. Candidate in Academy of Arts and Design and Design Management Research Lab of Tsinghua University, China PR. His research area includes human-centred design processes and methods, lifestyle design research. Current projects include the user value design research based on cross-cultural lifestyle.

Salu Ylirisku is Associate Professor in SDU Design Research at the University of Southern Denmark. For the last 15 years he has worked in the field of user-centred concept design and interaction design. His theoretical focus is currently on project-specific learning in conceptual designing.

Index